IEE ELECTROMAGNETIC WAVES SERIES 39

Series Editors: Professor P. J. B. Clarricoats
Professor Y. Rahmat-Samii
Professor J. R. Wait

MICROWAVE HORNS and FEEDS

Other volumes in the IEE Electromagnetic Waves Series:

Volume 1 **Geometrical theory of diffraction for electromagnetic waves** G. L. James
Volume 2 **Electromagnetic waves and curved structures** L. Lewin, D. C. Chang and E. F. Kuester
Volume 3 **Microwave homodyne systems** R. J. King
Volume 4 **Radio direction-finding** P. J. D. Gething
Volume 5 **ELF communications antennas** M. L. Burrows
Volume 6 **Waveguide tapers, transitions and couplers** F. Sporleder and H. G. Unger
Volume 7 **Reflector antenna analysis and design** P. J. Wood
Volume 8 **Effects of the troposphere on radio communications** M. P. M. Hall
Volume 9 **Schumann resonances in the earth-ionosphere cavity** P. V. Bliokh, A. P. Nikolaenko and Y. F. Flippov
Volume 10 **Aperture antennas and diffraction theory** E. V. Jull
Volume 11 **Adaptive array principles** J. E. Hudson
Volume 12 **Microstrip antenna theory and design** J. R. James, P. S. Hall and C. Wood
Volume 13 **Energy in electromagnetism** H. G. Booker
Volume 14 **Leaky feeders and subsurface radio communications** P. Delogne
Volume 15 **The handbook of antenna design, Volume 1** A. W. Rudge, K. Milne, A. D. Olver, P. Knight (Editors)
Volume 16 **The handbook of antenna design, Volume 2** A. W. Rudge, K. Milne, A. D. Olver, P. Knight (Editors)
Volume 17 **Surveillance radar performance prediction** P. Rohan
Volume 18 **Corrugated horns for microwave antennas** P. J. B. Clarricoats and A. D. Olver
Volume 19 **Microwave antenna theory and design** S. Silver (Editor)
Volume 20 **Advances in radar techniques** J. Clarke (Editor)
Volume 21 **Waveguide handbook** N. Marcuvitz
Volume 22 **Target adaptive matched illumination radar** D. T. Gjessing
Volume 23 **Ferrites at microwave frequencies** A. J. Baden Fuller
Volume 24 **Propagation of short radio waves** D. E. Kerr (Editor)
Volume 25 **Principles of microwave circuits** C. G. Montgomery, R. H. Dicke, E. M. Purcell (Editors)
Volume 26 **Spherical near-field antenna measurements** J. E. Hansen (Editor)
Volume 27 **Electromagnetic radiation from cylindrical structures** J. R. Wait
Volume 28 **Handbook of microstrip antennas** J. R. James and P. S. Hall (Editors)
Volume 29 **Satellite-to-ground radiowave propagation** J. E. Allnutt
Volume 30 **Radiowave propagation** M. P. M. Hall and L. W. Barclay (Editors)
Volume 31 **Ionospheric radio** K. Davies
Volume 32 **Electromagnetic waveguides: theory and application** S. F. Mahmoud
Volume 33 **Radio direction finding and superresolution** P. J. D. Gething
Volume 34 **Electrodynamic theory of superconductors** S.-A. Zhou
Volume 35 **VHF and UHF antennas** R. A. Burberry
Volume 36 **Propagation, scattering and dissipation of electromagnetic waves** A. S. Ilyinski, G. Ya. Slepyan and A. Ya. Slepyan
Volume 37 **Geometrical theory of diffraction** V. A. Borovikov and B. Ye. Kinber
Volume 38 **Analysis of metallic antennas and scatterers** B. D. Popović and B. M. Kolundžiya

MICROWAVE HORNS and FEEDS

A D Olver, P J B Clarricoats
A A Kishk and L Shafai

IEE
The Institution of Electrical Engineers

IEEE PRESS
The Institute of Electrical and Electronics Engineers, Inc., New York

Copublished by: The Institution of Electrical Engineers,
London, United Kingdom, and the Institute of Electrical
and Electronics Engineers, Inc., New York, USA

Distribution in North and South America is by the IEEE Press
ISBN 0 7803 1115 9
IEEE Order No. PC4689

Distribution elsewhere is by the IEE
ISBN 0 85296 809 4

The Institution of Electrical Engineers,
Michael Faraday House,
Six Hills Way, Stevenage,
Herts. SG1 2AY, United Kingdom

British Library Cataloguing in Publication Data

A CIP catalogue record for this book
is available from the British Library

ISBN 0 85296 809 4

Printed in England by Bookcraft, Bath

Contents

Preface **x**

1 Introduction **1**
A.D. Olver and P.J.B. Clarricoats
1.1 Preliminary remarks 1
1.2 History of reflector antennas and feeds 1
1.3 Performance requirements of feeds and reflectors 5
1.3.1 Spatial characteristics 5
1.3.2 Impedance characteristics 9
1.3.3 Frequency characteristics 10
1.4 Organisation of the book 10
1.5 Choice of type of feed 14
1.6 References 15

2 Reflector antennas **16**
P.J.B. Clarricoats
2.1 Introduction 16
2.1.1 General remarks 16
2.1.2 Reflector types 16
2.1.3 Constraints 17
2.2 Theory and performance of reflector antennas 18
2.2.1 The scattered field from a reflector 18
2.2.2 The aperture-field method 22
2.2.3 Radiation from a symmetrical paraboloidal reflector 22
2.2.4 Features of the secondary-radiation pattern 24
2.2.5 The dual-reflector antenna 29
2.2.6 The offset-reflector antenna 32
2.2.7 The dual-offset-reflector antenna 40
2.2.8 Focal-field analysis 46
2.2.9 Antenna return loss 51
2.2.10 Radome-covered reflectors 53
2.2.11 The effect of reflector surface errors on antenna performance 57
2.2.12 The effect of positional errors on antenna performance 60
2.2.13 Analysis and synthesis of shaped-beam antennas 61
2.2.14 Dual-gridded reflector antennas 71
2.2.15 Reconfigurable reflector antennas 72
2.2.16 Dichroic reflectors 74
2.3 References 77

3 Radiation from apertures **79**
A.D. Olver
3.1 Introduction 79
3.2 Radiation equations from aperture fields 80

3.3 Radiation characteristics from apertures 84
3.3.1 Solution of radiation equations 84
3.3.2 Line-source apertures 85
3.3.3 Compound aperture distributions 87
3.3.4 Rectangular apertures 89
3.3.5 Apertures with phase tapers 90
3.3.6 Circular apertures 92
3.3.7 Gaussian aperture distributions 95
3.4 Directivity of apertures 96
3.5 Near fields 97
3.6 References 98

4 Numerical analysis using cylindrical and spherical waves in horns 100
A.D. Olver
4.1 Introduction 100
4.2 Fundamentals of modal analysis 101
4.3 Cylindrical-wave analysis 102
4.4 Spherical-wave analysis 105
4.5 Principles of modal-matching technique 106
4.5.1 Scattering matrix of a uniform section 110
4.5.2 Scattering matrix of a junction 110
4.6 Modal-matching technique for cylindrical waveguides 113
4.6.1 Numerical procedure 114
4.6.2 Number of modes and sections 115
4.7 Spherical-modal-matching technique for conical waveguides 117
4.8 References 122

5 Numerical analysis of small axisymmetric feeds 123
A.A. Kishk and L. Shafai
5.1 Introduction 123
5.2 Basic formulation 125
5.3 Moment-method solution 129
5.3.1 Co-ordinates of bodies of revolution 129
5.3.2 Application of moment method 131
5.3.3 Excitation vectors 136
5.3.4 Computation of the radiation patterns 139
5.3.5 Different integral-equation formulations 143
5.3.6 Solution verification 145
5.4 References 148

6 Pure-mode horns 149
A.D. Olver
6.1 Principles of pure-mode horns 149
6.2 Fields and modes in cylindrical waveguides 150
6.3 Conical horns 152
6.3.1 Radiation patterns 154
6.3.2 Gain 157
6.3.3 Optimum designs 157
6.3.4 Phase centre 159
6.4 Pyramidal horns 160
6.4.1 Radiation patterns 162
6.4.2 Gain 165
6.4.3 Design of pyramidal horns 167
6.5 Other shaped-aperture horns 168

6.6 Wideband ridge-loaded horns 172
6.6.1 Radiation patterns 174
6.6.2 Ridge-loaded input section 174
6.7 References 178

7 Prime-focus waveguide feeds 181
L. Shafai and A.A. Kishk
7.1 Introduction 181
7.2 Phase-centre determination 182
7.3 Circular waveguide feeds 185
7.4 Circular waveguide with rectangular corrugated flange 200
7.5 Circular waveguide with trapezoidally corrugated flange 203
7.6 Phase-error effects 208
7.7 Coaxial waveguide feeds 209
7.8 Feeds for deep reflectors 218
7.9 Dualband feeds 222
7.10 Small-horn feeds 225
7.11 References 227

8 Multimode horns 229
A.D. Olver
8.1 Introduction 229
8.2 Multiple modes in cylindrical waveguides 230
8.2.1 Circular waveguides 232
8.2.2 Rectangular waveguides 241
8.2.3 Other waveguide cross-sections 241
8.3 Excitation of higher-order modes 244
8.4 Shaped-pattern horns 246
8.4.1 Horns for pattern symmetry and low crosspolarisation 246
8.4.2 Horns for increased aperture efficiency 249
8.4.3 Matched feed horns for offset reflectors 253
8.5 Tracking horns 253
8.6 References 255

9 Corrugated horns 258
A.D. Olver
9.1 Introduction 258
9.2 Principles of operation of corrugated horns 259
9.3 Design procedure for corrugated horns 261
9.4 Analytical techniques for predicting performance 265
9.4.1 Balanced hybrid model 266
9.4.2 Impedance techniques: cylindrical 266
9.4.3 Impedance techniques: spherical 271
9.4.4 Space-harmonic techniques 274
9.4.5 Modal-matching techniques: cylindrical 275
9.4.6 Modal-matching techniques: spherical 278
9.4.7 Gaussian-beam techniques 282
9.5 Types of corrugated horn 286
9.5.1 Standard narrow-flare-angle horns 286
9.5.2 Small-aperture horns 287
9.5.3 Millimetre-wave horns 287
9.5.4 Wide-flare-angle horns 288
9.5.5 Broadband corrugated horns 296
9.5.6 Ring-loaded slots 298

9.5.7 Dual-depth corrugated horns 300
9.5.8 Profile corrugated horns 305
9.5.9 Rectangular corrugated horns 313
9.6 References 315

10 Dielectric-loaded horns **318**
A.D. Olver
10.1 Introduction 318
10.2 Principle of operation 321
10.3 Prediction using modal-matching technique 322
10.3.1 Dielectric-filled circular waveguide 323
10.3.2 Power-coupling integrals 328
10.4 Design considerations 328
10.5 Parametric studies on linear horn 330
10.5.1 Permittivity of dielectric 330
10.5.2 Variation of aperture diameter 331
10.5.3 Airgap 332
10.5.4 Change in flare angle 332
10.5.5 Design of throat region 332
10.6 Horn-design examples 336
10.6.1 Dielectric-foam cone in metal horn 337
10.6.2 Metallised dual-dielectric cone-loaded horn 338
10.6.3 Millimetre-wave horns using a solid-dielectric cone 342
10.6.4 Millimetre-wave horns using two solid dielectrics 342
10.7 Compact profiled dielectric-loaded horns 346
10.7.1 Comparison with linear horn 347
10.7.2 Reduction in length 350
10.7.3 Influence of shape of profile 352
10.8 Dielectric losses in a dielectric-loaded conical horn 356
10.8.1 Perturbation expression for the attenuation coefficient 357
10.8.2 Losses in dielectric-loaded horn 358
10.9 Tolerance on relative permittivity of dielectric 361
10.10 Control of crosspolarisation and efficiency with a lens 364
10.10.1 Lens profile 364
10.10.2 Design of lens to improve efficiency 365
10.10.3 Design of lens to reduce crosspolarisation 366
10.11 References 368

11 Lens-corrected horns **370**
A.D. Olver
11.1 Introduction 370
11.2 Theory of lens design 371
11.3 Plastics-foam lenses 373
11.4 Shape of lens 373
11.5 Zoned lenses 378
11.6 Aperture amplitude distribution 379
11.7 Radiation characteristics 383
11.8 Experimental lenses 387
11.9 Uses of lens-corrected horns 389
11.10 References 390

12 Dielectric feeds **391**
L. Shafai and A.A. Kishk
12.1 Introduction 391

12.2 Dielectric-rod feed 392
12.2.1 Dielectric-diameter effect 397
12.2.2 Dielectric constant 397
12.3 Dielectric-rod feed with ring loading 401
12.4 Dielectric rod with cavity loading 404
12.5 Profiled dielectric rods 407
12.6 Waveguides with dielectric rings 409
12.7 Dualband feed 411
12.8 References 414

13 Dipole feeds **416**
L. Shafai and A.A. Kishk
13.1 Introduction 416
13.2 Ring-loaded dipole-disc feed 417
13.3 Cone-dipole feed 419
13.4 Cavity-dipole feed 423
13.5 References 426

14 Microstrip and resonator type feeds **427**
L. Shafai and A.A. Kishk
14.1 Introduction 427
14.2 Microstrip-antenna feed 428
14.3 Effect of mode excitation on the crosspolarisation 431
14.4 Effect of conducting chokes 432
14.5 Feed performance with the reflector antenna 434
14.6 Patch-ring feed 436
14.7 Dielectric-resonator feed 438
14.8 References 443

15 Array feeds **444**
P.J.B. Clarricoats
15.1 Introduction 444
15.2 Mutual coupling in an array 448
15.2.1 Fundamentals 448
15.2.2 Mutual coupling in an array of conical horns 448
15.2.3 Mutual-coupling predictions for conical-horn arrays 452
15.3 Effect of mutual coupling on secondary patterns 461
15.3.1 Prediction methods 461
15.3.2 Predicted primary and secondary patterns 464
15.3.3 Effect of mutual coupling on secondary patterns when the array feed is displaced 469
15.4 Mutual coupling for elements of rectangular and elliptical cross-sections 472
15.5 Mutual coupling for a microstrip patch array 475
15.6 References 480
15.7 Appendix: Coefficients of the admittance matrix for an array of circular apertures 480

Index **483**

Preface

The book is devoted to describing the theory, design, performance and application of microwave horns and feeds for reflector antennas. The idea for a text covering horns and feeds arose from the large amount of knowledge which has been collected by the authors over more than two decades of research into horns and feeds. The book is a collaboration between two research groups. The microwave antenna group at Queen Mary and Westfield College (University of London) has researched, designed, built and tested many large and medium-size, high-performance horns and reflector antennas. The electromagnetic group at the University of Manitoba (Winnipeg, Canada) has pioneered analysis techniques and developed many small horns, microstrip and other antennas with desirable radiation characteristics. Thus the knowledge of the two groups complements each other. The result is a book which covers all the main types of feeds in use today.

The book contains material which should be of value both to the designer and the research worker. Parts of the book have been published previously in papers and conference proceedings, but the text also contains many new results. Some of these have been developed in conjunction with the research staff and students in both universities. To these people we offer our warmest thanks.

Chapter 1
Introduction

A.D. Olver and P.J.B. Clarricoats

1.1 Preliminary remarks

This book is concerned with feeds for reflector antennas and as such is believed to be the first general treatment of its kind. The aim is to describe the design principles and methods of analysis so that readers will be able to understand and develop feeds for reflector antennas. In the past 40 years a number of excellent texts on microwave antennas have been published and one of the earliest, 'Microwave antenna theory and design', edited by the late Samuel Silver appears as a reprint in the IEE Electromagnetic series [1] to which this text belongs. Within that classic work, rather little attention is given to feeds *per se* as the waveguide and horn were the only forms readily available to the antenna designer of the 1940s other than waveguides terminated in dipoles. Much of the design of that period was empirical, although many of the theories we currently use were first developed then. In the nearly halfcentury following the Second World War, antenna technology has burgeoned, stimulated by communications, radar, radio astronomy and remote sensing. There has been a continual requirement to develop improved feeds and reflectors and for an improved understanding of the factors influencing their performance. Today the antenna engineer has available a wide variety of feeds ranging from high-performance corrugated horns to simple waveguide-fed dipoles. The theory and practice of these and many other types of feed are described in the book.

1.2 History of reflector antennas and feeds

The origins of the modern reflector antenna lie in the optical reflecting telescope invented by Isaac Newton in 1672. Newton believed that it was not possible to correct chromatic aberration which occurred in the refracting telescopes of the day and thus he conceived of a telescope involving mirrors which was free from chromatic aberration. The Newtonian reflector, which is shown in Figure 1.1*a*, comprises a concave mirror which directs incoming light rays to an eyepiece after reflection in a small plane mirror. The eyepiece performs the function of a feed in a modern reflector antenna. Figures 1.1*b* and 1.1*c* identify other dual-reflector telescopes devised in the 17th century. Figure 1.1*b* shows the arrangement proposed by Cassegrain while Figure 1.1*c* shows Gregory's design.

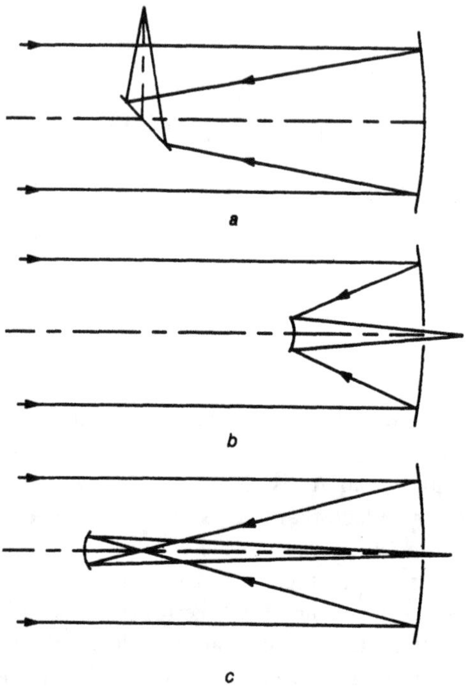

Figure 1.1 Reflector telescopes

a Newtonian reflecting telescope and plane subreflector
b Cassegrain dual-reflector telescope
c Gregorian dual-reflector telescope

Both have counterparts in modern antennas.

Two centuries elapsed before Hertz conducted experiments at Karlsruhe in Germany in 1887 to demonstrate a number of fundamental properties of electromagnetic waves [2]. His experiments were performed at 450 MHz using, among other components, a cylindrical reflector fed by means of a dipole antenna (Figure 1.2). Soon after the discoveries of Hertz, other researchers used microwave frequencies to explore the properties of radio waves. Bose [3] and Lodge [4] made pyramidal horns as simple radiators of electromagnetic waves in the 1890s. They were experimenting with the radiated waves and were able to investigate the fundamental properties by setting up microwave optical benches. The theoretical properties of modes in waveguides was described by Lord Rayleigh and published in 1897 [5]. No further work on reflectors or feeds is known to have occurred until the 1930s, when Marconi used a paraboloid reflector while investigating UHF-propagation phenomena from his ship SS *Electra* in the Mediterranean. A microwave radio-relay link across the English Channel was demonstrated in 1931 using 10 ft-diameter (~3 m) parabolic reflectors. In 1934 in the USA, Karl Jansky built the first radiotelescope to use a

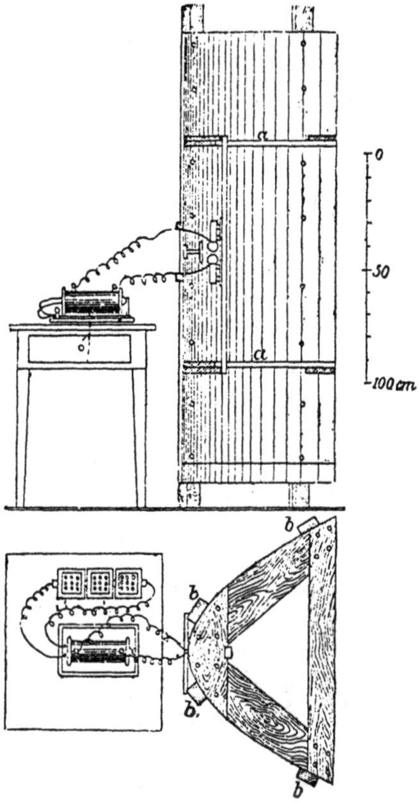

Figure 1.2 Hertz's cylindrical reflector and dipole feed

paraboloidal reflector. With it he received signals coming from the vicinity of the Milky Way. These experiments gave birth to the subject of radioastronomy, which gained considerable momentum following the end of the Second World War.

The first detailed experimental investigation of horns was undertaken by Southworth [6] and Barrow [7] in 1936 followed by detailed theoretical treatment of the pyramidal horn which appeared in a paper by Barrow and Chu [8] in January 1939. A series of classic papers was published in the late 1930s and 1940s which described the basic theory and principles of operation of pyramidal and conical horns. This included papers by Southworth and King [9] and Chu [10].

The years of the Second World War saw the development of both pencil-beam and shaped-beam antennas for radar. The latter were especially important in applications calling for different elevation- and azimuth-radiation patterns. These early designs made use of ray-tracing techniques but in more recent times synthesis methods for shaped beams evolved using both geometric optics and diffraction methods.

In the 1950s reflector antennas became widely used in terrestrial microwave-communication systems and the first large radiotelescopes were designed and constructed, including the telescope at Jodrell Bank in England, which monitored signals from the first earth-orbiting satellite *Sputnik*. The early radiotelescopes had feeds at the prime focus of a paraboloid reflector, a design which has been sustained over the years. More recently, telescope designs have also employed dual reflectors, which have among their advantages a more convenient location for the low-noise receivers. In parallel with these developments, plans evolved for the launch of the first telecommunications satellite. *Telstar* commenced operation in 1962 and live trans-Atlantic television pictures were exchanged for the first time. Antennas using both paraboloidal and horn-paraboloid reflectors were employed at earth stations in the UK, USA and France. *Telstar* was a low-earth-orbit satellite which called for continuous tracking in order to maintain communication. In 1965 the first Intelsat geostationary satellite was launched, to be followed by more complex satellite types, with the *Intelsat VI* satellite launched in 1990 to be followed by *Intelsat VII* and *VIII*. As an example *Intelsat VI* can support around 86 000 telephone channels, which compares with only the 450 channels of *Intelsat I*. The trend is towards increasingly more sophisticated antennas on the satellite, providing for both regional and continental earth coverage. These reflectors are in general offset paraboloids fed by means of multiplexed arrays. The most recent designs also carry singly fed shaped reflectors which provide a most efficient means of illuminating a region. Some designs permit in-orbit antenna-pattern reconfiguration.

Satellite communications and radioastronomy required high-performance feeds. The corrugated horn was proposed by Kay [11] in 1962 as a means of radiating a symmetrical radiation pattern. Subsequently this horn was shown to radiate very low levels of crosspolarisation, which is essential for dual-polarised systems. In consequence, the corrugated horn has become the prime choice for high-performance reflector antennas. It is discussed within this text, new material being introduced to supplement that presented by two of the present authors whose text 'Corrugated horns for microwave antennas' appeared as Volume 18 in this series in 1984 [12]. The potential of dielectric feeds was first realised in 1966 by Bartlett and Mosely [13] who developed a dielguide horn. Dielectrics have since been used in various configurations to produce feeds which radiate both shaped-beam and high-quality pencil-beam patterns.

Feeds have developed in two directions. There are the high-performance horns which feed large satellite-communication antennas, radar antennas and radioastronomy antennas. These are characterised by aperture diameters in the range three to eight wavelengths diameter. On the other hand, there are small-diameter feeds required for small-size, prime-focus reflectors used in terrestrial communications and direct-broadcast-by-satellite (DBS) satellite-receiving systems. During the last two decades much effort has gone into producing high performance small feeds. Microstrip feeds have also come to the fore and although normally configured as direct-radiating arrays they are nonetheless useful as feeds for reflectors, particularly when low cost is a factor. However, to date, the small size restricts the performance of these feeds.

As satellite-borne reflectors have evolved and the signal intensity received on the ground from the satellite has increased, the size and cost of earth-station antennas has significantly decreased. The Intelsat system began with 30 m diameter antennas but the size has now decreased to antennas a few metres in diameter for modern VSAT systems. Direct-broadcast satellites launched in the late 1980s enable television signals to be received directly in the home using either a small-diameter offset-paraboloid antenna with a circular waveguide feed or a small array comprising either microstrip radiators or slots.

From the beginning of the 1960s the computer played an increasingly important role in the design process, and methods which would have been impossible to implement without them have become central to their analysis and synthesis.

In the last decade, precision telescopes have been built to operate at millimetre wavelengths and have been based at high-altitude observatories to reduce atmospheric attenuation. An example is the James Clerk Maxwell Telescope owned jointly by the UK, the Netherlands and Canada and operated in Hawaii. The telescope permits observations down to below 1 mm wavelength and requires a high-precision reflector and feed assembly.

This short review shows the reflector as an important antenna component for communication, broadcast and scientific purposes. These applications all call for efficient feeds if the benefit of the reflector aperture is to be realised. Many systems require the simultaneous use of two senses of polarisation and in some cases there is a call for multiband operation. These requirements are particularly challenging for the feed designer.

1.3 Peformance requirements of feeds and reflectors

The performance of feeds and reflectors is specified by a number of criteria. These are the same as those for most antennas, though the pencil-beam radiation patterns mean that the emphasis is different. The shape and characteristics of the radiation pattern of a feed are the most important parameters because these will directly influence the fields which are directed at a reflector, thus determining the radiation characteristics of the complete antenna. The main performance requirements and specifications are briefly described here. They can be divided into the spatial characteristics, the impedance characteristics and the frequency characteristics.

1.3.1 Spatial characteristics

1.3.1.1 Radiation patterns

A feed-and-reflector antenna acting as a transmitter radiates spherical waves into space with different amplitudes and phases in different directions. The

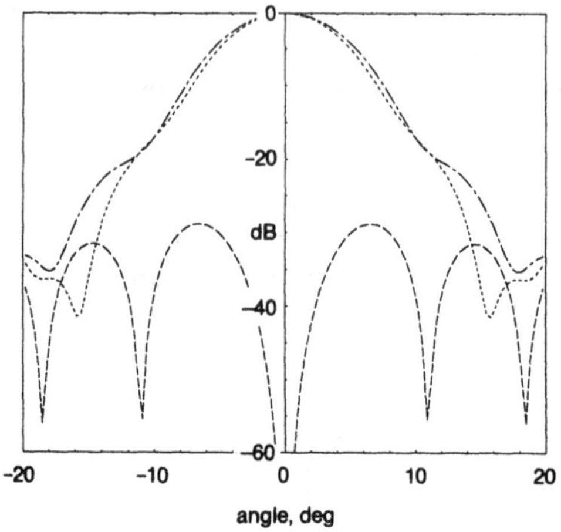

Figure 1.3 Radiation pattern most commonly used in reflector and feed studies

------- E-plane copolar pattern
—·—·— H-plane copolar pattern
– – – crosspolar pattern in 45° plane

representation of the relative power in the different directions is the radiation pattern. The universal form used in reflector antennas is the classic rectangular plot of radiated power in decibels against spherical angle θ from boresight (Figure 1.3). These plots are either in the E-plane or the H-plane or the $\phi = 45°$ and 135° planes. Other forms of plot are contour plots showing curves of constant power on a grid of $(\sin\theta \sin\phi)$ against $(\sin\theta \cos\phi)$, and isometric plots showing a 3-dimensional sketch of a contour plot. The shape of the main beam of a pencil pattern, as shown in Figure 1.3, is characterised by its shape (for instance Gaussian), the −3 dB beamwidth, and the sidelobe levels.

1.3.1.2 Polarisation

The polarisation of a radiated wave at a single frequency describes the shape and orientation of the locus of the extremities of the field vectors as a function of time. The radiated waves may be considered locally to be a plane wave in which the electric and magnetic fields are related by the intrinsic admittance of free space $Y_0 = \sqrt{(\varepsilon_0/\mu_0)}$. In these circumstances, it is sufficient when describing the polarisation of a wave to specify the polarisation of the electric-field vector. A wave may be described as 'linear polarised', 'circular polarised' or 'elliptical polarised'. In the study of horns and feeds, linear polarisation is the most common term used, with circular polarisation in some specific cases. The linear-polarised field is broken into a copolar component and a crosspolar component. There is no unique definition of crosspolarisation, but the definition universally used in the analysis and specification of feeds is Ludwig's third definition [14]. This relates the copolar electric field $E_p(\theta, \phi)$ and the crosspolar electric field $E_q(\theta, \phi)$ to the far-field spherical electric fields E_θ, E_ϕ by

$$\left.\begin{aligned} E_p(\theta, \phi) &= \sin\phi\, E_\theta + \cos\phi\, E_\phi \\ E_q(\theta, \phi) &= \cos\phi\, E_\theta - \sin\phi\, E_\phi \end{aligned}\right\} \tag{1.1}$$

where the electric field is aligned along the y direction. If the electric field is aligned along the x direction, then eqn. 1.1 becomes

$$\left.\begin{aligned} E_p(\theta, \phi) &= \cos\phi\, E_\theta - \sin\phi\, E_\phi \\ E_q(\theta, \phi) &= \sin\phi\, E_\theta + \cos\phi\, E_\phi \end{aligned}\right\} \tag{1.2}$$

Ludwig's third definition is convenient because it corresponds exactly to the fields which are measured when a horn or reflector is mounted on a polarisation-over-azimuth turntable.

A circular-polarised field has a right-hand and a left-hand component given by

$$\left.\begin{aligned} E_{RH} &= \frac{1}{\sqrt{2}}(E_\theta - jE_\phi)\exp(-j\phi) \\ E_{LH} &= \frac{1}{\sqrt{2}}(E_\theta + jE_\phi)\exp(-j\phi) \end{aligned}\right\} \tag{1.3}$$

In this case the copolar component is the wanted component (RH or LH) and the crosspolar component is the unwanted component.

The characteristic of linear crosspolar patterns radiated from pencil-beam-type feeds is that the power radiated along the E-plane and H-plane is zero and peaks along the 45° and 135° planes (Figure 1.3), thus producing four lobes. The first peaks of these lobes invariably fall within the angle subtended by the reflector at the feed, so this level provides a measure of how much crosspolarisation is radiated by both the horn and the reflector antenna.

1.3.1.3 Gain and directivity

The gain is the ability of a feed to concentrate power into a narrow angular region of space. The power gain in a specified direction (θ, ϕ) is defined as

$$G(\theta, \phi) = 4\pi \frac{\text{power radiated per unit solid angle in direction } \theta, \phi}{\text{total power accepted from source}} \tag{1.4}$$

This expression includes the dissipative losses in the antenna. These are normally specified by the 'insertion loss' which cannot easily be predicted and so is a measured parameter. Instead the 'directivity' is calculated which has a definition similar to eqn. 1.4 except that the denominator is replaced by the total power radiated. In practice, the term gain tends to be used to mean either the true gain or the directivity. The distinction is that the gain is measured whereas the directivity is calculated. In terms of the radiated electric fields the directivity is

$$D(\theta, \phi) = \frac{|E(\theta_0, \phi_0)|^2}{\frac{1}{4\pi}\int_0^{\pi}\int_0^{2\pi} |E(\theta, \phi)|^2 \sin\theta \, d\theta \, d\phi} \tag{1.5}$$

The 'effective area' A of an antenna is a fictional area which, if used in receive mode, collects enough power to produce a gain G. For an aperture antenna, this is given by

$$G = \frac{4\pi A}{\lambda^2} \tag{1.6}$$

1.3.1.4 Efficiency

The 'aperture efficiency' of an antenna is the ratio of the effective radiating area of an antenna to the physical radiating area of the antenna. Particularly for horns, it is a useful measure of how well the actual antenna makes use of the physical area available. It can be calculated by predicting the directivity, then using eqn. 1.6 to find the effective area and dividing by the physical radiating area of the antenna.

The aperture efficiency of a feed-and-reflector combination is made up of four separate components. These are the 'illumination efficiency' η_I, the 'spillover efficiency' η_S, the 'phase efficiency' η_P, and the 'crosspolar efficiency' η_X;

$$\eta_a = \eta_I \eta_S \eta_P \eta_X \tag{1.7}$$

The illumination efficiency is a measure of the nonuniformity of the field across the aperture of a reflector caused by the tapered radiation pattern and the unequal path loss from the feed to different parts of the reflector surface. It can be calculated by computing the power in the angular region subtended by the reflector as a proportion of the total power:

$$\eta_I = 2\cot^2\frac{\theta_0}{2} \frac{\left|\int_0^{2\pi}\int_0^{\theta_0} E(\theta, \phi)\tan(\theta/2) \, d\theta \, d\phi\right|^2}{\int_0^{2\pi}\int_0^{\theta_0} |E(\theta, \phi)|^2 \sin\theta \, d\theta \, d\phi} \tag{1.8}$$

where θ_0 is the subtended angle of the reflector. When a feed illuminates a reflector, only a proportion of the power from the feed will intercept the reflector, the remainder being the spillover power. The 'spillover efficiency' is then given by

$$\eta_S = \frac{\int_0^{2\pi}\int_0^{\theta_0} |E(\theta, \phi)|^2 \sin\theta \, d\theta \, d\phi}{\int_0^{2\pi}\int_0^{\pi} |E(\theta, \phi)|^2 \sin\theta \, d\theta \, d\phi} \tag{1.9}$$

The remaining two efficiencies, the phase efficiency and the crosspolarisation efficiency, are very small for feeds and reflectors because if the value is large enough to be a problem, the radiation pattern will already be unsatisfactory. The phase efficiency measures the uniformity of the spherical phase across the aperture, and the crosspolar efficiency measures the amount of power lost in the crosspolar-radiation pattern.

In remote-sensing work the 'beam efficiency' is commonly used as a measure of how much energy is being captured by the main beam of the antenna. The beam efficiency is the proportion of power radiated within a fixed angle from boresight, so the definition is the same as eqn. 1.9 except that the angle θ_0 is the angle away from boresight. This was originally defined by the position of the first null but is now more usually specified as 1.25 times the -3 dB halfbeamwidth.

Related to the efficiency of the antenna is the 'noise temperature' 'seen' by a receiving antenna. The noise is detected by the sidelobes and spillover lobes of the antenna and sets a lower limit to the sensitivity of a receiving system. This parameter is only of interest in high-gain reflector antennas with low-noise receivers where the noise arising through the spillover lobes is greater than that due to the front end of the receiver. The noise is usually measured by the ratio of the gain to the noise temperature, or the G/T ratio.

1.3.1.5 Phase centre

All horns and feeds have a 'phase centre'. This is the theoretical point along the axis of the horn which is the centre of curvature of the phase fronts of the spherical waves. In most feeds there is a phase-centre region rather than a single point because the radiation consists of a summation of spherical waves which emanate from the aperture. The approximate phase centre is found in computation by postulating that the radiation in a particular angular direction comes from one point. It can be found experimentally by rotating a feed around various points along its axis and finding the point which gives the most constant phase pattern. Formulas for the computation in both the far field and the near field are given in Reference 15. A detailed discussion is in Chapter 7.

1.3.2 Impedance characteristics

From the point of view of the microwave circuit behind an antenna, the antenna can be represented as a complex load impedance whose characteristics depend on the radiation patterns of the antenna and hence on the design of the antenna. The goal of a good design is to match the impedance of the antenna to the impedance of the waveguide or other transmission line connecting the antenna to the transmitter or receiver. The impedance match may be specified by one of the 'input-voltage-reflection coefficient' ρ_L, the 'return loss' $-20\log_{10}|\rho_L|$ (decibels) or the 'voltage standing-wave ratio' (VSWR) $(1+|\rho_L|)/(1-|\rho_L|)$. Horns and feeds usually have relatively good impedance characteristics and, for this reason, the impedance is not normally an initial design parameter. The impedance influences the final design and so should be computed as part of the complete design process.

The impedance referred to above means any mismatch between the antenna and the circuit behind the antenna. This is sometimes called the input impedance. There is also a possible mismatch between a feed antenna and the air surrounding the antenna. This is usually small and only significant for small-aperture antennas where the impedance of the antenna cannot be easily matched to the impedance of free space (377 Ω).

1.3.3 Frequency characteristics

The performance of the antenna will change with frequency. A reflector antenna is usually designed to operate at one frequency and the frequency characteristics refer to the way in which the performance varies over a given frequency band. The performance can be assessed by any of the spatial characteristics or impedance characteristics of the antenna. The frequency characteristics will, in general, be different for different performance criteria. For feeds and reflectors the most common criteria are in terms of the variation of the input impedance, the beamwidth and the peak crosspolar level.

Feeds may be specified as narrowband or wideband (or equivalently broadband). There is no precise definition of these terms. In feed work, narrowband generally means less than 10% bandwidth, where there is usually little problem in meeting a specification. Wideband can mean any frequency band from 10% to 100% or even larger. The actual definition tends to have evolved for individual applications. Sometimes there is a real challenge for the feed designer to try to design a wideband feed.

1.4 Organisation of the book

The book is divided into three parts:

(i) reflector antennas: Chapter 2
(ii) techniques for the analysis and design of feeds: Chapters 3–5
(iii) design and operation of horns and feeds: Chapters 6–15.

Most of the horns and feeds described in later chapters of this book have the prime purpose of feeding a reflector antenna. Thus the design of a feed cannot be understood without a firm knowledge of the characteristics of reflector antennas. Chapter 2 gives a comprehensive review of the theory and performance of reflector antennas. Techniques for the prediction of performance are described, followed by explanations of the design principles of current types of reflector antenna which are used in communication, radar and remote sensing. Satellite communication has been a significant motivation for the development of new types of reflector antenna. Examples are shaped-beam reflectors, dual-gridded reflectors, reconfigurable reflectors and dichroic reflectors. The performance of these systems is described at the end of the chapter.

The second section of the book, on the techniques for the analysis and design of feeds, describes three common prediction techniques: aperture techniques in Chapter 3, numerical analysis using cylindrical or spherical modes in Chapter 4, and the moment method for small feeds in Chapter 5. In each case the intention is to explain the principles behind the methods. Aperture techniques using Fourier-transform theory are the most common and convenient methods of computing a far-field-radiation pattern from a knowledge of the electric and magnetic fields across an aperture plane. For horns, the aperture plane is well defined and a range of techniques exists for obtaining the internal fields over the plane. Chapter 4 is concerned with methods of predicting these internal fields for horns which have one or more modes travelling along the horn. The fields in the horn can be represented using either cylindrical waves or spherical waves. The former are easier to handle but the latter are a better representation of the real fields inside the horn. The main part of Chapter 4 is given over to an exposition of the powerful modal-matching technique which divides a horn up into a series of short waveguide sections and matches all modes across the junctions between sections. This technique has been developed during the last decade as an accurate method of predicting performance which gives radiation patterns that are usually in good agreement with experimental measurements.

Modal techniques can be used for horns with sizes of a couple of wavelengths upwards, but become increasingly unreliable for small feeds where the currents on all parts of the radiating structure are important. A numerically powerful technique for predicting the performance of small feeds is the method of moments. This is the subject of Chapter 5. The chapter starts by developing integral equations which can be used to generate a set of field equations. These are then reduced to a matrix equation to determine the surface currents on the feed. These currents are then used to predict the radiation characteristics. The numerical details of the method of moments applied to feeds is complicated and these are discussed in Chapter 5. The theory applies both to feeds composed of conductors and feeds composed of dielectrics and conductors.

Each chapter of the third, and main, section of the book describes the theory and design of one type of antenna used to feed reflectors. The emphasis is on explaining the characteristics of the performance through the use of the analytical techniques described in Chapters 3, 4 and 5. The first type of feed described in Chapter 6 is the pure-mode horn such as the conical and pyramidal horn. This is the simplest category of horn antenna because of the use of a single mode to form the radiated field. These horn types were also the earliest to be developed. However, their performance is relatively poor so they are not often used in modern systems. The chapter also deals with wideband ridge-loaded horns which are designed to operate over very broad bandwidths, albeit with reduced performance.

Chapter 7 studies the analysis and design of small waveguide horns so that they can be used as feeds in prime-focus reflector antennas. A feed for a prime-focus reflector requires a broad radiation pattern with good pattern symmetry. The currents on the waveguide flange are a significant factor in these feeds and methods of controlling the currents using rings and slots are of considerable importance in the design. Several examples of prime-focus feeds are also described including coaxial feeds and dual-band feeds which radiate at two separate frequency bands.

Large-aperture horns are capable of supporting a number of modes in addition to the fundamental mode. Chapter 8 deals with multimode horns by explaining how the higher-order modes can be used to control the radiation patterns and then describing examples of horns intended to radiate shaped patterns for optimising the performance of reflector antennas. Multimode horns can also be used for receiving tracking information through a radiation pattern which has a null along the boresight direction.

The next two chapters deal with high-performance feeds which radiate the best characteristics obtainable from horns. These use hybrid modes and produce excellent radiation patterns for feeding dual-reflector antennas. Chapter 9 describes corrugated horns and Chapter 10 describes dielectric-cone-loaded horns. The corrugated horn has become the leading choice for a high-performance feed because of its excellent and reliable performance. The design methods have evolved to the stage where stable numerical techniques are available which usually produce good predictions of practical performance. Chapter 9 builds on the book on corrugated horns [12] by two of the present authors. There have been many developments since this was published in 1984 and these are described in the chapter.

The dielectric-cone-loaded horns discussed in Chapter 10 are a more recent development, but can radiate patterns with performance as good as that of corrugated horns. The dielectric-loaded horn consists of a metal horn containing a cone of dielectric which nearly fills the horn and can be either solid or foam dielectric. The principle of operation and design considerations are described. Various types of horn are analysed and the factors which influence the performance for use as a feed are studied.

A different use of dielectric is to form a lens which is placed in front of a smooth-wall or corrugated horn. This is done to improve the efficiency and increase the directivity of the horn. Chapter 11 describes the design procedure for lens-corrected horn antennas. Lenses made from solid dielectric and foam dielectric are described.

Small dielectric rods without a metal horn can also be used as a controllable radiator of energy and hence as a useful feed. These are the subject of Chapter 12 which describes the performance of various shapes of dielectric rod designed to optimise the radiation pattern and produce an efficient feed. The chapter also considers how dielectrics can be used as a means of modifying the radiation patterns of small waveguide feeds.

Attention then turns to small feeds which can be analysed using the numerical methods developed in Chapter 5. Chapter 13 studies dipole feeds. This was the first category of feed to be used because of the simplicity of construction and also the wide use at submicrowave frequencies. However, a single dipole has a poor radiation pattern with a large difference between the

Table 1.1 Quality assessment of types of feed

Type of feed	*Shape of pattern*	*Pattern symmetry*	*Peak crosspolar level*	*Gain*	*Efficiency*	*Bandwidth*	*VSWR*	*Size and volume*
Pure-mode horn	Gaussian	Medium	High	Medium	Medium	Medium	Good	Medium
Prime-focus waveguide feed	Broad	Medium	Medium	Low	High	Medium	Medium	Small
Multimode horn	Shaped	Medium to good	Medium to low	Medium	Medium	Narrow	Good	Medium
Corrugated horn	Gaussian	High	Very low	Medium to high	Low	Wide	Good	Large
Dielectric-loaded horn	Gaussian	High	Very low	Medium	Medium	Wide	Medium	Large
Dielectric-rod feed	Broad	Medium	Medium	Low	High	Medium	Poor	Small
Dipole feed	Broad	Poor	High	Low	Medium	Narrow	Poor	Small
Microstrip feed	Broad	Medium to poor	High	Low	High	Narrow	Medium	Small

principal-plane radiation patterns. For use as a feed, some form of small reflector is needed to direct the energy towards the main reflector. A small disc is one choice, and this and other options are analysed in Chapter 13 with the aim of showing that, by using good numerical techniques, dipole feeds can be designed to give good performance.

The feeds discussed in the earlier chapters are of traditional design based on waveguides, horns and dipoles. Their design and performance have improved to the extent that a wide choice of feeds is available. Chapter 14 studies the relatively new type of microstrip and resonator antennas. Microstrip feeds are becoming increasingly popular because they are easy to fabricate, low in cost and have compact profiles. Early microstrip antennas had poor bandwidth and asymmetric radiation characteristics, but these have now improved to the extent that microstrip antennas can be used as feeds in a reflector antenna. Chapter 14 studies microstrip antennas both as single units for use in prime-focus reflectors and as an element in an array to obtain a narrower radiation pattern.

The final chapter, Chapter 15, discusses the theory and performance of arrays of radiating elements to feed multiple- and shaped-beam reflector antennas. The feed element could, in principle, be any of the antennas described in the rest of the book. A practical array feed must be compact to occupy the smallest physical space, and this then introduces the problem of mutual coupling between feed elements. This is a controlling factor on performance and is studied in detail for conical waveguide feeds in Chapter 15, with examples of performance drawn from satellite communications.

1.5 Choice of type of feed

Many factors will determine the type of feed which should be used for a specific application. The system specifications for the communication, radar or remote-sensing system will determine the class of reflector antenna which will be needed. This will divide into two broad classes: prime-focus reflectors using a single-reflector antenna and dual reflectors employing a subreflector. The class of reflector antenna will then lead to a narrowing down of the choices of which feed to use. A single-reflector antenna usually requires a feed which radiates a broad beam, whilst a dual-reflector antenna needs a feed with a narrow beam of radiation. Feeds radiating narrower beams have wider apertures and generally give higher-quality radiation characteristics. Other electrical factors which may be relevant to the choice of feed are pattern symmetry, peak crosspolar level, gain, efficiency, bandwidth and input impedance. Often one or more of these is very important, in which case that factor may dominate the choice of feed. There are also some mechanical aspects which might be relevant such as the size, volume and weight of the feed. A summary of the relative merits of the main types of horn and other feeds described in this book is given in Table 1.1. This is inevitably a snapshot and there will be specific varieties of feed within each type which do not fit with the qualitative measures given in the Table. However, it can be treated as a general guide and qualitative assessment of the merits of each type of feed.

1.6 References

1 SILVER, S. (Ed.): 'Microwave antenna theory and design' (McGraw–Hill, 1949; reprinted Peter Peregrinus, 1984), IEE Electromagnetic Waves series 19
2 HERTZ, H.: 'Electric waves' (Dover Books, 1960; Macmillan, London, 1984)
3 BOSE, J.C.: 'Collected physical papers' (Longman, Green and Co., New York, 1927)
4 LODGE, O.J.: 'Signalling through space without wires' (The Electrician Printing and Publishing Co, London, 1898)
5 RAYLEIGH, LORD: 'On the passage of electric waves through tubes', *Philos. Mag.*, 1897, **44**, p. 28
6 SOUTHWORTH, G.C.: 'Hyperfrequency waveguides–general considerations and experimental results', *Bell Syst. Tech. J.*, 1936, **15**, p. 284
7 BARROW, W.L.: 'Transmission of electromagnetic waves in hollow tubes of metal', *Proc. Inst. Radio Eng.*, 1936, **24**, p. 1298
8 BARROW, W.L., and CHU, L.J.: 'Theory of the electromagnetic horn', *Proc. Inst. Radio Eng.*, 1939, **27**, pp. 51–64
9 SOUTHWORTH, G.C. and KING, A.P.: 'Metal horns as directive radiators of ultra-short waves', *Proc. Inst. Radio Eng.*, 1939, **27**, pp. 95–102
10 CHU, L.J.: 'Calculation of the radiation properties of hollow pipes and horns', *J. Appl. Phys.*, 1940, **11**, pp. 603–610
11 KAY, A.F.: 'A wide flare angle horn. A novel feed for low noise broadband and high aperture efficiency antennas'. US Airforce Cambridge Research Laboratories, report 62–757, 1962
12 CLARRICOATS, P.J.B., and OLVER, A.D.: 'Corrugated horns for microwave antennas' (Peter Peregrinus, 1984), IEE Electromagnetic Waves series 18
13 BARTLETT, H.E. and MOSELEY, R.E.: 'Dielguides–high efficiency low noise antenna feeds', *Microw. J.*, 1966, **9**, pp. 53–58
14 LUDWIG, A.C.: 'The definition of cross-polarisation', *IEEE Trans*, 1973, **AP–21**, pp. 116–119
15 SLETTEN C.J. (Ed): 'Reflector and lens antennas' (Artech House, Norwood, MA, 1988)

Chapter 2
Reflector antennas

P.J.B. Clarricoats

2.1 Introduction

2.1.1 General remarks

Although this book is concerned primarily with the characteristics of feeds, since feeds are the source of illumination for reflector antennas in the transmit mode and conversely on receive, it is important to understand the main characteristics of reflector antennas. This is the purpose of Chapter 2 which begins with a brief survey of reflector types and the principles of antenna design before considering the factors which underlie the theory and performance of reflector antennas in rather more detail.

2.1.2 Reflector types

Reflector antennas can be classified in various ways as shown in Table 2.1. There the antenna pattern, the reflector type and the type of feed establish a number of categories. The Table is intended to be read as a multiply connected diagram. For example, a satellite antenna may comprise an offset (off-axis)

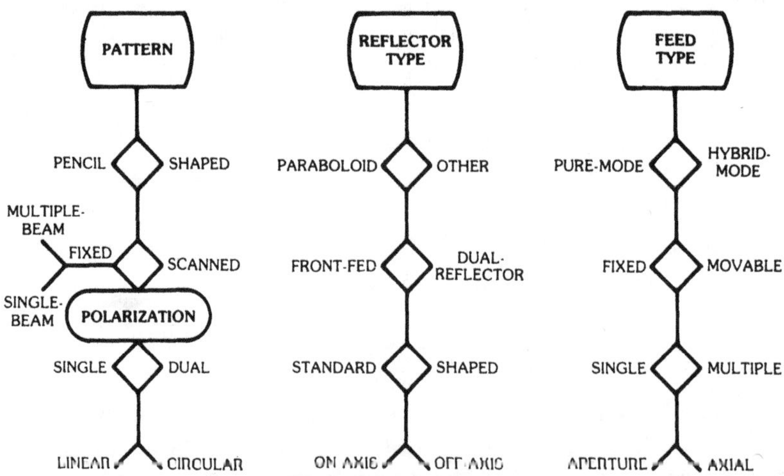

Table 2.1 Classification of reflector antennas

Reproduced by permission of Van Nostrand Reinhold

paraboloidal reflector fed through an array (multiple feed) so as to generate scanned pencil beams. These may be excited with (dual) orthogonal linear polarisation. As another example, the present radiotelescope at Arecibo in Puerto Rico utilises a spherical (other type) reflector with line (axial) feed. The pencil beam can be scanned by movement of the feed. Another illustration lies in the next generation of Arecibo radio telescope where the line feed is replaced by a dual reflector with a single feed; together with the spherical main reflector the antenna forms a trireflector offering improved efficiency, wide bandwidth and good crosspolar performance.

2.1.3 Constraints

The frequency of operation of an antenna, its bandwidth, beamwidth, gain and scanning requirements are among the key parameters which determine whether a reflector antenna constitutes a suitable choice for a particular system. A paraboloidal-reflector antenna operates efficiently, provided that it has dimensions exceeding about 10 wavelengths and has a surface roughness less than 1/16th wavelength RMS. A paraboloid reflector of 10 wavelengths diameter has a gain of about 25 dB, while at 100 wavelengths diameter the gain is 45 dB. At 450 MHz an antenna of 10 wavelengths diameter is 66 m across and this size constraint determines the lowest frequency at which reflectors can reasonably be employed. Usually reflectors have diameters in the range 40–100 wavelengths and operate at frequencies above 1 GHz. At 300 GHz the 15 m James Clerk Maxwell Telescope has a gain of about 88 dB and a beamwidth of less than 0.01°, but to achieve this performance an RMS surface accuracy of about 60 μm is required. Further, the profile of the antenna must be precisely maintained, and with large antennas this calls for massive structural support. Also, as the shape of a reflector antenna deforms under gravity, its shape will change if the pointing direction is changed. To maintain their performance, all large radiotelescopes are homologically mounted on their bearings so that they will always maintain a paraboloidal shape irrespective of the pointing direction, albeit that the focal length changes in a known way.

Terrestrial-communication antennas normally operate over a bandwidth of about 15%, but earthstation antennas can have their transmit and receive bands separated by 50%. The thrust of achieving this bandwidth lies with the feed, whose pattern must not change significantly over the band if efficient illumination of the reflector is to be maintained. Radiotelescopes can be required to operate over several octaves of frequency and this performance is achieved by the use of multiple feeds. These are usually mounted in a rotary turret arranged so that any feed can be brought to the focal point as required. Some communication antennas have extended operating bands by the use of an additional dichroic reflector and multiple feeds.

Early antennas operated with a single polarisation which was most usually linear, although circular polarisation was used in radars to reduce rain clutter. However, to accommodate a doubling of communication capacity, most satellite systems and many terrestrial systems now operate with dual polarisation. The choice between linear and circular polarisation depends on the application and the propagation path. Dual-polarisation operation places a stringent requirement on the feed design especially if it is necessary to maintain low levels of crosspolarisation, typically below −35 dB over a wide field of view and up to 50% bandwidth.

Relatively few antennas employ mechanically scanned feeds, but many satellite antennas use array feeds. This implies that most elements of the array do not lie at the reflector focus. Consequently, the scan properties of reflectors are important and have been the subject of investigation. This information is also relevant in antennas which use arrays either to furnish multiple beams or single- or multiple-scanned beams. Many satellite antennas are required to generate shaped beams and minimum gain contours will then be specified, usually together with a crosspolarisation specification. In a shaped reflector, the reflector diameter in wavelengths serves to determine the precision with which the contours can be defined and the gain falloff outside coverage. The maximum gain is set by the subtended solid angle within the coverage area.

Of prime concern in satellite systems is the antenna mass, and rigorous constraints are always applied. This influences the feed design; for example, a shaped reflector and a single feed are lighter than an array-fed reflector. However, the shaped reflector is unsuited when multiple regional coverages are required and to date array feeds have been widely used. Mass is also of importance in large earth-station antennas and in radiotelescopes. Mass sets one constraint on the maximum antenna diameter; cost and tolerances set the others. Antenna mass plays a rôle in terrestrial-communication systems where tower strength is dictated by this factor together with the requirement to tolerate thrust due to wind. Wind effects can be reduced by enclosing the antenna in a radome whose electrical effects on the antenna performance must be known.

The reflection coefficient or return loss of an antenna is another significant performance parameter in communication and radar systems. There are two contributors, namely the feed reflections and the reflector, and radome-reflections. Return loss is normally set below −30 dB.

It will be clear from the above that the feed plays a vital rôle in antenna performance and a proper understanding of feed design is vital to the antenna manufacturer.

2.2 Theory and performance of reflector antennas

2.2.1 The scattered field from a reflector

The purpose of the reflector in an ideal reflector antenna is to transform the angular distribution of energy emerging from the feed into a spatial distribution across the aperture which leads to the desired far-field radiation pattern. The converse applies on receive, but it is generally more convenient to treat the antenna as a transmitting device. Reciprocity ensures that the radiation pattern is identical on receive. For some special purposes it is useful to consider the antenna in a receiving mode, as is discussed in Section 2.2.8.

The theory of the reflector antenna was developed mainly during the 1940s [1] when, in particular, the aperture-field method was used to predict the radiation patterns of parabolic cylinder reflectors and paraboloidal reflectors. Although the theory of the physical optics (PO) method was developed, the absence of computers restricted its application. For radar applications, geometric optics (GO) techniques were used to design reflectors shaped in one dimension. By the 1960s, computers were available and physical optics could be used to analyse simple reflectors and later the geometrical theory of diffraction (GTD) and variants [2] enabled the radiation pattern to be predicted more accurately, especially away from boresight. For small reflectors, the method of moments provided a powerful tool in both pattern and scattering prediction. By the 1980s, computer power was sufficient to enable physical optics to be used for arbitrarily shaped reflectors of all but the largest sizes and the techniques which are currently available for analysis are given in Figure 2.1*a* with comments by Rahmat–Samii [3] in Figure 2.1*b*.

Figure 2.1a Analysis techniques for reflector antennas

GO (geometrical optics): Diffraction effect not considered; need to search for the specular points
GTD (geometrical theory of diffraction): suffer from caustic singularities; need to search for the diffraction points
GO/AF (geometrical optics/aperture field): Predict symmetric far-fields patterns for offset paraboloid with an onfocus feed
PO (physical optics): May not be accurate for far angular regions and crosspolarised fields
MoM (method of moments): Most accurate, but not easily applicable to large reflectors
PTD (physical theory of diffraction): Effectively modifies PO; flexible, and applicable to large reflectors

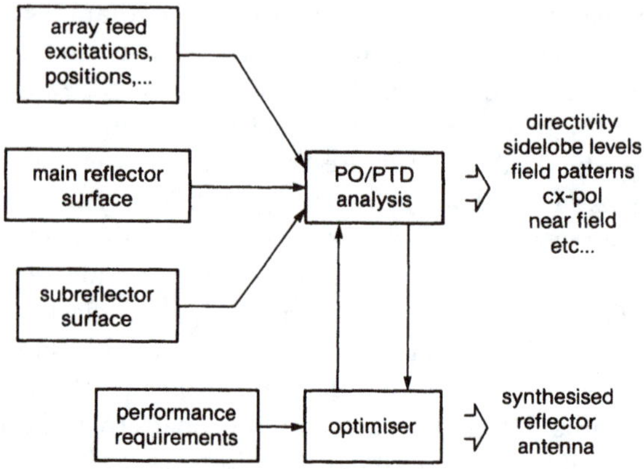

Figure 2.1b Techniques of reflector analysis [3]

Figure 2.2 shows an arbitrary reflector and an observation point which is assumed to lie in the far field. The electic component of the field radiated by the feed causes a current $\boldsymbol{J}$ to flow in the reflector surface which, for a solid metallic reflector, is assumed with great accuracy to be perfectly conducting. Then the scattered electric field $\boldsymbol{E}$ is given by:

$$\boldsymbol{E} = \frac{-j\omega\mu_0}{4\mu R} \exp(-jkR) \int_s \{\boldsymbol{J} - (\boldsymbol{J}.\hat{\boldsymbol{u}}_R)\hat{\boldsymbol{u}}_R\} \exp(jk\boldsymbol{r}'.\hat{\boldsymbol{u}}_R)\, ds \tag{2.1}$$

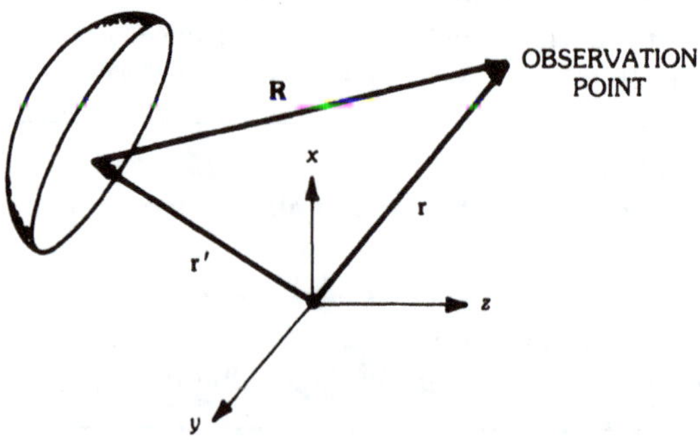

Figure 2.2 Co-ordinate system for radiation from reflector
Reproduced by permission of Van Nostrand Reinhold

where, to a good approximation, $\boldsymbol{J}$ and the incident $\boldsymbol{H}$ field are related through

$$\boldsymbol{J} = 2\hat{\boldsymbol{u}}_n \times \boldsymbol{H} \tag{2.2}$$

$$= C\{G_f(\theta, \phi)\}^{1/2} \frac{\exp(-jk\boldsymbol{r}')}{\boldsymbol{r}'} \tag{2.3}$$

G_f is the feed gain function which depends, in general, on both θ and ϕ although for an important class of feed the pattern can be made nearly independent of ϕ. Eqn. 2.1 is widely used to predict the radiation pattern of reflector antennas although a number of techniques, such as the Jacobi–Bessel expansion method [4], have been developed to reduce the computational load. In this method, the integral of eqn. 2.1 is rearranged in terms of a modified Jacobi polynomial whose coefficients are independent of observation angles. A schematic of the method due to Rahmat-Samii is in Table 2.2.

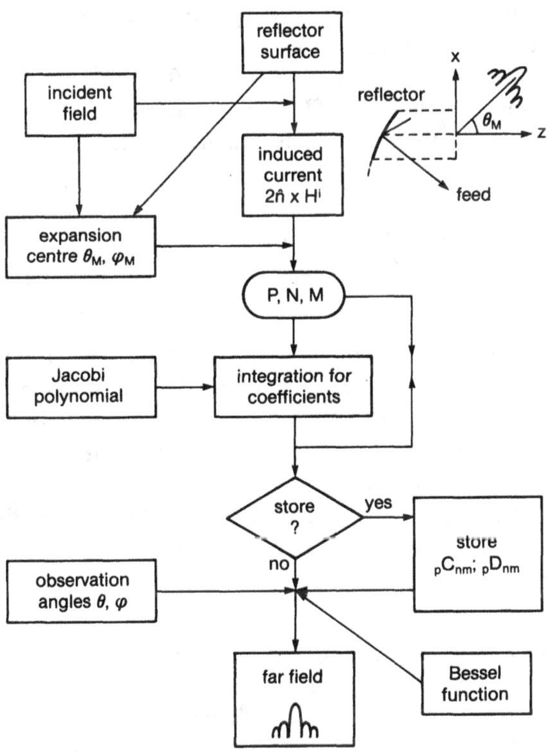

Table 2.2 Jacobi–Bessel method for determining radiation from a reflector [3]

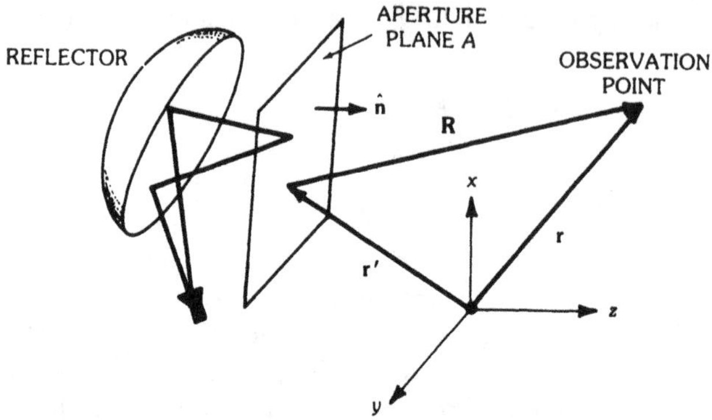

Figure 2.3 Co-ordinates for aperture-field method
Reproduced by permission of Van Nostrand Reinhold

2.2.2 *The aperture-field method*

The aperture-field method [1] is straightforward and is well suited to reflector antennas where the field in the reflector aperture has either a constant phase or nearly constant phase. From Figure 2.3 the power pattern of the feed is transformed into the aperture plane of the reflector using geometrical optics which can be supplemented by GTD and the far field of the antenna computed using an integration of the electric and magnetic fields over the aperture, as in eqn. 2.4:

$$\boldsymbol{E}=\frac{-jk^2\exp(-jkR)}{(\omega\varepsilon_0)4\pi R}\int_A\left[\hat{\boldsymbol{u}}_n\times\boldsymbol{H}_a-\{(\hat{\boldsymbol{u}}_n\times\boldsymbol{H}_a)\,.\,\hat{\boldsymbol{u}}_r\}\hat{\boldsymbol{u}}_r-\left(\frac{\varepsilon_0}{\mu_0}\right)^{1/2}(\hat{\boldsymbol{u}}_n\times\boldsymbol{E}_a\times\hat{\boldsymbol{u}}_r)\right]\exp(jk\hat{\boldsymbol{u}}_r.\boldsymbol{r}')\,\mathrm{d}A \tag{2.4}$$

General expressions for the far field are given in Chapter 3 and these reveal that the far field is essentially the Fourier transform of the aperture distribution which can be expressed either as in eqn. 2.4 or in terms of only $\boldsymbol{E}_a$ or $\boldsymbol{H}_a$. This observation is especially helpful in understanding, in physical terms, the role of tapered aperture distributions and the effect of aperture blockage. The aperture-field method gives very accurate results in angular directions near to boresight and is quite accurate over the first few sidelobes of high-gain antennas.

2.2.3 *Radiation from a symmetrical paraboloidal reflector*

Figure 2.4 shows a symmetrical paraboloidal reflector fed at its focus. The incident electric field at the reflector due to radiation from the feed is given by

$$\boldsymbol{E}_i=\hat{\boldsymbol{u}}_\theta E_\theta\cos\phi-\hat{\boldsymbol{u}}_\phi E\phi\sin\phi \tag{2.5}$$

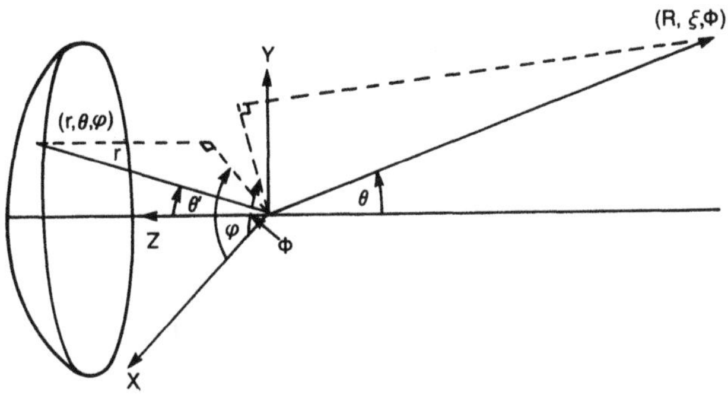

Figure 2.4 Co-ordinates for radiation from a symmetrical paraboloid

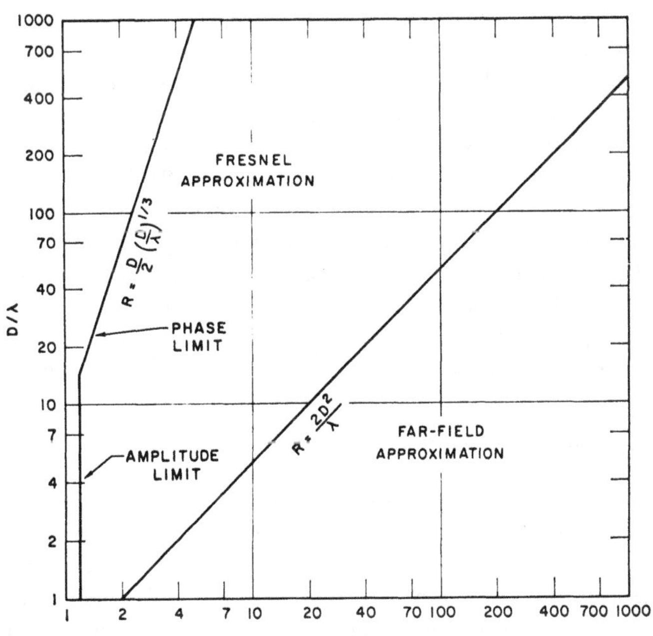

Figure 2.5 Far-field and Fresnel approximation for circular aperture of diameter D [6]

The electric field at R in the far field of the reflector is then given by a physical-optics integration over the reflector as

$$\boldsymbol{E}_R = \frac{-jkf}{R}\int_0^{\theta_0}\left\{\begin{matrix}\hat{\boldsymbol{u}}_x\{(\boldsymbol{E}_\theta - \boldsymbol{E}_\phi)J_o(u_1) + (E_\theta + E_\phi)J_2(u_1)\cos 2\Phi\} \\ +\hat{\boldsymbol{u}}_y(E_\theta + E_\phi)J_2(u_1)\sin 2\Phi - \hat{\boldsymbol{u}}_z j2E_\theta J_1(u_1)\tan(\theta'/2)\sin\Phi\end{matrix}\right\}$$

$$\times \exp(-ju_2)\tan\theta'/2\, d\theta'$$

where

$$u_1 = 2kf\tan\frac{\theta'}{2}\sin\theta$$

and

$$u_2 = kf\sec^2\frac{\theta'}{2}(1+\cos\theta\cos\theta') \tag{2.6}$$

Eqn. 2.6 shows that the radiation pattern contains both copolar and crosspolar components, the latter being especially important in dual-polarisation antennas. Notice that when $E_\theta = E_\phi$, the feed pattern is circularly symmetric and the crosspolar term in the far field vanishes. This condition implies that the feed is free from crosspolarisation and feeds with these properties are discussed in later chapters.

In nearly all systems, the receiving antenna lies in the far field of the transmitting antenna. However, the concept of the far-field distance of a planar aperture is important when the radiation properties of antennas have to be measured. Figure 2.5 [6] shows, for a circular aperture of diameter D, the conditions under which both the far-field approximation and the Fresnel approximation are valid. Outdoor antennas ranges normally aim to satisfy the far-field approximation, while indoor compact-antenna ranges aim to simulate the condition. As examples of the far-field distance $2D^2/\lambda$, an antenna of 5 m diameter operating at 6 GHz has a far-field distance of 1 km while a 30 m earth-station antenna operating at the same frequency has a far-field distance of 18 km. This latter distance is obviously impracticable to implement in a conventional outdoor range but by using a beacon on a satellite it is possible to measure the gain of an earth-station antenna. If the system has sufficient dynamic range, the beamwidth and first sidelobe can also be measured.

2.2.4 Features of the secondary-radiation pattern

The features of the secondary-radiation pattern of pencil-beam antennas will now be discussed. These antennas form a very important class, especially for telecommunication systems. They are found in microwave (point-to-point) radio systems where the major objective is to maximise gain, consistent with satisfying crosspolar constraints when they apply. Similarly, pencil beams are required in earth-station antennas and in radioastronomy, where reduction of antenna noise temperature is also a consideration. Antennas such as that of

Figure 2.4 and the related offset paraboloidal reflector, have a nearly constant phase front across their aperture. The constant-phase condition only departs because of reflector-surface inaccuracies, feed positional errors and changes in the feed phase centre with elevation angle. If the feed pattern is symmetric then, following Hansen, the normalized far-field radiation pattern depends on the parameter $\pi u = ka \sin \theta$, where a is the radius of the aperture, θ is the angle subtended by the far-field point with respect to boresight, and the feed taper is c. The latter is defined as the feed gain at the rim of the reflector relative to the axial gain. For a circular aperture, the generalised expression for the antenna pattern $F(u)$ is then related to a normalised aperture parameter $q = \pi(r/a)$ through

$$F(u) = \int_0^{\pi} g(q) J_0(qu) q \, dq \tag{2.7}$$

For certain distributions eqn. 2.7 can be solved exactly. For example, if

$$g\left(\frac{r}{a}\right) = c + (1 - c)\left\{1 - \left(\frac{r}{a}\right)^2\right\}^p \tag{2.8}$$

where c defines the edge taper T_E, through

$$T_E = 20 \log c \tag{2.9}$$

The case $p = 0$ corresponds to a uniform aperture distribution and leads to a $\{J_1(\pi u)\}/(\pi u)$ pattern. Simple closed-form expressions are available for integer values of p. Results for $p = 1$ and $p = 2$ have been provided, for example, by Hansen [6] and Rahmat–Samii [5] and Figure 2.6*a* shows the radiation pattern for a -10 dB edge taper with $p = 1$. Table 2.3 summarises results for $p = 1$ and $p = 2$ where it can be seen that the edge taper affects the beamwidth, sidelobes and aperture efficiency, the latter being referred to the uniform illumination condition for which $\eta_a = 1$.

As the edge-taper parameter c decreases, the beamwidth increases, the first sidelobe level falls and the efficiency also falls; c has to be less than 1 for a single-reflector antenna because all feeds have radiation patterns which decrease monotonically with angle from their boresight direction. By selecting the feed characteristics such as aperture size and flare angle, the edge taper 20 log c can be varied from typically -3 dB down to about -20 dB. The first value leads to considerable spillover beyond the main reflector but may be considered necessary in, for example, an array-fed reflector; the second value leads to very low sidelobes (in an unblocked aperture) but the aperture efficiency η_a of 69% gives a gain loss of 1.6 dB. A commonly chosen value of edge taper is -10 dB which is a compromise between spillover loss η_s and aperture efficiency η_a. The corresponding pattern is in Figure 2.6*a*.

In antennas for microwave radio systems, large earth-station antennas and radioastronomy antennas, the reflector is usually symmetrical and the feed

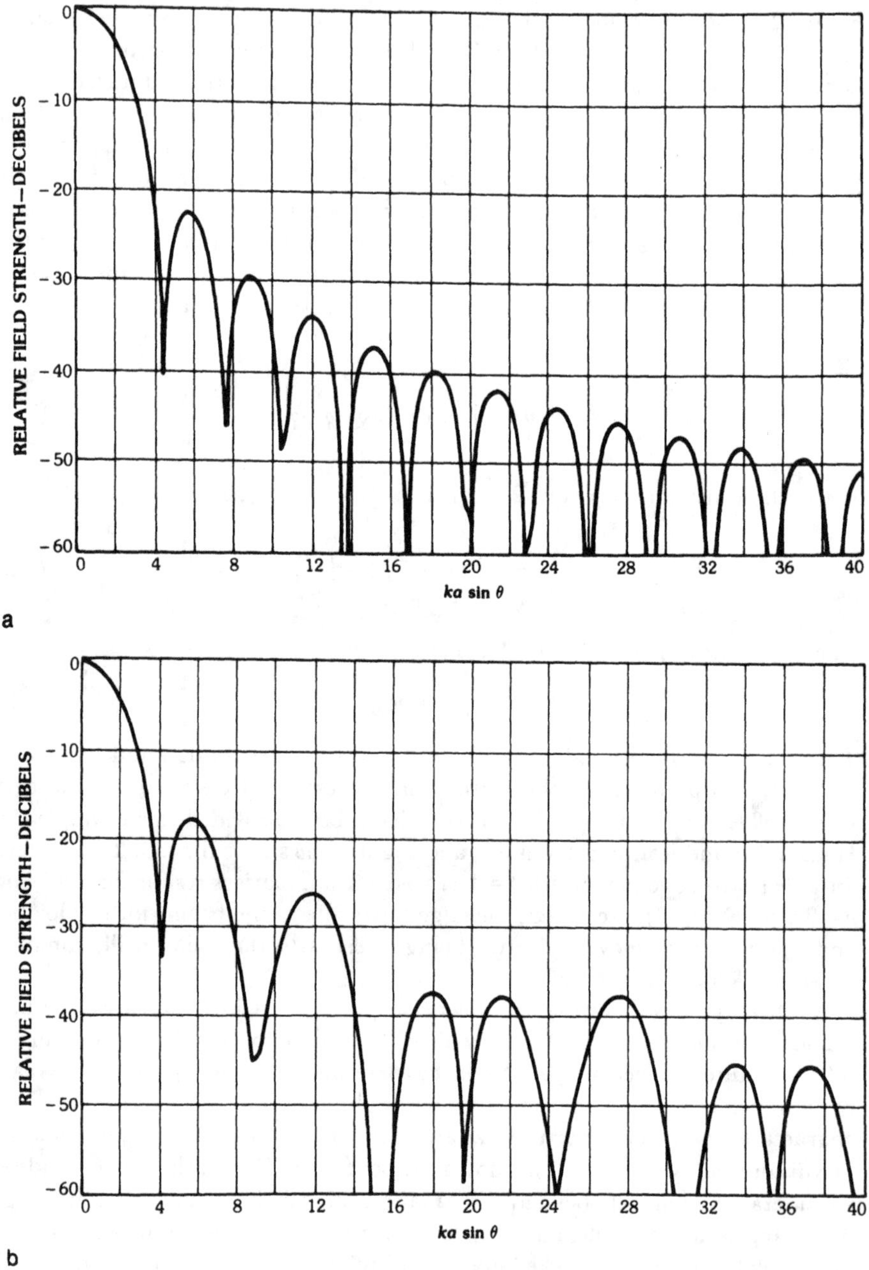

Figure 2.6 Radiation pattern of constant phase aperture with − 10 dB edge taper [5]

$p = 1$
a Without blockage
b With 20% blockage
Reproduced by permission of Van Nostrand Reinhold

Table 2.3 Table of radiation characteristics of circular aperture [5]

Edge illumination		$p=1$			$p=2$		
T_E	C	Halfpower beamwidth	Sidelobe level	η_t	Halfpower beamwidth	Sidelobe level	η_t
dB		rad	dB		rad	dB	
0	1	$1.01\lambda/2a$	−17.6	1	$1.01\lambda/2a$	−17.6	1
−8	0.398	$1.12\lambda/2a$	−21.5	0.942	$1.14\lambda/2a$	−24.7	0.918
−10	0.316	$1.14\lambda/2a$	−22.3	0.917	$1.17\lambda/2a$	−27.0	0.877
−12	0.251	$1.16\lambda/2a$	−22.9	0.893	$1.20\lambda/2a$	−29.5	0.834
−14	0.200	$1.17\lambda/2a$	−23.4	0.871	$1.23\lambda/2a$	−31.7	0.792
−16	0.158	$1.19\lambda/2a$	−23.8	0.850	$1.26\lambda/2a$	−33.5	0.754
−18	0.126	$1.20\lambda/2a$	−24.1	0.833	$1.29\lambda/2a$	−34.5	0.719
−20	0.100	$1.21\lambda/2a$	−24.3	0.817	$1.32\lambda/2a$	−34.7	0.690

obstructs the radiating aperture of the antenna and so introduces blockage. In a dual-reflector antenna to be discussed later, the blockage is usually entirely due to the subreflector. In these cases, the aperture integration of eqn. 2.7 does not have zero as its lower bound but the integral may still be evaluated in closed-form for values of $g(r/a)$ given in eqn. 2.8, provided that p is integer. A radiation pattern for a −10 dB edge taper with a blockage ratio D_0/D of 0.2 is shown in Figure 2.6*b*. The main effect of blockage is to reduce the gain in proportion to the loss of aperture area and to raise the sidelobe level due to truncation of the aperture field. For $D_0/D=0.2$, the first sidelobe lies at −17.5 dB (compared with −22.7 dB for $D_0/D=0$) while for $D_0/D=0.1$ the value is −20.7 dB. D_0 normally lies between 20 and 40 wavelengths and D_0/D between 0.1 and 0.2.

The main factors contributing to the antenna efficiency of an unblocked aperture are the aperture efficiency η_a and the spillover efficiency η_s leading to an overall efficiency $\eta=\eta_a\eta_s$. The manner in which η depends on the edge taper is shown in Figure 2.7. The shallow maximum of η around the value of $T_E=-10$ dB is evident and the corresponding maximum value of efficiency is 84%, as first noted by Silver. Other factors such as polarisation loss and surface errors also reduce the gain of a practical antenna, as does blockage, but the above two parameters are the main ones to be influenced by feed design. A very complete treatment of low-sidelobe-aperture distributions for blocked and unblocked circular apertures has been given by Ludwig [7]. He shows that very low sidelobes and reasonable efficiency are possible with certain idealised distributions but the designer of a single-refector antenna is greatly constrained by the feed distributions which can be realised. Nonetheless, the use of a second shaped reflector in conjunction with a shaped main reflector considerably enhances the range of realisable patterns.

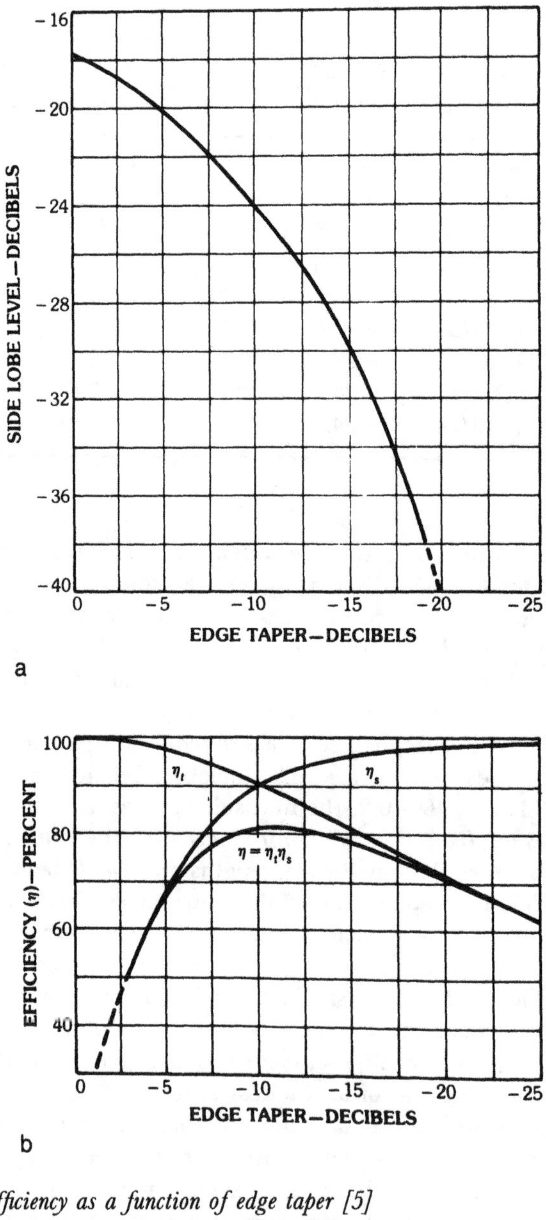

Figure 2.7 Efficiency as a function of edge taper [5]
Reproduced by permission of Van Nostrand Reinhold

2.2.5 The dual-reflector antenna

The dual-reflector concept was founded in the 17th century by the inventors of the optical reflecting telescope. Figure 2.8 shows examples of symmetrical Cassegrainian and Gregorian designs. Offset equivalents also exist and will be discussed later. The dual reflector offers a number of advantages over the front-fed single reflector. First, it enables the feed to be imaged so as to increase the effective focal length of the antenna, which is important in beam scanning. Secondly, by shaping the subreflector a more uniform illumination of the main reflector can be achieved consistent with a rapid falloff of energy at the reflector

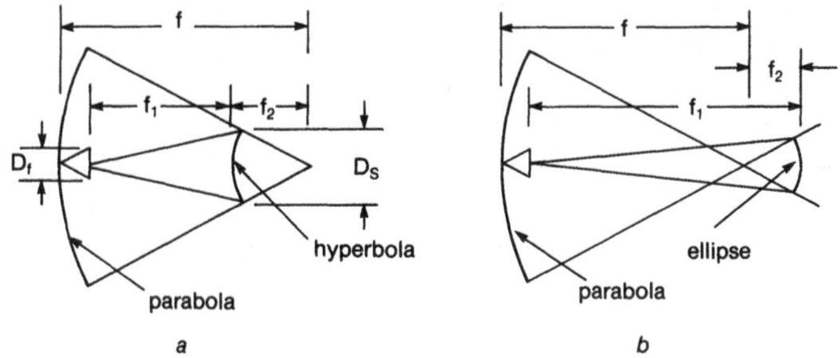

Figure 2.8 Geometries of Cassegrain and Gregorian reflectors [6]

a Cassegrain system
b Gregorian system. Parabola $f_\theta = mf = (f_1/f_2)f$

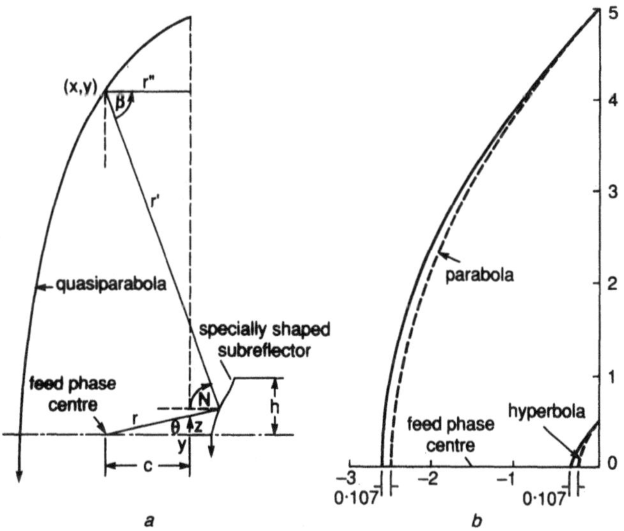

Figure 2.9 Dual-shaped reflector design after Galindo and Williams [16]

rim. To compensate for the phase error caused by this process, slight shaping of the main reflector is also required. Thirdly, it is possible to reduce the energy coupled back into the feed by suitably shaping the subreflector near to its vertex. Fourthly, the need for long feeder waveguides is eliminated, thereby reducing loss.

Figure 2.9 shows the geometrical-optics method adopted by Galindo [8] in the 1960s to enhance the efficiency of a dual Cassegrainian earth-station antenna. The technique relies on maintaining a constant path length between the feed phase centre and the antenna aperture, while simultaneously ensuring that the energy distribution over the aperture is constant. Details of the method have been given by Clarricoats and Poulton [16] and by Wood [10].

Although the geometrical-optics method is quite successful with large reflectors (of order 400 wavelengths diameter) the gain loss owing to diffraction is still about 0.5 dB. This value increases as the diameter of the reflector decreases and the designer is driven to diffraction methods. A discussion of diffraction-synthesis methods applied to symmetric dual reflectors is in Reference 16. Of these, the method which employs field correlation over the subreflector is both efficient and effective. The concept, which was developed by Wood [10], involves matching across the subreflector the radiated field of the feed and the incoming wave from the main reflector, as in Figure 2.10. Subject to an assumption, equivalent to the physical-optics approximation at the subreflector surface, the overall efficiency of the antenna η is

$$\eta = \boldsymbol{J} \cdot \boldsymbol{J}^* \tag{2.10}$$

where $\boldsymbol{J}$ is given either by

$$\boldsymbol{J} = \int_s \boldsymbol{E}_i \times \boldsymbol{H}_t . \mathrm{d}\boldsymbol{S} \tag{2.11}$$

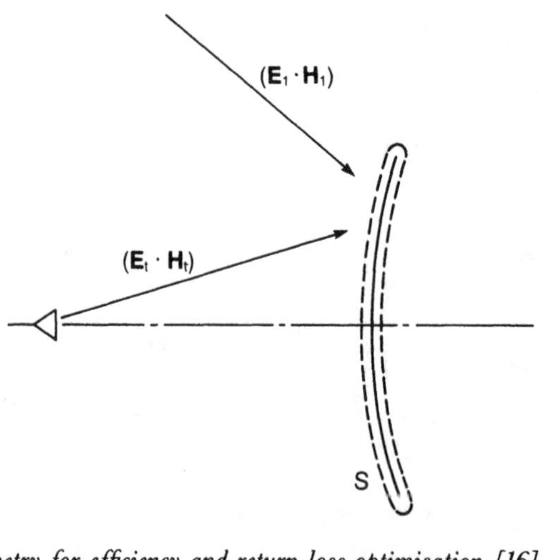

Figure 2.10 Geometry for efficiency and return-loss optimisation [16]

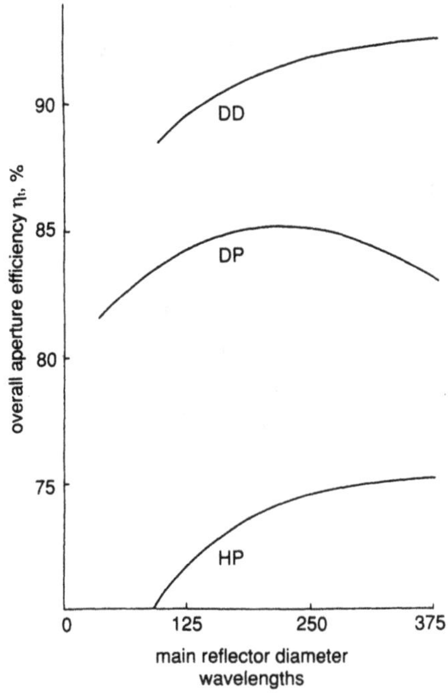

Figure 2.11 Efficiency of shaped Cassegrain antenna of 150-wavelength diameter [10]
HP = hyperboloid/paraboloid
DP = partially shaped antenna (Galindo and Williams)
DD = fully shaped antenna (diffraction analysis)

or

$$\boldsymbol{J} = \int_s \boldsymbol{E}_t \times \boldsymbol{H}_i . \mathrm{d}\boldsymbol{S} \tag{2.12}$$

The fields are defined in Figure 2.10.

A variety of methods can be used to calculate the incoming fields from the main reflector which are incident on the subreflector. GTD is one technique described by Poulton [9] while Wood [10] uses a spherical-wave expansion. The radiated fields from the feed can be obtained using various methods which are discussed later and also include the spherical-wave-expansion method. It is important to note that the subreflector may well lie in the near field of the feed and account should be taken of this. Once the above fields are known the subreflector shape can be selected to optimise efficiency. Figure 2.11. shows results due to Wood [10] for a Cassegrain antenna, with a main reflector diameter of 375 wavelengths. The diffraction-optimised subreflector is seen to give the highest efficiency. Figure 2.12 shows the scatter pattern of the 20-wavelength subreflector which shows that energy is steered away from the blocked portion of the main reflector aperture.

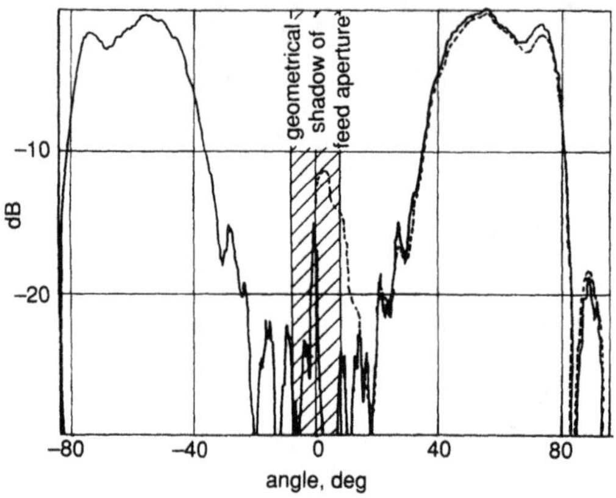

Figure 2.12 Subreflector radiation pattern for diffraction-optimised dual reflector (DPS) [10]

--- *predicted*
—— *measured*

2.2.6 *The offset-reflector antenna*

The offset-reflector antenna of Figure 2.13 has important advantages over the symmetrically fed reflector for certain applications. Subject to proper design, the feed system does not block the main reflector aperture so the aperture efficiency can be much higher than for the symmetrical counterpart. Furthermore, the absence of the feed and feed supports from the main aperture significantly reduces scattering away from the beam axis and reduces the far-out sidelobes. The absence of blockage is especially important when the feed is in the form of a large array of elements, as is the case for many satellite-communication or satellite-broadcast antennas.

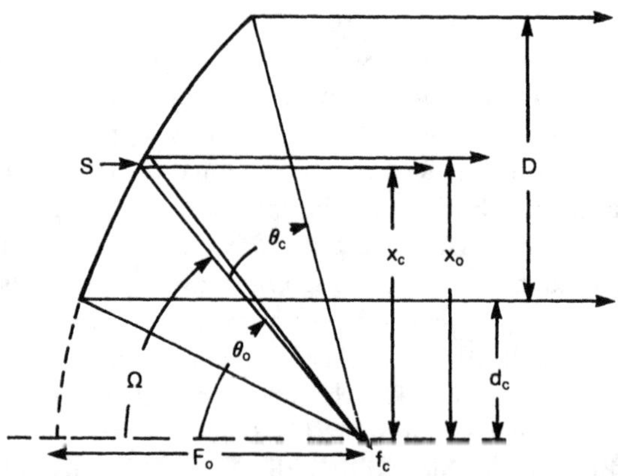

Figure 2.13 Co-ordinates of offset paraboloid reflector

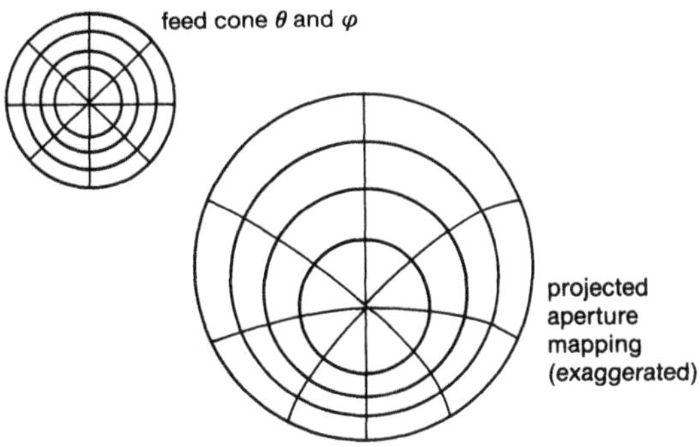

Figure 2.14 Mapping of feed co-ordinates onto reflector aperture

The main disadvantage of the offset reflector arises because, as shown in Figure 2.14, a circularly symmetric feed pattern does not map symmetrically into the reflector aperture, and although this has little influence on the copolar pattern there is inherent crosspolarisation even with a feed having intrinsic zero crosspolarisation (Figure 2.15) [11]. With an ideal linearly polarised feed, the

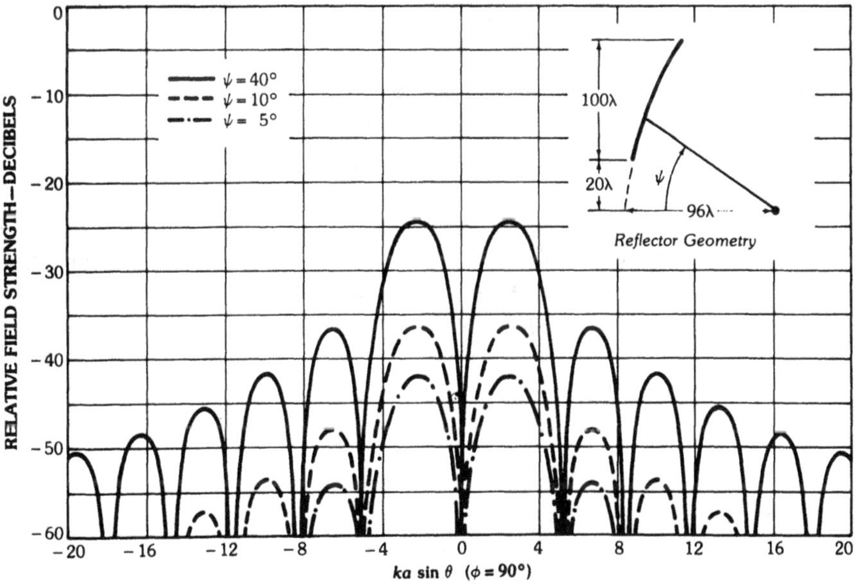

Figure 2.15 Crosspolar-radiation pattern of offset-reflector antenna (see inset for reflector geometry) [11]

Reproduced by permission of Van Nostrand Reinhold

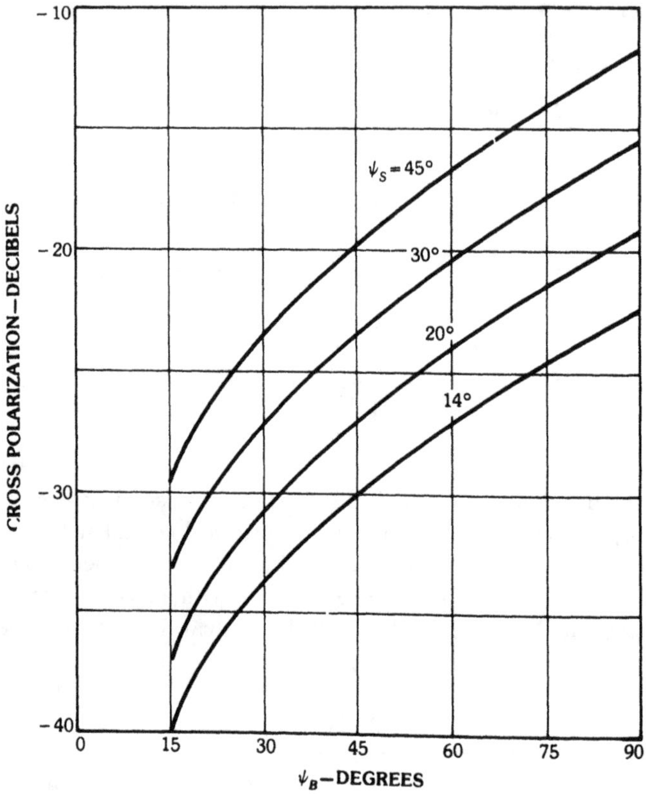

Figure 2.16 Crosspolar maximum as a function of offset angle ψ_B [11]
Parameter half subtended angle of reflector
Reproduced by permission of Van Nostrand Reinhold

level of secondary crosspolar radiation in the plane normal to the offset, depends on the tilt angle as in Figure 2.16. The maximum rises to a value of the order of −25 dB which is typical of practical offset antennas. The crosspolarisation can be compensated for by using a circularly polarised feed but this then leads to a beam squint θ_s given by

$$\sin\theta_s = \pm\frac{\sin\psi_B}{4\pi f/\lambda} \tag{2.13}$$

where the ± signs correspond to right- and left-handed circular polarisation, respectively. Figure 2.17 shows the normalised beam squint of an offset paraboloidal reflector illuminated by a circularly polarised feed at the focus.

Analysis of the fields in the focal region of an offset paraboloidal reflector due to an incident linearly polarised wave, shows that a feed supporting both a dominant mode and a higher-order-mode in the correct amplitude and phase

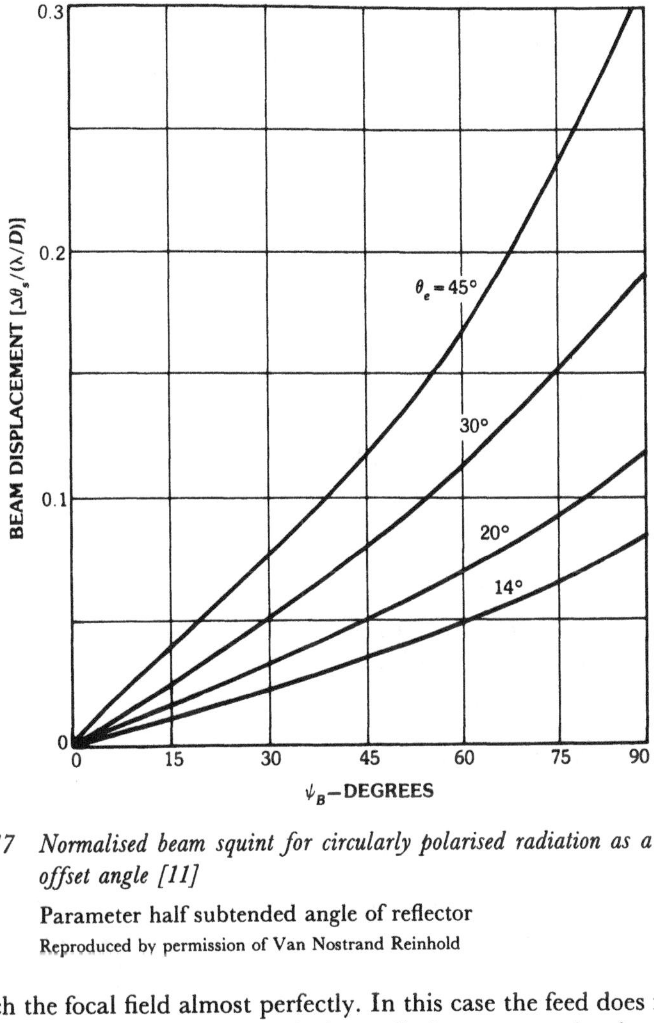

Figure 2.17 Normalised beam squint for circularly polarised radiation as a function of offset angle [11]

Parameter half subtended angle of reflector

Reproduced by permission of Van Nostrand Reinhold

can match the focal field almost perfectly. In this case the feed does not couple to the crosspolar components and the radiation pattern is almost free of crosspolarisation. This strategy was devised by Rudge and Adatia [12] to produce a so-called matched feed and, for applications requiring only a modest bandwidth, this type of feed is very effective.

The radiation pattern of an offset paraboloidal reflector fed by a matched feed is shown in Figure 2.18 [12] where the improvement over a conventional dual-mode Potter horn is evident.

As mentioned above, the main advantage of the offset paraboloidal reflector lies in the opportunity to mount a large feed array without blocking the antenna aperture. Obviously not all of the elements of the array can lie at the reflector focus and it is thus important to understand how feed displacement affects the radiation pattern. When the feed is displaced from the focus, the phase in the aperture plane of an otherwise perfect reflector system is no longer constant. A

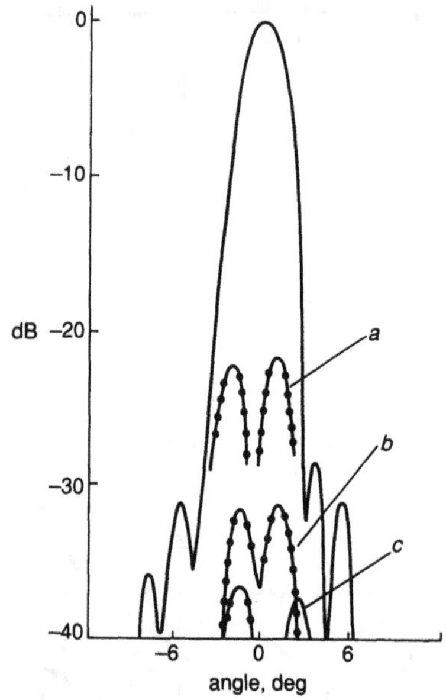

Figure 2.18 Copolar and crosspolar radiation of offset-reflector antenna fed by linearly polarised matched feed [12]

a Dual-mode hybrid feed
b Matched feed
c Optimised matched feed

linear phase dependence (Figure 2.19*a*) is desired, for this directs the radiation away from boresight, as may be required in a multibeam satellite antenna. However, this is not the only phase perturbation and Figure 2.19*b–e* shows the principal aberrations which occur. In principle, quadratic and spherical aberrations can be compensated for by axial feed movement but coma and astigmatic aberrations are not easily corrected, other than by special reflector designs.

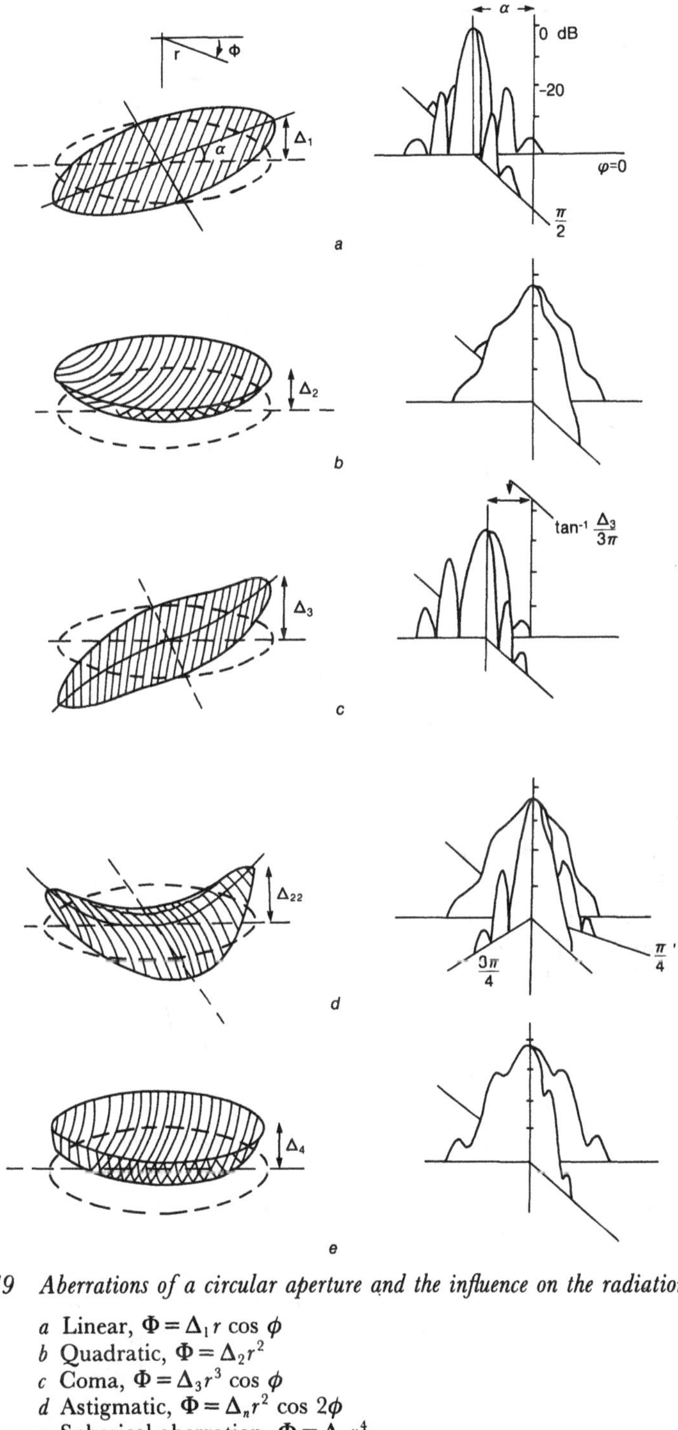

Figure 2.19 Aberrations of a circular aperture and the influence on the radiation pattern

a Linear, $\Phi = \Delta_1 r \cos \phi$
b Quadratic, $\Phi = \Delta_2 r^2$
c Coma, $\Phi = \Delta_3 r^3 \cos \phi$
d Astigmatic, $\Phi = \Delta_n r^2 \cos 2\phi$
e Spherical aberration, $\Phi = \Delta_4 r^4$

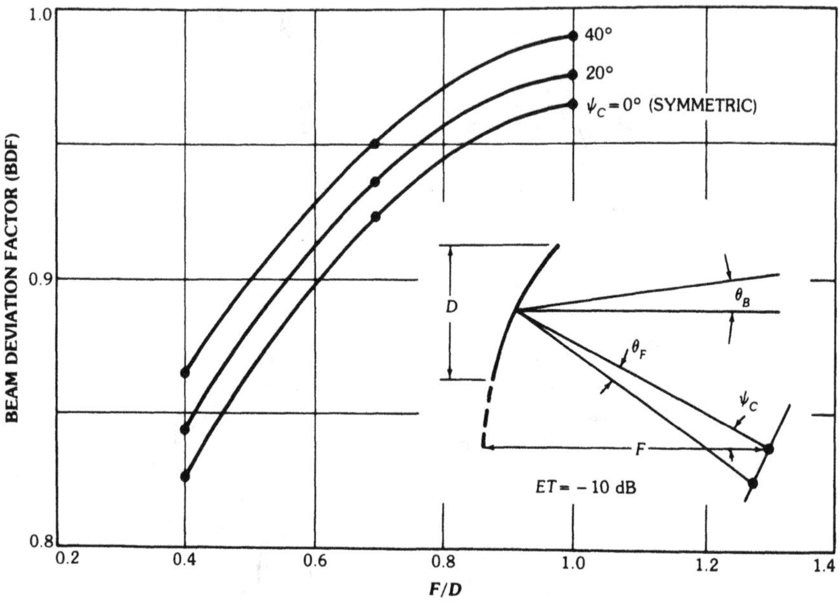

Figure 2.20 Beam-deviation factor for 100λ reflector with F/D = 0.96
Reproduced by permission of Van Nostrand Reinhold

An important parameter in describing the scanning characteristics of a reflector is the beam-deviation factor (BDF) which measures the angular extent θ_b by which the main beam is scanned for an angular displacement θ_f of the feed. The beam-deviation factor, which is defined by the ratio θ_b/θ_f, depends on the parameters F/D, the offset angle and the feed edge taper. Figure 2.20 shows the BDF [5] as a function of the first two parameters for an edge taper of −10 dB. In practice, it is found that the aberrations depend more on the ratio of the focal length to the diameter of the parent paraboloid rather than on the ratio for the actual paraboloid. Figure 2.21*a–c* illustrates the radiation patterns of a 100-wavelength diameter offset paraboloid for various feed displacements. The appearance of coma and astigmatism is evident and, because the feed is not tilted as it is displaced, the edge of the reflector becomes poorly illuminated. A considerable improvement in the pattern is possible if the feed can be tilted to point at the reflector centre as it is displaced but this is usually not possible in an array. To supplement our understanding, Figure 2.22 shows a far-field contour plot for a 50.5-wavelength diameter reflector with a 4-wavelength feed displacement. This produces a 2.5-beamwidth main-beam shift. We observe that, although the beam is circular at the −3 dB contour, there is distortion at lower gain levels owing to both coma and astigmatic phase errors. The crosspolar contours are also shown.

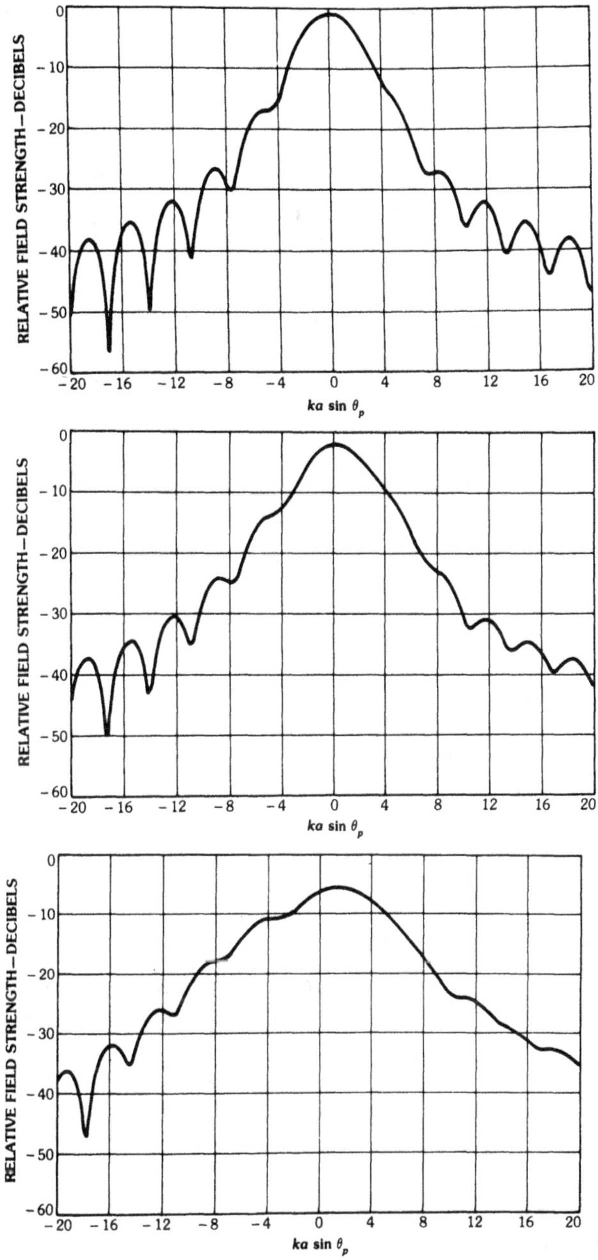

a

b

c

Figure 2.21 *Far-field-radiation pattern as a function of feed displacement [5]*

$F/D = 0.96$, $T_E = -10$ dB

a Four beamwidths

b Six beamwidths

c 10 beamwidths

Reproduced by permission of Van Nostrand Reinhold

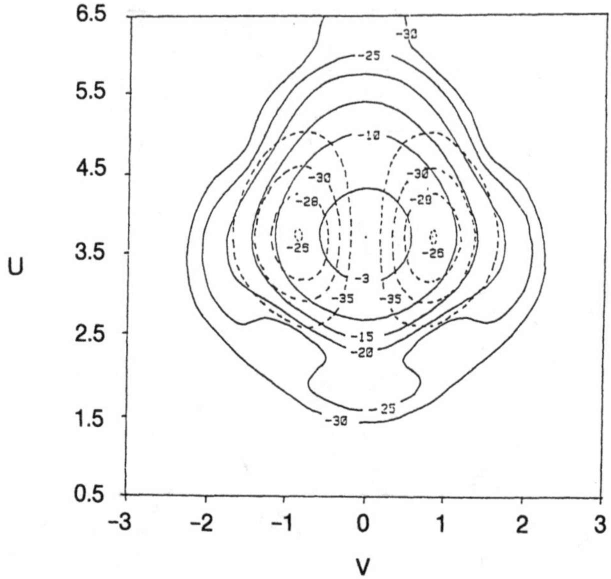

Figure 2.22 *Copolar and crosspolar contours for 50.5λ reflector of focal length 50λ with 40° offset and 4λ feed displacement*

$U = D \sin\theta \cos\phi$
$V = D \sin\theta \sin\phi$
—— copolar
– – – crosspolar

2.2.7 The dual-offset-reflector antenna

The potential absence of blockage in an offset reflector renders it an ideal candidate for a dual-reflector configuration. The most common form is that of the Cassegrainian design of Figure 2.23 or the Gregorian equivalent. If the main reflector is of paraboloidal shape, the subreflector is either hyperboloidal (Cassegrain) or ellipsoidal (Gregorian). The offset principle has been extended to other main reflector shapes such as spherical or quasiparaboloidal, in which case the subreflector is no longer a conic surface. Although the rim projection of the subreflector is elliptical, it is still customary to illuminate the antenna with a feed exhibiting a circularly symmetric radiation pattern. However, by judicious choice of the offset angles the high level of crosspolarisation, inherent in a single offset paraboloid, can be eliminated (within the limits of geometrical optics). The condition obtains when

$$\tan\frac{\theta_0}{2} = \frac{1}{M}\tan\frac{\alpha}{2} \qquad (2.14)$$

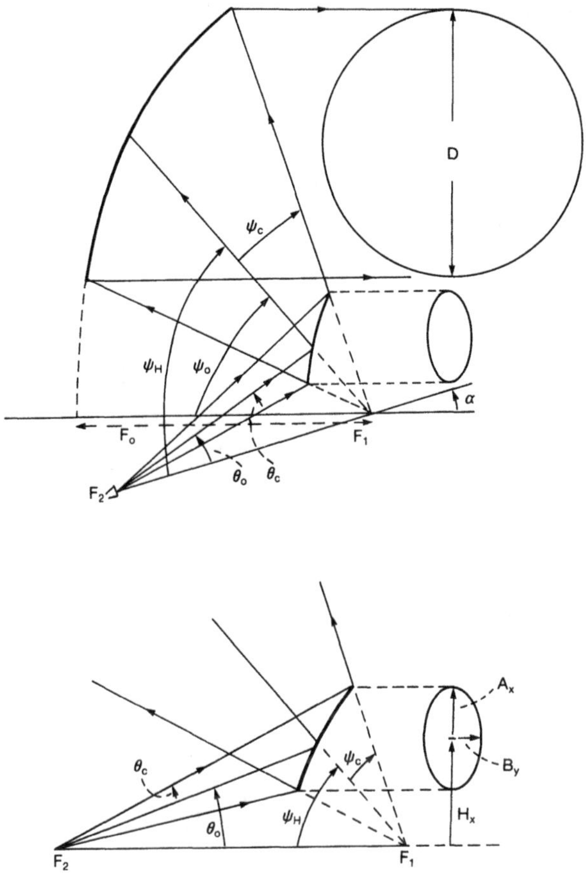

Figure 2.23 Geometry of offset Cassegrain reflector

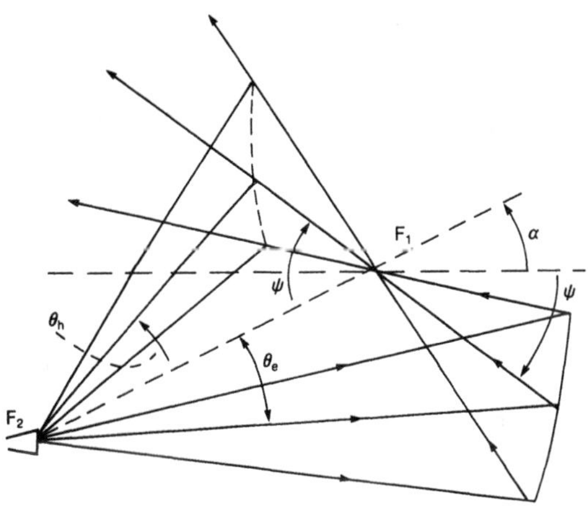

Figure 2.24 Equivalence between Gregorian and Cassegrain subreflectors

where M is the magnification which relates to the subreflector eccentricity e through

$$M = \frac{e+1}{e-1} \tag{2.15}$$

Within the geometrical–optics approximation there is an equivalence between Cassegrain and Gregorian reflectors as illustrated in Figure 2.24.

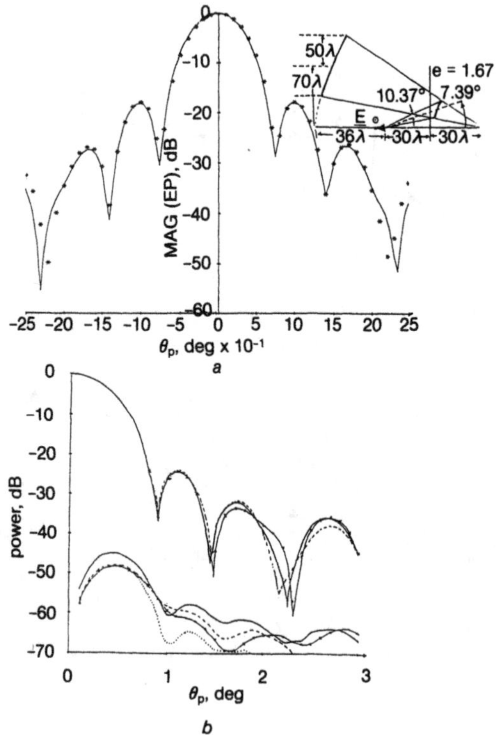

Figure 2.25 *Radiation patterns of dual-offset-reflector antenna (see inset) predicted by physical optics on the main reflector and different techniques for determining main-reflector currents*

a Copolar pattern
Feed = isotropic
—— GO + UAT
* * * GO + GTD

b Crosspolar pattern
Feed taper = − 10 dB
Subreflector scatterings:
— — GTD
—*— GTD + slope
—— PO
……. GO only

The radiation pattern of the subreflector can be determined by a variety of means and Figure 2.25*a* shows a comparison of the secondary pattern for a main reflector diameter of 50 wavelengths in a Cassegrainian configuration (see inset). The feed pattern is assumed to be circularly symmetric with a −10 dB taper at the subreflector edge. Evidently the copolar secondary pattern is only slightly dependent on the method of analysis whereas there is a much larger difference for the crosspolar pattern (Figure 2.25*b*). Of main note is the 20 dB improvement in the level of peak crosspolarisation, compared with the single offset reflector.

Offset-reflector antennas can also be used to form the optics of a more complex feed arrangement such as the beam-waveguide-fed 64 m deep-space antenna of the Jet Propulsion Laboratory (Figure 2.26*a*) [13]. There the beam-waveguide system, shown in Figure 2.26*b*, enables the low-noise receiver to be located in a stationary position as the antenna pointing angle varies. Three

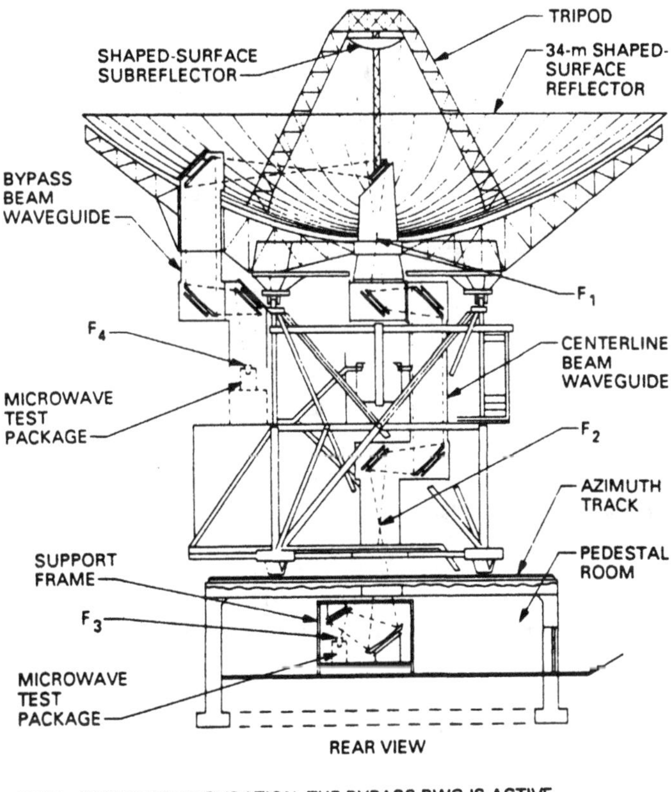

a

Figure 2.26 *Beam waveguides*

a JPL 64 m beam-waveguide-fed deep-space antenna. In this configuration the bypass beam waveguide is active

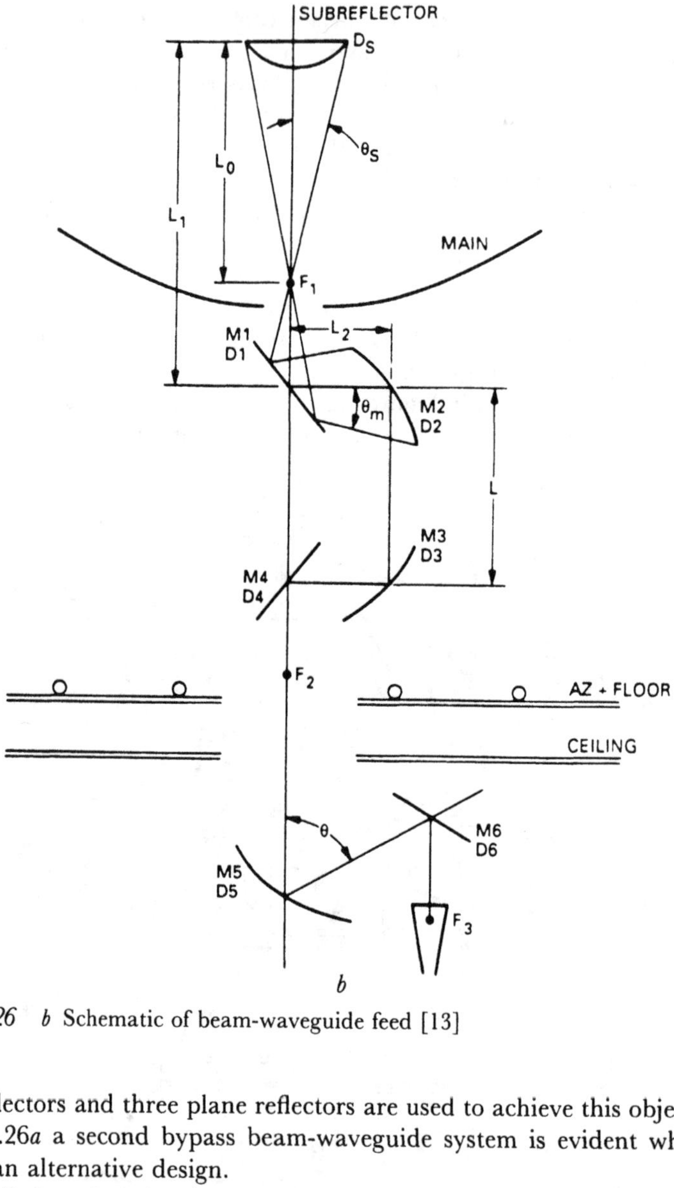

b

Figure 2.26 *b* Schematic of beam-waveguide feed [13]

offset reflectors and three plane reflectors are used to achieve this objective. In Figure 2.26*a* a second bypass beam-waveguide system is evident which was used as an alternative design.

A detailed analysis has been made of a spherical offset reflector, see Figure 2.27 [14]. The spherical surface offers ease of manufacture when antennas are required in large quantities since the surface can be made using a spinning process. To correct the inherent spherical aberration, a specially shaped subreflector lacking axial symmetry is required but, since it is much smaller in size, it could be produced as a pressed surface. Figure 2.28 shows the measured and predicted H-plane radiation pattern of such an antenna with an approximate 100-wavelength aperture. The subreflector was fed by means of 20°-flare-

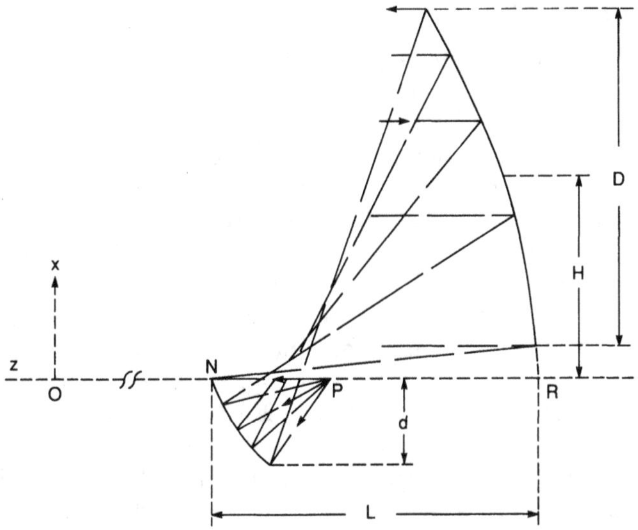

Figure 2.27 Offset spherical-reflector antenna with Gregorian geometry

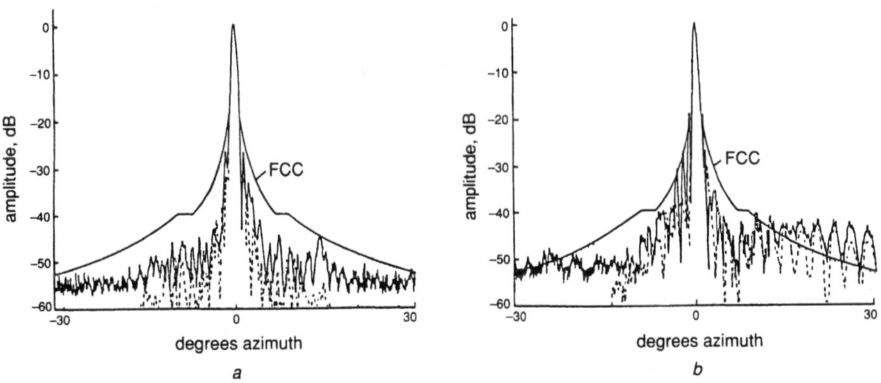

Figure 2.28 Predicted and measured copolar-radiation pattern of spherical offset antenna of 100λ diameter

a H-plane
b E-plane

angle corrugated horn and the measurements were performed at 37.5 GHz. The very low sidelobe envelope is of interest since this would lend itself to small earth-station applications. With closer separation of satellites in the geostationary arc, a lower sidelobe envelope is demanded to safeguard against interference between adjacent satellites.

2.2.8 Focal-field analysis

In previous sections the antenna has been treated as a transmitting device and attention given to the copolar and crosspolar radiation patterns. For some purposes it is helpful to view the antenna as a receiving system and to consider the fields created in the focal-region when a plane-wave is incident. In principle, it would be possible to determine the excitation of the appropriate mode or modes in the feed by a field-matching procedure across the feed aperture and so determine the antenna gain for a prescribed direction of incidence of the plane wave. The process could be repeated for other directions of incidence and the radiation pattern of the antenna constructed. Such a method would be laborious compared with the techniques presented above, but the advantage of the focal-field approach lies in the conceptual advantages it offers in initial design stages, especially when the reflector is fed by an array. It is also valuable in the design of matched feeds, as mentioned in Section 2.2.6.

Figure 2.29 shows a schematic of a paraboloidal reflector receiving a plane wave of electric-field intensity E_0 propagating along the z axis. The incident magnetic-field components of the wave excite currents in the reflector surface which radiate and lead to the fields observed in the focal region located in the vicinity of the geometrical-optics focus at the point (0, 0, 0). For a paraboloidal reflector, the electric and magnetic fields are given by

$$\boldsymbol{E} = j\omega\mu_0 H_0\{\hat{\boldsymbol{u}}_x I_1 + I_2(\cos 2\phi_f \hat{\boldsymbol{u}}_x + \sin 2\phi_f \hat{\boldsymbol{u}}_y) + \hat{\boldsymbol{u}}_z jI_3 \cos\phi_f\} \tag{2.16}$$

$$\boldsymbol{H} = j\omega\mu_0 \frac{H_0}{Z_0}\{\hat{\boldsymbol{u}}_y I_1 + I_2(\sin 2\phi_f \hat{\boldsymbol{u}}_x - \cos 2\phi_f \hat{\boldsymbol{u}}_y) + \hat{\boldsymbol{u}}_z jI_3 \sin\phi_f\} \tag{2.17}$$

$$I_1 = f\int_0^{\theta_0} \sin\theta J_0(kr_f \sin\theta)\, \mathrm{d}\theta \exp(-j\Phi_0) \tag{2.18}$$

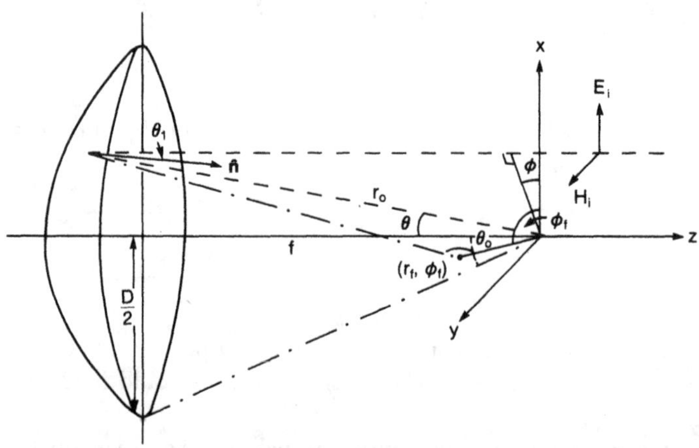

Figure 2.29 Paraboloid-reflector configuration for determining fields in the focal region

$$I_2 = f\int_0^{\theta_0} \sin\theta \tan^2\frac{\theta}{2} J_2(kr_f \sin\theta)\, d\theta \exp(-j\Phi_0) \tag{2.19}$$

$$I_3 = 4f\int_0^{\theta_0} \sin^2\frac{\theta}{2} J_1(kr_f \sin\theta)\, d\theta \exp(-j\Phi_0) \tag{2.20}$$

$$\Phi_0 = k(2f + z) \tag{2.21}$$

The above equations were originally derived by Minnett and Thomas [15] who used them to examine the focal fields of the Australian 64 m radiotelescope at Parkes.

In general, the integrals of eqns. 2.18–2.20 are complex but in the focal plane, where $z=0$, they are real and for a reflector of long focal length

$$I_1 = l_1\frac{J_1(u)}{u} \tag{2.22}$$

$$I_2 \sim 0 \tag{2.23}$$

and

$$I_3 = l_2\frac{J_2(u)}{u} \tag{2.24}$$

$l_{1,2}$ are constants and $u = kr_f \sin\theta_0$.

Under these conditions the x-directed field in the focal region is given by

$$E_x = 2E\left(\frac{\pi D^2}{4f\lambda}\right)\frac{J_1(kr_f\theta_0)}{(kr_f\theta_0)} \tag{2.25}$$

Eqn. 2.25 was first derived by Airy in his 19th-century investigations of the distribution of light in the focal region of a lens. He found the focal region to consist of a central bright disc surrounded by annular rings of decreasing intensity with interspersed dark rings corresponding to the nulls of the function $J_1(U)$. Figure 2.30 [15] shows the electric-field directions in one quadrant of the focal plane of a paraboloidal reflector with $F/D=0.4$. One quadrant is shown since the distribution is symmetric about the axis. The central region is essentially linearly polarised, as is the field in the illuminated rings, only the direction of the field alternating with distance from the focal point. The intensity varies as in Figure 2.31(i). Figure 2.30 shows that, along planes lying at 45° with respect to the principal planes, there are regions of crosspolarisation. When the incidence angle is changed, for small deviations, the pattern of Figure 2.30 shifts and only minor changes in form occur. For example, scanning along one of the principal planes, either $x=0$ or $y=0$, causes intensity fluctuations as

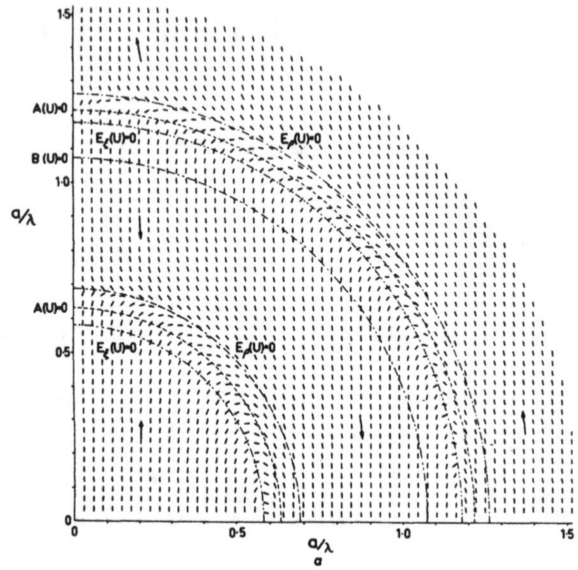

Figure 2.30 Electric-field distribution in focal region of reflector due to a plane-polarised incident wave [15]. Paraboloid-opening angle $\theta_0 = 63°$, $F/D = 0.4$

shown in Figure 2.31(*i*) [15] with nulls occurring as the 'dark rings' cross the feed. The polarisation remains pure linearly polarised throughout this scan. By contrast, scanning along a 45° plane produces a similar result for the copolar field but for an arbitrary feed the level of crosspolarisation rises to a maximum as the focal crosspolar field of Figure 2.30 overlays the feed aperture. The term arbitrary feed indicates that there can be a class of feed for which, in principle, the crosspolarisation can remain zero throughout a 45° scan. This condition is obtained with a so-called hybrid feed whose properties are examined in later chapters. The proof of this condition follows.

To calculate the power coupled into a feed a theorem based on reciprocity gives

$$P = \left| \left\{ \frac{1}{2} \int_s (\boldsymbol{E}_1 \times \boldsymbol{H}_2 + \boldsymbol{H}_1 \times \boldsymbol{E}_2) \cdot \mathrm{d}\boldsymbol{S} \right\} \right|^2 \tag{2.26}$$

where the fields $\boldsymbol{E}_1$, $\boldsymbol{H}_1$ arise from unit incident power in the plane wave of Figure 2.29 while $\boldsymbol{E}_2$, $\boldsymbol{H}_2$ arise from unit power in the feed. If the focal field is resolved into transverse fields of two types, based on eqns. 2.16 and 2.17, we have

$$\boldsymbol{E}_A \propto \hat{\boldsymbol{u}}_x I_1 \tag{2.27}$$

$$\boldsymbol{H}_A \propto \hat{\boldsymbol{u}}_y I_1 \tag{2.28}$$

$$\boldsymbol{E}_B \propto \hat{\boldsymbol{u}}_x I_2 \cos 2\phi + \hat{\boldsymbol{u}}_y I_2 \sin 2\phi \tag{2.29}$$

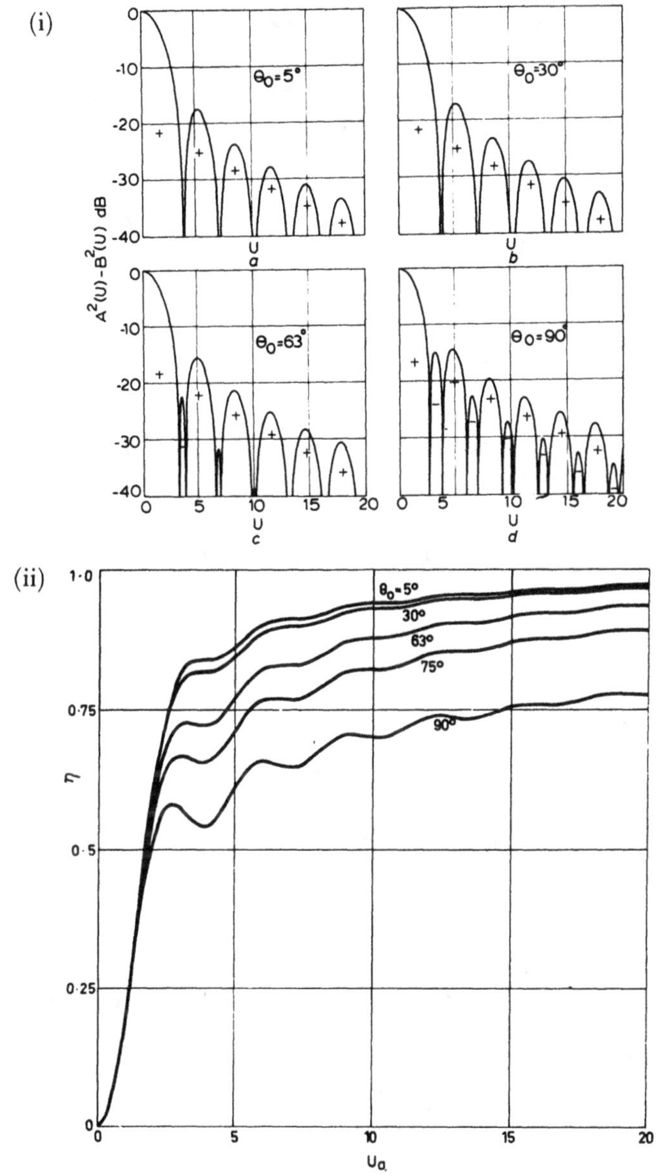

Figure 2.31 *Intensity of electric field in focal region and aperture efficiency for different opening angles θ_0 [15]*

(i) Intensity of electric field
a $\theta_0 = 5°$
b $\theta_0 = 30°$
c $\theta_0 = 63°$
d $\theta_0 = 90°$

(ii) Aperture efficiency η as a function of $u_a = kr_f \sin \theta_0$. r_f denotes the radial distance from geometrical-optics focal point

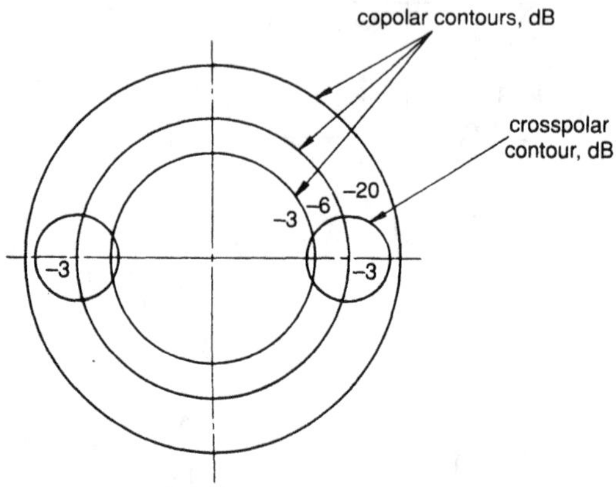

Figure 2.32 Focal-field contours for offset paraboloid illuminated by a linearly polarised plane wave [17]

$$\boldsymbol{H}_B \propto -\hat{\boldsymbol{u}}_y I_2 \cos 2\phi + \hat{\boldsymbol{u}}_x I_2 \sin 2\phi \tag{2.30}$$

If

$$\boldsymbol{E}_2 = \hat{\boldsymbol{u}}_x F(r, \phi)$$
$$\boldsymbol{H}_2 = -\hat{\boldsymbol{u}}_y F(r, \phi) Y_0$$

where

$$Y_0 = \frac{1}{Z_0} = \left(\frac{\varepsilon_0}{\mu_0}\right)^{1/2}$$

corresponding to a 'balanced-hybrid' mode travelling in the negative z direction, substitution of eqns. 2.27–2.30 into eqn. 2.26 gives

$$P = \int_0^{2\pi} \int_0^a 2Y_0 I_1(r) F(r, \phi) r \, dr \, d\phi \tag{2.31}$$

and the contribution from the crosspolar 'B' fields of eqns. 2.29 and 2.30 disappears. This important observation was first made by Poulton [see Reference 16].

In a similar manner, the focal field of an offset antenna can be analysed. Figure 2.32 [17] shows the focal-field contours for an offset paraboloid. The crosspolar maxima now lie in the H-plane. The matched feed referred to in Section 2.2.6 generates a field which is the conjugate of the focal field and again crosspolarisation can be effectively eliminated. With a hybrid-mode feed this can be accomplished by exciting the HE_{21} mode in addition to the dominant HE_{11} mode.

Focal-field analysis permits us to determine the power flowing in the focal plane $\boldsymbol{S}_z$

$$S_z = -\frac{1}{2} \frac{(kfE_0)^2}{Z_0} \frac{\theta_0}{2} (|I_1|^2 - |I_2|^2) \tag{2.32}$$

If $\boldsymbol{S}_z$ is integrated over an aperture in the focal plane the total energy P is obtained from which the normalised aperture efficiency η_a is also obtained. Figure 2.31(*ii*) shows it as a function of $U_a = ka \sin \theta_0$. The power incident on the aperture of the paraboloid is P_0 where

$$P_0 = 2\pi \left(f \tan \frac{\theta_0}{2} \right)^2 \frac{E_0^2}{Z_0} \tag{2.33}$$

thus $\eta = P/P_0$ represents the fraction of the total power which passes through a regon of radius a. For small θ_0, corresponding to a reflector with long focal length,

$$\eta = 1 - J_0^2(U_a) - J_1^2(U_a) \tag{2.34}$$

This result was originally derived by Lord Raleigh.

If all the power flowing through the central region of the focal plane is intercepted, an efficiency of 84% is achieved. To improve on this value calls for the use of dual-shaped reflectors as described in Section 2.2.5.

2.2.9 *Antenna return loss*

In a number of applications, especially microwave-communication systems, the power coupled back into the feed owing to scattering from the reflector is of importance and stringent limits on this parameter are set. Following the treatment by Silver [1], a reflector fed by a feed of peak gain G_f gives rise to a voltage-reflection coefficient Γ_R which in the asymptotic limit has a magnitude

$$|\Gamma_R| = G_f \frac{\lambda}{4\pi f} \tag{2.35}$$

The result can be interpreted as follows. Assume that the feed has an aperture A and is positioned at the focus which lies at a distance f from the reflector vertex. If the feed radiates a power P_T then that contained in the solid angle A/f^2 is

$$\frac{P_T}{4\pi} \frac{G_f A}{f^2}$$

and this is the power which is collimated back into the feed. On substituting for A where $A = G_f \lambda^2/(4\pi)$, into the above expression, the power-reflection coefficient is obtained as $\{G_f \pi/(4\pi f)\}^2$. This corresponds to the voltage-reflection coefficient of eqn. 2.35.

For typical parameters of communication antennas, the value of Γ_R expressed in decibels can be above -30 dB and a means to reduce this value is required. The technique which is widely used is to incorporate into the feed-support structure a flat plate which covers the vertex of the reflector, as in Figure 2.33*c* [1]. The technique, which is known as vertex-plate matching, relies on causing a portion of the reflected energy from the vertex region to arrive in antiphase with energy from the remainder of the reflector. If r defines the radial distance from the axis to a point on the paraboloidal surface, and $\nu = r^2/f^2$, then the real and imaginary parts of Γ, for Γ tending to infinity, are given by

$$R_\infty = \frac{\alpha G_0}{4\left(\alpha^2 + \frac{\pi^2 f^2}{\lambda^2}\right)} \tag{2.36}$$

$$I_\infty = \frac{\pi f G_0}{4\lambda\left(\alpha^2 + \frac{\pi^2 f^2}{\lambda^2}\right)} \tag{2.37}$$

In the derivation, a Gaussian-feed radiation pattern $G_0\, e^{-\alpha\nu}$ is assumed, which for a corrugated feed constitutes a good approximation. For large f/λ, Γ becomes independent of the feed taper.

Again, using the method described by Silver [1] and referring to Figure 2.33, the method of reflection compensation is as follows. The perpendicular bisector of the total reflection coefficient Γ_t is erected and its intersection point v_c on the

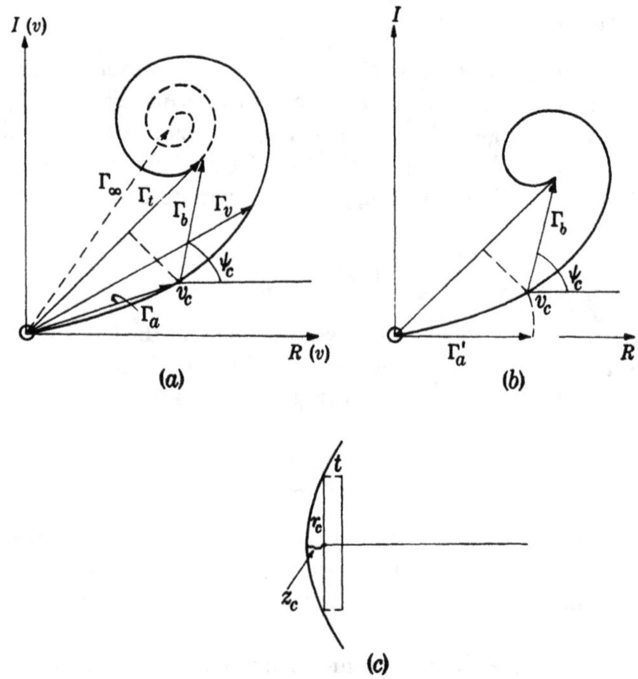

Figure 2.33 Complex reflection coefficient due to scattering from a paraboloid reflector, construction for vertex-plate matching and section through paraboloid showing vertex plate at distance r_c from vertex

a Complex reflection coefficient
b Construction
c Section

spiral determined. This divides the reflector into two zones, one within the radius r_c contributing a reflection coefficient Γ_a and that outside contributing Γ_b, where the magnitudes of Γ_a and Γ_b are equal. A vertex plate of radius r_c is now placed against the surface of the reflector. The path from the focus F to the edge is $(f+zc)$ while the path to the centre of the vertex plate is $(f-zc)$, the average path length is thus f. The plate thus rotates Γ_a onto the real axis R of the reflection coefficient, so as to bring it into coincidence with the contribution from the vertex-plate region. To bring Γ_a and Γ_b into antiphase, the vertex plate is made of thickness t such that

$$t=(2n+1)\frac{\lambda}{4}-\frac{\lambda\Psi_c}{4\pi} \qquad n=0,\ 1,\ 2,\ \ldots \tag{2.38}$$

Ψ_c is the angle between Γ_b and the R axis. As f/λ becomes very large, the parameters of the vertex plate become practically independent of the aperture and the feed taper. Furthermore, if the taper is not too large the initial portion of the spiral in Figure 2.33 [1] can be regarded as a circle of constant radius (Γ_∞) with centre on the I axis. Then, according to Silver,

$$r=\frac{1}{2}\left(\frac{4f\lambda}{3}\right)^{1/2} \tag{2.39}$$

and

$$t=(2n+1)\frac{\lambda}{4}-\frac{5\lambda}{24} \tag{2.40}$$

A reflector operating at 6 GHz with $f=12\lambda$ has $r=2\lambda=10$ cm.

Since the vertex-matching technique is frequency dependent, it is customary to design for midband. But, in many microwave-communication systems, it is still possible to ensure that Γ lies below -30 dB over the required band of frequencies.

In symmetric dual-reflector antennas it is possible to minimise reflection by perturbing the subreflector shape near to vertex. Poulton [9] has developed an optimisation process which simultaneously maximises antenna gain and minimises return loss. The details of the method are to be found in Reference 9 and the results are shown in Figure 2.34. There a substantial improvement in efficiency is shown to be possible compared with a standard Cassegrain reflector while maintaining a low value of return loss.

For offset-reflector antennas in general, there is minimal scattering into the feed and the return loss is dominated by the reflection coefficient of the feed. Matching of feeds is discussed later.

2.2.10 Radome-covered reflectors

In many microwave-communication systems the reflector antenna is protected by a radome which is relatively transparent to electromagnetic waves at the frequency of operation. Curved radomes not only provide protection from the

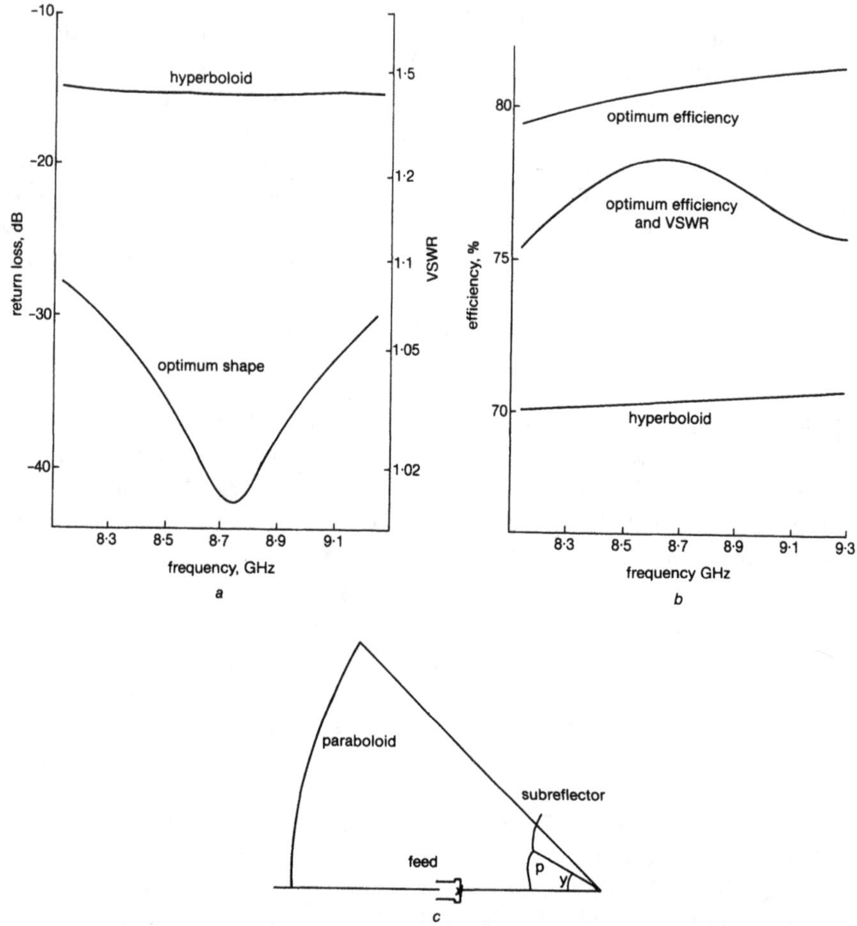

Figure 2.34 *Return loss and efficiency of antenna illustrated in Figure 2.34c, and Cassegrain antenna with shaped reflector*

a Return loss of 8.7 GHz dual-reflector antenna of Figure 2.34*c* when main reflector is a paraboloid of 2 m diameter and subreflector is either hyperboloidal or of optimum shape to reduce return loss [9]
b Efficiency of dual-reflector antenna of Figure 2.34*c* under different optimisation conditions
c Cassegrain antenna with shaped subreflector

adverse effects of precipitation; they also significantly reduce wind loading, leading to substantial savings in the cost of the tower required to support the antenna. Microwave antennas with plane radomes serve principally to protect the reflector from weather effects. Usually, the radome is made from a hydrophobic cloth which has enough flexibility in the wind to shed most forms of precipitation. Some large military radar antennas and some military earth-station antennas are covered by a space-frame radome which serves to protect

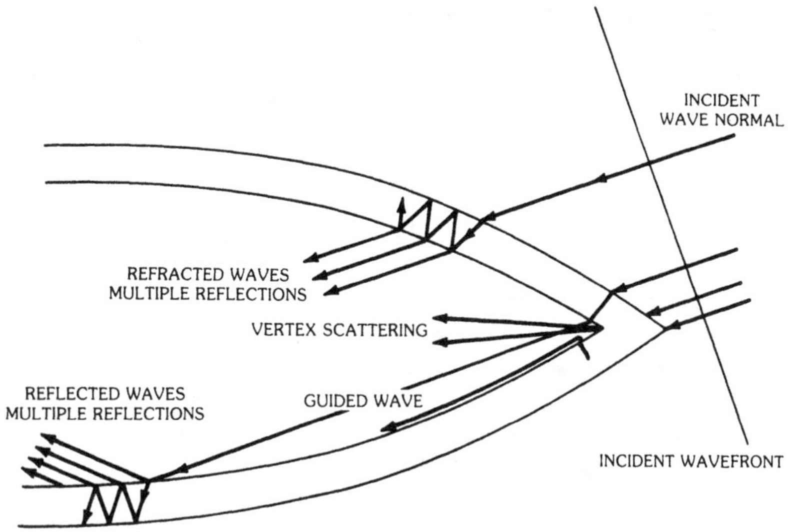

Figure 2.35 ***Waves in a pointed radome due to incident plane wave [18]***
Reproduced by permission of Van Nostrand Reinhold

the system and inhibit visual surveillance. Airborne and missile radars are always radome covered for aerodynamic reasons. Millimetre-wavelength radiotelescope antennas are covered to prevent wind from influencing the direction of antenna pointing. For example, the James Clerk Maxwell 15 m radiotelescope on Mauna Kea is protected by a rotatable housing with a narrow slit opening covered by a polypropylene sheet.

In communication applications, the main adverse effect of the radome is to increase the return loss of the antenna although transmission losses may not be negligible. In airborne applications, beam-pointing errors can be significant. There solid radomes are used and a number of waves can be excited in the radome, as Figure 2.35 shows [18]. Such radomes can be made from halfwavelength-thick materials or in the form of an A-sandwich construction. An A-sandwich comprises two thin layers of high permittivity surrounding a core of lower-permittivity material. The scheme offers a reasonable match over a broad operating-frequency band. A-sandwich radomes are also lighter than solid radomes which, in general, are one halfwavelength thick, a factor of importance in airborne applications. Figure 2.36 [18] shows the power transmission through an A-sandwich wall. The incidence angle is relevant in an application where the antenna rotates relative to the radome housing. A comprehensive treatment of radomes for airborne applications is in Reference 18.

Microwave-communication-system antennas using curved radomes often employ paraboloidal shapes since they can be made using a mandrel which is itself used to form a reflector but of a different focal length. These radomes, as in Figure 2.37*a* and *b*, are normally made from glass fibre since this offers strength,

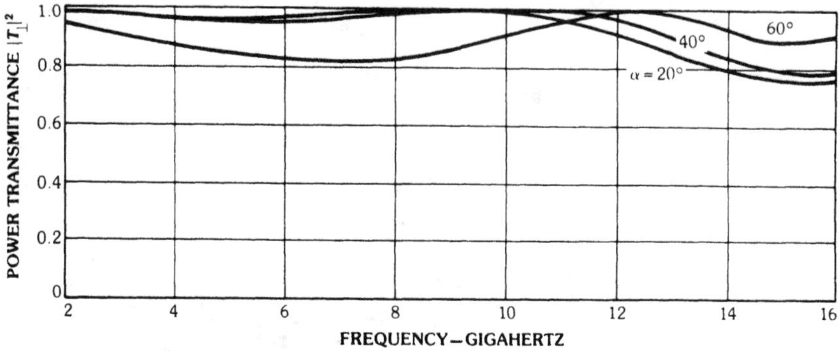

Figure 2.36 *Energy transmission through an A-sandwich radome wall*

Skin thicknesses: 0.035 in and 0.225 in (core)
Relative permittivities: 3.2 and 1.1 (core)
Loss tangents: 0.015 and 0.005 (core)
Reproduced by permission of Van Nostrand Reinhold

low mass and relatively low cost. An investigation of the return-loss and transmission effects of such radomes has been made by Cory and Clarricoats [19]. Following a method similar to that originally described by Silver for determining the return loss of reflectors [1], they find that the return loss of a paraboloidal reflector, of focal length f, covered by a radome of focal length f', is given by

$$|\Gamma_{total}| = \frac{Gf\lambda}{4\pi f}\left[1 + \frac{f'}{f} R(0)\ \exp\{-2k(f-d+f')\}\right] \tag{2.41}$$

where $R(0)$ is the voltage-reflection coefficient of the radome at normal incidence, and $d = D(\cot\theta_0 + \cot\theta_0')/2$ where θ_0 and θ_0' are the subtended angles of the reflector rim at the respective foci where $d=0$ for a confocal system. The condition $d=0$ is, however, an undesirable arrangement as rays striking the radome are directly returned to the feed.

Eqn. 2.41 has a simple physical interpretation; the term $Gf\lambda/4\pi$ is the reflection due to the reflector alone (as given in eqn. 2.35). This is modulated by the first two reflections occurring along the antenna boresight. In the asymptotic evaluation, the radome reflection coefficient is polarisation-independent. The derivation neglects feed blockage since the feed radius is much less than the radius of the first Fresnel zone which is $(2\lambda/f)^{1/2}\{1 + (\lambda/2 + d)/f\}$.

If, as is usual, the reflector incorporates a vertex plate, assumed here to be at distance d' from the reflector focus, then the second term in eqn. 2.41 must be multiplied by $f^2/\{(2d'-d) - f'^2\}$.

To reduce the return loss, the above authors have proposed the use of a modified radome surface: see Figure 2.37*b* which shows a Cassegrain form of geometry (a Gregorian form is also possible and in fact has some additional advantages). In a simple geometrical-optics analysis, the subradome deflects rays away from the feed so as to reduce the radome contribution. The

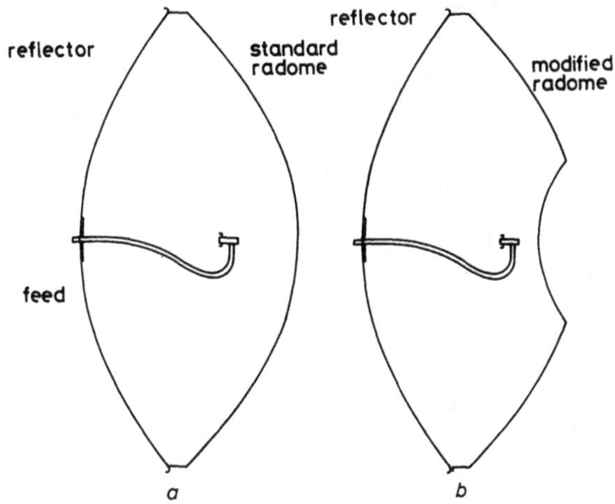

Figure 2.37 Radome-covered reflectors [19]

a Paraboloidal
b Modified paraboloidal

calculation of the return loss follows a similar procedure to that given above and a 10 dB improvement has been found both theoretically and experimentally for an antenna of 2.4 m diameter operating at a frequency near 6 GHz. The above authors have also studied the influence of both types of radome on the radiation pattern of the antenna. The prime effect of the radome is to cause multiple reflections (in a geometrical optics sense) which increase the sidelobe level. The modified radome brings about worthwhile improvements.

2.2.11 The effect of reflector surface errors on antenna performance

In our discussion of reflector antennas we have assumed that the reflector surface is precisely that desired. In practice there are two main classes of distortion, namely random errors and systematic errors. The effect of random errors was analysed by Ruze [20] in the 1950s. He first characterised the errors through two statistical parameters, the transverse extent of the perturbations, and the mean-square phase error δ^2 (radians) over the disturbed aperture. Complete correlation is assumed to exist over a region of diameter $2c$. Then, following Rusch [21], for uniform illumination of a reflector with a circular aperture of radius a, the antenna power pattern $G(\theta)$ is given by

$$G(\theta) = \eta\left(\frac{2\pi a}{\lambda}\right)^2 2\frac{J_1(u)}{u}\exp(-\delta^2) + \left(\frac{2\pi c}{\lambda}\right)^2 2\frac{J_1(u_c)}{u_c}\{1-\exp(-\delta^2)\} \tag{2.42}$$

$$u = Ka\sin\theta$$

$$u_c = Kc\sin\theta$$

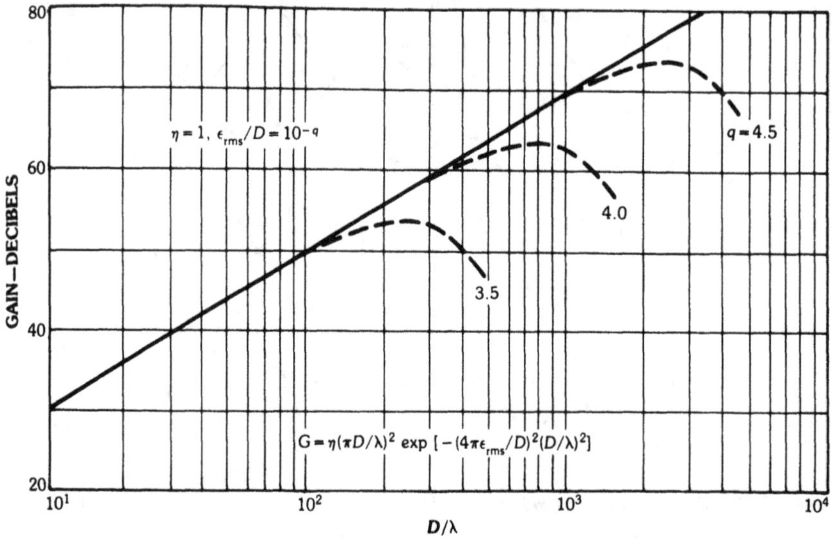

Figure 2.38 *Gain loss owing to random irregularities (defined in Figure) as a function of D/λ [22]*

$G = (\pi D/\lambda)^2 \exp\{-(4\pi\varepsilon_{RMS}/D)^2(D/\lambda)^2\}$
$\eta = 1$, $\varepsilon_{RMS}/D = 10^{-q}$

Reproduced by permission of Van Nostrand Reinhold

Eqn. 2.42 shows that the principal effect is to reduce the gain of the unperturbed aperture by a factor determined by the phase error. The second perturbing term is the pattern of a reflector of diameter $2c$.

For the more practical case in which Gaussian shaped distortions are assumed together with small correlation intervals, the peak gain reduction of a reflector of diameter D is given by the well known expression

$$G = \eta\left(\frac{\pi D}{\lambda}\right)^2 \exp - \left(\frac{4\pi\kappa\varepsilon_{RMS}}{\lambda}\right)^2 \tag{2.43}$$

where

$$\kappa = \frac{4F}{D}\left[\ln\left\{1 + \frac{1}{(4FD)^2}\right\}\right]^{1/2}$$

Figure 2.38, due to Love [22], shows the gain as a function of a normalised parameter q where $\varepsilon_{rms}/D = 10^{-q}$ and the antenna diameter D is expressed in wavelengths.

From eqn. 2.43, if $\eta = 1$, the gain increases with the square of frequency until the tolerance effect becomes significant. The shortest operating wavelength, denoted by the maximum in Figure 2.38 [22], is then given by

$$\lambda_s = 4\pi\varepsilon_{RMS} \tag{2.44}$$

where in Figure 2.38 the factor κ has been taken as unity. At a wavelength equal to λ_S, the RMS error is approximately $\lambda_S/16$, although at this wavelength the sidelobes of the antenna may become unacceptable. As an illustration, a tolerance of 60 μm (60×10^{-3} mm) enables a radiotelescope to operate down to about 1 mm wavelength (300 GHz) while satisfying the above criterion.

The above analysis does not account for tapered aperture distributions and nonuniform distortions. These effects have been analysed in detail by Rahmat–Samii [23].

In addition to random errors, systematic errors in a reflector surface influence performance. These are characterised by large-scale deviations such as may occur in a spun reflector if the metal deforms after manufacture or is mishandled during assembly. Large reflectors, such as those used in radiotelescopes, are usually made from precision panels. Although individual panels may have a very high surface accuracy their mutual alignment when fastened to the antenna-backing structure poses a problem to the designer. It is customary to use optical methods such as twin theodolites and reference targets to achieve the desired accuracy. Subsequently, microwave or millimetre-wave holography can be used to check the overall surface accuracy. The maintenance of this accuracy over a long period of time, and with changes in temperature, constitutes one reason for systematic errors. These normally manifest themselves in loss of gain, increased sidelobes and pattern distortion. Large radiotelescope antennas also experience gravitational distortion when the telescope-pointing direction is changed but, if a holomorphic design is chosen, the shape of the reflector maintains paraboloidal shape as it deforms although its focal length alters. Compensation is achieved by moving the feed position. In spacecraft antennas, thermal-distortion effects arise because of solar radiation, which is especially severe if part of the reflector is in shadow, and these are particularly important for antennas operating at frequencies above 20 GHz.

Various techniques to compensate for systematic distortion are being evaluated. These procedures first involve detecting and estimating the distortion by, for example, electromagnetic, mechanical or optical techniques; then, either by use of actuators attached to the reflector or by the use of an array feed with optimised excitation, an appropriate compensation is achieved.

So far we have addressed errors in the surface of high-gain reflector antennas such as are used to create spot beams or to receive efficiently from a narrow angular region. However, shaped-beam antennas (Section 2.2.13) are required in a number of applications and the effect of distortion on performance is no less important. A study has been undertaken by Pontoppidan [24] who shows that the effect of errors on gain is more sensitive than in a spot-beam antenna. His physical explanation is based on the concept that, in a shaped-beam antenna, the reflector is intrinsically deformed to reduce the 'axial gain' and enhance the gain away from boresight. As a consequence, within the shaped-beam coverage region there is substantial cancellation of fields coming from different regions of the reflector surface, which contrasts with the essential addition of contributions in the boresight direction for a spot-beam antenna. As a consequence, the gain of a shaped-beam antenna is more sensitive to random errors and Pontoppidan shows that these are linearly related to the RMS deviation. This contrasts with the square-law dependence for spot-beam antennas (as in eqn. 2.43).

2.2.12 The effect of positional errors on antenna performance

Imperfections in the surface of a reflector are not the only reasons why the antenna may perform below specification. Errors in the feed model are a common source of discrepancy but discussion of this will be deferred to later chapters. Errors arising from incorrect positioning of the feed, of the subreflector in a dual-reflector antenna, or both, constititute additional reasons. An understanding of these effects allows the designer to create a structure which minimises these defects and to propose realistic tolerances for antenna manufacture.

The main effect of feed misalignment in a paraboloidal reflector is to cause a phase variation over the reflector aperture. The principal aberrations are given in Figure 2.19. Axial defocusing generates a quadratic error which is independent of azimuthal angle. This error manifests itelf as a symmetrical increase in sidelobes and filling of the antenna-pattern nulls. Spherical aberration is a 4th-order phase error with similar symmetry to the quadratic error. Both can be compensated for by axial-feed movement, assuming that the feed has a unique phase centre, an issue to which we shall return.

Coma is associated with a cubic phase error and exhibits a cos ϕ dependence. It arises from lateral feed displacement. Apart from gain loss, the sidelobes are symmetric in height. Astigmatism also arises from lateral displacement of the feed and is associated with a quadratic phase error and a cos 2ϕ dependence. The sidelobe distortion therefore carries a $\pi/4$ cycle. The cos 2ϕ dependence of the phase error also interacts with the cos 2ϕ dependence of the crosspolar radiation from the field and exacerbates crosspolar effects.

Figure 2.39 [25] shows the effect of transverse defocusing for the case of a 25-wavelength-diameter reflector. The initial taper gives a first sidelobe level of about -28 dB for the perfectly focused reflector. A feed deviation causing a one-beamwidth scan is sufficient to raise one near-in sidelobe by 8 dB while leaving the sidelobe level in the direction of scan unchanged. By four beamwidths of scan there is a 2 dB gain loss and the sidelobe, relative to the peak, lies close to 10 dB, the first null in the direction of scan has disappeared and the pattern is significantly distorted. The process continues with further transverse movement of the feed.

The phase centre of a feed is defined as the local centre of the spherical wavefront of the radiated field from the feed. Consider a wide-angle corrugated-horn feed operating at its design frequency. At angles from boresight not significantly greater than the flare angle of the feed, the centre of curvature of the phase front is the same irrespective of the plane of the observation point, e.g. E-plane or H-plane. However, away from the design frequency and in general for many other feeds, the E-plane phase centre and H-plane phase centre do not coincide along the feed axis. When such a feed illuminates a reflector, there is

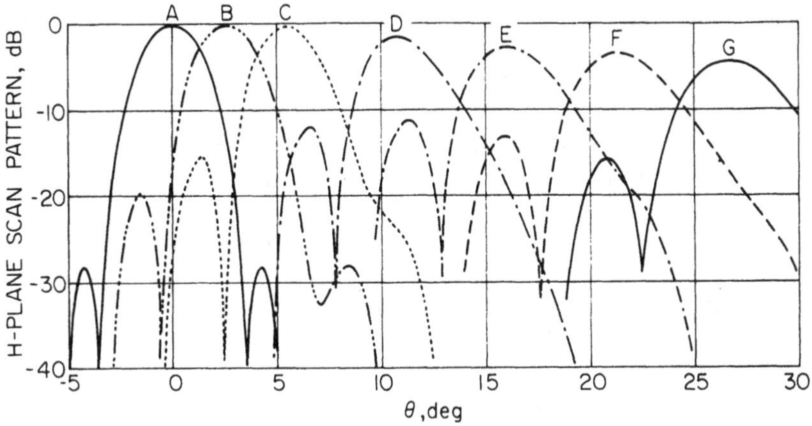

Figure 2.39 H-plane radiation patterns of paraboloid with $D = 25\lambda$ when illuminated by defocused circular waveguide feed excited in TE_{11} mode [25]

Curves $A-G$ correspond, respectively, to halfbeamwidth scans of 0, 1, 2, 4, 6, 8 and 10

inevitable axial defocusing. This reduces the antenna gain, raises the sidelobe level in a symmetrical way and, most importantly in some applications, raises the peak level of crosspolarisation in the 45° planes. This problem is especially significant with small feeds and the phase-centre location becomes an important design parameter. For example a circular waveguide supporting the dominant TE_{11} mode has a phase-centre difference between E-plane and H-plane of about $\lambda/20$ at midband. Apart from the difference in the amplitude patterns of the feed in E- and H-planes, the phase-centre difference contributes to raise the crosspolar peak.

2.2.13 Analysis and synthesis of shaped-beam antennas

Although the antennas described above are ideal for point-to-point communication, target-locating radars and for radioastronomy, there are many other applications where a circularly symmetric pattern is not optimum. Examples are to be found in search radars and in satellite communication and satellite broadcasting. In the former, a narrow antenna beam is required in azimuth but a broad coverage is required in elevation. This allows good azimuthal target discrimination to be achieved while providing adequate illumination of targets at different ranges, whose elevation angle with respect to the radar antenna will vary. In the latter, both satellite-communication and satellite-broadcast applications generally achieve optimum performance if the spacecraft-antenna pattern has a shape which closely matches the countries or regions to be served. In both applications it is desired to minimise radiation outside the desired coverage and in radar, there is often a need for very low sidelobes so as to minimise spurious target returns. In satellite applications, energy falling outside the coverage region represents energy wasted and the associated signals are a source of potential interference to other users.

When a single-point feed or line feed is used to create a shaped beam, the angular distribution of energy relies on shaping the reflector. If the coverage solid angle is large, a reasonable design can be made using geometrical optics, as with radar antennas made in the 1940s. They were restricted to beam shaping in one plane, and a discussion of these GO techniques is to be found in a number of references [1, 26]. A general GO-synthesis method was developed more recently by Jørgensen [27] and by Westcott [28] whose work will be discussed in this section. For GO to work accurately the reflector needs to be large in terms of wavelengths.

The development of satellite-communication systems began in the 1960s. At first restrictions on payload mass led to simple reflectors providing global coverage, but with improvements in satellite launchers the payload mass could be increased. Then larger antennas could be deployed generating shaped beams leading to more efficient illumination of the coverage region thereby reducing earth-station costs. These designs utilised array-fed offset paraboloidal reflectors with the array excited through a microwave beamforming network (BFN). Some antennas now have the potential for inorbit reconfiguration by virtue of telemetry control of a microwave switch in the BFN. Although the array-fed reflector offers advantages if regional coverage is required and is essential if multiple regions or spot beams form part of the specification, they suffer the disadvantage of increased mass compared with a singly fed shaped reflector and the BFN also gives rise to loss. In general, the array-fed reflector has higher spillover losses. The shaped reflector can offer an excellent solution if a fixed continental coverage is required, e.g. North America or Europe, but it is unsuitable when regions subtending a small solid angle or closely separated regions are specified. The potential for shaped-reflector reconfiguration is being explored.

We now consider the design of an array-fed offset-paraboloidal reflector as indicated in Figure 2.40. Each element of the array produces a spot beam in a direction which is approximately determined by the element position in the focal plane and the focal length of the antenna (see Section 2.2.6). The constituent spot beams with correct amplitude weighting can be combined to produce a number of specific coverage patterns. A first stage in the synthesis procedure requires a choice of the reflector size and the number of elements of the array. In practice both will be constrained by the maximum payload mass and, in general, by the size of the payload bay on the launcher. However, it is useful to understand the influence of these constraints on performance. From Figure 2.41 and a knowledge of the required angular coverage, the corresponding focal-field area is defined through

$$\sin\theta_p \exp(i\psi_p) = -\frac{(x_0 + jy_0)}{f_0} \tag{2.45}$$

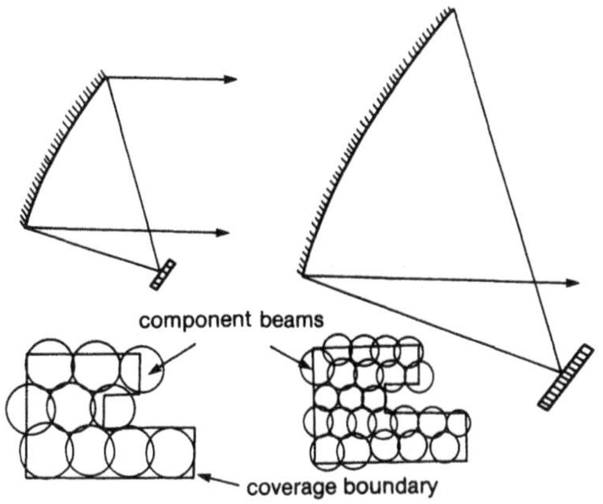

Figure 2.40 Configuration of array-fed offset reflector

which essentially relates the far-field beam angle θ_p, to the focal-field position through

$$\theta_p = \frac{r}{f_0} \tag{2.46}$$

for small values of θ_p.

The size of the focal-field area depends on f_0 while the reflector diameter D determines the beamwidth of the constituent beams for a given feed illumination. This in turn depends on the size of the feed aperture. Neglecting the effect of mutual coupling between feed horns, the finite size of the feed horn determines the closest feed packing attainable and the minimum spacing between component beams.

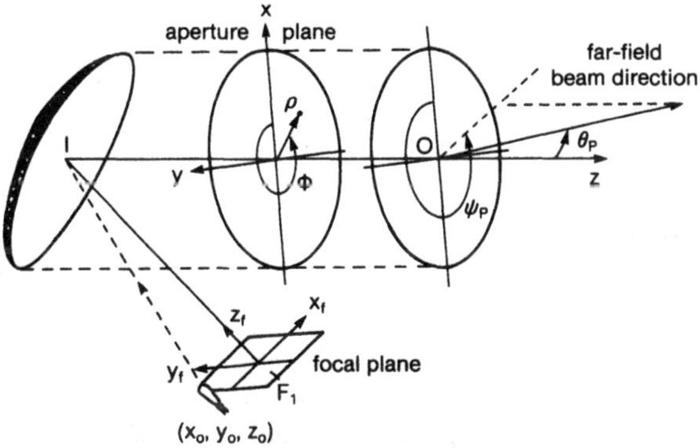

Figure 2.41 Beam orientation for displaced feed in focal plane

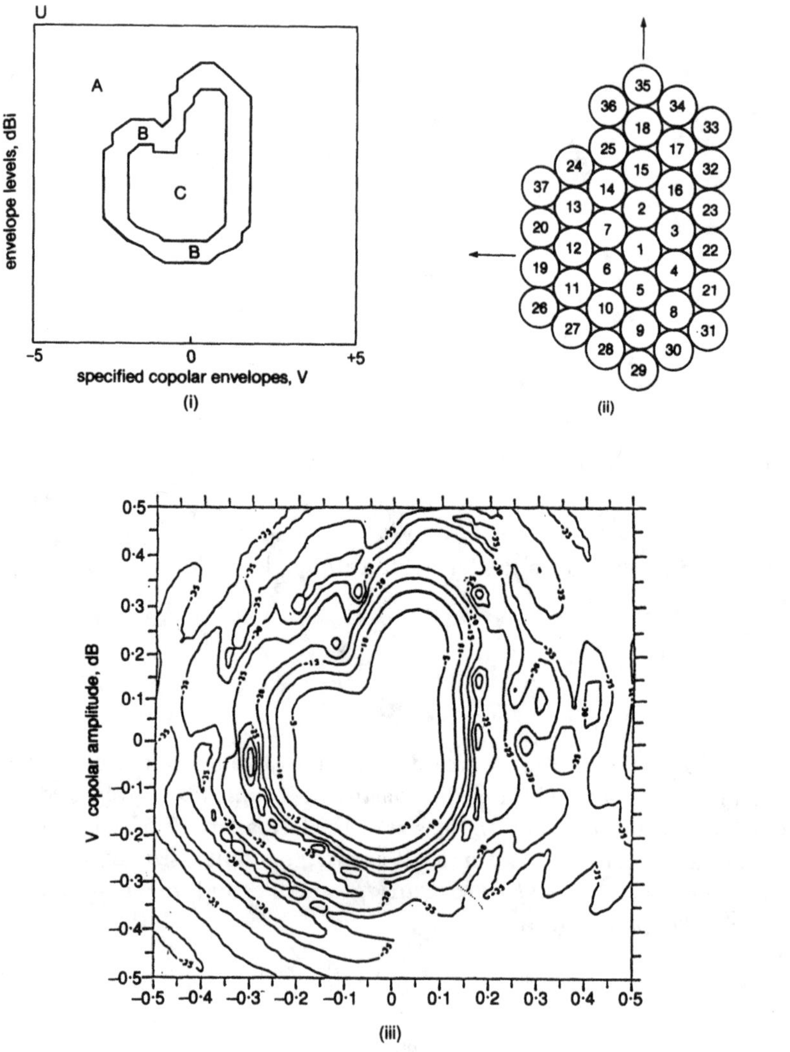

Figure 2.42 Template of shaped coverage, array feed for coverage of Figure 2.42(i) and predicted copolar contour of this feed

(i) Template of shaped coverage

Region	*A*	*B*	*C*
${}^{up}T_i^{co}$	2	32	32
${}^{lo}T_i^{co}$	$-\infty$	$-\infty$	30

(ii) Array feed for coverage of Figure 2.42(i) using single-offset reflector of 50λ diameter; radius $a = 0.51\lambda$*; separation* $= 1.1\lambda$*; flare angle* $= 5°$

(iii) Predicted copolar contour of antenna using optimally excited array feed of Figure 2.42(ii)

Since a shaped-beam antenna illuminating an angular reigon Ω_c has an upper limit on coverage gain

$$G_{max} = 10 \log \frac{4\pi}{\Omega_c} \qquad (2.47)$$

the principal merit of increasing the antenna aperture is to improve the pattern resolution by decreasing the beamwidths of component beams, as in Figure 2.40. For the larger antenna in Figure 2.40 a greater number of feed elements is needed to provide the same coverage, thus requiring a more complex feed network. The f/D ratio also influences the scanning performance, as evidenced by the aberrations discussed in Section 2.2.12. However, in most cases, the spacecraft-antenna size is dictated by the launcher. Once the focal length and diameter have been determined, the angular coverage can be translated directly into a focal-field area and the number of elements determined according to the desired resolution, with an upper bound determined by the smallest acceptable feed-aperture size. When multiple spot beams are required, an increase in antenna aperture gives a higher gain and, if the spots are widely spaced, each can be generated by a single horn placed at the correct position in the focal field. The feed aperture must be sufficient to provide the appropriate edge taper, thereby controlling energy lost in the far-out sidelobes and the possibility that the sidelobes may give rise to out-of-coverage interference. For far-out scanned beams it is possible, in principle, to use a small array of elements to excite the beam which, if properly weighted, can compensate for aberrations. On the other hand, if spot beams are very closely spaced it is impossible to achieve a low crossover level between the beams without strong edge illumination and high sidelobes. Once the far-field patterns of the antenna for each element in the feed array have been computed and stored, the patterns can be summed with the correct weightings to provide the desired shaped-beam pattern. This procedure avoids the need continually to repeat the time-consuming pattern predictions made using PO (with appropriate corrections if required). The optimisation program continuously adjusts the weights associated with each beam until a coverage is achieved which best matches the original template. Each complex excitation coefficient of the N array elements is represented by two real variables corresponding to the real and imaginary parts. The optimisation is then performed in a $2N$-variable space along the steepest descent path to a minimum in a predetermined error function. Different error functions can be used such as least-squares, least pth, minimax, etc. To start the optimisation process, it is usual to choose a uniform amplitude excitation for the feed elements with equal phase.

Figure 2.42(i) shows a typical template with upper- and lower-bound gains set relative to isotropic. The reflector is a 50.5-wavelength-diameter offset paraboloid with a focal length of 50 wavelengths and an offset angle of 40°. The array in Figure 2.42(ii) comprises 37 single-mode horns of 5° semiflare angle, with a radius of 0.51 wavelengths and a separation of 1.1 wavelengths (equivalent to the horns touching with a 0.05 wavelength wall thickness). Figure 2.42(iii) shows the copolar gain contours relative to the peak gain of 37 dB with the specification being adequately met. By exciting the feeds with circular polarisation, the inherent crosspolarisation of the offset configuration is avoided. The peak crosspolarisation is −29 dB below the copolar peak, principally on account of strong mutual coupling exciting the TE_{21} mode in the horn. If the aperture size were to be increased, the excitation level would decrease and the crosspolar maximum would decrease, however, with fewer elements, the copolar contour resolution would change. An excellent example of a practical application of contoured-beam design, including the effects of mutual coupling, is provided by Bird [29] who describes the design of an antenna to cover the total territory of the USA in which the offset paraboloidal reflector is fed through 26 rectangular horns. He shows that neglect of mutual coupling can lead to a 0.76 dB gain error in the worst case. A detailed discussion of mutual-coupling effects is in Chapter 15.

An alternative approach to the production of a shaped-beam radiation pattern is to utilise a shaped reflector (or reflectors for dual-reflector antenna) and either a single high-performance feed, such as a corrugated horn, or a small array-feed cluster. First attempts [27, 28] to synthetise a shaped beam were made using geometrical optics to establish either the far field directly or a desired aperture distribution from which the far field could be computed, the latter being more accurate when the coverage region occupies a small solid angle and diffraction effects are pronounced. The amplitude can be expressed by

$$F(r) = a + b(1 - r^2)^p \tag{2.48}$$

and the phase by $\phi(x, y) = Ax + By + Cx^2 + Dy^2 + Exy + \Sigma f_m(x) f_n(y)$ where the aperture co-ordinates are normalised to $(-\pi/2, \pi/2)$ to allow $f(x)$ to be expressed as

$$f_m(x) = 1,\ \sin x,\ \cos x,\ \sin 2x,\ \cos 2x \ldots \quad \text{and similarly for } f_n(y) \tag{2.49}$$

for $m, n = 1, 2, 3, 4, 5, \ldots.$

Most phase distributions over an aperture for shaped-beam applications are bowl shaped and the inclusion of quadratic terms in the phase expansion reduces the number of terms in the Fourier series. The terms a, b and p are chosen to give an appropriate edge taper.

The coefficients are varied in a quasi-Newton search in which the far-field pattern is sampled and compared with the specification at Nyquist sample points within the coverage region. In the example of Figure 2.43, the search minimises a one-sided sum of the fourth-power gain differences

$$F = \sum_i F_i^4 \tag{2.50}$$

where $f_i = \{(G_c - G_t)/\Delta G\}$ and $F_i = f_i$ if $f_i < 0$, otherwise $F_i = 0$. C_c, G_t and ΔG are the calculated gain, target gain and tolerance respectively.

Once the amplitude and phase of the aperture field are known, geometrical optics can be used to determine the co-ordinates of the reflector. In our studies,

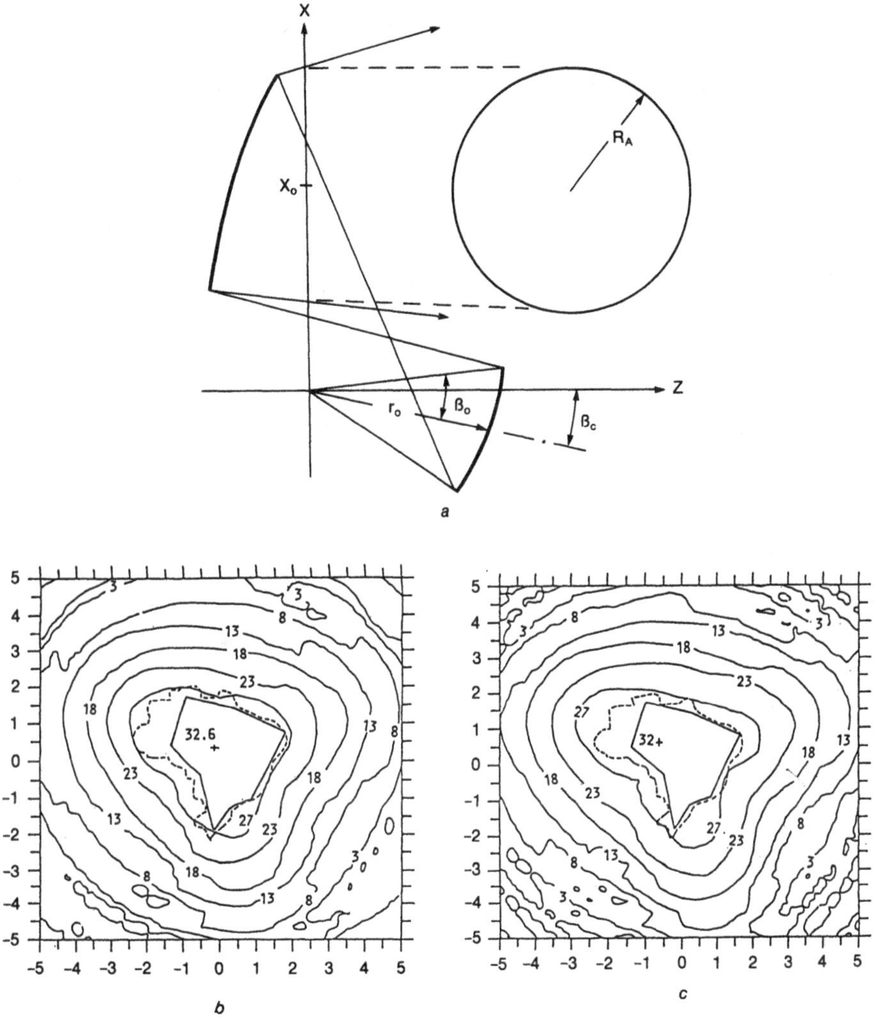

Figure 2.43 *Gregorian dual reflector and copolar contours*

a Gregorian dual reflector for coverage of Brazil
b Copolar contours corresponding to desired reflector-aperture field
c Copolar contours corresponding to achieved aperture using GO synthesis

we have used the method of Westcott and his coworkers; their procedures use power conservation and Snell's laws. They lead to a 2nd-order nonlinear partial-differential equation of the Monge–Ampere type. The output of the GO synthesis leads typically to between 600 and 1000 points defining the surfaces of the subreflector and main reflector. The discrete reflector points are fitted to provide continuous second derivatives of the surface, as are required for a PO analysis. Small differences are observed in predicted far-field patterns; Figure

2.43*a* and *b* shows results obtained from the prototype aperture field and the PO analysis performed on both subreflector and main reflector. The differences arise principally from subreflector diffraction. Nonetheless, the coverage has been achieved with a margin. For this optimised Gregorian configuration, the crosspolar maximum lies 38 dB below the copolar maximum.

In the above method of synthesis, four steps are required, as in Table 2.4. A more direct approach is that of synthesis by direct optimisation, so-called diffraction synthesis [30]. This is the method now used almost universally by the designers of shaped reflectors for satellite applications.

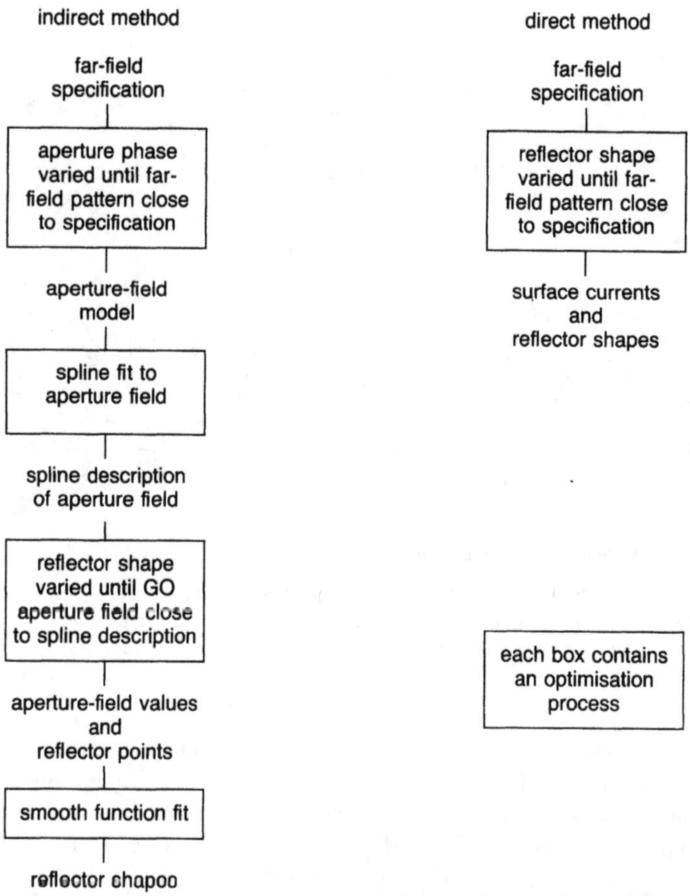

Table 2.4 Procedures used for GO synthesis (indirect) and PO synthesis (direct)

In the direct approach, also shown in Table 2.4, the height of the reflector surface z above a plane is characterised by a polynomial and a Fourier series thus:

$$z = ax + by + cx^2 + dxy + ey^2 + \sum_{k}^{Nx} \sum_{l}^{Ny} c_{kl} f_k(x) f_l(y) \tag{2.51}$$

with the same normalisation as in eqn. 2.48.

The reflector shape is controlled by the coefficients in the above equation and these are varied so that the gain of the antenna conforms to the specification within a tolerance expressed as in eqn. 2.50. The gain of the antenna is calculated using the physical-optics currents on the main reflector. PO can also be used on the subreflector in a dual-reflector antenna, although it is computationally more efficient to trace rays using GO onto the main reflector. Account should be taken where necessary of the possibility that the subreflector may lie in the near-field of the feed. Figure 2.44 shows the procedure followed in the direct method. Bergman, Brown, Clarricoats and Zhou [30] have compared both the indirect and direct methods for a European-coverage specification corresponding to *Eutelsat 2* and they found that the average gain within the coverage region differed by only 0.01 dB.

Alternative design procedures have been used;.for example Stirland, Fox and Malik [31] follow some of the stages of each of the above methods. They use the method of successive approximations to generate an aperture field correspond-

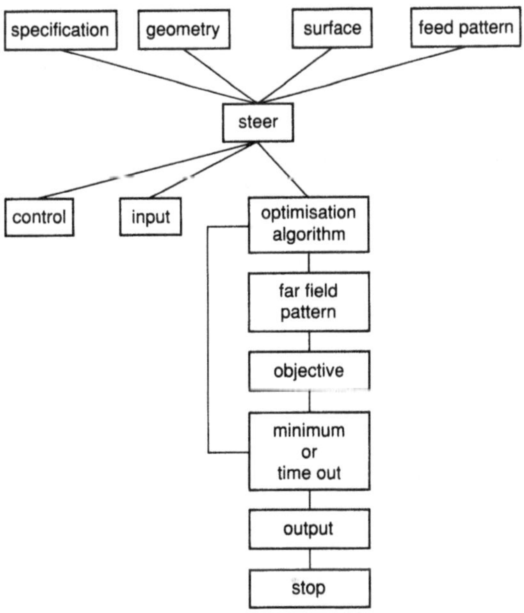

Figure 2.44 Schematic of diffraction-optimisation procedure [30]

ing to the desired far field, and then use PO optimisation to generate a subreflector shape to give the desired amplitude distribution over the main-reflector aperture. The main-reflector shape is then generated to give the correct aperture phase. The same authors have compared designs using array-fed reflectors and single-fed shaped reflectors. Table 2.5 compares the performance of an antenna to illuminate Germany. The shaped reflector offers a mass saving of 2.4 kg with improvements in edge-of-coverage (EOC) gain. Similar improvements are found for other cases they have studied. They conclude that there are performance advantages associated with a shaped reflector compared with an array-fed reflector, provided that only single coverage without reconfigurability is required. The key advantages are:

(i) higher directivity following from reduced spillover
(ii) elimination of about 1 dB of loss associated with the BFN
(iii) ease of manufacture compared with dual-gridded reflectors
(iv) 30% mass saving for the same performance
(v) improvement in EOC gain between 0.7 and 2.0 dB depending on the coverage and antenna-accommodation constraints.

Table 2.5 Comparison of performance of array-fed single offset reflector of 1.8 m projected aperture and shaped reflector with corrugated-horn feed [31]

Parameter	Multifeed antenna	Shaped-reflector antenna
Antenna components	1 reflector 8 conical horns 8 polarisers 1 BFN	1 reflector 1 corrugated horn 1 polariser
Total mass (kg)	14.8	12.4
Maximum VSWR	1.2	1.2
RF loss (dB)	0.72	0.50
Crosspolar w.r.t. peak (dB)	<−40	<−40
Minimum EOC gain (dBi, net of pointing error)		
B_1 (45–57°N, 5–20°E)	35.64	36.42
B_2 (Austria, Germany, Switzerland)	39.14	39.84

Frequency: 11.7–12.2 GHz
Polarisation: Left-hand circular

2.2.14 *Dual-gridded reflector antennas*

One of the principal requirements in satellite communications is to provide the means to use the same frequency band by transmitting and receiving in orthogonal linear polarisation, a technique known as polarisation frequency re-use. The standard means of achieving this with a single-offset-reflector geometry is to employ two orthogonal offset gridded reflectors whose aperture planes are slightly shifted in angle so that their foci are displaced. In the most general case, shown in Figure 2.45*a* [32], the reflector (shell) can be paraboloidal and fed from an array feed to effect the appropriate coverage, or it can be shaped so that a single high-performance feed can be used. This is the arrangement used in the *Orion* spacecraft where the array-fed reflector provides multiple regional coverage and the shaped reflector continental coverage. Figure 2.45*b* shows the dual-shell construction with ribs, known as intercostals, which are required to provide structural support. This is especially critical in the launch phase of the spacecraft.

In the *Orion* reflector, the reflecting surface is in the form of conducting aluminium strips. Initially, aluminium is vapour deposited on to the whole of a Kevlar shell; then the gridded reflector strips are produced by etching using an excimer laser. The grid dimensions are

pitch	1.20 mm
width	0.30 mm
thickness	0.35 mm

Both front and rear shells are gridded in order to achieve better than −38 dB crosspolar isolation. The 2.3 m-diameter shells are rotated by 8° with respect to each other, which is the maximum allowed by the constraints of the particular

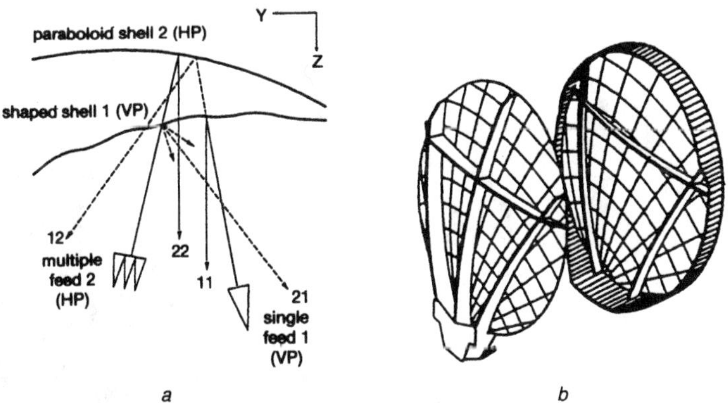

Figure 2.45 Dual-gridded reflectors and shells [32]

a Dual-gridded-reflector concept. Shell 1 is for continental coverage in vertical polarisation and the design uses a single corrugated-horn feed. Shell 2 is for regional 'spot' beams in horizontal polarisation and uses an array feed

b Artist's concept of dual gridded shells when separated showing the intercostal supports launch-vehicle-spacecraft bay. Although different manufacturers utilise different manufacturing techniques, the above specifications are generally representative.

2.2.15 Reconfigurable reflector antennas

The shaped reflector has been seen to offer advantages over the array-fed reflector but it suffers the disadvantage that, if the reflector is of rigid construction, once the antenna is installed, especially if on board a spacecraft, there is no means to alter the coverage pattern. This situation contrasts with the array-fed reflector where, in principle, means can be found to alter the excitation by use of remotely controlled switches, power dividers or phase shifters in the BFN. To overcome this difficulty, a reconfigurable reflector has been proposed [33] where the surface shape is controlled through the use of actuators, as in Figure 2.46. The antenna is an offset paraboloidal reflector whose surface is made from gold-plated molybdenum mesh. In the prototype antenna, the surface is tethered at 52 control points by means of nylon cords attached to stepper motors which are operated under computer control. A mesh surface was chosen for the prototype antenna since the material was known to be space-qualified. The number of control points depends on the scheme used to adjust the reflector, and in later models a smaller number has been found satisfactory. By using the diffraction-optimisation method of Section 2.2.13, the reflector shape is determined and the required motor settings computed to create the desired antenna pattern. The motors are then driven and the antenna shape

Figure 2.46 850 mm-diameter reconfigurable mesh reflector for operation in band centred on 10 GHz [33]

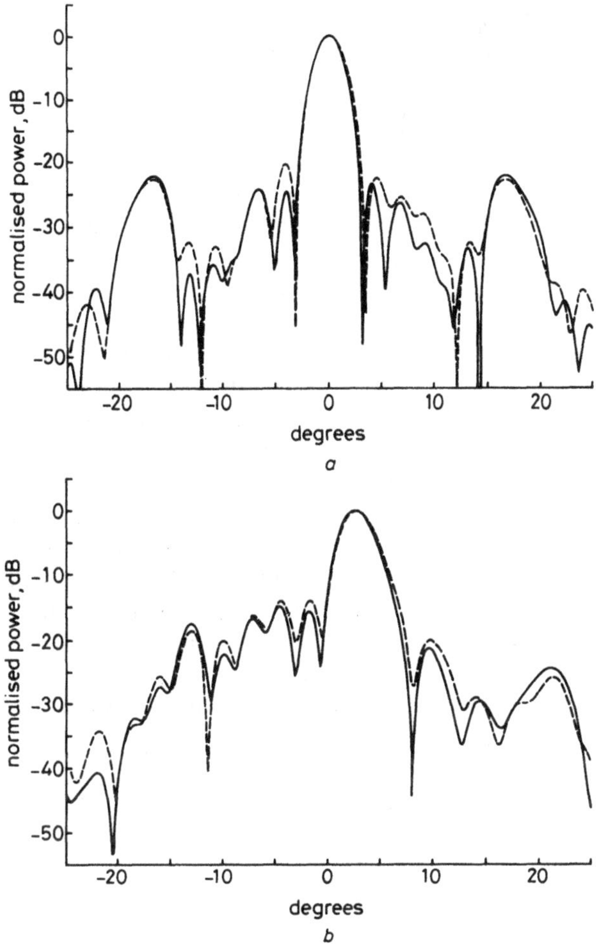

Figure 2.47 H-plane radiation patterns of reconfigurable antenna of Figure 2.46 [33]

—— predicted
– – – measured
a When set to produce a spot beam
b When set to produce a shaped beam

can be set in under a minute. Predicted and measured pattern cuts shown for a pencil-beam and a shaped beam are given in Figure 2.47*a* and *b*. The antenna, which is fed by means of a 30° corrugated horn, operates from 8 to 12 GHz. In an alternative design the antenna can be fed from a small array. Then, if the antenna is configured as an offset paraboloid, the array can be used to generate multiple spot beams, but when configured as a shaped reflector the array can be excited symmetrically and the antenna can then generate a reconfigurable continental beam. The reconfigurable antenna could also be used as the back shell of a dual-gridded reflector. Then if the front shell uses an array feed with an adjustable BFN, full reconfigurability of dual linear polarisation is possible.

2.2.16 *Dichroic reflectors*

As a means of increasing the communication capacity of an antenna, especially that of a spacecraft antenna, the use of more than one operating frequency band is possible and one way this can be achieved is to fit the antenna with one or more dichroic reflectors. The example of Figure 2.48 [34] depicts a 3-band antenna where the prime focus and two secondary foci of the antenna are available by virtue of the use of subreflectors which reflect at one frequency band while being almost transparent at another.

Dichroic or frequency-selective surfaces (FSS) are made by covering a transparent support structure such as Kevlar with an array of frequency-selective elements. The elements can take the form of gratings but more usually comprise dipoles, crossed dipoles, resonant rings, concentric rings, square loops, concentric loops or Jerusalem crosses, among a wide variety of configurations dependent on the nature of the surface, the frequency band of operation and the transmission properties.

Following Parker [35], three principal techniques are available for the analysis and design of a FSS. The first is a method of modal analysis in which the distribution of the current induced in the conducting elements (or fields in the slots) is represented as a series of basis functions, such as sinusoidal functions or waveguide modes. Local incident fields are expanded as a set of Floquet modes and the two sets matched by applying boundary conditions. The resulting integral equation is solved by the method of moments.

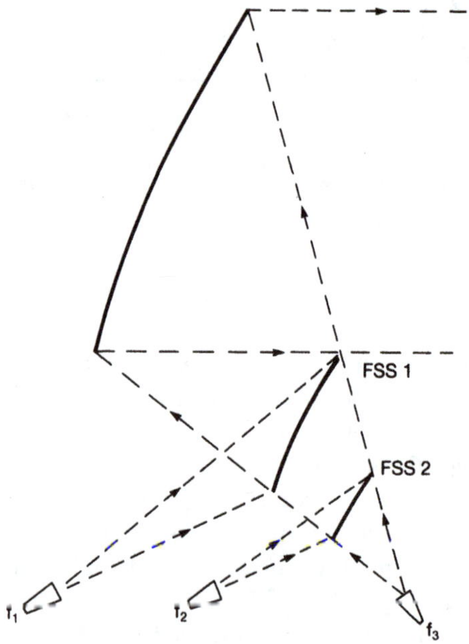

Figure 2.48 Concept of 3-band dichroic (FSS) reflector

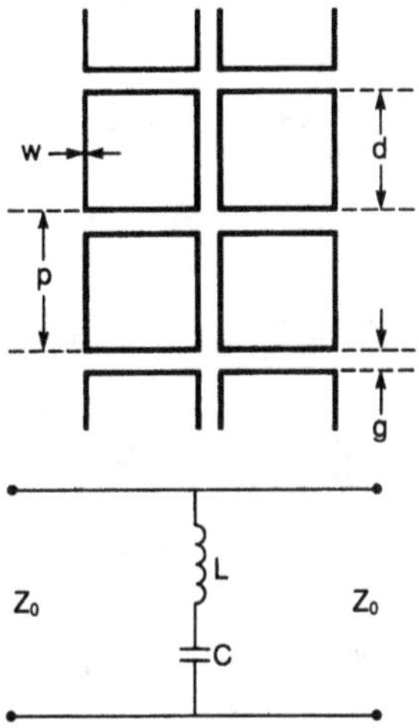

Figure 2.49 Square-grid elements and their equivalent circuit [35]

The second technique is based on an equivalent-circuit approach where the array is modelled by means of lumped impedances on a transmission line as in Figure 2.49 [35]. The third technique is iterative. A first estimate of the element current induced by the incident electric field is used to calculate the local field distribution which is forced to zero on the conductors. The modified field is then used to recalculate the element current distribution and so on. The first two methods assume that the surface, and hence the arrays, is of infinite extent, the third method is not so restricted. A review of the methods of analysis has been given by Mittra, Chan and Cwik [36].

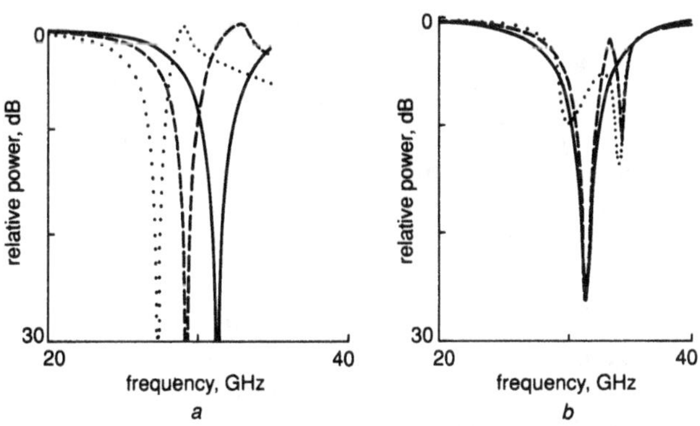

Figure 2.50 Response of crossed-dipole array in square lattice [35]
See Figure 2.51 for different incidence angles

a TE polarisation
b TM polarisation

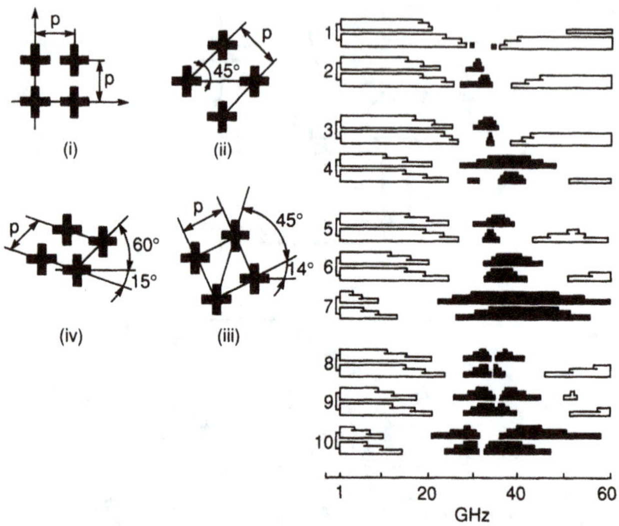

Figure 2.51 Transmission and reflection bands for crossed dipoles with different parameters and in different lattice configurations [35]

(i) Lattices 1, 2 with $L/p = 0.70$, 0.82
(ii) Lattices 3, 4 with $L/p = 0.70$, 1.05
(iii) Lattices 5, 6 with $L/p = 0.88$, 1.05, 1.60
(iv) Lattices 7, 8 with $L/p = 0.93$, 1.05, 1.33

Many satellite antennas operate with dual polarisation; thus there is considerable interest in those types of FSS elements which perform satisfactorily when illuminated with either sense of polarisation. Rings and crossdipoles are examples. Consider crossdipoles arranged on a square lattice as in Figure 2.51(i). The response characteristics depend on not only the incidence angle of the incoming plane wave but the direction of polarisation of the electric- and magnetic-field components (TE or TM). For the TE-mode case, the response is shown in Figure 2.50*a*, while that for TM mode is shown in Figure 2.50*b*. There a second null appears above the main reflection resonance owing to current flow in the orthogonal arm of the crossed element, the structure then being excited in an asymmetrical mode. There is a general tendency for more complicated structures to exhibit more complicated transmission responses. This can offer both advantages and disadvantages. It can favourably modify the transmission-band/reflection-band ratios but, on the other hand, subsidiary responses are undesirable if the objective is to design a surface with a single transmission band. A very comprehensive view of the possibilities with crossed dipoles due to Parker is displayed in Figure 2.51. It shows transmission and reflection bands for crossed dipoles of length 4.2 mm printed on a 0.75 mm thick substrate of relative permittivity 2.3 and for various lattice configurations. Besides crossed dipoles, many other elements have also been considered.

For example, convoluted dipoles devised by Parker offer a means to pack more resonance length into a unit cell. This decreases sensitivity to incidence angle and renders easier fabrication on curved surfaces. Layers of resonant elements have also been investigated to enhance reflection. In general, the solution depends on the application. For a dual-band dual-reflector antenna, circular rings prove superior to dipole elements. In one example, the subreflector was required to be transparent in the band 10.95–12.25 GHz while being reflective in the band 18.2–20.2 GHz. The antenna exhibits a gain loss of less than 0.4 dB due to the FSS in both bands. Sidelobes associated with the FSS array give levels below −25 dB in transmission and below −20 dB in reflection while the associated crosspolarisation lies below −25 dB and −36 dB, respectively.

Dichroic reflectors continue as an area of detailed study both in terms of configurations and packaging. The above is intended to provide the reader with a brief introduction to the subject.

2.3 References

1 SILVER, S. (Ed.): 'Microwave antenna theory and design' (McGraw-Hill, 1949; reprinted Peter Peregrinus, 1984), IEE Electromagnetic Waves series 19

2 JAMES, G.L.: 'Geometric theory of diffraction for electromagnetic waves' (Peter Peregrinus, 1976), IEE Electromagnetic Waves series 1

3 RAHMAT-SAMII, Y.: 'Reflector antenna design and antenna modelling for satellite and radar applications'. Lecture 2, antenna symposium, Queen Mary and Westfield College, University of London, 1993

4 RAHMAT-SAMII, Y., and GALINDO V.: 'Shaped reflector antenna analysis using the Jacobi–Bessel series', *IEEE Trans.*, 1980, **AP-28**, pp. 425–435

5 RAHMAT-SAMII, Y.: *in* LO, Y.Y., and LEE, S.W. (Eds.): 'Antenna handbook' (Van Nostrand and Reinhold Book Co, 1988)

6 HANSEN, R.C.: 'Microwave scanning antennas, Vol. 1' (Academic Press, 1964)

7 LUDWIG, A.C.: 'Low sidelobe aperture distributions for blocked and unblocked circular apertures', *IEEE Trans.*, 1982, **AP-30**, pp. 923–946

8 GALINDO, V.: 'Design of dual reflector antennas with arbitrary phase and amplitude', *IEEE Trans.*, 1964, **AP-12**, pp. 403–408

9 POULTON, G.T.: 'The design of Cassegrain antennas for high efficiency and low VSWR'. Proceedings of the 5th European Microwave Conferences, 1975, pp. 61–65

10 WOOD, P.J.: 'Reflector antenna analysis and design' (Peter Peregrinus, 1980), IEE Electromagnetic Waves series 7

11 CHU, T.S., and TURRIN, R.H.: 'Depolarisation properties of offset reflector antennas', *IEEE Trans.*, 1973, **AP-21**, pp. 339–345

12 RUDGE, A.W., and ADATIA, N.A.: 'New class of primary-feed antennas for use with offset parabolic reflector antennas', *Electron. Lett.*, 1975, **11**, pp. 597–599

13 IMBRIALE, W.: 'Antennas for satellite systems' Lecture 4, antenna symposium, Queen Mary and Westfield College, University of London, 1991

14 RAMANUJAM, P., CLARRICOATS, P.J.B., and BROWN, R.C.: 'Offset spherical reflector with a low sidelobe radiation', *IEE Proc. H*, 1987, **134**, (2), pp. 199–204

15 MINNETT, H.C., and THOMAS, B.M.: 'Fields in the image space of symmetrical focusing reflectors', *Proc. IEE*, 1968, **115**, pp. 1419–1430

16 CLARRICOATS, P.J.B., and POULTON G.T.: 'High efficiency microwave reflector antennas—a review', *Proc. IEEE*, 1977, **65**, (10) pp. 1473–1475

17 BEM, D.J.: 'Electric field distribution in the focal field of an offset paraboloid', *Proc. IEE*, 1974, **116**, pp. 579–584

18 TRICOLES, G.P.: *in* LO, Y.T., and LEE, S.W. (Eds.): 'Antenna handbook' (Van Nostrand and Reinhold Book Co., 1988)

19 CORY, H., and CLARRICOATS, P.J.B.: 'Reflection coefficient at the feed of a radome-covered reflector antenna', *IEE Proc. H*, 1982, **129** (4), pp. 145–160
20 RUZE, J.: 'Antenna tolerance theory – a review', *Proc. IEEE*, 1966, **54**, (4) pp. 633–640
21 RUSCH, W.V.T., and POTTER, P.D.: 'Analysis of reflector antennas' (Academic Press, 1970)
22 LOVE, A.W.: 'Some highlights in reflector antenna development', *Radio Sci.*, 1976, **11**, pp. 671–684
23 RAHMAT-SAMII, Y.: 'Effects of deterministic surface distortions on reflector antenna performance', *Ann. Telecommun.*, 1985, **40**, pp. 350–360
24 PONTOPPIDAN, K., and SORENSEN, S.B.: 'Reflector surface accuracy for shaped-beam antennas'. IEE colloquium on Reflector Antennas for the 90s, 1992
25 RUSCH, W.V.T.: *in* RUDGE, A.W., MILNE, K., OLVER, A.D., and KNIGHT, P. (Eds.): 'The handbook of antenna design Vol. 1' (Peter Peregrinus, 1982), IEE Electromagnetic Waves series 15
26 COLLIN, R.E., and ZUCKER, F.J.: 'Antenna theory, Part 2' (McGraw-Hill Book Co., 1969), Chap. 16
27 JØRGENSEN, R.: 'Coverage shaping of contoured beam antennas by aperture field synthesis', *IEE Proc. H*, 1980, **127**, pp. 210–208
28 WESTCOTT, B.S.: 'Shaped reflector antenna design' (Research Studies Press, 1983)
29 BIRD, T.S.: 'Design of the Ku-band antennas for the Galaxy HS601C satellites', *J. Elec. Electron. Eng. Aust.*, 1992, pp. 267–273
30 BERGMANN, J., BROWN, R.C., CLARRICOATS, P.J.B., and ZHOU, H.: 'Synthesis of shaped-beam reflector antenna patterns', *IEE Proc. H*, 1988, **135**, (1), pp. 48–53
31 STIRLAND, S.J., FOX, G.P.D., and MALIK, D.P.S.: 'Comparison between multi-feed and shaped reflector satellite antennas for contoured beams'. Proceedings of 14th AIAA conference on Communication and Satellite Systems 1990
32 MALIK, D.P.S.: 'Dual-gridded reflector antenna design'. Lecture 3, antenna symposium, Queen Mary and Westfield College, University of London, 1993
33 CLARRICOATS, P.J.B., and ZHOU, H.: 'Design and performance of a reconfigurable mesh reflector antenna', *Proc. IEEE*, 1991, **138**, (6), pp. 485–492
34 COMPTESSE, L.E., LANGLEY, R.J., PARKER, E.A., and VARDAXOGLOU, J.C.: 'Frequency selective surfaces in dual and triple band offset reflector antennas'. Proceedings of 17th European Microwave Conference, 1987, pp. 208–213
35 PARKER, E.A.: 'The gentleman's guide to frequency selective surfaces'. Lecture 1, antenna symposium, Queen Mary and Westfield College, University of London, 1991
36 MITTRA, R., CHAN, C.H., and CWIK, T.: 'Techniques for analysing frequency selective surfaces—a review', *Proc. IEEE*, 1988, **76**, pp. 1593–1615

Chapter 3

Radiation from apertures

A.D. Olver

3.1 Introduction

Horn and feed antennas can be divided into two categories for the purpose of the analysis of the radiation characteristics:

(i) antennas which consist of a well defined aperture with imposed electric and magnetic fields over the aperture and which are a few wavelengths in cross-section. This category includes most of the horns and open-ended waveguides used as feeds. The electromagnetic process of determining the fields across an aperture can be separated from the process of determining the radiation characteristics of the fields in the aperture;
(ii) antennas which are small in size and in which the currents on the antenna structure determine the radiation characteristics directly.

This chapter is concerned with the process of predicting the radiation characteristics from antennas which fall into category (i). Antennas falling into category (ii) will be studied in Chapter 5. Antennas where the radiation characteristics are determined solely from the electric and magnetic fields across an aperture are called 'aperture antennas'. This embraces not only most horn and feed antennas but also reflector antennas and lens antennas (Figure 3.1). In all these cases, a plane aperture can be constructed which divides the analysis problem neatly into the 'internal problem' of determining the fields across the aperture and the 'external problem' of predicting the radiated fields from the aperture. The internal problem can be solved by a knowledge of the electromagnetic fields and boundary conditions appropriate to the particular antenna structure. For horns this is usually done by solving the waveguide modal fields and is described in Chapter 4. Here we are concerned solely with the radiation from the plane aperture without regard to how the fields across the aperture come to be formed. The theory outlined in this chapter applies to all aperture antennas, but the focus of attention is obviously on horn and feed antennas rather than reflectors or lenses (although a feed might incorporate a small lens). The size of aperture will normally be between about one wavelength and ten wavelengths in equivalent diameter.

The concept of predicting the radiation from an aperture is particularly important because of the fact that, under certain assumptions, the far-field-

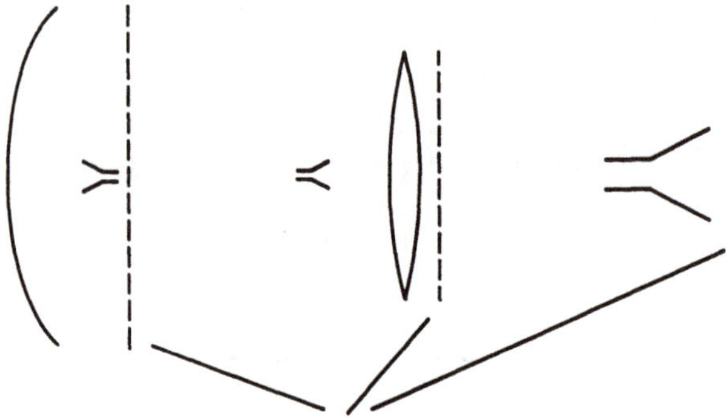

Figure 3.1 Plane apertures in front of a reflector, a lens and a horn

radiation characteristics and aperture electric or magnetic fields are the Fourier transform of each other. Fourier transforms are well known in many branches of science and engineering, and as a result their properties are well understood and can be used to explain all the basic principles of radiation from apertures.

The principles of radiation from apertures have been dealt with in many textbooks (see for example, Stutzman and Thiele [1], Collin and Zucker [2] and Jull [3]). There are two basic approaches to developing the radiation equations. In the first approach Huygen's principle or the field-equivalence principle together with the radiation integral [1, 2] are used to determine the relation between the aperture fields and the radiation patterns. In the second approach the fields from a plane aperture are assumed to consist of a spectrum of plane waves. This approach is associated with Booker and Clemmow [4], but was also developed by others [5–7]. The two approaches lead to identical equations in the far field and it is a matter of choice which approach is preferred. In this chapter the field-equivalence principle is used to develop the basic Fourier-transform relationship because this illustrates the various approximations which must be applied to obtain the radiation from apertures.

3.2 Radiation equations from aperture fields

The field-equivalence principle, as applied to planar apertures, states that if a region of sources ($z<0$) causes radiated fields, $\boldsymbol{E}, \boldsymbol{H}$ in the region $z>0$ (Figure 3.2*a*), then the same radiated fields would be obtained by replacing an imaginary surface s at $z=0$ by a conducting surface on which an electric current density $\boldsymbol{J}_s$ and a magnetic current density $\boldsymbol{J}_m$ flow (Figure 3.2*b*). The current densities are given by

$$\left.\begin{aligned} \boldsymbol{J}_s &= \hat{\boldsymbol{u}}_z \times \boldsymbol{H}_a \\ \boldsymbol{J}_m &= -\hat{\boldsymbol{u}}_z \times \boldsymbol{E}_a \end{aligned}\right\} \tag{3.1}$$

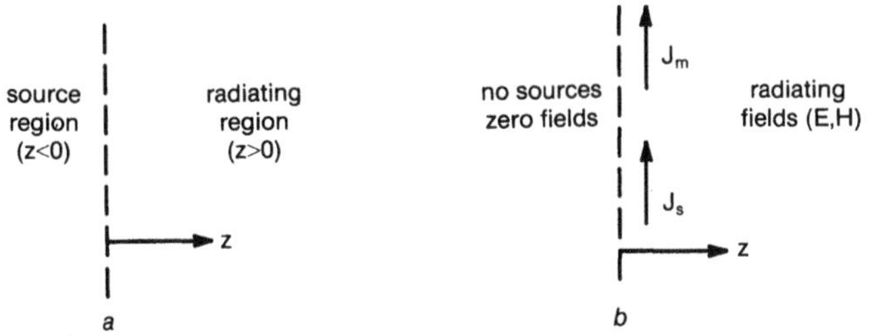

Figure 3.2 Field-equivalence principle applied to a plane aperture

where $\boldsymbol{E}_a$ and $\boldsymbol{H}_a$ are, respectively, the tangential electric and magnetic fields in the aperture. $\boldsymbol{E}_a$ and $\boldsymbol{H}_a$ can be computed by solving the internal problem and then the radiated fields can be derived using the current densities as explained below. Note that it is implicit within the field-equivalence principle that the fields in the region $z<0$ do not need to be known.

The above is only one of many field-equivalence relations. For the radiation from apertures problem, there are two other versions which are useful. Since the region $z<0$ contains no fields, it is possible to introduce an imaginary surface without altering $\boldsymbol{J}_s$ or $\boldsymbol{J}_m$. If a perfect electric conductor is introduced behind the surface s (Figure 3.3), image theory indicates that there will be an image of $\boldsymbol{J}_m$ and an opposite image of $\boldsymbol{J}_s$. If the conductor is removed and the images remain, the result will be a surface s on which $\boldsymbol{J}_s=0$ and $\boldsymbol{J}_m=2(\hat{\boldsymbol{u}}_z\times\boldsymbol{E}_a)=\hat{\boldsymbol{u}}_z\times 2\boldsymbol{E}_a$. This form of the field-equivalence principle is particularly useful because it involves only the aperture electric field. It is sometimes referred to as the E-field model [8]. A third form of the principle is the dual of the last version. If a perfect magnetic conductor is introduced behind surface s, the radiated fields can be determined from $\boldsymbol{J}_m=0$ and $\boldsymbol{J}_s=-\hat{\boldsymbol{u}}_z\times 2\boldsymbol{H}_a$.

The surface s in Figures 3.2 or 3.3 is infinite in the plane transverse to z and the currents are known everywhere on the plane. In an aperture antenna, the electric and/or magnetic fields (and hence currents) are only known over the aperture (Figure 3.4). Outside the aperture, assumptions must be made about the extent of the surface s and the currents flowing on it. These assumptions are different for the three versions described above and consequently lead to three

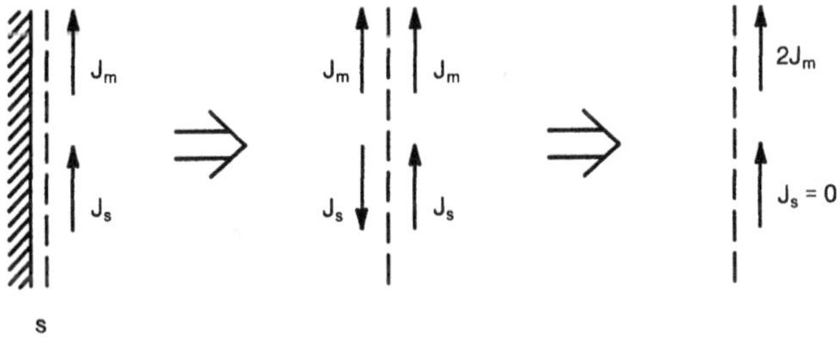

Figure 3.3 Image plane with field-equivalence principle

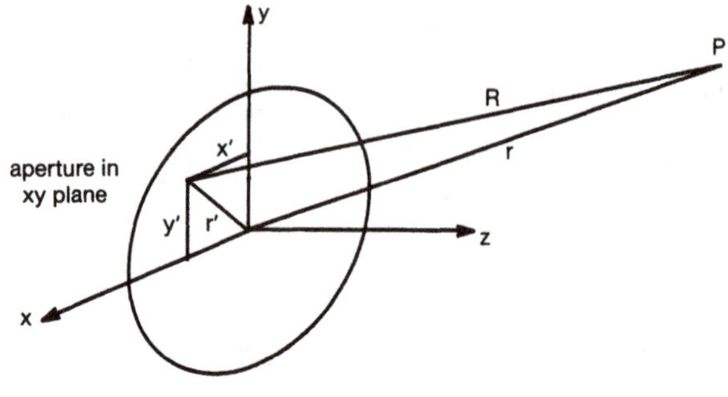

Figure 3.4 Aperture in xy plane

versions of the radiation equations. In all cases there will either be no actual conducting surface (e.g. a horn with no flange), or a finite conducting surface (e.g. a corrugated horn). In addition the condition $\hat{\boldsymbol{u}}_z \times \boldsymbol{E}_a = (\boldsymbol{J}_m) = 0$ is not a valid boundary condition on a practical conducting sheet and the assumption that the aperture electric fields are undisturbed by the presence of the aperture is approximate.

The radiation equations will now be derived for the first version using both $\boldsymbol{E}_a$ and $\boldsymbol{H}_a$. The magnetic vector potential $\boldsymbol{A}$ gives the radiation from an electric current density $\boldsymbol{J}_s$,

$$\boldsymbol{A} = \int_s \boldsymbol{J}_s \frac{e^{-jkR}}{4\pi R} \, ds \tag{3.2}$$

Similarly, the electric vector potential $\boldsymbol{A}_m$ gives the radiation from an electric current density $\boldsymbol{J}_m$,

$$\boldsymbol{A}_m = \int_s \boldsymbol{J}_m \frac{e^{-jkR}}{4\pi R} \, ds \tag{3.3}$$

In the far field $R = r - r' \cos\theta$ (Figure 3.4), and if r' is expressed in terms of the spherical co-ordinates θ and ϕ, eqn. 3.2 becomes

$$\boldsymbol{A} = \frac{e^{-jkR}}{4\pi R} \hat{\boldsymbol{u}}_z \times \int_{x'} \int_{y'} E_a(x', y') \exp\{jk(x' \sin\theta \cos\phi + y' \sin\theta \sin\phi)\} \, dx' \, dy' \tag{3.4}$$

and similarly for eqn. 3.3. Eqn. 3.4 is a 2-dimensional Fourier transform and its components are written down as

$$\left.\begin{aligned} f_x &= \int_{x'} \int_{y'} E_{ax}(x', y') \exp\{jk(x' \sin\theta \cos\phi + y' \sin\theta \sin\phi)\} \, dx' dy' \\ f_y &= \int_{x'} \int_{y'} E_{ay}(x', y') \exp\{jk(x' \sin\theta \cos\phi + y' \sin\theta \sin\phi)\} \, dx' \, dy' \end{aligned}\right\} \tag{3.5}$$

A comparable set of Fourier transforms relates the aperture magnetic field to the radiated field:

$$\left.\begin{aligned} g_x &= \int_{x'}\int_{y'} H_{ax}(x', y') \exp\{jk(x' \sin\theta \cos\phi + y' \sin\theta \sin\phi)\}\, dx'\, dy' \\ g_y &= \int_{x'}\int_{y'} H_{ay}(x', y') \exp\{jk(x' \sin\theta \cos\phi + y' \sin\theta \sin\phi)\}\, dx'\, dy' \end{aligned}\right\} \quad (3.6)$$

In the far field the electric and magnetic fields are related by the TEM relation $\boldsymbol{E} = \eta_0 \boldsymbol{H} \times \hat{\boldsymbol{u}}_r$ where η_0 is the free-space wave impedance of 377 Ω. The radiated electric fields are then given by

$$\boldsymbol{E} = -j\omega(\boldsymbol{A} + \eta_0 \boldsymbol{A}_m \times \hat{\boldsymbol{u}}_r) \quad (3.7)$$

Combining the above equations and transforming the electric fields into spherical co-ordinates gives the radiated electric fields using J_s and J_m as

$$\left.\begin{aligned} E_\theta &= \frac{jk_0 \exp(-jk_0 r)}{4\pi r}\{f_x \cos\phi + f_y \sin\phi + \eta_0 \cos\theta(g_y \cos\phi - g_x \sin\phi)\} \\ E_\phi &= \frac{jk_0 \exp(-jk_0 r)}{4\pi r}\{\cos\theta(f_y \cos\phi - f_x \sin\phi) - \eta_0(g_y \sin\phi + g_x \cos\phi)\} \end{aligned}\right\} \quad (3.8)$$

A common approximation is to assume that the fields in the aperture are TEM-type fields so that $E_{ax} = \eta_0 H_{ay}$ and $E_{ay} = -\eta_0 H_{ax}$. In this case eqns. 3.8 simplify to

$$\left.\begin{aligned} E_\theta &= \frac{jk_0 \exp(-jk_0 r)}{4\pi r}\{(1 + \cos\theta)(f_x \cos\phi + f_y \sin\phi)\} \\ E_\phi &= \frac{jk_0 \exp(-jk_0 r)}{4\pi r}\{(1 + \cos\theta)(f_y \cos\phi - f_x \sin\phi)\} \end{aligned}\right\} \quad (3.9)$$

If an E-field model is used with only $2\boldsymbol{J}_m$ in the aperture, the far-field radiated fields are

$$\left.\begin{aligned} E_\theta &= \frac{jk_0 \exp(-jk_0 r)}{2\pi r}(f_x \cos\phi + f_y \sin\phi) \\ E_\phi &= \frac{jk_0 \exp(-jk_0 r)}{2\pi r}\cos\theta(f_y \cos\phi - f_x \sin\phi) \end{aligned}\right\} \quad (3.10)$$

If an H-field model is used with only $2\boldsymbol{J}_s$ in the aperture, the far-field radiated fields are

$$\left.\begin{aligned} E_\theta &= \frac{jk_0 \exp(-jk_0 r)}{2\pi r}\cos\theta(g_y \cos\phi - g_x \sin\phi) \\ E_\phi &= \frac{-jk_0 \exp(-jk_0 r)}{2\pi r}(g_x \cos\phi + g_y \sin\phi) \end{aligned}\right\} \quad (3.11)$$

For horns and feeds, the linear polarised copolar and crosspolar fields are normally required. Using Ludwig's third definition [9], these can be obtained from the spherical components by

$$\left.\begin{aligned} E_{co} &= E_\theta \sin\phi + E_\phi \cos\phi \\ E_{xp} &= E_\theta \cos\phi - E_\phi \sin\phi \end{aligned}\right\} \tag{3.12}$$

This assumes that the aperture electric field is directed along the y direction.

If circular polarisation is required, the right-hand (RH) and left-hand (LH) components are given by

$$\left.\begin{aligned} E_{RH} &= \frac{\exp(-j\phi)}{\sqrt{2}}(E_\theta - jE_\phi) \\ E_{LH} &= \frac{\exp(-j\phi)}{\sqrt{2}}(E_\theta + jE_\phi) \end{aligned}\right\} \tag{3.13}$$

The question naturally arises as to which of eqns. 3.8, 3.9, 3.10 and 3.11 should be used. There is no single answer to this question, because it depends on how well a particular aperture antenna can be approximated by one of the models. For large apertures $\cos\theta \simeq 1$, in which case eqns. 3.9 and 3.10 are the same. For large antennas with nearly linear aperture fields (e.g. a corrugated horn), all three models give the same results around the main beam. In practice, the differences between models is most significant for wide radiation angles and small apertures. The potentially most accurate model is eqn. 3.8 since it uses both electric and magnetic fields. However, experience has tended to show that when the aperture is small the currents on the outside of the antenna structure influence the radiation (and these are not included in the model), so that no particular model is always better at predicting the correct result. For accurate predictions, a moment-method approach as described in Chapter 4 must be used. The surprising feature is how reasonably good the models are at predicting the radiation, even for relatively small apertures. The most common model is to use either the pure E-field model (eqn. 3.10) or the modified E-field and H-field model (eqn. 3.9).

3.3 Radiation characteristics from apertures

3.3.1 Solution of radiation equations

The radiation equations which were derived in Section 3.2 will now be used to describe the basic properties of the radiation patterns from apertures. The Fourier-transform relationship between the transverse aperture fields and the far-field radiation patterns means that the deductions have considerable applicability to practical horns even though the results are expressed in terms of idealised aperture fields. The aim, therefore, is to understand the factors which control the radiation patterns so that designers of horns can choose the most appropriate type of horn for a given requirement.

From this point of view, the factors of most interest to the feed designer are the shape of the copolar-radiation pattern, the beamwidth of the main beam and the level of the first sidelobe. The beamwidth is normally quoted in terms of the aperture size which will produce a beamwidth at the halfpower point, i.e. the −3 dB level. However, for feed design it is also useful to know the size of aperture which will produce a beamwidth at the −10 dB level. This is because −10 dB is the most common edge taper chosen for the illumination of a reflector system and that which produces approximately optimum efficiency.

The radiation equations can always be solved numerically and for certain aperture distributions they can be solved analytically. The discussion below describes some of the analytical results because these illustrate the main features. Apart from the simple aperture distributions it is not generally worthwhile using the analytical results because numerical integration is easy and relatively quick using a method such as Simpson's rule. The Fourier-transform relation lends itself to solution using an FFT approach. This is potentially quicker but the relatively small aperture dimensions of horns do not lead to long computation times by any method.

3.3.2 Line-source apertures

Line source apertures (Figure 3.5) are artificial but convey all the main features of radiation from apertures and are the basic building block of rectangular apertures. For a line source of length L in the x direction, the Fourier-transform relation (eqn. 3.5) becomes

$$f_x = \int_{-L/2}^{L/2} E_a(x') \exp(jkx' \sin\theta \cos\phi)\, dx' \tag{3.14}$$

The solution to this equation in the $\phi = 0$ plane depends on the form of $E_a(x')$.

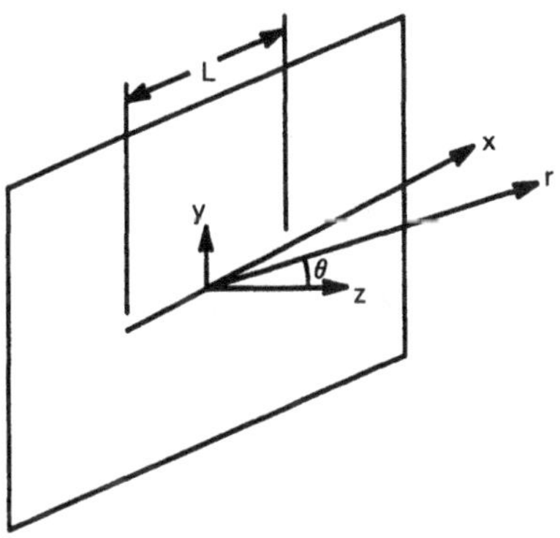

Figure 3.5 Line-source aperture of length L

The simplest is a uniform distribution $E_a(x) = 1(|x| < L/2)$. This gives the radiated field as

$$f_x = \frac{\sin u}{u} \tag{3.15}$$

where the patterns have been normalised to $u = kL/2 \sin\theta$ $(k = 2\pi/\lambda)$. This is the well known $\sin u/u$ antenna pattern. The radiation pattern is plotted in Figure 3.6, which has a first-sidelobe level of −13.2 dB. This is a high level and leads to a low efficiency with a reflector since too much power is going into the sidelobes instead of into the main beam. The radiated power in the sidelobes does not hit the reflector and is lost. However, the high sidelobes lead to a high aperture efficiency which is reflected in the −3 dB beamwidth. This occurs when $\sin u/u = 1/\sqrt{2}$. The full beamwidth, $\Delta\theta_{B3}$ is then

$$\Delta\theta_{B3} = 2\sin^{-1}\left(0.44\frac{\lambda}{L}\right) \tag{3.16}$$

For long sources $(L \gg \lambda)$ this approximates to $\Delta\theta_{B3} = 0.88\,\lambda/L$ where $\Delta\theta_{B3}$ is in radians. This is a widely quoted and useful design equation. Note, however, that the angular value predicted by the equation, and similar values quoted below for other aperture distributions, are for medium to large aperture dimensions. This means that, for many horns where the beamwidth is quite broad, these simple formulae should be treated as only estimates of the beamwidth.

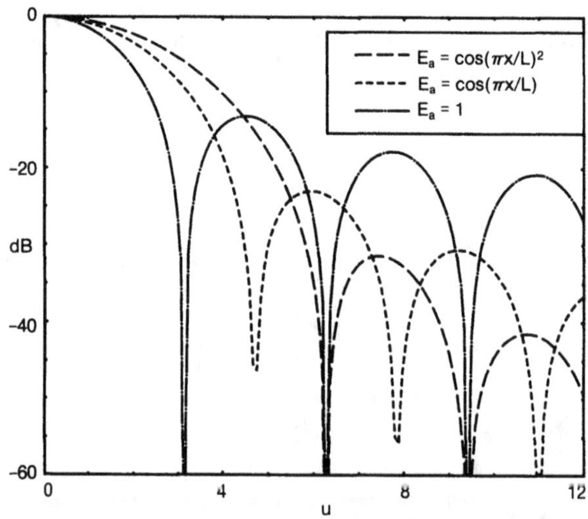

Figure 3.6 Radiation patterns for line-source aperture

Using similar reasoning, the -10 dB beamwidth works out as $\Delta\theta_{B10} = 1.48\ \lambda/L$.

The phase pattern for the sin u/u radiation pattern is $\theta = \pi/2$ for

$$2(n-1)\pi < u < (2n-1)\pi \qquad (n=0, 1, 2, \ldots)$$

and $\theta = -\pi/2$ for

$$(2n-1)\pi < u < 2n\pi \qquad (n=1, 2, 3, \ldots)$$

Thus the phase is constant over a lobe and switches by 180° at each null.

Most feeds have aperture distributions which are tapered away from the axis (the exceptions are feeds with a horn-plus-lens combination which generate a nearly uniform aperture distribution). A representative aperture distribution of many horns is a cosine function

$$E_a(x') = \cos^n\left(\frac{\pi x'}{L}\right) \qquad |x'| < \frac{L}{2} \tag{3.17}$$

where n is 1 or 2. The results for these cases are given in Table 3.1 and the radiation patterns shown in Figure 3.6, these having been computed using the E-field model (eqn. 3.10). By comparison with the uniform distribution, the tapered apertures have a broader beamwidth but a lower sidelobe level.

There are some situations where a horn is required to provide tracking information from a radiation pattern which has a null on axis. This can be obtained with an aperture distribution of the form

$$E_a(x') = \sin\left(\frac{\pi x'}{L}\right) \qquad |x'| < \frac{L}{2} \tag{3.18}$$

which gives the radiation equation display in Table 3.1 and the radiation pattern in Figure 3.7.

3.3.3 Compound aperture distributions

A compound distribution is one which combines a tapered source with a uniform distribution, sometimes called a taper on a pedestal (Table 3.1). This is a better estimate of a practical aperture distribution because there will always be a finite field at the edge of a horn. The total aperture field can be represented as the sum of two fields $E_{a1}(x') + E_{a2}(x')$. Eqn. 3.14 then becomes

$$f_{x1} + f_{x2} = \int_{-L/2}^{L/2} \{E_{a1}(x') + E_{a2}(x')\} \exp(jkx' \sin\theta \cos\phi)\,\mathrm{d}x' \tag{3.19}$$

where f_{x1} is the pattern of $E_{a1}(x')$ and f_{x2} is the pattern of $E_{a2}(x')$. In other words, the result is the sum of the Fourier transforms of the separate parts of the aperture distribution. The results for a cosine on a pedestal are given in Table

Table 3.1 *Radiation characteristics of line sources [$u=(kL/2)\sin\theta$]*

Distribution	Aperture field	Normalised pattern $f(u)$	−3 dB full beamwidth	−10 dB full beamwidth	First-sidelobe level (dB)
Uniform	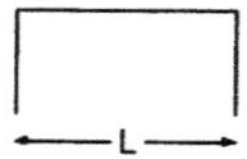	$\frac{\sin u}{u}$	0.88 λ/L	1.48 λ/L	−13.2
$\cos\left(\frac{\pi x'}{L}\right)$		$\frac{\cos u}{1-\left(\frac{2u}{\pi}\right)^2}$	1.2 λ/L	2.04 λ/L	−23.1
$\cos^2\left(\frac{\pi x'}{L}\right)$		$\frac{1}{1-\left(\frac{u}{\pi}\right)^2}\frac{\sin u}{u}$	1.44 λ/L	2.51 λ/L	−31.5
$\sin\left(\frac{\pi x'}{L}\right)$	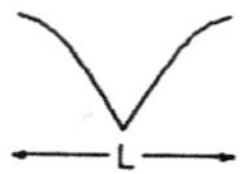	$\frac{u\cos u}{1-\left(\frac{2u}{\pi}\right)^2}$	—	—	
$C+(1+C)\cos\left(\frac{\pi x'}{L}\right)$		$\dfrac{C\frac{\sin u}{u}+(1-C)\frac{2}{\pi}\frac{\cos u}{1-\left(\frac{2u}{\pi}\right)^2}}{C+(1-C)\frac{2}{\pi}}$	1.03 λ/L (C = −10 dB) 1.09 λ/L (C = −15 dB)	1.75 λ/L 1.86 λ/L	−20.0 −21.9

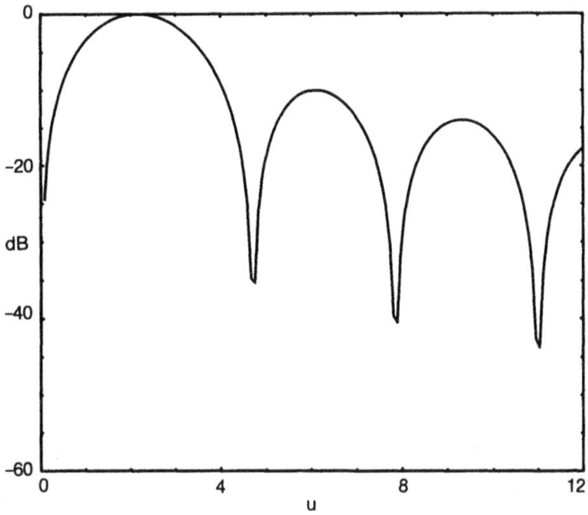

Figure 3.7 Radiation pattern of a sin (x) line-source aperture

3.1 for two edge tapers of −10 dB and −15 dB. By comparison with the cosine distributions alone, the compound distribution gives a narrow beamwidth with only a slightly higher first-sidelobe level.

3.3.4 Rectangular apertures

Most rectangular apertures (Figure 3.8) have aperture distributions $E_a(x', y')$ which are separable in x and y so that $E_a(x', y') = E_{a1}(x')E_{a2}(y')$. If this is not the case, the problem cannot be solved analytically because the electromagnetic-wave equation and Maxwell's equation can only be solved for separable co-

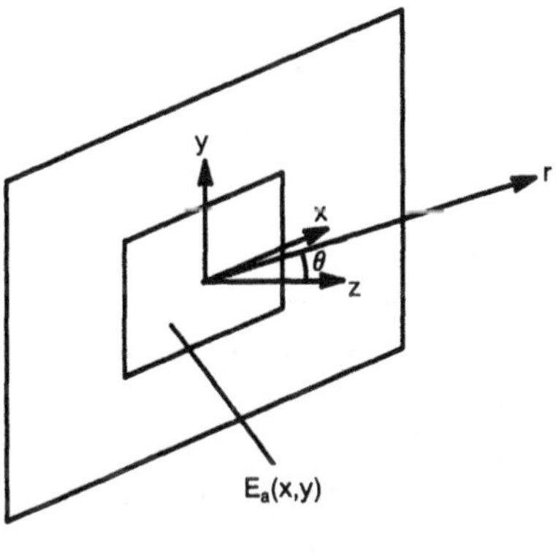

Figure 3.8 Rectangular aperture

ordinate systems. For separable cases, the radiation equation for f_x in eqn. 3.14 becomes

$$f_x = f_{x1} f_{x2} \tag{3.20}$$

where

$$\left.\begin{aligned} f_{x1} &= \int_{-L/2}^{L/2} E_{a1}(x') \exp(jkx' \sin\theta \cos\phi)\, dx' \\ f_{x2} &= \int_{-L/2}^{L/2} E_{a2}(y') \exp(jky' \sin\theta \sin\phi)\, dy' \end{aligned}\right\} \tag{3.21}$$

The integrations in these equations are over the x' or y' dimensions only so that the radiation pattern is given by the multiplication of two linear-source apertures and the results of Section 3.3.3 are directly applicable to rectangular apertures. For instance, an open-ended rectangular waveguide with the dominant TE_{10} mode has a cosine distribution in one plane and a constant field in the other plane. If

$$u_x = k\frac{L_x}{2} \sin\theta \cos\phi$$

with $E_a(x') = \cos\left(\dfrac{\pi x'}{a}\right)$

$$v_y = k\frac{L_y}{2} \sin\theta \sin\phi$$

with $E_a(y') = 1$

Then, from Table 3.1,

$$f = \frac{\cos u_x}{1 - \left(\dfrac{2u_x}{\pi}\right)^2} \frac{\sin v_y}{v_y} \tag{3.22}$$

3.3.5 *Apertures with phase tapers*

The linear and rectangular apertures described above produce real Fourier transforms from real aperture distributions. In most practical feeds, the electromagnetic fields across the aperture are produced by a horn which has a finite flare angle. In these circumstances the aperture field can be represented by the field due to an open-ended waveguide plus a spherical phase factor which accounts for the influence of the finite flare angle. This procedure is approximate and only works well for narrow flare angles, but this includes most of the horns used in reflector antennas. The phase taper can be simulated in the

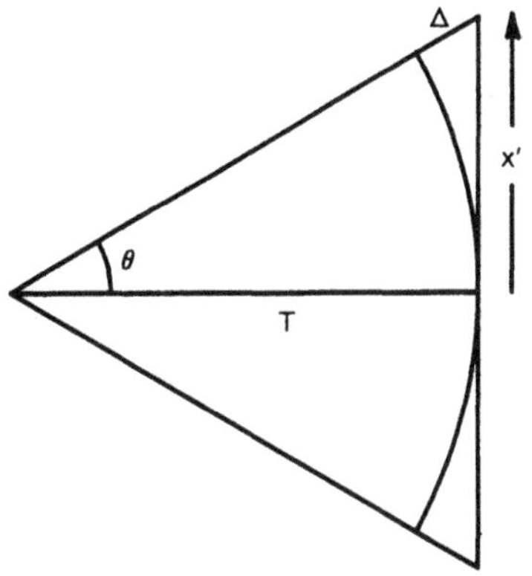

Figure 3.9 Spherical phase taper on a line-source aperture

idealised aperture by multiplying the aperture electric field by a phase factor as follows.

The fields across the aperture of a tapered horn (Figure 3.9) are generated by spherical waves with the origin at the apex. The spherical waves can be approximated in the aperture by plane waves with an additional spherical phase factor Δ to account for the extra path length travelled by waves at an arbitrary point x across the aperture. The phase factor for narrow flare angles is given by

$$\Delta = R - R\cos\theta = x'\frac{(1-\cos\theta)}{\sin\theta} = x'\tan\left(\frac{\theta}{2}\right) \approx x'\frac{x'}{2T} \tag{3.23}$$

The Fourier-transform relation for the far fields of a line source then becomes

$$f_x = \int_{-L/2}^{L/2} E_a(x')\exp(jkx'\sin\theta\cos\phi)\exp(jk\Delta)\,\mathrm{d}x' \tag{3.24}$$

This equation does not have the simplicity of the original form, although it is still possible to obtain closed-form solutions in certain cases [10] using Fresnel integrals. Generally, it is better to solve the transform numerically. The integral now has both real and imaginary parts, meaning that the far-field radiation pattern has a phase term which is not constant over a lobe.

Figure 3.10 shows the radiation patterns for a 6λ line source with uniform aperture distribution and phase tapers Δ/λ of 1/16, 1/8 and 1/4 (or 22.5°, 45° and 90°). These correspond to horn semiflare angles of 2.4°, 4.8° and 9.6° for an aperture of width 6λ. Larger flare angles would violate the assumptions in eqn. 3.23, and can only be approximately computed with the Fourier-transform method. The phase factor has the consequence that the nulls in the amplitude radiation pattern are filled in. For large phase tapers, the separate nulls disappear and the radiation pattern has a continuous decay with angle. In this case the pattern approximates to that produced by an ideal Gaussian aperture field (see below).

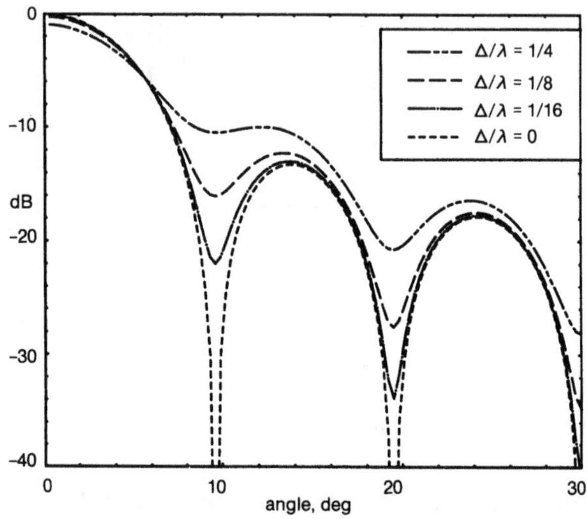

Figure 3.10 Radiation patterns of uniform line source with phase tapers

3.3.6 Circular apertures

Conical horns are widely used as feeds so that the radiation characteristics of circular apertures are of considerable interest (Figure 3.11). For circular geometry the distance factor in the radiation integral (eqn. 3.4) becomes

$$\hat{\boldsymbol{u}}_r \cdot \boldsymbol{r}' = r' \sin\theta \cos(\phi - \phi') \tag{3.25}$$

The radiation equation is now

$$f = \int_0^{2\pi} \int_0^a E_a(r', \phi') \exp\{jkr' \sin\theta \cos(\phi - \phi')\} r' \, dr' \, d\phi' \tag{3.26}$$

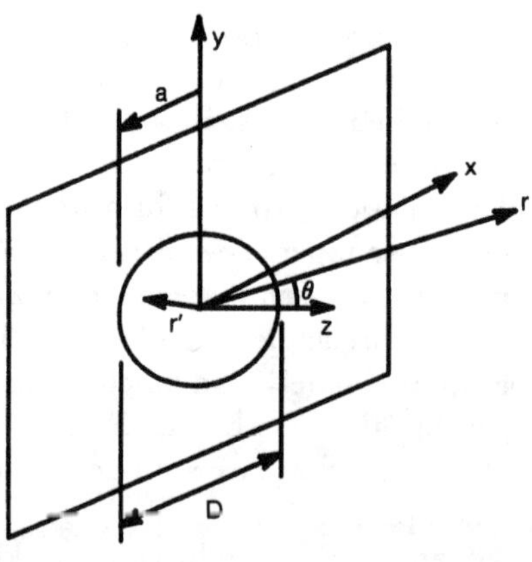

Figure 3.11 Circular aperture

In many cases the aperture electric field can be assumed to be independent of the azimuthal co-ordinate, ϕ, and then the integral in eqn. 3.36 can be evaluated using

$$\int_0^{2\pi} \begin{pmatrix} \cos n\phi' \\ \sin n\phi' \end{pmatrix} \exp\{jA \cos(\phi - \phi')\} d\phi' = 2\pi j^n \begin{pmatrix} \cos n\phi' \\ \sin n\phi' \end{pmatrix} J_n(A) \quad (3.27)$$

where $J_n(A)$ is a Bessel function of order n. The radiation equation becomes

$$f = \int_0^a E_a(r') J_0(kr' \sin \theta) r' \, dr' \quad (3.28)$$

If the aperture distribution is uniform, so that $E_a(r') = 1$, the normalised radiation equation becomes

$$f = 2\frac{J_1(u)}{u} \quad (3.29)$$

where $u = kD/2 \sin \theta$. This is to be compared with the $\sin(u)/u$ pattern produced by line sources. The full beamwidth is $1.03\lambda/D$ and the first-sidelobe level is at -17.6 dB. The uniformly illuminated circular aperture thus produces a broader main beam than the equivalent linear aperture but has a lower sidelobe level. Generally this is a better pattern, particularly since it is valid in both orthogonal planes whereas a square aperture is more difficult to design.

Practical circular apertures will have tapered transverse aperture fields. Some representative results are shown in Table 3.2, which can be compared with the linear apertures in Table 3.1 and seen to produce similar results. The amplitude-radiation patterns are plotted in Figure 3.12.

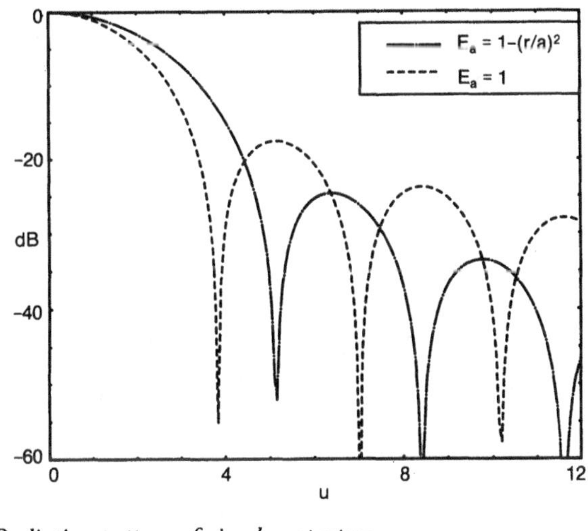

Figure 3.12 Radiation pattern of circular aperture

Table 3.2 Radiation characteristics of circular apertures [$u=(ka)\sin\theta$, $D=2a$]

Distribution	Normalised pattern	−3 dB full beamwidth	−10 dB full beamwidth	First-sidelobe level (dB)
$\left\{1-\left(\frac{r}{a}\right)^2\right\}^n$	$f(\theta, n)=\frac{2^{n+1}(n+1)!J_{n+1}(u)}{u^{n+1}}$			
$n=0$ (uniform)	$2\frac{J_1(u)}{u}$	1.03 λ/D	1.74 λ/D	−17.6
$n=1$	$8\frac{J_2(u)}{u^2}$	1.27 λ/D	2.19 λ/D	−24.6
$C+(1+C)\cos\left\{1-\left(\frac{r}{a}\right)^2\right\}^n$	$f(\theta, n, C)=\frac{Cf(\theta, n=0)+\frac{1-C}{n+1}f(\theta, n)}{C+\left(\frac{1-C}{n+1}\right)}$			
$n=1$, $C=-10$ dB		1.14 λ/D	1.95λ/D	−22.3
$n=1$, $C=-15$ dB		1.18 λ/D	2.03 λ/D	−23.7

3.3.7 Gaussian aperture distributions

An aperture distribution which is of particular interest to feed designers is the Gaussian distribution $E_a(r') = \exp(-r'^2)$. The radiation equation for an infinite circular aperture with a Gaussian field which is independent of ϕ is

$$f = \int_0^\infty \exp(-r'^2) J_0(kr' \sin\theta) r' \, dr' \tag{3.30}$$

The solution to this equation is [11]

$$f = \frac{1}{2} \exp\left\{ -\left(\frac{k}{2} \sin\theta \right)^2 \right\} \tag{3.31}$$

Hence the radiation pattern is a Gaussian function of sin θ, and the Gaussian aperture distribution has been transformed into a Gaussian function. It has the property of a continuous decay with angle and no sidelobes.

This feature is only true for an infinite aperture. In practice, a horn with a Gaussian aperture distribution will be truncated at the edge of the finite aperture. Then the radiation pattern is given by

$$f = \int_0^a \exp\left\{ -C\left(\frac{r'}{a} \right)^2 \right\} J_0(kr' \sin\theta) r' \, dr' \tag{3.32}$$

where C is the edge taper in decibels. The radiation pattern now has a Gaussian-shaped main beam but then has finite sidelobes because of the truncated aperture distribution. Figure 3.13 shows radiation patterns for edge tapers of 6 dB and 40 dB. The pattern for a 6 dB taper shows virtually no Gaussian characteristics and is dominated by the nearly uniform aperture field

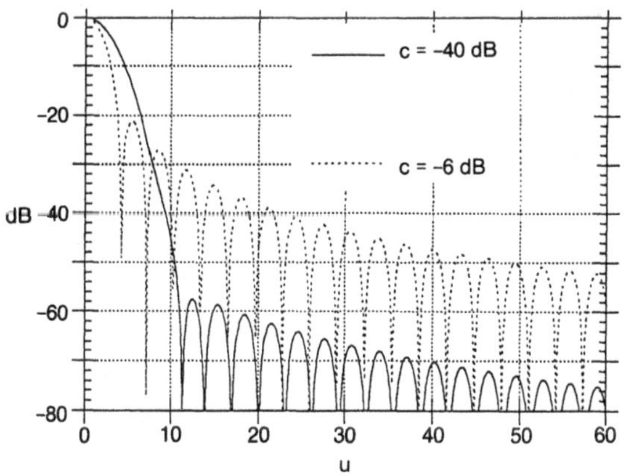

Figure 3.13 Radiation patterns of tapered Gaussian circular aperture distribution

so that the pattern is nearly a $J_1(u)/u$ pattern. By contrast the 40 dB edge-taper pattern has the first three nulls filled-in and the main beam gradually decays as expected from a Gaussian pattern. For larger angular values of u, sidelobes appear. There are some horns, for instance a wide-angle corrugated horn, where, under the appropriate conditions, the field at the edge of the aperture has decayed to nearly zero so that a patttern similar to Figure 3.12 is produced.

3.4 Directivity of apertures

The 'directivity' D in direction θ, ϕ is defined as

$$D(\theta, \phi) = \frac{4\pi \times \text{power radiated per unit solid angle in direction } \theta, \phi}{\text{total power radiated by antenna}} \tag{3.33}$$

This is often referred to as the 'gain', but the directivity and the gain are different since the gain is normalised against the total power accepted from the source. Thus the directivity is a function of the radiation characteristics, while the gain includes the dissipative losses in the antenna. In general, the gain is a measured quantity and the directivity is a computed quantity.

The general expression given in eqn. 3.33 can be written in terms of the far-field electric field $E(\theta, \phi)$ for the radiation from an aperture as

$$D(\theta_0, \phi_0) = \frac{4\pi |\boldsymbol{E}(\theta_0, \phi_0)|^2}{\int_0^{2\pi} \int_0^{\pi/2} |\boldsymbol{E}(\theta, \phi)|^2 \sin\theta \, \mathrm{d}\theta \, \mathrm{d}\phi} \tag{3.34}$$

Usually the peak directivity is required which is in the boresight direction $\theta_0 = 0$, $\phi_0 = 0$. To evaluate this expression the far fields, as computed using the radiation equations in Section 3.2, are inserted into eqn. 3.34 and evaluated. For a few cases the denominator can be evaluated analytically, but it is more easily done using numerical integration.

An alternative expression for the directivity, which is applicable only to apertures, can be obtained by noting that all the radiated power must flow through the aperture. Thus the denominator in eqn. 3.34 can be derived by integrating the power in the aperture. When the aperture fields are TEM so that $\boldsymbol{E} = \eta_0 \boldsymbol{H} \times \boldsymbol{u}_z$, the peak directivity D becomes

$$D = \frac{4\pi (|f_x|^2 + |f_y|^2)}{\lambda^2 \int_x \int_y |E_a(x, y)|^2 \, \mathrm{d}x \, \mathrm{d}y} \tag{3.35}$$

where f_x and f_y are evaluated along the boresight direction. This equation is expressed only in terms of the aperture fields. It is less accurate than eqn. 3.34 because of the assumption that the electric and magnetic fields in the aperture are related through the free-space wave impedance.

3.5 Near fields

The radiation equations derived in Section 3.2 and the results presented in Section 3.3 apply only to the far field. Sometimes the radiation pattern in the near-field, or Fresnel, region is required. This might be because a reflector or lens is required to be placed close to a horn and the distances from different parts of the aperture to the reflector or lens cannot be considered to be constant. In this case, the radiation equations must be modified for the near-field region.

The far-field region, or Fraunhofer region, is the region where the spherical waves radiating from the aperture can be assumed to be plane waves with no variation in the plane transverse to the direction of travel. The change in field behaviour between the far-field region and the near-field region is clearly gradual, so an arbitrary boundary is drawn at the distance where the phase error on the spherical wavefront differs by $\lambda/16$ from a plane wavefront (Figure 3.14). On the assumption that R is some distance from the aperture of diameter D, this 'far-field distance' can be expressed as

$$R = \frac{2D^2}{\lambda} \tag{3.36}$$

For a horn with a 150 mm diameter at 12 GHz (6λ), the far-field distance is 1.8 m. This distance can easily be greater than the distance between a feed and a subreflector. Fortunately the far-field distance is particularly noncritical for feeds. This is partly because most feeds have well behaved radiation patterns which change only gradually with distance and partly because good feeds have Gaussian shaped patterns which necessarily remain Gaussian at all distances from the aperture. There are, however, some cases where the near-field patterns must be computed.

The near-field region extends from the point outside the aperture at which the reactive fields can be assumed to be zero. This is normally taken to be when

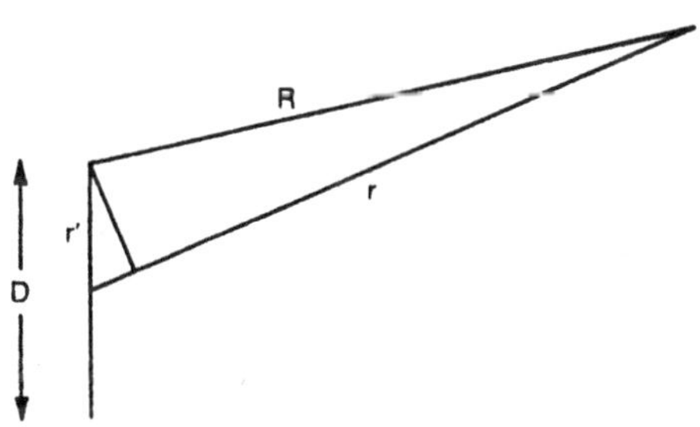

Figure 3.14 Geometry of near field

$r > \lambda$. In the near-field region, the radiation equations (eqns. 3.2 and 3.3) must be evaluated without approximating the distance R. Using the notation of Figure 3.14, R can be expanded in terms of r by the first terms of a Taylor expansion:

$$R = r - \hat{\boldsymbol{u}}_r \boldsymbol{r}' + \frac{1}{2r}\{r'^2 - (\hat{\boldsymbol{u}}_z \cdot \boldsymbol{r}')^2\} + \cdots \tag{3.37}$$

In the far field, only the first two terms are used. For points in the near field which are near the main beam, the fourth term $(\hat{\boldsymbol{u}}_r \cdot \boldsymbol{r}')^2 = (r' \sin\theta)^2$ is small and can be ignored. The near field is then represented by a quadratic term $r'^2/2r$. The radiation equations (eqn. 3.5) now have an extra multiplying term $\exp(jkr'^2/2r)$. For instance, the radiation from a circular aperture (eqn. 3.26) becomes

$$f = \int_0^{2\pi} \int_0^a E_a(r', \phi') \exp\{jkr' \sin\theta \cos(\phi - \phi')\} \exp(jkr'^2/2r) r' \, dr' \, d\phi' \tag{3.38}$$

where r' is the distance from the centre of the aperture to the observation point. This equation is no longer a Fourier transform. In optics it is called a Fresnel transform. There are certain cases where it can be evaluated analytically, but it is not generally worthwhile and is better done numerically.

The extra term can be recognised as identical in form to that used in Section 3.3.5 to describe an aperture with a phase taper. This is not surprising since the method of derivation is similar. The results in Figure 3.10 therefore also give the behaviour of the radiation patterns in the near-field region. This shows that the nulls of the amplitude-radiation pattern will fill in and the main beam will broaden as the distance r from the antenna is reduced. The pattern for $\Delta/\lambda = 1/16$ corresponds to the far-field distance in eqn. 3.36 and shows that, for a uniformly illuminated line source, the pattern measured at the far-field distance has a minimum first null of -22 dB instead of $-\infty$ dB.

3.6 References

1 STUTZMAN, W.L., and THIELE, G.A.: 'Antenna theory and design' (Wiley, New York, 1981)
2 COLLIN, R.E., and ZUCKER, F.J.: 'Antenna theory. Part 1' (McGraw–Hill, New York, 1969)
3 JULL, E.V.: 'Aperture antennas and diffraction theory' (Peter Peregrinus, 1981), IEE Electromagnetic Waves series 10
4 BOOKER, H.G., and CLEMMOW, P.C.: 'The concept of an angular spectrum of plane waves and its relation to that of polar diagram and aperture distribution', *Proc. IEE*, Pt III, 1950, **97**, pp. 11–17
5 WOODWARD, P.M., and LAWSON, J.D.: 'The theoretical precision with which an arbitrary radiation pattern may be obtained from a source of finite size', *Proc. IEE*, Pt III, 1948, **95**, pp. 363–370
6 CLEMMOW, P.C.: 'The plane wave representation of electromagnetic fields' (Pergamon, London, 1966)

7 RHODES, D.R.: 'On a fundamental principle in the theory of planar antennas', *IEEE Trans.*, 1966, **AP–12**, pp. 1013–1021
8 RUDGE, A.W., MILNE, K., OLVER, A.D., and KNIGHT, P. (Eds.): 'The handbook of antenna design' (Peter Peregrinus, 1982), IEE Electromagnetic Wave series 15 and 16
9 LUDWIG, A.C.: 'The definition of cross-polarisation', *IEEE Trans.*, 1973, **AP–21**, pp. 116–119
10 SLETTEN, C.J. (Ed.): 'Reflector and lens antennas' (Artech House, USA, 1988), pp. 63–67
11 ABROMOWITZ, M., and STEGUN, I.A.: 'Handbook of mathematical functions' (Dover, 1965), p. 486

Chapter 4

Numerical analysis using cylindrical and spherical waves in horns

A.D. Olver

4.1 Introduction

Chapter 3 dealt with the analysis procedure for predicting the radiation pattern from apertures. The transverse electric and magnetic fields across the aperture were assumed to be known and the method by which these fields were obtained was not a consideration in the analysis. In other words, the aperture theory deals with the external region of a horn and predicts the radiation characteristics of a horn. This chapter now deals with the internal region of a horn where the transverse electric and magnetic fields across the aperture plane are formed (Figure 4.1). The results of the internal analysis predict the aperture fields and possibly also the reflection coefficient at the input terminals of the horn. If the internal fields are known accurately, the radiation characteristics and the impedance characteristics can be computed more accurately. Thus the aim should be to analyse the internal characteristics of the horn as accurately as possible.

The classical method of studying the internal region is to predict the modal characteristics which describe the fields in the waveguide. For simple geometries and homogeneous waveguides, the results are simple functions of cylindrical or spherical waves. This method works well when the fields are stable along

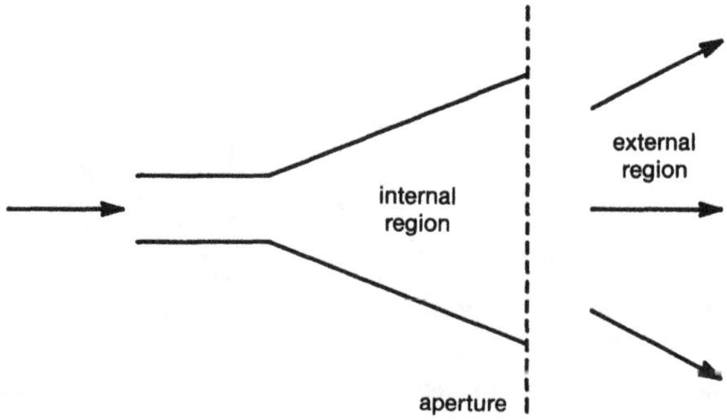

Figure 4.1 Internal and external regions in a horn

the horn and the aperture fields are composed of either one or two discrete modes. It has been used over many years to predict successfully the aperture fields of all types of horn [1–3]. There are two restrictions on the method. First, it does not predict the consequences of discontinuities along the horn which will lead to the excitation of multiple modes. Secondly it gives no information of the impedance properties of the horn. Both these restrictions can be overcome by using the modal-matching technique which aims to predict precisely how the fields vary at all points along the horn. However, this is only done at the expense of a much more complicated analysis and lengthy computational process. Thus the designer has the choice of a simple but approximate method which gives a good physical insight into the operation of the electromagnetic fields in the horn, or a more complicated method which is cumbersome but produces accurate predictions. The choice will depend on how well the geometry can be modelled by an electromagnetic analysis and how much computational effort is considered justifiable in obtaining the fields. This chapter will outline the principles of both the simple modal approach and the modal-matching technique. The detailed application to particular horns will be dealt with in subsequent chapters.

4.2 Fundamentals of modal analysis

The fundamentals of modal analysis can be found in most electromagnetic-wave textbooks. A detailed exposition is given by Collin [3]. The general aim is to predict the electromagnetic fields in the horn or waveguide. This can be summarised in a series of steps.

(i) The form of the field components and the relationship between the field components is derived from Maxwell's equations.
(ii) The form of the propagating waves is found from solving the Helmholtz wave equation for the co-ordinate system appropriate to the geometry of the structure.
(iii) The boundary conditions are used to find expressions for the propagation coefficients and amplitude coefficients which apply to the specific cross-section of the horn. For hollow rectangular and circular cross-sections, the propagation coefficients are closed-form expressions. For other cases, the propagation coefficient must usually be derived by solving an equation by iterative techniques.
(iv) The field components are obtained. It is convenient to divide these into the components along the axis of the horn (the z direction) and the components transverse to the axis of the horn. The transverse components make up the aperture fields and are used to predict the radiation characteristics.

The horn or waveguide will support two orthogonal mode sets. For waveguides which have perfectly conducting smooth walls and are filled with a homogeneous material (i.e. hollow or completely filled with dielectric), the two mode sets consist of 'transverse magnetic' (TM) modes with $H_z=0$ and 'transverse electric' (TE) modes with $E_z = 0$. In general, there are an infinite number of possible propagating or evanescent modes. The number of finite propagating modes will depend on the relative size of the cross-section of the waveguide. These are designated TM_{nm} and TE_{nm} modes where n and m are integers and refer to the solutions to the wave equation in two co-ordinate directions. When the cross-section of the waveguide is inhomogeneous (for example two dielectrics or corrugated), the TM and TE mode are coupled together and in principle all field components exist. The modes are now called 'hybrid' modes. Two orthogonal hybrid-mode sets still exist and these are usually designated HE_{nm} and EH_{nm} modes.

The modal characteristics of a horn can be found either by using 'cylindrical waves' or by using 'spherical waves'. Cylindrical waves are characterised by modes which travel along one co-ordinate direction (the z direction) with a sinusoidal dependence $Ae^{-\gamma z} + Be^{\gamma z}$ where A and B are the forward and backward travelling functions, respectively, of the transverse co-ordinates. The propagation coefficient is $\gamma = \alpha + j\beta$, where α is the attenuation coefficient and β the phase coefficient. The transverse-field pattern for cylindrical waves repeats every half wavelength along the waveguide and therefore the actual physical length of a waveguide is not significant as far as prediction of the transverse fields is concerned. The electrical length is only relevant if there is more than one mode present in the structure when the transverse fields are the linear combination of two or more modes.

Spherical waves are characterised by modes which spread outwards from the centre of a spherical-co-ordinate system along the radial R direction. The waves have a dependence $(1/\gamma R)(Ae^{-\gamma R} + Be^{\gamma R})$ where the denominator describes the spherical spreading factor of the power and the other terms have the same meaning as for cylindrical waves.

Most horns used as feeds are conical in cross-section so it may be thought that a spherical-wave analysis will be most appropriate, particularly because the radiation fields are composed of spherical waves. However, spherical waves are much more complicated than cylindrical waves. The boundary conditions cannot, in general, be applied exactly so that in only a few cases can horns be analysed using spherical waves. They have been used successfully for studying conical corrugated horns with wide flare angles, but not for many other cases. Thus the more obvious spherical model has to be weighed against the approximations which need to be built into the model in order to solve the equations. Consequently, most horns are analysed using cylindrical waves in which the horn is approximated by a cylinder and the mode functions depend on the cross-section of the horn.

4.3 Cylindrical-wave analysis

Cylindrical waves can be used to study the propagation characteristics of horns with cross-sections which are rectangular, circular, elliptical or arbitrary. The horns can have metal or dielectric boundaries and can be homogeneous (e.g.

hollow) or inhomogeneous (e.g. coaxial, corrugated, dielectric loaded). The same general principles apply in all cases.

The electric and magnetic field in the homogeneous waveguide can be divided into the axial components E_z and H_z and the transverse components E_t and H_t. The forward travelling fields are

$$\left.\begin{aligned} \boldsymbol{E}&=\boldsymbol{E}_t(u_1, u_2)\ \mathrm{e}^{-\gamma z}+E_z(u_1, u_2)\ \mathrm{e}^{-\gamma z} \\ \boldsymbol{H}&=\boldsymbol{H}_t(u_1, u_2)\ \mathrm{e}^{-\gamma z}+H_z(u_1, u_2)\ \mathrm{e}^{-\gamma z} \end{aligned}\right\} \tag{4.1}$$

where u_1 and u_2 refer to the transverse co-ordinates of a general co-ordinate system. The axial-field components satisfy the 2-dimensional scalar electromagnetic-wave equations

$$\left.\begin{aligned} \nabla_t^2 E_z+K^2 E_z=0 \\ \nabla_t^2 H_z+K^2 H_z=0 \end{aligned}\right\} \tag{4.2}$$

For homogeneous waveguides, one of the axial-field components will be zero ($H_z=0$ for TM modes, and $E_z=0$ for TE modes), but for inhomogeneous waveguides both axial components exist. In this latter case, either the E_z- or the H_z-wave equation can be used. The wavenumber K is given by

$$K^2=\omega^2\varepsilon\mu+\gamma^2 \tag{4.3}$$

where for the most general medium filling the waveguide, $\varepsilon\mu=\varepsilon_0\varepsilon_r\mu_0\mu_r$. In most horns $\mu_r=1$ and the dielectric is air so that $\varepsilon_r=1$. Then $K^2=\omega^2\varepsilon_0\mu_0+\gamma^2=k_0^2+\gamma^2$.

The transverse Laplacian operator ∇_t^2 depends on the co-ordinate system. For rectangular co-ordinates it is

$$\nabla_t^2=\frac{\partial^2}{\partial x^2}+\frac{\partial^2}{\partial y^2} \tag{4.4}$$

The solutions are products of trigonometric functions.

The Laplacian operator for circular co-ordinates is

$$\nabla_t^2=\frac{1}{r}\frac{\partial}{\partial r}\left(\frac{r\partial}{\partial r}\right)+\frac{1}{r^2}\frac{\partial^2}{\partial\phi^2} \tag{4.5}$$

The solutions are products of Bessel functions and trigonometric functions.

The transverse-field components are obtained from

$$\left.\begin{aligned} \boldsymbol{E}_t&=-\frac{\gamma}{\boldsymbol{K}^2}\nabla_t\boldsymbol{E}_z+\frac{j\omega\mu}{\boldsymbol{K}^2}\hat{\boldsymbol{u}}_z\times\nabla_t\boldsymbol{H}_z \\ \boldsymbol{H}_t&=-\frac{\gamma}{\boldsymbol{K}^2}\nabla_t\boldsymbol{H}_z-\frac{j\omega\varepsilon}{\boldsymbol{K}^2}\hat{\boldsymbol{u}}_z\times\nabla_t\boldsymbol{E}_z \end{aligned}\right\} \tag{4.6}$$

The characteristic equation for the waveguide is obtained by applying the boundary conditions for the waveguide under consideration. The general boundary conditions between two regions 1 and 2 are

$$\left.\begin{aligned}\hat{\boldsymbol{u}}_n \times (\boldsymbol{E}_1 - \boldsymbol{E}_2) = 0\\ \hat{\boldsymbol{u}}_n \times (\boldsymbol{H}_1 - \boldsymbol{H}_2) = 0\end{aligned}\right\} \tag{4.7}$$

If one of the regions is a perfect conductor, then

$$\left.\begin{aligned}\hat{\boldsymbol{u}}_n \times \boldsymbol{E}_1 = 0\\ \hat{\boldsymbol{u}}_n \times \boldsymbol{H}_1 = \boldsymbol{J}_s\end{aligned}\right\} \tag{4.8}$$

where $\boldsymbol{J}_s$ is the surface current density. The electric boundary condition in eqns. 4.8 states that the tangential components of the electric field are zero at the conductor.

If the waveguide is homogeneous (e.g. air-filled) and the walls are metal, the boundary conditions lead to simple functions for E_z and H_z. If the waveguide is inhomogeneous (e.g. corrugated walls or containing a dielectric rod), then there are boundary conditions at each interface and the resultant form of the electric and magnetic fields is more complicated. The field components will be discussed in more detail in the appropriate chapter later in the book. For the purposes of illustration, the solutions for hollow rectangular and circular waveguides will now be given.

For rectangular waveguides, the wavenumber is given by

$$K^2 = \left(\frac{n\pi}{a}\right)^2 + \left(\frac{m\pi}{b}\right)^2 \tag{4.9}$$

where a is the broad dimension and b the narrow dimension of the waveguide. The order of cuton of the modes is a function of the ratio a/b. For TM_{nm} modes

$$E_z(x, y) = A_{nm} \sin\left(\frac{n\pi}{a} x\right) \sin\left(\frac{m\pi}{b} y\right) \tag{4.10}$$

and for TE_{nm} modes

$$H_z(x, y) = A_{nm} \cos\left(\frac{n\pi}{a} x\right) \cos\left(\frac{m\pi}{b} y\right) \tag{4.11}$$

The lowest-order mode is the TE_{10} mode and this is the mode used in standard rectangular waveguides.

The wavenumber for a circular waveguide is different for TM and TE modes. For TM_{nm} modes the axial electric field and wavenumber are given by

$$E_z(r, \phi) = A_n J_n(Kr)\, e^{jn\phi} \tag{4.12}$$

and

$$J_n(Ka) = 0 \tag{4.13}$$

where a is the radius of the circular waveguide. The wavenumber is equal to the roots of the Bessel function of order n. For TE_{nm} modes the axial magnetic field and wavenumber are given by

$$H_z(r, \phi) = B_n J_n(Kr)\, e^{jn\phi} \tag{4.14}$$

and

$$J'_n(Ka) = 0 \tag{4.15}$$

The wavenumber is now equal to the roots of the derivative of the Bessel function of order n. The lowest-order mode in the circular waveguide is the TE_{11} mode.

4.4 Spherical-wave analysis

Spherical-wave analysis uses spherical modal functions in a spherical (r, θ, ϕ) co-ordinate system to represent the fields in a horn. Spherical-wave theory is a powerful method of analysing horns, but it is also much more complicated than cylindrical-wave analysis. Its major use is in the prediction of the radiation characteristic of antennas, where it represents a useful alternative to the aperture theory described in Chapter 3. The theory of spherical waves is discussed in detail by Stratton [4], Rusch and Ludwig [5] and Wood [6]. Its major use in horns has been to predict the radiation patterns of wide-flare-angle corrugated horns. This application is described by Clarricoats and Olver [2], Clarricoats and Saha [7] and Mahmoud and Clarricoats [8].

The process of solution is similar to that of cylindrical waves described in Section 4.3. The fields in the radial direction are specified by equations of the form

$$\left.\begin{aligned} E_R &= A\,\frac{v(v+1)}{jkR^2}\, h_v^{(2)}(kR) P_v^m(\cos\theta)\cos m\phi \\ H_R &= B\,\frac{v(v+1)}{jkR^2}\, h_v^{(2)}(kR) P_v^m(\cos\theta)\sin m\phi \end{aligned}\right\} \tag{4.16}$$

where $P_v^m(\cos\theta)$ is an associated Legendre function of order m and degree v [23], and $h_v^{(2)}(kR)$ is a spherical Hankel function of order v. The eigenfunction v must be determined from the boundary conditions.

An advantage of the spherical-mode theory is that the fields in the aperture of the horn are represented as a sum of spherical waves which can then be projected into the radiation sphere and the fields computed at any distance in front of the horn. Unlike the aperture theory, there are no approximations involved in computing the radiation patterns in the near field of the horn.

4.5 Principles of modal-matching technique

The modal-matching technique is a powerful computer method of analysing horn antennas in which the actual profile of the horn is replaced by a series of uniform waveguide sections. The waveguides can have any cross-section and can be propagating either cylindrical or spherical modes. Examples of horns which can be studied using the modal-matching technique are shown in Figures 4.2–4.4. For the corrugated horn in Figure 4.2, the uniform waveguide sections are either hollow cylindrical waveguides between slots and ridges, or conical waveguides with impedance walls representing the corrugations. For the homogeneously filled conical horn in Figure 4.3, or the inhomogeneously filled dielectric horn of Figure 4.4, the uniform waveguide sections are sections of cylindrical waveguide.

The modal-matching technique involves matching the total modal field at each junction between uniform sections so that conservation of power is maintained. From this process, the amplitudes of the separate modes at the output of a junction can be deduced in terms of the amplitudes of the mode spectrum at the input to the junction. The number of propagating and evanescent modes which are needed to represent the total power must be found by trial and error. The power of the modal-matching technique stems from the fact that the amplitudes of the modes can be expressed as the components of a scattering matrix. Each junction along the horn has its own scattering matrix. The matrices for all junctions can be cascaded and an overall scattering matrix derived for the horn. The process of computing the overall scattering matrix can be decoupled from the process of obtaining the elements of a particular scattering matrix. The latter will depend on the geometry of the waveguide, but the former is quite general. Thus the basic technique can readily be extended to different geometries. The overall scattering matrix for the horn contains the input reflection coefficients and the output transmission coefficients from which the aperture fields are computed.

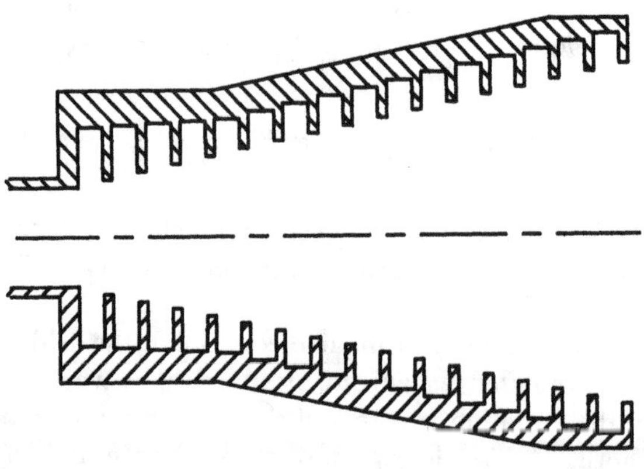

Figure 4.2 Corrugated horn

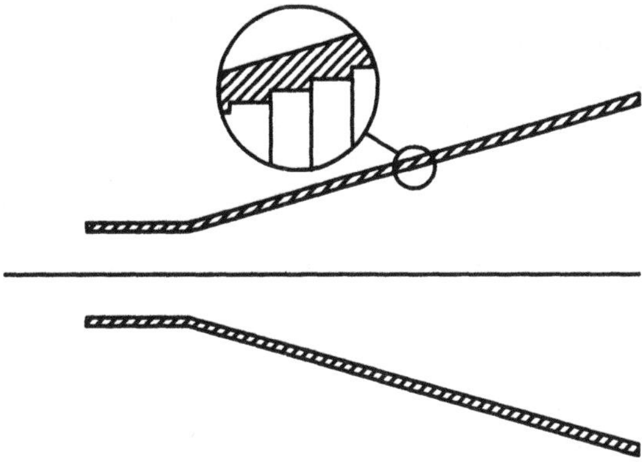

Figure 4.3 Smooth-wall horn

The concept of matching modes at a junction was first attempted in the late 1960s and early 1970s by Wexler [9], Masterman and Clarricoats [10] and English [11]. However, the process of computing the coefficients is lengthy and because of the limited computer power available it was not possible to do more than simple computations. The computational emphasis was on reducing the number of modes to the minimum so that a numerical solution could be obtained. It was the arrival of powerful computers which enabled the concept to be applied to the analysis of complete horns. It was first developed for corrugated horns by James [12] and Kuhn and Hombach [13] and has been adapted for the more complicated case of dielectric-loaded horns by Raghavan *et al.* [14, 15], Olver *et al.* [16] and also by Tun *et al.* [17]. Coaxial structures have been studied by Dubrovka *et al.* [18]. The description of the technique given here follows the procedure of James [12]. The modal-matching technique involves a large amount of computation because there will be a large number of

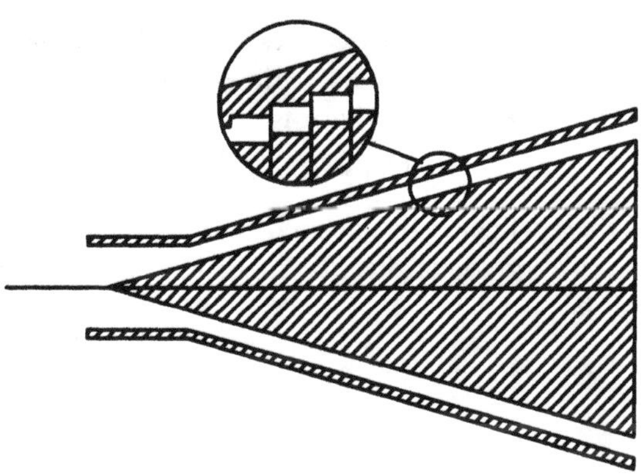

Figure 4.4 Dielectric-cone-loaded horn

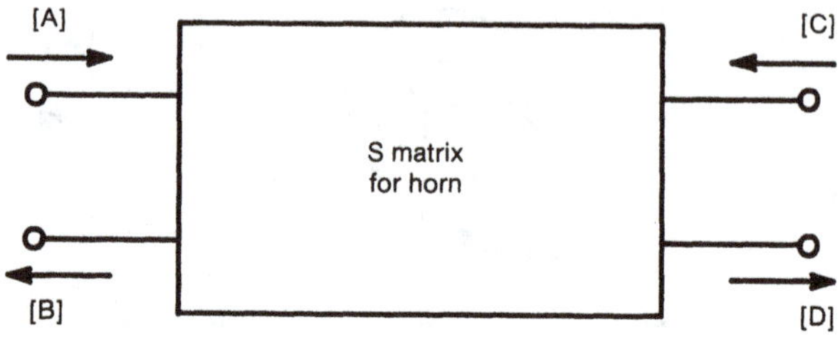

Figure 4.5 Horn represented as a scattering matrix [S]

modes to be matched across each junction and a large number of junctions along the horn.

The rest of this section describes the principles of the modal-matching technique in general terms which can then be applied to any specific waveguide and horn system.

The modal-matching technique can be considered as a method of obtaining the overall transmission and reflection properties of a horn. The horn is represented as a box (Figure 4.5), where $[A]$ and $[B]$ are column matrices containing the forward and reflection coefficients of all the modes looking into the horn from the source side. Similarly, $[C]$ and $[D]$ represent column matrices containing the forward and reflected coefficients of all the modes looking into the aperture of the horn from the outside. The characteristics of the horn are then given by a scattering matrix $[S]$,

$$\begin{bmatrix} [B] \\ [D] \end{bmatrix} = [S] \begin{bmatrix} [A] \\ [C] \end{bmatrix} \tag{4.17}$$

where the scattering matrix $[S]$ is

$$[S] = \begin{bmatrix} [S_{11}] & [S_{12}] \\ [S_{21}] & [S_{22}] \end{bmatrix} \tag{4.18}$$

The elements of $[S]$ are square matrices describing the power coupling between all modes at the input with all modes at the output of the horn.

The reflection coefficient for the horn is

$$[B] = [S_{11}]\ [A] \tag{4.19}$$

For many horns there will only be one mode at the input (for instance, the TE_{11} mode in a circular waveguide or the TE_{10} mode in a rectangular waveguide). In this case $[A]$ will be

$$[A] = \begin{bmatrix} 1 \\ 0 \\ 0 \\ \cdot \\ \cdot \end{bmatrix} \tag{1.20}$$

The length of the column matrix depends on the number of modes at the output of the horn.

The transmission coefficient of the horn, from which the aperture fields are determined, is

$$[D] = [S_{21}]\,[A] \tag{4.21}$$

The horn is divided into a number of sections and junctions. For instance, the corrugated horn shown in Figure 4.2 has 33 junctions and 32 uniform sections. Each junction or section can be represented by its own scattering matrix with a box similar to Figure 4.5 and an equation similar to eqn. 4.17. $[S]$ now represents the scattering properties of the individual junction or section. The scattering matrix of the complete horn is made up of a series of scattering matrices, as shown in Figure 4.6.

Scattering matrices are particularly useful because it is straightforward to cascade two scattering matrices. If these two scattering matrices have elements

$$[S^a] = \begin{bmatrix} [S^a_{11}] & [S^a_{12}] \\ [S^a_{21}] & [S^a_{22}] \end{bmatrix} \quad \text{and} \quad [S^b] = \begin{bmatrix} [S^b_{11}] & [S^b_{12}] \\ [S^b_{21}] & [S^b_{22}] \end{bmatrix} \tag{4.22}$$

then the cascaded scattering matrix is

$$[S^c] = \begin{bmatrix} [S^c_{11}] & [S^c_{12}] \\ [S^c_{21}] & [S^c_{22}] \end{bmatrix} \tag{4.23}$$

where

$$\left.\begin{aligned} [S^c_{11}] &= [S^a_{12}]\,[[I]-[S^b_{11}]\,[S^a_{22}]]^{-1}[S^b_{11}]\,[S^a_{21}]+[S^a_{11}] \\ [S^c_{12}] &= [S^a_{12}]\,[[I]-[S^b_{11}]\,[S^a_{22}]]^{-1}\,[S^b_{12}] \\ [S^c_{21}] &= [S^b_{21}]\,[[I]-[S^a_{22}]\,[S^b_{11}]]^{-1}\,[S^a_{21}] \\ [S^c_{22}] &= [S^b_{21}]\,[[I]-[S^a_{22}]\,[S^b_{11}]]^{-1}[S^a_{22}]\,[S^b_{12}]+[S^b_{22}] \end{aligned}\right\} \tag{4.24}$$

where $[I]$ is a unit matrix and $[\;]^{-1}$ represents the inverse matrix. The cascading process has the advantage that the exact number of junctions and sections does not have to be known at the start of the analysis as the process proceeds from the input to the output in a logical fashion.

The next stage is to determine the scattering matrices of the separate sections. These can be of two forms, either a uniform section or a junction between two uniform sections.

Figure 4.6 Cascading of scattering matrices

4.5.1 Scattering matrix of a uniform section

The scattering matrix elements for a uniform section of guide are

$$\left.\begin{aligned}[S_{11}]&=[S_{22}]=[0]\\ [S_{12}]&=[S_{21}]=[V]\end{aligned}\right\} \tag{4.25}$$

where $[V]$ is an $N\times N$ diagonal matrix, with elements $V_{nn}=\exp(-\gamma_n l)$ where l is the length of the section and γ_n is the propagation constant for the nth mode in the waveguide $(1<n<N)$. In principle, the waveguide sections could contain lossy material so that γ_n is complex. However, this would lead to extensive computation and is generally unnecessary since the influence of lossy materials can usually be adequately accounted for by a perturbation approach. The propagation coefficient is normally either purely imaginary $(\gamma_n=j\beta_n)$ for travelling modes or is purely real $(\gamma_n=\alpha_n)$ for evanescent modes. In both cases the elements of the matrices are real. A substantial number of evanescent modes must be included in the analysis. This is because the uniform sections will be relatively short in length so that the amplitude of a decaying wave may still be significant by the time the wave reaches the next junction.

4.5.2 Scattering matrix of a junction

The derivation of the scattering matrix at a junction is more complicated, since it involves the matching of the total power in all modes on both sides of the junction. The number of modes on the left-hand side of the junction and the number of modes on the right-hand side of the junction can in general be arbitrary. However, it simplifies the analysis, and the computational procedure if the number of modes N is the same on both sides of the junction. This will be assumed in the following analysis.

Figure 4.7 shows a junction between two sections of cylindrical waveguide. Each uniform waveguide section contains travelling waves in which the transverse electric fields can be represented as a spectrum of N nodes. The transverse electric and magnetic modal functions on the left-hand side of the junction are represented by the subscript L (i.e. e_{nL}, and h_{nL}), and those on the right-hand side of the junction by the subscript R.

The electric and magnetic fields on the left-hand side are

$$\left.\begin{aligned}\boldsymbol{E}_L&=\sum_{n=1}^{N}\{A_n\exp(-\gamma_n z)+B_n\exp(\gamma_n z)\}\boldsymbol{e}_{nL}\\ \boldsymbol{H}_L&=\sum_{n=1}^{N}\{A_n\exp(-\gamma_n z)-B_n\exp(\gamma_n z)\}\boldsymbol{h}_{nL}\end{aligned}\right\} \tag{4.26}$$

where A_n and B_n are the forward and reflected amplitude coefficients of mode n

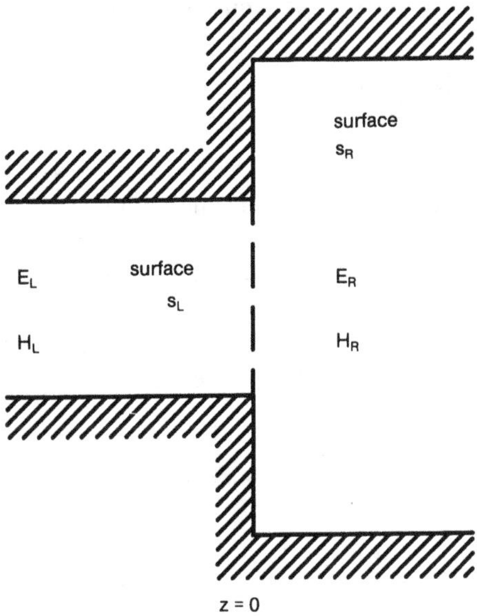

Figure 4.7 Fields at a junction

on the left-hand side of the junction.

On the right-hand side of the junction, the fields have the form

$$\left.\begin{aligned} \boldsymbol{E}_R &= \sum_{n=1}^{N} \{C_n \exp(-\gamma_n z) + D_n \exp(\gamma_n z)\} \boldsymbol{e}_{nR} \\ \boldsymbol{H}_R &= \sum_{n=1}^{N} (C_n \exp(-\gamma_n z) - D_n \exp(\gamma_n z) \boldsymbol{h}_{nR} \end{aligned}\right\} \tag{4.27}$$

where C_n and D_n are the forward and reflected amplitude coefficients of mode n on the right-hand side of the junction, looking into the junction.

The total transverse fields must match across the junction. If the junction is at $z=0$, then

$$\left.\begin{aligned} \sum_{n=1}^{N} (A_n + B_n)\boldsymbol{e}_{nL} &= \sum_{n=1}^{N} (C_n + D_n)\boldsymbol{e}_{nR} \\ \sum_{n=1}^{N} (A_n - B_n)\boldsymbol{h}_{nL} &= \sum_{n=1}^{N} (C_n - D_n)\boldsymbol{h}_{nR} \end{aligned}\right\} \tag{4.28}$$

If the cross-section area of the waveguide on the left-hand side of the junction is s_L and that on the right-hand side of the junction is s_R, the boundary conditions give that the transverse electric fields over the area $(s_R - s_L)$ will be zero. The fields over the area s_L will be continuous. The continuity of fields and the orthogonality relationship between modes leads to a pair of simultaneous matrix equations

$$\begin{aligned}[P]\ [[A]+[B]]&=[Q]\ [[C]+[D]]\\ [P]^T\ [[D]-[C]]&=[R]\ [[A]-[B]]\end{aligned} \tag{4.29}$$

where $[A]$ and $[B]$ are N-element column matrices in the section on the left-hand side of the junction containing the unknown modal coefficients A_1 to A_N and B_1 to B_N. Similarly, $[C]$ and $[D]$are N-element column matrices for the right-hand side of the junction containing the unknown modal coefficients C_1 to C_N and D_1 to D_N.

The matrix $[P]$ is an $N\times N$ square matrix whose elements are integrals representing the mutual coupled power between mode i on the left-hand side and mode j on the right-hand side,

$$P_{ij}=\int_{s_L}(\boldsymbol{e}_{iL}\times\boldsymbol{h}_{jR}).\mathrm{d}s_L \tag{4.30}$$

The matrix $[P]^T$ is the transpose of $[P]$, i.e. the rows and columns are interchanged. The matrix $[Q]$ is an $N\times N$ diagonal matrix describing the self-coupled power between modes on the right-hand side of the junction. The elements are integrals over the area s_R,

$$Q_{jj}=\int_{s_R}(\boldsymbol{e}_{jR}\times\boldsymbol{h}_{jR}).\mathrm{d}s_R \tag{4.31}$$

Similarly the matrix $[R]$ is an $N\times N$ diagonal matrix describing the self-coupled power between modes on the left-hand side of the junction. The elements are integrals over the area s_L,

$$R_{ii}=\int_{s_L}(\boldsymbol{e}_{iL}\times\boldsymbol{h}_{iL}).\mathrm{d}s_L \tag{4.32}$$

The power-coupling integrals in eqns. 4.30–4.32 contain the information about the type of waveguide on either side of the junction. They must be evaluated for the appropriate homogeneous or inhomogeneous cross-section. This can be done either analytically or numerically. Numerical evaluation reduces the amount of mathematical effort but increases the computational time since the integrals need to be evaluated for all modal combinations at each junction. Analytical evaluation is only possible in some cases and may involve considerable mathematics, but where it is possible the saving in computer time can be considerable.

Eqn. 4.29 needs to be rearranged into the scattering-matrix form. This gives the elements of $[S]$ from eqn. 4.18 as

$$\left.\begin{aligned}[S_{11}]&=[[R]+[P]^T[Q]^{-1}[P]]^{-1}[[R]-[P]^T[Q]^{-1}[P]]\\ [S_{12}]&=2[[R]+[P]^T[Q]^{-1}[P]]^{-1}[P]^T\\ [S_{21}]&=2[[Q]+[P][R]^{-1}[P]^T]^{-1}\ [P]\\ [S_{22}]&=-[[Q]+[P]\ [R]^{-1}\ [P]^T]^{-1}\ [[Q]-[P]\ [R]^{-1}\ [P]^T]\end{aligned}\right\} \tag{1.33}$$

and the inverse matrix $[Q]^{-1}=[I]/[Q]$.

The analysis above assumes that the area s_R is greater than the area s_L. If this is not the case, the elements of $[S]$ in eqn. 4.18 become

$$[S]=\begin{bmatrix}[S_{22}] & [S_{21}]\\ [S_{12}] & [S_{11}]\end{bmatrix} \tag{4.34}$$

This completes the description of the general modal-matching technique for analysing any combination of junctions and waveguide sections.

4.6 Modal-matching technique for cylindrical waveguides

The modal-matching technique described in Section 4.5 can be implemented in a straightforward manner for cylindrical waveguides. The power-coupling integrals in eqns. 4.30–4.32 need to be evaluated for the specific geometry and internal waveguide. This will be illustrated by the case of the junction of two hollow circular waveguides (Figure 4.8). This problem is at the heart of the solution of the conical corrugated and conical smooth-wall horns sketched in Figures 4.2 and 4.3.

The power-coupling integrals involve the evaluation of integrals of the form

$$\int_s (\boldsymbol{e}\times\boldsymbol{h}).\mathrm{d}s=\int_0^{2\pi}\int_0^a (\boldsymbol{E}_r\boldsymbol{H}_\phi^*-\boldsymbol{E}_\phi\boldsymbol{H}_r^*)r\,\mathrm{d}r\,\mathrm{d}\phi \tag{4.35}$$

where the fields are evaluated for each mode and for either the left-hand side or the right-hand side of the junction. The field components are obtained from eqns. 4.12, 4.14 and 4.6. The integrals can be evaluated numerically or analytically. In this case the closed-form solutions are standard and should therefore be used. In more complicated cases, for instance of a coaxial

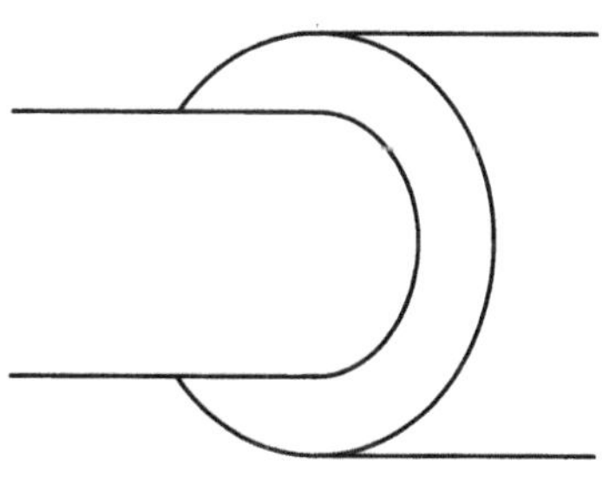

Figure 4.8 Junction between two cylindrical sections

waveguide, numerical evaluation may be preferable. The penalty will be a considerable increase in the computational time.

Inserting the field components into the self-coupling power integrals Q_{ij} and R_{ii} and evaluating the integrals of products of Bessel functions gives for TE modes

$$Q_{TE}=\frac{\pi}{2}\,CZ_0a^4\bar{\beta}K^2\{J_{n-1}^2(ka)-J_{n-2}(Ka)J_n(Ka)+J_{n+1}^2(Ka)-J_n(Ka)J_{n+2}(Ka)\} \quad (4.36)$$

where Z_0 is the impedance of free space, $\bar{\beta}=\beta/k_0=\lambda_0/\lambda_g$ and C is a constant. For TM modes $Q_{TM}=Q_{TE}/Z_0^2$.

The mutual coupling-power integrals P_{ij} are as follows.

(*a*) Left-hand side TE mode–right-hand side TE mode:

$$P_{TE\text{-}TE}=2\pi C_1C_2abZ_0\bar{\beta}_RX_1 \quad (4.37)$$

(*b*) Left-hand side TE mode–right-hand side TM mode:

$$P_{TE\text{-}TM}=2\pi C_1C_2ab\,X_2 \quad (4.38)$$

(*c*) Left-hand side TM mode–right-hand side TE mode:

$$P_{TM\text{-}TE}=0 \quad (4.39)$$

(*d*) Left-hand side TM mode–right-hand side TM mode:

$$P_{TM\text{-}TM}=2\pi C_1C_2\frac{ab}{Z_0}\bar{\beta}_LX_1 \quad (4.40)$$

where

$$X_1=\frac{aK_LK_R}{K_L^2-K_R^2}\{K_RJ_{n-1}(K_La)J_{n-2}(K_Ra)-K_LJ_{n-2}(K_La)J_{n-1}(K_Ra)$$
$$+K_RJ_{n+1}(K_La)J_n(K_Ra)-K_LJ_n(K_La)J_{n+1}(K_Ra)\} \quad (4.14)$$

and

$$X_2=J_n(K_La)J_n(K_Ra) \quad (4.42)$$

where C_1 and C_2 are constants.

4.6.1 Numerical procedure

The theory outlined above can be implemented numerically using the following strategy:

<table>
<tr><td rowspan="7">For each section</td><td rowspan="4">For each mode</td><td>Wavenumber γ_n computed for waveguide section</td></tr>
<tr><td>Left-hand side of junction: self-coupling power coefficients (R_{ii})</td></tr>
<tr><td>Right-hand side of junction: self-coupling power coefficients (Q_{jj})</td></tr>
<tr><td>Crosscoupling power coefficients (P_{ij})</td></tr>
<tr><td colspan="2">Scattering coefficients for junction</td></tr>
<tr><td colspan="2">Scattering matrix of junction cascaded with matrix for section length</td></tr>
<tr><td colspan="2">Scattering matrix cascaded with scattering matrix computed at previous junction</td></tr>
<tr><td colspan="3">Aperture modal coefficients [D] and input reflection coefficients [B]</td></tr>
</table>

The main part of the computation is the four sections which have to be calculated for each section and for each mode. For hollow circular waveguide, the wavenumber γ_n is the roots of Bessel functions as given by eqns 4.13 and 4.15. Most conical feeds use input circular waveguides propagating the TE_{11} mode. If the structure contains no asymmetry, the modes will all have an azimuthal dependence of unity. Junctions will excite only TE_{1m} and TM_{1m} modes. The wavenumbers for the first 60 TE_{1m} and TM_{1m} modes are listed in Table 4.1.

4.6.2 Number of modes and sections

The time taken to compute the input and output coefficients for a horn depends on the number of modes and the number of sections. The time is proportional to the number of sections and approximately proportional to the square of the number of modes. It is important to be able to choose the number of sections and the number of modes correctly so that the desired accuracy is achieved without unnecessary computation.

The total number of modes depends on the relative diameters at each junction and must be chosen by repeated tests. The larger the change in diameter, the more modes will be excited locally and the higher the level of mode conversion. The smooth wall and dielectric loaded horns (Figures 4.3 and 4.4) require relatively few modes, but a large number of sections. The corrugated horn of Figure 4.2 can be made up of straight sections between the

Table 4.1 Table of roots of $J_1(x)$ and $J'_1(x)$ for first 60 modes

Mode $1m$	TE_{1m} mode	TM_{1m} mode
1	1.84118	3.83171
2	5.33144	7.01559
3	8.53632	10.17347
4	11.70600	13.32369
5	14.86359	16.47063
6	18.01553	19.61586
7	21.16437	22.76008
8	24.31133	25.90367
9	27.45705	29.04683
10	30.60192	32.18968
11	33.74618	35.33231
12	36.88999	38.47477
13	40.03344	41.61709
14	43.17663	44.75932
15	46.31960	47.90146
16	49.46239	51.04354
17	52.60504	54.18555
18	55.74757	57.32753
19	58.89000	60.46946
20	62.03235	63.61136
21	65.17462	66.75323
22	68.31683	69.89507
23	71.45899	73.03690
24	74.60110	76.17870
25	77.74316	79.32049
26	80.88519	82.46226
27	84.02719	85.60402
28	87.16916	88.74577
29	90.31110	91.88750
30	93.45302	95.02923

corrugations so the number of sections is twice the number of corrugations, but the number of modes is high. This can be seen by computing the crosspolar patterns for different numbers of modes and comparing the peak crosspolar level, which is a sensitive measure of accuracy since the crosspolarisation is the difference between two orthogonal copolar patterns. Figure 4.9 shows this comparison for the horn of Figure 4.2 with an aperture diameter of 3λ. Forty modes gives an accuracy of 0.02 dB, 30 modes gives an accuracy within 0.2 dB,

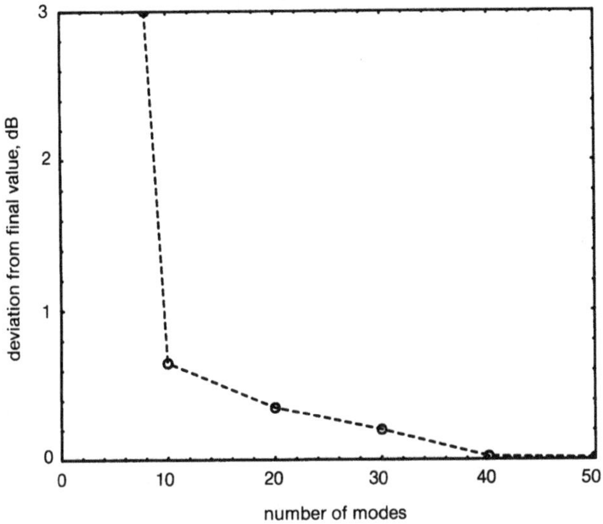

Figure 4.9 Accuracy against number of modes for horn of Figure 4.2

so that about 12 modes per wavelength-aperture diameter is required for good accuracy. However it is notable that the accuracy does not fall off rapidly until less than ten modes are used. Thus quick initial predictions are potentially possible, although this may not be true for different geometries.

The smooth wall horn of Figure 4.3, which is a Potter horn of 3λ aperture diameter and 7λ long (described in Chapter 9), needs only ten modes for good accuracy. In this example, it is the number of sections which is the main choice. Figure 4.10 shows the difference in the peak crosspolar level against the number of sections. Between 80 and 100 sections are needed, or about 14 sections/wavelength. In practical computations, it is easy to increase the number of sections because of the low number of modes required.

4.7 Spherical-modal-matching technique for conical waveguides

The general theory of modal matching can be applied to spherical modes. The analysis is both more complicated and more approximate than using cylindrical modes. In some cases it is a worthwhile exercise because spherical modes much more nearly match the true field picture inside many practical conical smooth-wall and corrugated horns. Consequently, many fewer modes are needed to represent the horn. For instance, a conical corrugated horn with a 5λ aperture might require 100 modes using cylindrical waves, whereas the same horn

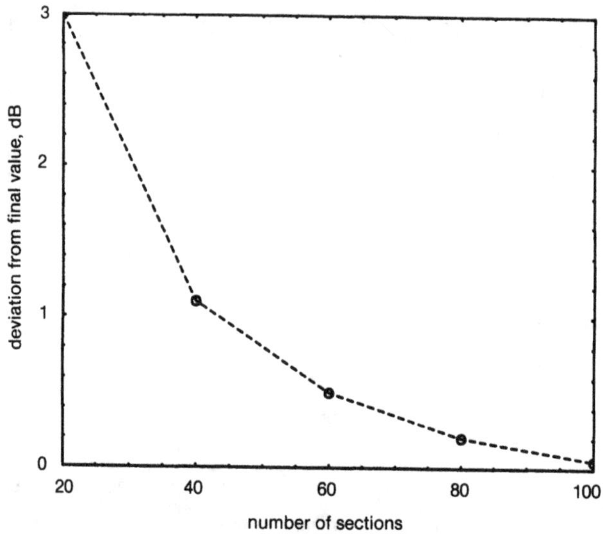

Figure 4.10 Accuracy against number of sections for horn of Figure 4.3

analysed using spherical waves might be adequately modelled with ten modes. The theory of spherical modal matching [20–22] will be outlined in this section, emphasising the differences from the general theory developed in Section 4.5.

A junction in a conical horn is shown in Figure 4.11 with the mode notation in Figure 4.12. The horn section on the left-hand side may be a smooth-wall circular waveguide, a smooth-wall conical horn or a conical corrugated horn. The horn section on the right-hand side may be either of the latter two types. The junction shown in Figure 4.11 does not have a single boundary. There is a crescent-shaped overlap region owing to the different radii of the two conical sections. The distance Δ can be made to be small so that the amplitudes of the fields over the two spherical caps will be nearly constant and the phase can be accounted for by adding a spherical phase factor to the fields on the left-hand side of the junction. Consequently the fields on the left-hand side, corresponding to eqn. 4.26, are

$$\left.\begin{aligned} \boldsymbol{E}_L &= \sum_{n=1}^{N} \{A_n \boldsymbol{e}_{nL} \exp(-jk\beta_n\Delta) + B_n \boldsymbol{e}'_{nL} \exp(jk\beta_n\Delta)\} \\ \boldsymbol{H}_L &= \sum_{n=1}^{N} \{A_n \boldsymbol{h}_{nL} \exp(-jk\beta_n\Delta) - B_n \boldsymbol{e}'_{nL} \exp(jk\beta_n\Delta)\} \end{aligned}\right\} \tag{4.43}$$

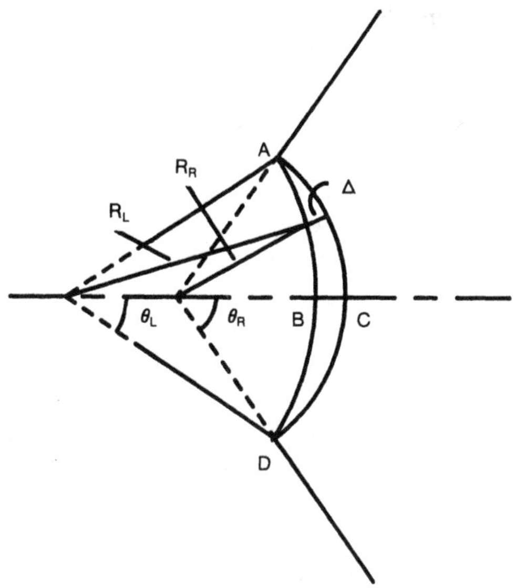

Figure 4.11 Junction between two conical sections

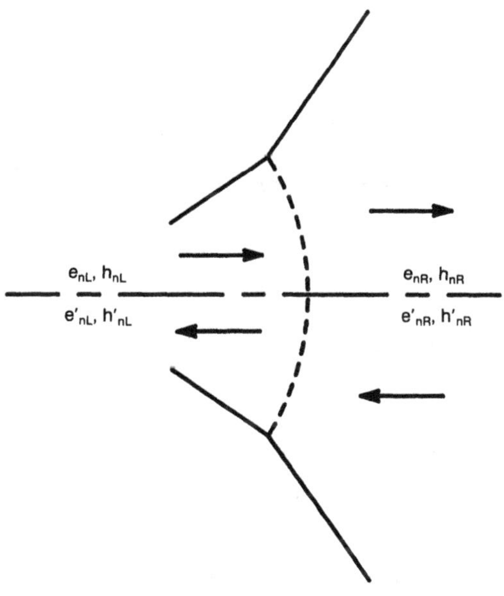

Figure 4.12 Fields at a conical junction

where $\boldsymbol{e}_{nL}$, $\boldsymbol{h}_{nL}$ are the incident fields and $\boldsymbol{e}'_{nL}$, $\boldsymbol{h}'_{nL}$ are the reflected fields. The distance between the two spherical caps $R=R_L$ and $R=R_R$ is given by

$$\Delta(\theta)=R_L\left\{1+\frac{2\sin\theta_L}{\sin\theta_R}(\cos\theta_L-\sin\theta_L\cot\theta_R)(\cos\theta-\cos\theta_R)\right\}^{1/2}-R_L \quad (4.44)$$

On the right-hand side of the junction, the fields are

$$\left.\begin{aligned}\boldsymbol{E}_R&=\sum_{n=1}^{N}(C_n\boldsymbol{e}_{nR}+D_n\boldsymbol{e}'_{nR})\\ \boldsymbol{H}_R&=\sum_{n=1}^{N}(C_n\boldsymbol{h}_{nR}-D_n\boldsymbol{h}'_{nR})\end{aligned}\right. \quad (4.45)$$

If $\theta_L<\theta_R$, a direct modal matching is strictly not applicable because neither modal expansion is valid in the segment bounded by ABD and ACD and there is no common boundary except at A and D. However, a rigorous formulation involves providing the segment with its own field representation which leads to considerable mathematical problems [19], so the approximate technique of using a spherical phase cap is a simple but effective solution. If $\theta_L>\theta_R$ this problem does not arise since two horn sections share a common area where both modal expansions are valid [19].

Expression for the electric and magnetic fields $\boldsymbol{e}_{nL}$, $\boldsymbol{h}_{nL}$ and $\boldsymbol{e}'_{nL}$, $\boldsymbol{h}'_{nL}$ depend on the geometry of the horn. Those for cylindrical waveguides are given in Section 4.3. For spherical TE and TM modes they are as follows.

TE_{1m} modes:

$$\left.\begin{aligned}e_\theta&=-\frac{j}{\sin\theta}P_v^1(\cos\theta)\cos\phi\\ e_\phi&=jP_v^{1\prime}(\cos\theta)\sin\phi\\ h_\theta&=-jY_0\bar{\beta}P_v^{\prime}(\cos\theta)\sin\phi\\ h_\phi&=-\frac{jY_0}{\sin\theta}\bar{\beta}P_v^1(\cos\theta)\cos\phi\end{aligned}\right\} \quad (4.46)$$

TM_{1n} modes:

$$\left.\begin{aligned}e_\theta&=-\bar{\beta}P_v^{1\prime}(\cos\theta)\cos\phi\\ e_\phi&=\frac{1}{\sin\theta}\bar{\beta}P_v^1(\cos\theta)\sin\phi\\ h_\theta&=-\frac{Y_0}{\sin\theta}P_v^1(\cos\theta)\sin\phi\\ h_\phi&=-Y_0P_v^{1\prime}(\cos\theta)\cos\phi\end{aligned}\right\} \quad (4.47)$$

where

$$\bar{\beta}=j\frac{h_v^{(2)\prime}(KR)}{h_v^{(2)}(KR)}$$

$$Y_0 = 1/Z_0$$

The expressions for spherical hybrid modes in corrugated waveguides where the corrugations are represented by a surface impedance which determines the value of the eigenfunction v are

$$\left.\begin{aligned}
e_\theta &= -\left\{\bar{\beta}P_v^{1'}(\cos\theta) + \bar{\Lambda}\frac{P_v^1(\cos\theta)}{\sin\theta}\right\}\cos\phi \\
e_\phi &= \left\{\bar{\Lambda}P_v^{1'}(\cos\theta) + \bar{\beta}\frac{P_v^1(\cos\theta)}{\sin\theta}\right\}\sin\phi \\
h_\theta &= -Y_0\left\{\bar{\beta}\bar{\Lambda}P_v^{1'}(\cos\theta) + \frac{P_v^1(\cos\theta)}{\sin\theta}\right\}\sin\phi \\
h_\phi &= -Y_0\left\{P_v^{1'}(\cos\theta) + \bar{\Lambda}\bar{\beta}\frac{P_v^1(\cos\theta)}{\sin\theta}\right\}\cos\phi
\end{aligned}\right\} \qquad (4.48)$$

where $\bar{\Lambda} = jZ_0H_R/E_R$ and E_R and H_r are given in eqn. 4.16.

The fields are matched across the junction and the scattering matrix is obtained. The result is more complicated than the cylindrical case because the eigenfunction v depends to some extent upon kR. This means that the spherical hybrid modes are not orthogonal and the power-coupling integrals Q_{ij} or R_{ij} are nonzero for $i \neq j$. This means that eqn. 4.33 must be replaced by

$$\left.\begin{aligned}
[S_{11}] &= -[[T_1]^{-1}[Q_1] + [T_2]^{-1}[Q_2]]^{-1}[[T_1]^{-1}[P_1] - [T_2]^{-1}[P_2]] \\
[S_{12}] &= [[T_1]^{-1}[Q_1] + [T_2]^{-1}[Q_2]]^{-1}[[T_1]^{-1}[R_1] + [T_2]^{-1}[R_2]] \\
[S_{21}] &= [[Q_1]^{-1}[T_1] + [Q_2]^{-1}[T_2]]^{-1}[[Q_1]^{-1}[P_1] + [Q_2]^{-1}[P_2]] \\
[S_{22}] &= -[[Q_1]^{-1}[T_1] + [Q_2]^{-1}[T_2]]^{-1}[[Q_1]^{-1}[R_1] - [Q_2]^{-1}[R_2]]
\end{aligned}\right\} \qquad (4.49)$$

where the power-coupling integrals are defined in spherical co-ordinates by

$$\langle \boldsymbol{e}, \boldsymbol{h}^* \rangle = \int_0^{2\pi}\int_0^{\theta_R} (\boldsymbol{e} \times \boldsymbol{h}^*) . \boldsymbol{u}_R . R^2 \sin\theta \, d\theta \, d\phi \qquad (4.50)$$

and the elements of the component matrices are

$$\left.\begin{aligned}
P_{1ij} &= \langle \boldsymbol{e}_{jL} \exp(-jk\bar{\beta}\Delta), \boldsymbol{h}_{iR}^* \rangle \\
P_{2ij} &= \langle \boldsymbol{e}_{iR}^*, \boldsymbol{h}_{jL} \exp(-jk\bar{\beta}\Delta) \rangle \\
Q_{1ij} &= \langle \boldsymbol{e}'_{jL} \exp(jk\bar{\beta}\Delta), \boldsymbol{h}_{iR}^* \rangle \\
Q_{2ij} &= \langle \boldsymbol{e}_{iR}^*, \boldsymbol{h}'_{jl} \exp(jk\bar{\beta}\Delta) \rangle \\
R_{1ij} &= \langle \boldsymbol{e}'_{jR}, \boldsymbol{h}_{iR}^* \rangle \\
R_{2ij} &= \langle \boldsymbol{e}_{iR}^*, \boldsymbol{h}'_{jR} \rangle \\
T_{1ij} &= \langle \boldsymbol{e}_{jR}, \boldsymbol{h}_{iR}^* \rangle \\
T_{2ij} &= T_{1ij}^*
\end{aligned}\right\} \qquad (4.51)$$

The remainder of the analysis is as in Section 4.5. The application of spherical modal matching to the study of corrugated horns will be described in Chapter 10.

4.8 References

1 LOVE, A.W., RUDGE, A.W., and OLVER, A.D., *in* RUDGE, A.W., MILNE, K., OLVER, A.D., and KNIGHT, P. (Eds.): 'The handbook of antenna design' (Peter Peregrinus, 1982), IEE Electromagnetic Waves series 15 and 16
2 CLARRICOATS, P.J.B., and OLVER, A.D.: 'Corrugated horns for microwave antennas' (Peter Peregrinus, 1984), IEE Electromagnetic Waves series 18
3 COLLIN, R.E.: 'Field theory of guided waves' (McGraw-Hill, New York, 1960, 2nd ed, IEEE Press, 1991)
4 STRATTON, J.A.: 'Electromagnetic theory' (McGraw–Hill, New York, 1941)
5 RUSCH, W.V.T., LUDWIG, A.C., and WONG, W.C., *in* RUDGE, A.W., MILNE, K., OLVER, A.D., and KNIGHT, P. (Eds.): 'The handbook of antenna design Vol 1' (Peter Peregrinus, 1982), IEE Electromagnetic Waves series 15, chapter 2
6 WOOD, P.J.: 'Spherical waves in antenna problems', *Marconi Rev.*, 1971, **34**, pp. 149–172
7 CLARRICOATS, P.J.B., and SAHA, P.K.: 'Propagation and radiation behaviour of corrugated feeds. Pt. 2, Corrugated conical horn feeds', *Proc. IEE*, 1971, **118**, pp. 187–199
8 MAHMOUD, S.F., and CLARRICOATS, P.J.B.: 'Radiation from wide flare-angle corrugated conical horns', *IEE Proc. H.*, 1982, **129**, pp. 221–228
9 WEXLER, A.: 'Solution of waveguide discontinuities by modal analysis', *IEEE Trans.*, 1967, **MTT–15**, pp. 508–517
10 MASTERMAN, P.H., and CLARRICOATS, P.J.B.: 'Computer field-matching solution of waveguide transverse discontinuities', *Proc. IEE*, 1971, **118**, pp. 51–63
11 ENGLISH, W.J.: 'The circular waveguide step-discontinuity mode transducer', *IEEE Trans.*, 1973, **MTT–21**, pp. 633–636
12 JAMES, G.L.: 'Analysis and design of TE11 to HE11 corrugated cylindrical waveguide mode converters', *IEEE Trans.*, 1981, **MTT–29**, pp. 1059–1066
13 KUHN, E., and HOMBACH, V.: 'Computer aided analysis of corrugated horns with axial or ring-loaded radial slots'. Proceedings of 3rd international conference on Antennas and Propagation, Norwich, 1983, *IEE Conf. Publ. 219*, Norwich, pp. 127–131
14 RAGHAVAN, K., OLVER, A.D., and CLARRICOATS, P.J.B.: 'Compact dual mode dielectric loaded horn', *Electron. Lett.*, 1986, **22**, pp. 1131–1132
15 RAGHAVAN, K., OLVER, A.D., and CLARRICOATS, P.J.B.: 'Computer-aided design of dielectric loaded horns'. Proceedings of 5th international conference on Antennas and Propagation, York, 1987, pp. 434–437
16 OLVER, A.D., CLARRICOATS, P.J.B., and RAGHAVAN, K.: 'Dielectric cone loaded horn antennas', *IEE Proc. H.*, 1988, **135**, pp. 158–162
17 TUN, S.M., PHILIPPOU, G.Y., and ADATIA, N.: 'Analysis and design of rotationally symmetric dielectric loaded waveguides and horns'. Proceedings of 5th international conference on Antennas and Propagation, York, 1987, pp. 438–442
18 DUBROVKA, F.F., OLVER, A.D., and RAGHAVAN, K.: 'Radiation characteristics of disc loaded horns'. Proceedings of 6th international conference on Antennas and Propagation, Warwick, 1989, pp. 149–153
19 LEWIN, L.: 'On the problem of applying mode matching techniques in some waveguide discontinuity problems', *IEEE Trans.*, 1970, **MTT–18**, pp. 364–372
20 XIANG, J.: 'Computer aided design of corrugated horns'. PhD thesis, University of London, 1986
21 OLVER, A.D., and XIANG, J.: 'Wide angle corrugated horns analysed using spherical modal-matching', *IEE Proc. H*, 1988, **135**, pp. 34–40
22 OLVER, A.D., and XIANG, J.: 'Design of profiled corrugated horns', *IEEE Trans.*, 1988, **AP–36**, pp. 936–940
23 BALANIS, C.A.: 'Advanced engineering electromagnetics' (Wiley, New York, 1989), pp. 951–953

Chapter 5

Numerical analysis of small axisymmetric feeds

A.A. Kishk and L. Shafai

5.1 Introduction

The desirable characteristics of reflector antenna feeds were discussed in Chapter 2. Two main requirements are the far-field pattern symmetry and low crosspolarisation, which are usually met by corrugated conical horns. Their geometry is symmetric and yields to analytical formulations [1] which facilitate their design and analysis, as discussed in Chapter 4. However, the requirement of pattern symmetry can also be met by other antenna configurations which may have additional useful characteristics. For instance, in prime-focus reflectors the feed blocks the reflector aperture and its size must be small. Also, since the feed must be held at the focus by the supporting struts, the reflector focal length should not be excessively large. This means that the feed-radiation pattern must be broad and remain symmetric over a larger angular range. The feed for prime-focus reflectors are therefore small in size. Their analysis using approximate or analytic methods thus becomes less accurate. Fortunately, the small size of the feed facilitates the use of numerical methods, and this chapter deals with the formulation of the problem. Initially, integral equations are used to generate a suitable set of field equations, which are then reduced to a matrix equation to determine surface currents. Far-field patterns are finally determined from the computed surface currents.

Figure 5.1 shows the geometry of four different feed types, which can be subdivided into two categories. Figure 5.1*a* and *b* identifies feeds which are made totally of a conducting material. Their analysis is therefore simpler, since one deals only with a single conducting surface to enforce the boundary conditions. Two other feeds, shown in Figure 5.1*c* and *d*, are made of both conducting and dielectric materials, and boundary conditions on both conducting and dielectric surfaces must be satisfied. Consequently, the formulation and numerical solution involving them become considerably more complex. In the following sections we first provide a general formulation for the field components and reduce it to a matrix equation for numerical solution. We then discuss various steps the user of the method must take to ensure the solution accuracy. We complete the chapter by presenting a convenient approach for verifying the solutions and include a few representative examples.

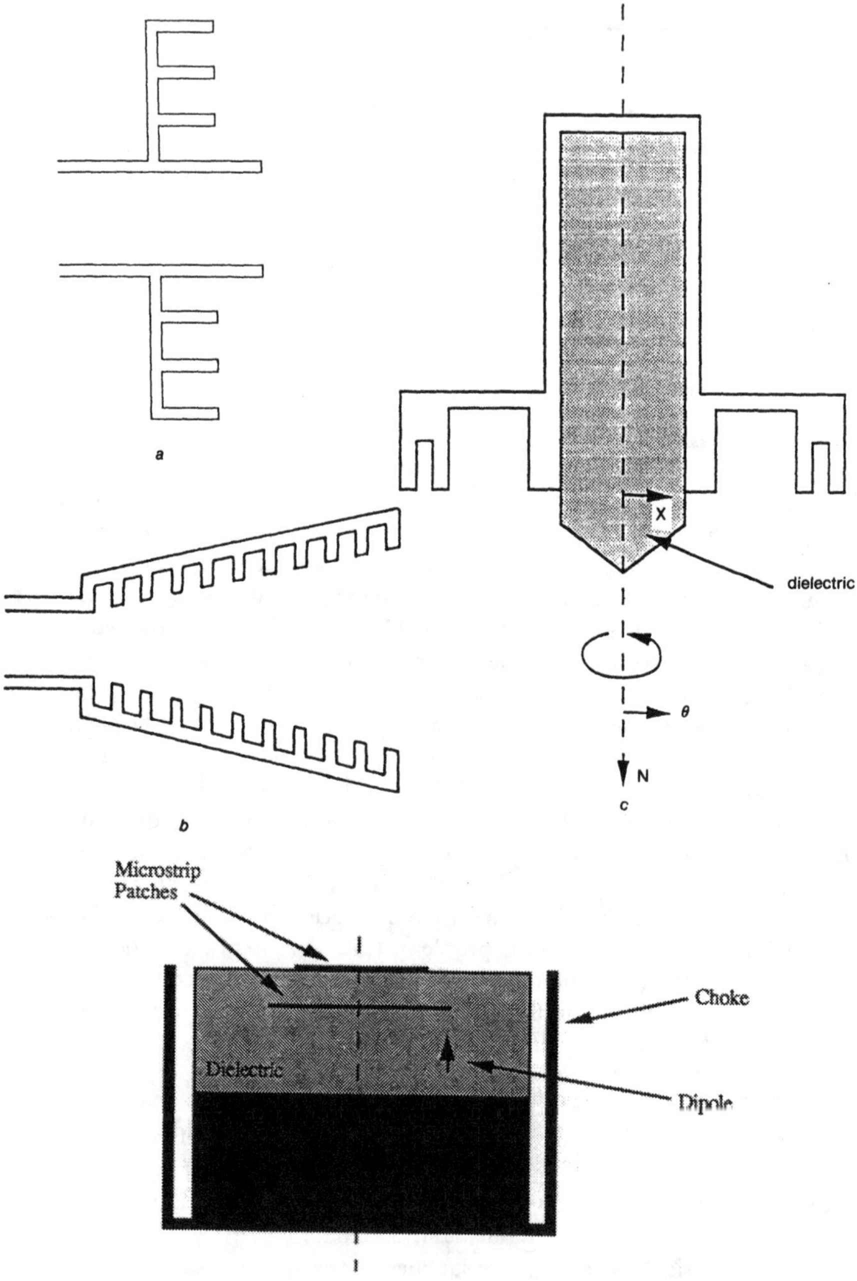

Figure 5.1 *Different feed-antenna geometries*

a Waveguide with a corrugated flange
b Small corrugated horn
c Dielectric- and cavity-loaded waveguide
d Microstrip-type feed

5.2. Basic formulation

In this section we provide a formulation for the field vectors of an antenna geometry that is made of conductors and dielectrics, such as Figure 5.1*c* and *d*. This is the general case, and formulation for the conducting antennas of Figure 5.1*a* and *b* can be obtained by removing parts involving the dielectric region. In practice, the excitation of these feeds may take a variety of forms, but most are by a circular waveguide or a coaxial probe. The modelling of the excitation region can therefore be specific for each antenna and may become excessively complex for some geometries. To generalise the formulation we have therefore assumed a special case of dipole excitation, which also simplifies formulation. As will be shown later, in investigation of different feed types, such an assumption is not restricted to a particular feed and multiple dipoles can be used to model complex excitations accurately. However, the use of dipoles does not provide information on the antenna input impedance, and when the antenna geometry is determined its impedance properties should be obtained by other means.

To present a formulation we consider the geometry of the antenna, in Figure 5.2*a*, which has both conducting and dielectric parts. Three distinct regions can be identified for this geometry, namely v_c for the conducting material, v_e for the exterior of the antenna and v_i for the dielectric material. The region for the conducting material can also be made of several parts as in Figure 5.1*d*. To identify the above regions, we define s_{cd} as the boundary surface between v_c and v_i and s_{ce} and s_d the boundary surfaces between v_c and v_e and between v_i and v_e, respectively. The medium of the exterior region is free space, and is therefore characterised by a real scalar permittivity ε_0, permeability μ_0 and a wave-number k_0. Similarly, the medium of the interior region is likewise linear, homogeneous, isotropic, lossy or lossless, and is therefore characterised by its complex constitutive parameters $\varepsilon=\varepsilon_r\varepsilon_0$, $\mu=\mu_r\mu_0$ and $k=k_0\sqrt{(\varepsilon_r\mu_r)}$, ε_r and μ_r are, respectively, the relative permittivity and permeability of the dielectric medium.

The excitation of the antenna is simulated by one or more infinitesimal electric dipoles each of current moment $\boldsymbol{I}l$ located in the dielectric region or in the exterior region (free-space region). The dipoles are arbitrarily oriented and located according to the antenna-design requirements. Each electric dipole produces the impressed field $(\boldsymbol{E}^{inc}, \boldsymbol{H}^{inc})$, where

$$\boldsymbol{E}_q^{inc} = -j\omega\boldsymbol{A}_q - \nabla\Phi_q \qquad q=i,\ e \tag{5.1}$$

$$\boldsymbol{H}_q^{inc} = \frac{1}{\mu_q q}\,\nabla\times A_q \tag{5.2}$$

In eqns. 5.1 and 5.2

$$\boldsymbol{A}_q = -j\,\frac{k_q\mu_q}{4\pi|\boldsymbol{r}-\boldsymbol{r}_0'|}\,\boldsymbol{I}l h_0^{(2)}(k_q|\boldsymbol{r}-\boldsymbol{r}_0'|) \tag{5.3}$$

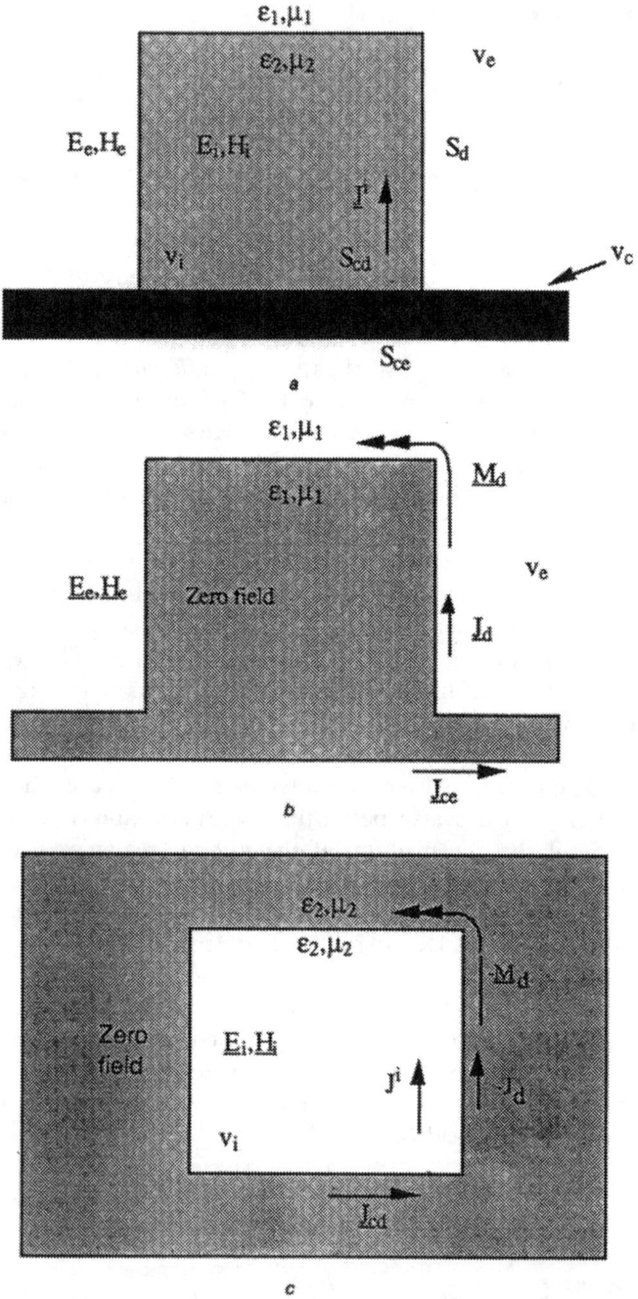

Figure 5.2 Problem representation by equivalent principle

a Original problem
b Equivalence for the exterior region
c Equivalence for the interior region

$$\Phi_q = \frac{\eta_q}{4\pi|\mathbf{r}-\mathbf{r}'_0|} \mathbf{I}l.\nabla h_0^{(2)}(k_q|\mathbf{r}-\mathbf{r}'_0|) \tag{5.4}$$

are, respectively, the magnetic vector and electric scalar potentials produced by the dipole. Here, $h_0^{(2)}$ is the spherical Hankel function of the second kind and zero order, $\mathbf{r}'_0$ is the position vector of the electric dipole, and $\eta_q = \eta_0 \eta_r$ is the intrinsic wave impedance of the medium in which the dipole is located, η_0 is the free-space intrinsic impedance and $\eta_r = \sqrt{(\mu_{rq}/\varepsilon_{rq})}$. It is important to note that the impressed field ($\mathbf{E}^{inc}$, $\mathbf{H}^{inc}$) is the field produced by the electric dipole in an unbounded homogeneous medium whose constitutive parameters are ε_q, μ_q and k_q.

A standard field-equivalence principle can be used to divide the problem into two separate problems for the exterior and interior regions as shown in Figure 5.2*b* and *c*, respectively. According to the field-equivalence principle (Reference 2, Section3–5), the total field ($\mathbf{E}_e$, $\mathbf{H}_e$) in the exterior region v_e is produced by an equivalent distribution of electric and magnetic currents $\mathbf{J}_e$ and $\mathbf{M}_e$ on $S_e = S_d + S_{ce}$ radiating in an unbounded free space, whereas the total field ($\mathbf{E}_i$, $\mathbf{H}_i$) in the interior region is produced by the electric dipole and an equivalent distribution of electric and magnetic currents $\mathbf{J}_i$ and $\mathbf{M}_i$ on $S_i = S_d + S_{cd}$ radiating in an unbounded homogeneous medium whose constitutive parameters are ε_i, μ_i and k_i, where

$$\mathbf{J}_{e,i} = \hat{\mathbf{u}}_{ne,i} \times \mathbf{H}_{e,i} \qquad \text{on } S_{e,i} \tag{5.5}$$

$$\mathbf{M}_{e,i} = \mathbf{E}_{e,i} \times \hat{\mathbf{u}}_{ne,i} \qquad \text{on } S_{e,i} \tag{5.6}$$

In eqns. 5.5 and 5.6, $\mathbf{J}_{e,i}$ means either $\mathbf{J}_e$ or $\mathbf{J}_i$ and $\hat{\mathbf{u}}_{ne}$ and $\hat{\mathbf{u}}_{ni}$ are the unit normal vectors pointing into v_e and v_i, respectively. The electric and magnetic fields in the exterior and interior regions can then be determined in terms of the equivalent electric and magnetic surface currents $\mathbf{J}_{e,i}$ and $\mathbf{M}_{e,i}$ as [3]

$$\theta(\mathbf{r})\mathbf{E}_e(\mathbf{r}) = \mathbf{E}_e^{inc}(\mathbf{r}) - L_e(\mathbf{J}_e) + K_e(\mathbf{M}_e) \tag{5.7}$$

$$\theta(\mathbf{r})\mathbf{H}_e(\mathbf{r}) = \mathbf{H}_e^{inc}(\mathbf{r}) - K_e(\mathbf{J}_e) + \frac{1}{\eta_0^2} L_e(\mathbf{M}_e) \tag{5.8}$$

and

$$\theta(\mathbf{r})\mathbf{E}_i(\mathbf{r}) = \mathbf{E}_i^{inc}(\mathbf{r}) - L_i(\mathbf{J}_i) + K_i(\mathbf{M}_i) \tag{5.9}$$

$$\theta(\mathbf{r})\mathbf{H}_i(\mathbf{r}) = \mathbf{H}_i^{inc}(\mathbf{r}) - K_i(\mathbf{J}_i) + \frac{1}{\eta_0^2} L_i(\mathbf{M}_i) \tag{5.10}$$

where

$$\theta(\mathbf{r}) = \left\{ \begin{array}{ll} 1 & \mathbf{r} \in v_e \text{ or } v_i \\ 0.5 & \mathbf{r} \in S_e \text{ or } S_i \\ 0 & \text{elsewhere} \end{array} \right\} \tag{5.11}$$

and the integro-differential operators $L_{e,i}$ and $K_{e,i}$ are given by

$$L_{e,i}(\mathbf{C}) = j\omega\mu_{e,i} \int_{S_{e,i}} \left[\mathbf{C}(\mathbf{r}') + \{\nabla'.\mathbf{C}(\mathbf{r}')\}\nabla \right] \Phi_{e,i}(|\mathbf{r}-\mathbf{r}'|)\, ds' \tag{5.12}$$

$$K_{e,i}(\boldsymbol{C}) = \int_{S_{e,i}} \boldsymbol{C}(\boldsymbol{r}') \times \nabla\Phi_{e,i}(|\boldsymbol{r}-\boldsymbol{r}'|)\, ds' \tag{5.13}$$

Here,

$$\Phi_{e,i} = \frac{1}{4\pi|\boldsymbol{r}-\boldsymbol{r}'|} \exp(-jk_{e,i}|\boldsymbol{r}-\boldsymbol{r}'|) \tag{5.14}$$

is the Green's function for unbounded homogeneous space, $\boldsymbol{r}$ and $\boldsymbol{r}'$ are the position vectors of the field and source points, respectively, and $k_e = k_0$ and $k_i = k_0\sqrt{(\varepsilon_{ri}\mu_{ri})}$. When $\boldsymbol{r}=\boldsymbol{r}'$, the integrals defining $L_{e,i}$ and $K_{e,i}$ are given by their Cauchy's principal values.

The boundary conditions are that the tangential components of the electric field must vanish at the conductor surface whereas, together with those of the magnetic field, they must be continuous across the dielectric interface S_d. It then follows that $\boldsymbol{J}_e = -\boldsymbol{J}_i = \boldsymbol{J}_d$ and $\boldsymbol{M}_e = -\boldsymbol{M}_i = \boldsymbol{M}_d$ on S_d since $\hat{\boldsymbol{u}}_{ne} = -\hat{\boldsymbol{u}}_{ni}$. Furthermore, only electric surface currents exist on the conductor surfaces, i.e., $\boldsymbol{J}_e = \boldsymbol{J}_{ce}$ on S_{ce} and $\boldsymbol{J}_i = \boldsymbol{J}_{cd}$ on S_{cd}. Enforcing the boundary conditions at the conductor surface and across the dielectric interface, there then results the following system of equations for the equivalent currents:

$$\{+L_e(\boldsymbol{J}_{ce}+\boldsymbol{J}_d) + K_e(\boldsymbol{M}_d)\}_{tan} = \boldsymbol{E}_e^{inc}|_{tan} \qquad \text{on } S_{ce} \tag{5.15}$$

$$\{+L_i(\boldsymbol{J}_{cd}+\boldsymbol{J}_d) + K_i(\boldsymbol{M}_d)\}_{tan} = -\boldsymbol{E}_i^{inc}|_{tan} \qquad \text{on } S_{cd} \tag{5.16}$$

$$\{L_e(\boldsymbol{J}_{ce}+\boldsymbol{J}_d) + L_i(\boldsymbol{J}_{cd}+\boldsymbol{J}_d) + K_e(\boldsymbol{M}_d) + K_i(\boldsymbol{M}_d)\}_{tan} = -\boldsymbol{E}_i^{inc}|_{tan} + \boldsymbol{E}_e^{inc}|_{tan} \qquad \text{on } S_d \tag{5.17}$$

$$\left[-K_e(\boldsymbol{J}_{ce}+\boldsymbol{J}_d) - K_i(\boldsymbol{J}_{cd}+\boldsymbol{J}_d) + \frac{1}{\eta_0^2} L_e(\boldsymbol{M}_d) + \frac{1}{\eta^2} L_i(\boldsymbol{M}_d)\right]_{tan} = -\boldsymbol{H}_i^{inc}|_{tan} + \boldsymbol{H}_e^{inc}|_{tan} \qquad \text{on } S_d \tag{5.18}$$

Solution of the system of eqns. 5.15–5.18 determines the equivalent electric and magnetic currents on the inner and outer surfaces of the antenna. The far-field components $E_{e\theta}$ and $E_{e\phi}$ can then be determined in terms of the currents on the outer surface of the antenna and the incident fields in the exterior region as

$$E_{e\theta}(\theta,\phi) = -j\frac{\omega\mu_0}{4\pi r} \exp(-jk_0 r) F_{e1}(\theta,\phi) + E_{e\theta}^{inc}(\theta,\phi) \tag{5.19}$$

$$E_{e\phi}(\theta,\phi) = -j\frac{\omega\mu_0}{4\pi r} \exp(-jk_0 r) F_{e2}(\theta,\phi) + E_{e\phi}^{inc}(\theta,\phi) \tag{5.20}$$

where $F_{e1}(\theta, \phi)$ and $F_{e2}(\theta, \phi)$ are given in terms of the equivalent surface currents by

$$F_{e1}\{\theta,\phi\} = \int_{S_e} \left(\boldsymbol{J}_e.\hat{\boldsymbol{u}}_{\theta'} + \frac{1}{\eta_0}\boldsymbol{M}_e.\hat{\boldsymbol{u}}_{\phi}'\right) \exp(jk_0\hat{\boldsymbol{u}}_r.\boldsymbol{r}')\, ds' \tag{5.21}$$

$$F_{e2}\{\theta,\phi\} = \int_{S_e} \left(\boldsymbol{J}_e.\hat{\boldsymbol{u}}_{\phi'} - \frac{1}{\eta_0}\boldsymbol{M}_e.\hat{\boldsymbol{u}}_{\theta}'\right) \exp(jk_0\hat{\boldsymbol{u}}_r.\boldsymbol{r}')\, ds' \tag{5.22}$$

In eqns. 5.21 and 5.22, $\hat{\boldsymbol{u}}_r$ is the unit vector in the direction of the field point, $\boldsymbol{r}'$ is a vector representing the source point on s_e, and $\hat{\boldsymbol{u}}_{\theta'}$ and $\hat{\boldsymbol{u}}_{\phi'}$ are the unit vectors at the source point in the direction of increasing θ' and ϕ', respectively. The electric-field components in eqns. 5.21 and 5.22 are complex quantities from which the amplitude and phase patterns are computed for the antenna relative to certain reference points on its axis.

Eqns. 5.15–5.18 provide one set of equations that can be used to determine the surface currents on the antenna. They are obtained by enforcing the boundary conditions on the tangential components of both electric and magnetic fields. In particular, as is evident from the equations the boundary conditions on the electric field is enforced on both conducting and dielectric surfaces. However, the boundary condition on the magnetic field is enforced only on the dielectric surface. In practice, one can generate the required set by applying the boundary conditions only on one of the field vectors, either the electric or magnetic. Similarly, one could reverse the role of field vectors in eqns. 5.15–5.18; i.e. apply the boundary conditions for the magnetic field on both conducting and dielectric surfaces, but limit that of the electric field only to the dielectric surface. Mathematically, each set should be adequate for determining the surface currents. However, in numerical computation they behave differently and care must be taken to ensure the solution accuracy. A discussion of these issues is provided later in this chapter.

5.3 Moment-method solution

The expressions provided in Section 2 did not specify the object shape and are valid for any shape conductors, dielectrics or their combinations. Since this chapter deals with application of numerical methods to axisymmetric horns and feeds, the remaining derivations will be limited to symmetric objects, i.e. bodies of revolution. A solution to eqns. 5.15–5.18 will be obtained using the moment methods [4].

5.3.1 Co-ordinates of bodies of revolution

Figure 5.3 shows the geometry of a body of revolution in which (ρ, ϕ, z) are cylindrical co-ordinates and (t, ϕ) form an orthogonal curvilinear co-ordinate system on the surface s of the body of revolution. Also, $\hat{\boldsymbol{u}}_t$ and $\hat{\boldsymbol{u}}_\phi$ are orthogonal unit vectors in the t and ϕ directions, respectively. The co-ordinate origin is on the axis of the body of revolution but not necessarily on the lower pole as shown in Figure 5.3. The outward normal to the surface of the structure is defined by the direction of the unit vector $\hat{\boldsymbol{n}}$ given by

$$\hat{\boldsymbol{u}}_n = \hat{\boldsymbol{u}}_\phi \times \hat{\boldsymbol{u}}_t \tag{5.23}$$

On the surface of the body we define a field point by its co-ordinates (t, ϕ) or (ρ, ϕ, z) and a source point by (t', ϕ') or (ρ', ϕ', z'). Their respective unit tangent vectors are $(\hat{\boldsymbol{u}}_t, \hat{\boldsymbol{u}}_\phi)$ and $(\hat{\boldsymbol{u}}_t', \boldsymbol{u}_\phi')$. The unit vector $\hat{\boldsymbol{u}}_\phi$ is orthogonal to the

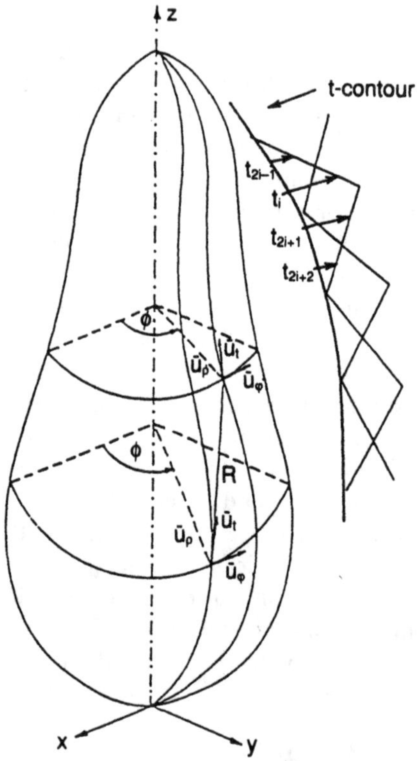

Figure 5.3 Geometry of a body of revolution

z axis, but $\hat{\boldsymbol{u}}_t$ and $\hat{\boldsymbol{u}}_z$ are at an angle ν. This angle is assumed to be positive if $\hat{\boldsymbol{u}}_t$ points away from the z axis. Similarly, ν' is the angle between $\hat{\boldsymbol{u}}'_t$ and the z axis at (t', ϕ'). The relationships among these unit vectors can be determined by an inspection of Figure 5.3, and are given by

$$\hat{\boldsymbol{u}}_t = \sin \nu \hat{\boldsymbol{u}}_\rho + \cos \nu \hat{\boldsymbol{u}}_z \tag{5.24}$$

$$\hat{\boldsymbol{u}}_{t'} = \sin \nu' \cos(\phi' - \phi) \hat{\boldsymbol{u}}_\rho + \sin \nu' \sin(\phi' - \phi) \hat{\boldsymbol{u}}_\phi + \cos \nu' \hat{\boldsymbol{u}}_z \tag{5.25}$$

$$\hat{\boldsymbol{u}}_{\phi'} = -\sin(\phi' - \phi) \hat{\boldsymbol{u}}_\rho + \cos(\phi' - \phi) \hat{\boldsymbol{u}}_\phi \tag{5.26}$$

In addition, if the positional vectors of points (t, ϕ) and (t', ϕ') are $\boldsymbol{r}$ and $\boldsymbol{r}'$, respectively, then

$$\boldsymbol{r} = \rho \hat{\boldsymbol{u}}_\rho + z \hat{\boldsymbol{u}}_z \tag{5.27}$$

and

$$\boldsymbol{r}' = \rho' \cos(\phi' - \phi) \hat{\boldsymbol{u}}_\rho + \rho' \sin(\phi' - \phi) \hat{\boldsymbol{u}}_\phi + z' \hat{\boldsymbol{u}}_z \tag{5.28}$$

where

$$|\boldsymbol{r} - \boldsymbol{r}'| = \left[(\rho - \rho')^2 + (z - z')^2 + 4\rho\rho' \sin^2 \frac{\phi - \phi'}{2} \right]^{1/2} \tag{5.29}$$

Also, by definition, the surface gradient of a scalar function Φ on the body of revolution is given by

$$\nabla_s \Phi = \hat{\boldsymbol{u}}_t \frac{\partial \Phi}{\partial t} + \hat{\boldsymbol{u}}_\phi \frac{\partial \Phi}{\rho \partial \phi} \tag{5.30}$$

and the surface divergence of a vector function $\boldsymbol{W}$ is defined as

$$\nabla_s . \boldsymbol{W} = \frac{1}{\rho} \frac{\partial}{\partial t} (\rho \boldsymbol{W} . \hat{\boldsymbol{u}}_t) + \frac{1}{\rho} \frac{\partial}{\partial \phi} (\boldsymbol{W} . \hat{\boldsymbol{u}}_\phi) \tag{5.31}$$

5.3.2 *Application of moment method*

The unknown currents $\boldsymbol{J}$ and $\boldsymbol{M}$ can now be decomposed into two components along the unit vectors $\hat{\boldsymbol{u}}_{t'}$ and $\hat{\boldsymbol{u}}_{\phi'}$ and expressed in the form [3–5]

$$\left.\begin{aligned} \boldsymbol{J}(\boldsymbol{r}') &= \hat{\boldsymbol{u}}_{t'} J^t(t', \phi') + \hat{\boldsymbol{u}}_{\phi'} J^\phi(t', \phi') \\ \boldsymbol{M}(\boldsymbol{r}') &= \hat{\boldsymbol{u}}_{t'} M^t(t', \phi') + \hat{\boldsymbol{u}}_{\phi'} M^\phi(t', \phi') \end{aligned}\right\} \tag{5.32}$$

where J^t, J^ϕ and M^t, M^ϕ are the current components along $\hat{\boldsymbol{u}}_{t'}$ and $\hat{\boldsymbol{u}}_{\phi'}$, respectively.

The surface is divided into annular rings and on the kth ring the surface currents are expanded in the form

$$\boldsymbol{J}_a(t', \phi') = \sum_{n,k} \left\{ I^t_{a,nk} \boldsymbol{J}^t_{a,nk}(t', \phi') + I^\phi_{a,nk} \boldsymbol{J}^\phi_{a,nk}(t', \phi') \right\} \text{ on } S_a \qquad a = ce,\ cd \text{ or } d \tag{5.33}$$

$$\boldsymbol{M}_d(t', \phi') = \eta_0 \sum_{n,k} \left\{ V^t_{d,nk} \boldsymbol{K}^t_{d,nk}(t', \phi') + V^\phi_{d,nk} \boldsymbol{K}^\phi_{d,nk}(t', \phi') \right\} \text{ on } S_d \tag{5.34}$$

where t and ϕ are the co-ordinates, respectively, along the generating arc and in the azimuthal direction of the body of revolution as defined above, and $n = 0$, ± 1, ± 2, ... A convenient choice of the expansion function is

$$\boldsymbol{J}^{t,\phi}_{a,nk}(t', \phi') = \boldsymbol{K}^{t,\phi}_{d,nk}(t', \phi') = \hat{\boldsymbol{u}}_{t',\phi'} f_k(t') \exp(jn\phi') \tag{5.35}$$

where the function $f_k(t')$ is defined for each segment along the generating curve.

The coefficients $I^t_{a,nk}$, $I^\phi_{a,nk}$, $V^t_{d,nk}$ and $V^\phi_{d,nk}$ are the current coefficients to be determined. Substituting eqns. 5.33 and 5.34 into eqns. 5.15–5.18 and taking the inner products with certain testing functions and integrating them over the surface will result in a matrix equation. The testing functions are defined as

$$\boldsymbol{W}^{t,\phi}_{a,lk} = \boldsymbol{J}^{t,\phi *}_{a,lk} = \boldsymbol{K}^{t,\phi *}_{d,lk} = \hat{\boldsymbol{u}}_{t,\phi} f_k(t) \exp(-jl\phi) \tag{5.36}$$

where the asterisk denotes the complex conjugate.

The expansion and testing functions, as defined above, are orthogonal over the period 0 to 2π in ϕ. This means, for example, that the inner products of $\boldsymbol{W}^{t,\phi}_{a,lk}{}^*$ and $\boldsymbol{J}^{t,\phi}_{a,nk}$ vanish for $l \neq n$ and give contributions only for different azimuthal modes n. The resulting equations therefore involve only a particular mode n. This is the major simplification which is introduced by the mode orthogonality in axisymmetric objects. Accordingly, the modes are decoupled and one obtains a separate matrix equation for each mode.

For explicit evaluation of the matrix elements, one must first choose $f_k(t')$. It is well known that subsectional expansions, i.e. either using flat-pulse or triangle-pulse functions, gives rise to well conditioned matrices. Thus, flat-pulse current expansions and point matching were first used to solve the scattering problems of a conducting sphere. It turns out that with such expansions the moment-method solutions for the surface currents will not converge rapidly to

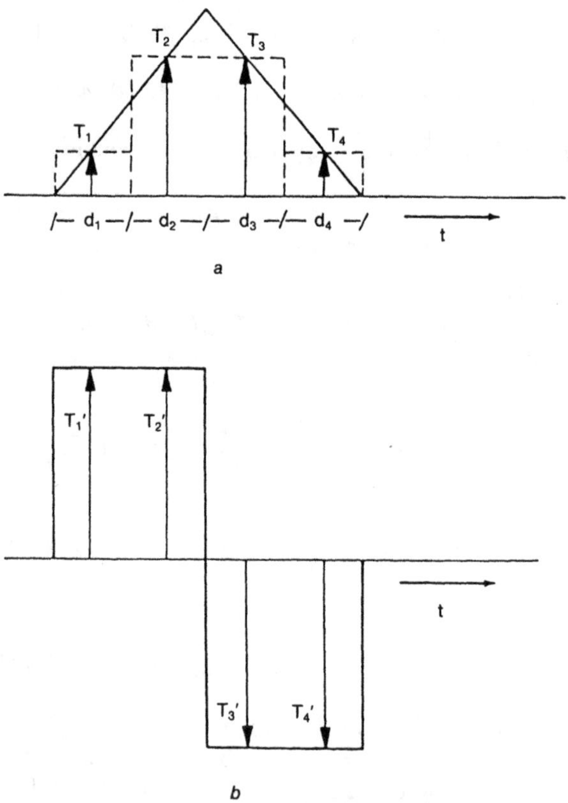

Figure 5.4 Triangle-function approximation and its derivative

a Triangle-function approximation
b Derivative

the exact eigenfunction solution. Even when a finite-difference approximation was used for the derivative operators on the Green's function, the solution accuracy was still unsatisfactory. For this reason the triangle-pulse functions are used instead of the flat-pulse currents for both the current expansion and testing functions. Here, we select $\rho(t')f_k(t')$ to be a triangle function, which is defined as

$$\rho(t')f_k(t') = \sum_{p=1}^{4} T_{p+4k-4}\delta(t' - t_{p+2k-2}) \tag{5.37}$$

where $\delta(t)$ is the unit impulse function and T is defined as shown in Figure 5.4, for $k=1$ with

$$\left.\begin{aligned} T_1 &= \frac{d_1^2}{2(d_1+d_2)} \\ T_2 &= \frac{(d_1+1/2d_2)d_2}{d_1+d_2} \\ T_3 &= \frac{(d_4+1/2d_3)d_3}{d_3+d_4} \\ T_4 &= \frac{d_4^2}{2(d_3+d_4)} \end{aligned}\right\} \tag{5.38}$$

and for the derivative we choose the 4 impulse approximation

$$\frac{\mathrm{d}}{\mathrm{d}t}\{\rho f_k(t')\} = \sum_{p=1}^{4} T'_{p+4k-4}\delta(t' - t_{p+2k-2}) \tag{5.39}$$

The derivative of the triangle function is shown in Figure 5.4, where the coefficients T' for $k=1$ are given by

$$\left.\begin{aligned} T'_1 &= \frac{d_1}{d_1+d_2} \\ T'_2 &= \frac{d_2}{d_1+d_2} \\ T'_3 &= \frac{-d_3}{d_3+d_4} \\ T'_4 &= \frac{-d_4}{d_3+d_4} \end{aligned}\right\} \tag{5.40}$$

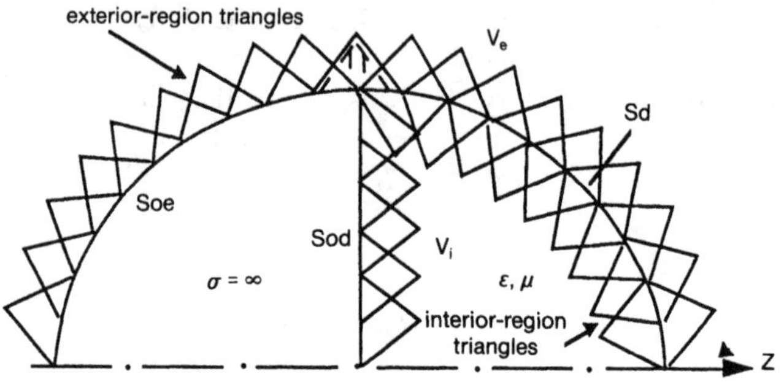

Figure 5.5 Overlapping triangles at intersections of conducting and dielectric surfaces of the object

* Exterior triangles and interior triangles must match each other on the dielectric surface

Now, all information needed to proceed with transferring the integral equations to a system of linear matrix equations is known. The following matrix equation is obtained for each Fourier mode n:

$$\begin{bmatrix} \mathbf{Z}^e_{ce,\,ce} & 0 & \mathbf{Z}^e_{ce,\,d} & -\mathbf{Y}^e_{ce,\,d} \\ 0 & -\eta_r \mathbf{Z}^i_{cd,\,cd} & \eta_r \mathbf{Z}^i_{cd,\,d} & -\mathbf{Y}^i_{cd,\,d} \\ \mathbf{Z}^e_{d,\,ce} & -\eta_r \mathbf{Z}^i_{d,\,cd} & \mathbf{Z}^e_{d,\,d} + \eta_r \mathbf{Z}^i_{d,\,d} & -\mathbf{Y}^e_{d,\,d} - \mathbf{Y}^i_{d,\,d} \\ \mathbf{Y}^e_{d,\,ce} & -\mathbf{Y}^i_{d,\,cd} & \mathbf{Y}^e_{d,\,d} + \mathbf{Y}^i_{d,\,d} & -\mathbf{Z}^e_{d,\,d} - \dfrac{1}{\eta_r}\mathbf{Z}^i_{d,\,d} \end{bmatrix}_n \begin{bmatrix} \tilde{\mathbf{I}}_{ce} \\ \tilde{\mathbf{I}}_{cd} \\ \tilde{\mathbf{I}}_{d} \\ \tilde{\mathbf{V}}_{d} \end{bmatrix}_n = \begin{bmatrix} \tilde{\mathbf{V}}^e_{ce} \\ \tilde{\mathbf{V}}^i_{cd} \\ \tilde{\mathbf{V}}^e_{d} - \tilde{\mathbf{V}}^i_{d} \\ \tilde{\mathbf{I}}^e_{d} - \tilde{\mathbf{I}}^i_{d} \end{bmatrix}_n \tag{5.41}$$

In eqn. 5.41, 0 is a null submatrix, whereas the impedance submatrices $\mathbf{Z}$ and admittance submatrices $\mathbf{Y}$ are given by

$$\mathbf{Z}^{e,\,i}_{\gamma,\,\beta} = [\langle \mathbf{W}^{t,\,\phi}_{\beta,\,nl}, L_{e,\,i}\{(\mathbf{C}^{t,\,\phi}_{\gamma,\,nk})\}_{tan}\rangle] \qquad \mathbf{C}_\gamma = \mathbf{J}_a, \mathbf{K}_d \tag{5.42}$$

$$\mathbf{Y}^{e,\,i}_{\gamma,\,\beta} = [\langle \mathbf{W}^{t,\,\phi}_{\beta,\,nl}, K_{e,\,i}\{(\mathbf{C}^{t,\,\phi}_{\gamma,\,nk})\}_{tan}\rangle] \qquad \mathbf{C}_\gamma = \mathbf{J}_a, \mathbf{K}_d \tag{5.43}$$

where the subscripts of the impedance and admittance submatrices identify the field and source surfaces, and the superscripts e and i correspond, respectively, to the exterior and interior regions. As a result of testing the left-hand sides of the set of eqns. 5.15–5.18 with the vectorial testing functions, every submatrix consists of four submatrices. Furthermore, the vectors $\tilde{\mathbf{I}}_a = [I^{t,\,\phi}_{a,\,nk}]$ and $\tilde{\mathbf{V}}_d = [\mathbf{V}^{t,\,\phi}_{d,\,nk}]$ are the vectors of the expansion coefficients, and

$$\tilde{\mathbf{V}}^q_\alpha = [\langle \mathbf{W}^{t,\,\phi}_{\alpha,\,nl}, \mathbf{E}^{inc}_{q\ tan}\rangle] \qquad \alpha = ce,\ cd,\ d \text{ and } q = i,\ e \tag{5.44}$$

$$\tilde{\mathbf{I}}^q_d = [\langle \mathbf{W}^{t,\,\phi}_{d,\,nl}, \mathbf{H}^{\mathrm{inc}}_{\mathrm{q\ tan}}\rangle] \tag{5.45}$$

are the excitation vectors (excitation-vector expressions will be given in the next section), and ⟨., .⟩ is the inner product of two vectors on the surface of the antenna. By definition, the inner product of two vectors $\boldsymbol{A}$ and $\boldsymbol{B}$ is given by the dot product of $\boldsymbol{A}$ and $\boldsymbol{B}$ integrated over the surface of the antenna, i.e.

$$\langle \boldsymbol{A}, \boldsymbol{B} \rangle = \int_s \int \boldsymbol{A} . \boldsymbol{B} \, ds' \tag{5.46}$$

The matrix elements of eqn. 5.41 consist of

$$[Y_n^q] = \begin{bmatrix} Y_n^{q,\,tt} & Y_n^{q,\,t\phi} \\ Y_n^{q,\,\phi t} & Y_n^{q,\,\phi\phi} \end{bmatrix} \qquad q = e, i \tag{5.47}$$

and

$$[Z_n^q] = \begin{bmatrix} Z_n^{q,\,tt} & Z_n^{q,\,t\phi} \\ Z_n^{q,\,\phi t} & Z_n^{q,\,\phi\phi} \end{bmatrix} \tag{5.48}$$

and their elements have the form

$$\begin{bmatrix} (Y_n^{q,\,tt})_{ij} \\ (Y_n^{q,\,\phi t})_{ij} \\ (Y_n^{q,\,t\phi})_{ij} \\ Y_n^{q,\,\phi\phi})_{ij} \end{bmatrix} = \sum_{p=1}^{4} T_{p'} \sum_{l=1}^{4} T_{l'} \begin{bmatrix} (Y_1)_{ij} \\ (Y_2)_{ij} \\ (Y_3)_{ij} \\ (Y_4)_{ij} \end{bmatrix} \tag{5.49}$$

$$(Y_1)_{ij} = \left\{ \begin{array}{ll} k_q\{(\rho_j - \rho_i) \cos \nu_j - (z_j - z_i) \sin \nu_j\} G_2 - k_q \rho_i G_1 \cos \nu_j & i \neq j \\ \dfrac{\pi}{k_q^2 d_i \rho_i} - k_q \rho_i G_1 \cos \nu_i & i = j \end{array} \right\} \tag{5.50}$$

$$(Y_2)_{ij} = jk_q(\rho_j \sin \nu_i \cos \nu_j - \rho_i \sin \nu_j \cos \nu_i - (z_j - z_i) \sin \nu_i \sin \nu_j) G_3 \tag{5.51}$$

$$(Y_3)_{ij} = jk_q(z_j - z_i) G_3 \tag{5.52}$$

$$(Y_4)_{ij} = \left\{ \begin{array}{ll} k_q\{(\rho_j - \rho_i) \cos \nu_i - (z_j - z_i) \sin \nu_i\} G_2 - k_q \rho_i G_1 \cos \nu_i & i \neq j \\ \dfrac{\pi}{k_q^2 d_i \rho_i} + k_q \rho_i G_1 \cos \nu_i & i = j \end{array} \right\} \tag{5.53}$$

$$(Z_n^{q,\,tt})_{ij} = j \sum_{p=1}^{4} \sum_{l=1}^{4} T_{p'} T_{l'} (G_5 \sin \nu_{i'} \sin \nu_{j'} + G_4 \cos \nu_i \cos \nu_j') - T'_{p'} T'_{l'} G_4 \tag{5.54}$$

$$(Z_n^{q,\,\phi t})_{ij} = -\sum_{p=1}^{4} \sum_{l=1}^{4} T_{p'} T_{l'} G_6 \sin \nu_{j'} + \frac{n}{k\rho_i'} T_{p'} T'_{l'} G_4 \tag{5.55}$$

$$(Z_n^{q,\,t\phi})_{ij} = \sum_{p=1}^{4} \sum_{l=1}^{4} T_{p'} T_{l'} G_6 \sin \nu_{i'} + \frac{n}{k\rho_j'} T'_{p'} T_{l'} G_4 \tag{5.56}$$

$$(Z_n^{q,\phi\phi})_{ij}=j\sum_{p=1}^{4}\sum_{l=1}^{4}T_{p'}T_{l'}\left(G_5-\frac{n^2}{k^2\rho'_i\rho'_j}G_4\right) \tag{5.57}$$

and

$$\left.\begin{aligned}p'&=p+4i-4\\ l'&=l+4j-4\\ i'&=p+2i-2\\ j'&=l+2j-2\end{aligned}\right\} \tag{5.58}$$

where

$$G_1=2\int_0^{\pi}\mathrm{d}\phi'\,G'_q\sin^2\left(\frac{\phi'}{2}\right)\cos(n\phi') \tag{5.59}$$

$$G_2=\int_0^{\pi}\mathrm{d}\phi'\,G'_q\cos\phi'\cos(n\phi') \tag{5.60}$$

$$G_3=\int_0^{\pi}\mathrm{d}\phi'\,G'_q\sin\phi'\sin(n\phi') \tag{5.61}$$

$$G_4=\int_0^{\pi}\mathrm{d}\phi'\,G_q\cos(n\phi') \tag{5.62}$$

$$G_5=\int_0^{\pi}\mathrm{d}\phi'\,G_q\cos\phi'\cos(n\phi') \tag{5.63}$$

$$G_6=\int_0^{\pi}\mathrm{d}\phi'\,G_q\sin\phi'\sin(n\phi') \tag{5.64}$$

with

$$G_q=\frac{\exp(-jk_qR)}{k_qR} \tag{5.65}$$

$$G'_q=-\frac{1+jk_qR}{k_q^3R^3}\exp(-jk_qR) \tag{5.66}$$

$$R=\left\{(\rho-\rho')^2+(z-z')^2+4\rho\rho'\sin^2\left(\frac{\phi'}{2}\right)\right\}^{1/2} \tag{5.67}$$

at $t=t'$, R is approximated by

$$R_e=\left\{\left(\frac{d_{i'}}{4}\right)^2+4\rho_{i'}^2\sin^2\left(\frac{\phi'}{2}\right)\right\}^{1/2} \tag{5.68}$$

5.3.3 *Excitation vectors*

The excitation vectors, as in eqns. 5.40 and 5.41, are the result of the inner products of the testing function with $\boldsymbol{E}^{inc}$ and $\boldsymbol{H}^{inc}$. For the dipole excitations $\boldsymbol{E}^{inc}$ and $\boldsymbol{H}^{inc}$ are given in eqns. 5.1 and 5.2 in terms of the magnetic vector potential and the scalar electric potentials which are given in terms of the spherical Hankel function of second kind and zero order in eqns. 5.3 and 5.4. If the Hankel function is represented by

$$h_0^{(2)}(k_q|\vec{\boldsymbol{r}}-\vec{\boldsymbol{r}}'|)=\sum_{-\infty}^{\infty} G_m \exp\{jm(\phi-\phi')\} \tag{5.69}$$

with

$$G_m = G_{-m} = \left\{ \begin{array}{ll} \sum\limits_{n=m}^{\infty} a_n h_n^{(2)}(k_q r') j_n(k_q r) P_n^m(\cos\theta) P_n^m(\cos\theta') & r' > r \\ \sum\limits_{n=m}^{\infty} a_n h_n^{(2)}(k_q r) j_n(k_q r') P_n^m(\cos\theta) P_n^m(\cos\theta') & r' < r \end{array} \right\} \tag{5.70}$$

where $a_n = (2n+1)(n-m)!/(n+m)!$ then, using eqn. 5.1 for $\boldsymbol{E}^{inc}$, its excitation matrix elements can be calculated from

$$(V_m^q)_{ix}^t = \frac{\eta_q|\boldsymbol{I}l|_x}{4}\left[-k_q^2\int_0^{tu} \mathrm{d}t\,\rho f_i(t)\sin\nu(G_{m+1}+G_{m-1}) + \int_0^{tu} dt\frac{\mathrm{d}}{\mathrm{d}t}\{\rho f_i(t)\}\left\{\frac{\partial G_{m+1}}{\partial r}+\frac{\partial G_{m-1}}{\partial r}+\frac{(m+1)G_{m+1}-(m-1)G_{m-1}}{\rho}\right\}\right] \tag{5.71}$$

$$(V_m^q)_{ix}^{\phi} = \frac{\eta_q|\boldsymbol{I}l|_x}{4}\left[jk_q^2\int_0^{tu} \mathrm{d}t\,\rho f_i(t)\sin\nu(G_{m+1}-G_{m-1}) - jm\int_0^{tu}\mathrm{d}t\,f_i(t) \times\left\{\frac{\partial G_{m+1}}{\partial r}+\frac{\partial G_{m-1}}{\partial r}+\frac{(m+1)G_{m+1}-(m-1)G_{m-1}}{\rho}\right\}\right] \tag{5.72}$$

$$(V_m^q)_{iy}^t = \frac{\eta_q|\boldsymbol{I}l|_y}{4}\left[-jk_q^2\int_0^{tu} \mathrm{d}t\,\rho f_i(t)\sin\nu(G_{m+1}-G_{m-1}) + j\int_0^{tu}\mathrm{d}t\frac{\mathrm{d}}{\mathrm{d}t}\{\rho f_i(t)\} \times\left\{\frac{\partial G_{m+1}}{\partial r}-\frac{\partial G_{m-1}}{\partial r}+\frac{(m+1)G_{m+1}+(m-1)G_{m-1}}{\rho}\right\}\right] \tag{5.73}$$

$$(V_m^q)_{iy}^{\phi} = \frac{\eta_q|\boldsymbol{I}l|_y}{4}\left[-k_q^2\int_0^{tu} \mathrm{d}t\,f_i(t)\sin\nu(G_{m+1}+G_{m-1}) + m\int_0^{tu}\mathrm{d}t\,f_i(t) \times\left\{\frac{\partial G_{m+1}}{\partial r}-\frac{\partial G_{m-1}}{\partial r}+\frac{(m+1)G_{m+1}+(m-1)G_{m-1}}{\rho}\right\}\right] \tag{5.74}$$

$$(V_m^q)_{iz}^t = \frac{\eta_q|\boldsymbol{I}l|_z}{2}\left[-k_q^2\int_0^{tu} \mathrm{d}t\,\rho f_i(t)\cos\nu G_m - \int_0^{tu}\mathrm{d}t\frac{\mathrm{d}}{\mathrm{d}t}\{\rho f_i(t)\}\frac{1}{\rho}\frac{\partial G_m}{\partial\theta}\right] \tag{5.75}$$

$$(V_m^q)_{iz}^{\phi} = \frac{\eta_q|\boldsymbol{I}l|_z}{2}\left\{jm\int_0^{tu}\mathrm{d}t\frac{f_i(t)}{\rho}\frac{\partial G_m}{\partial\theta}\right\} \tag{5.76}$$

where tu is the t co-ordinate of the upper end of segment i of the generating curve. Similarly, using eqn 5.2 for $\boldsymbol{H}^{inc}$, its matrix element can be calculated from

$$(I_m^q)_{ix}^t = \frac{-k_q|\boldsymbol{I}l|_x}{4}\left(\int_0^{tu} \mathrm{d}t\, \rho f_i(t) \cos\nu \left(\frac{\partial G_{m+1}}{\partial r} - \frac{\partial G_{m-1}}{\partial r}\right) + \int_0^{tu} \mathrm{d}t\, f_i(t)\right.$$

$$\left.\times\left[\sin\nu \left(\frac{\partial G_{m+1}}{\partial \theta} - \frac{\partial G_{m-1}}{\partial \theta}\right) + \cos\nu\{(m+1)G_{m+1} + (m-1)G_{m-1}\}\right]\right) \tag{5.77}$$

$$(I_m^q)_{ix}^\phi = \frac{jk_q|\boldsymbol{I}l|_x}{4}\left\{\int_0^{tu} \mathrm{d}t\, \rho f_i(t) \left(\frac{\partial G_{m+1}}{\partial \theta} + \frac{\partial G_{m-1}}{\partial \theta}\right)\right. \tag{5.78}$$

$$(I_m^q)_{iy}^t = \frac{-jk_q|\boldsymbol{I}l|_y}{4}\left(\int_0^{tu} \mathrm{d}t\, \rho f_i(t) \left[\cos\nu \left(\frac{\partial G_{m+1}}{\partial r} + \frac{\partial G_{m-1}}{\partial r}\right) + \frac{\sin\nu}{\rho}\right.\right.$$

$$\left.\left.\left(\frac{\partial G_{m+1}}{\partial \theta} + \frac{\partial G_{m-1}}{\partial \theta}\right) + \frac{\cos\nu}{\rho}\{(m+1)G_{m+1} - (m-1)G_{m-1}\}\right]\right) \tag{5.79}$$

$$(I_m^q)_{iy}^\phi = \frac{-k_q|\boldsymbol{I}l|_y}{4}\left\{\int_0^{tu} \mathrm{d}t\, f_i(t) \left(\frac{\partial G_{m+1}}{\partial \theta} - \frac{\partial G_{m-1}}{\partial \theta}\right)\right\} \tag{5.80}$$

$$(I_m^q) = \frac{-jk_q|\boldsymbol{I}l|_z}{4}\left\{\int_0^{tu} \mathrm{d}t\, f_i(t) \sin\nu G_m\right\} \tag{5.81}$$

$$(I_m^q)_{iz}^\phi = \frac{jk_q|\boldsymbol{I}l|_z}{2}\left\{\int_0^{tu} \mathrm{d}t\, \rho f_i(t) \frac{\partial G_m}{\partial r}\right\} \tag{5.82}$$

In eqn 5.70 one is free to choose the origin except that it must be placed on the z axis. In order to have a sum which converges as rapidly as possible the ratio r/r' must be either large or small depending on whether $r > r'$ or $r < r'$, although neither r nor r' must vanish. A simple way to ensure that we satisfy this condition is to replace the origin so that $\cos\theta = 0$. Then r has its minimum value and eqn. 5.70 becomes

$$G_m = G_{-m} = \left\{\begin{array}{ll} \displaystyle\sum_{n=m}^{\infty} \frac{(2n+1)(n-m)!}{(n+m)!} h_n^{(2)}(kr') j_n(k\rho) P_n^m(0) P_n^m(\cos\theta') & r' > \rho \\ \displaystyle\sum_{n=m}^{\infty} \frac{(2n+1)(n-m)!}{(n+m)!} h_n^{(2)}(k\rho) j_n(kr') P_n^m(0) P_n^m(\cos\theta') & r' < \rho \end{array}\right\} \tag{5.83}$$

The quantities $\partial G_m/\partial r$ and $\partial G_m/\partial\theta$ are obtained by differentiating eqn. 5.70

partially with respect to r and θ, respectively, and setting $r=\rho$ and $\cos\theta=0$. If the dipole is not located at $\phi'=0$, the incident-field vectors in eqns. 5.71–5.82 should be multiplied by a factor $\exp(-jm\phi')$.

5.3.4 Computation of the radiation patterns

To obtain the far-field radiation patterns it is necessary to calculate the integrals in eqns. 5.21 and 5.22. Referring to the field point, the vectors $\hat{\boldsymbol{u}}_\theta$ and $\hat{\boldsymbol{u}}_\phi$ can be written as

$$\hat{\boldsymbol{u}}_\theta=\cos\theta_0\cos\phi_0\hat{\boldsymbol{u}}_x+\cos\theta_0\sin\phi_0\hat{\boldsymbol{u}}_y-\sin\theta_0\hat{\boldsymbol{u}}_z \tag{5.84}$$

$$\hat{\boldsymbol{u}}_\phi=-\sin\phi_0\hat{\boldsymbol{u}}_x+\cos\phi_0\hat{\boldsymbol{u}}_y \tag{5.85}$$

and the dot product in the exponential term of eqns. 5.21 and 5.22 can be shown to be

$$\begin{aligned}\hat{\boldsymbol{r}}_0.\boldsymbol{r}'&=(x'\cos\phi_0+y'\sin\phi_0)\sin\theta_0+z'\cos\theta_0\\&=\rho'\cos(\phi'-\phi_0)\sin\theta_0+z'\cos\theta_0\end{aligned} \tag{5.86}$$

where (ρ', ϕ', z') are the cylindrical co-ordinates of the point (x', y', z'). If we now substitute for J and M in eqns. 5.21 and 5.22, and evaluate the dot products $(\hat{\boldsymbol{u}}_{t'}.\hat{\boldsymbol{u}}_\theta)$ and $(\hat{\boldsymbol{u}}_{t'}.\hat{\boldsymbol{u}}_\phi)$ as

$$\hat{\boldsymbol{u}}_{t'}.\hat{\boldsymbol{u}}_\theta=\sin\nu'\cos\theta_0\cos(\phi'-\phi_0)-\cos\nu\sin\theta_0 \tag{5.87}$$

$$\hat{\boldsymbol{u}}_{t'}.\hat{\boldsymbol{u}}_\phi=\sin\nu'\sin(\phi'-\phi_0) \tag{5.88}$$

and

$$\hat{\boldsymbol{u}}_{\phi'}.\hat{\boldsymbol{u}}_\theta=-\sin\theta_0\sin(\phi'-\phi_0) \tag{5.89}$$

$$\begin{aligned}\hat{\boldsymbol{u}}_{\phi'}.\hat{\boldsymbol{u}}_\phi&=\sin\phi'\sin\phi_0+\cos\phi'\cos\phi_0\\&=\cos(\phi'-\phi_0)\end{aligned} \tag{5.90}$$

then $F_{e1}(\theta_0, \phi_0)$ and $F_{e2}(\theta_0, \phi_0)$ take the form

$$\begin{aligned}F_{e1}(\theta_0,\phi_0)=\int_0^{t_u}\int_0^{2\pi}\sum_{m=-M_0}^{M_0}\sum_{j=1}^{N}\\&\times f_j(t')\exp(jm\phi')\exp(jk_0\{\rho'\cos(\phi'-\phi_0)\sin\theta_0+z'\cos\theta_0\})\\&\times\Bigg[I^t_{mj}\{\sin\nu'\cos\theta_0\cos(\phi'-\phi_0)-\cos\nu'\sin\theta_0\}\\&-I^\phi_{mj}\cos\theta_0\sin(\phi'-\phi_0)+\frac{1}{\eta_0}V^t_{mj}\sin\nu'\sin(\phi'-\phi_0)\end{aligned}$$

$$+\frac{1}{\eta_0} V^{\phi}_{mj} \cos(\phi' - \phi_0)\Bigg]\, \rho'\, d\phi'\, dt' \tag{5.91}$$

and

$$F_{e2}(\theta_0, \phi_0) = \int_0^{t_u}\int_0^{2\pi} \sum_{m=-M_0}^{M_0} \sum_{j=1}^{N}$$

$$\times f_j(t') \exp(jm\phi') \exp(jk_0\{\rho' \cos(\phi' - \phi_0) \sin\theta_0 + z' \cos\theta_0\})$$

$$\times\Bigg[I^t_{mj} \sin\nu' \sin(\phi' - \phi_0) + I^{\phi}_{mj} \cos(\phi' - \phi_0)$$

$$+\frac{1}{\eta_0} V^t_{mj}\{\sin\nu' \cos\theta_0 \cos(\phi' - \phi_0) - \cos\nu \sin\theta_0\}$$

$$+\frac{1}{\eta_0} V^{\phi}_{mj} \cos\theta_0 \sin(\phi' - \phi_0)\Bigg]\, \rho' d\phi'\, dt' \tag{5.92}$$

Changing the order of summations and integrations and using the integral representation for the Bessel function of the first kind, $J_n(\rho)$ can be written as

$$J_n(\rho) = \frac{j^n}{2\pi}\int_0^{2\pi} \exp(-j\rho \cos\alpha) \exp(-jn\alpha)\, d\alpha \tag{5.93}$$

the integration in eqns. 5.91 and 5.92 involving the azimuthal co-ordinate ϕ' can be evaluated as

$$\int_0^{2\pi} \exp(jm\phi') \exp\{jk_0\rho' \cos(\phi' - \phi_0) \sin\theta_0\}$$

$$\times d\phi' = 2\pi j^m \exp(jm\phi_0) J_m(k_0\rho' \sin\theta_0) \tag{5.94}$$

and

$$\int_0^{2\pi} \begin{Bmatrix} \cos(\phi' - \phi_0) \\ \sin(\phi' - \phi_0) \end{Bmatrix} \exp(jm\phi') \exp\{jk_0\rho' \cos(\phi' - \phi_0) \sin\theta_0\}\, d\phi'$$

$$= \pi j^m \exp(jm\phi_0) \begin{Bmatrix} j\{J_{m+1}(k_o\rho' \sin\theta_0) - J_{m-1}(k_0\rho' \sin\theta_0)\} \\ \{J_{m+1}(k_0\rho' \sin\theta_0) + J_{m-1}(k_0\rho' \sin\theta_0)\} \end{Bmatrix} \tag{5.95}$$

The far-field functions F_{1e} and F_{2e} can be represented as

$$F_{1e}(\theta_0, \phi_0) = F_{1J}(\theta_0, \phi_0) + F_{1M}(\theta_0, \phi_0) \tag{5.96}$$

$$F_{2e}(\theta_0, \phi_0) = F_{2J}(\theta_0, \phi_0) + F_{2M}(\theta_0, \phi_0) \tag{5.97}$$

$$F_{1J}(\theta_0, \phi_0) = \sum_{m=-M_0}^{M_0} \sum_{j=1}^{N_c+N_d} \{(R^{t\theta, J}_m)_j I^t_{mj} + (R^{\phi\theta, J}_m)_j I^{\phi}_{mj}\} \tag{5.98}$$

$$F_{2J}(\theta_0, \phi_0) = \sum_{m=-M_0}^{M_0} \sum_{j=1}^{N_c+N_d} \{(R^{t\phi, J}_m)_j I^t_{mj} + (R^{\phi\phi, J}_m)_j I^{\phi}_{mj}\} \tag{5.99}$$

$$F_{1M}(\theta_0, \phi_0) = \sum_{m=-M_0}^{M_0} \sum_{j=N_c+N_{d+1}}^{N} \{(R_m^{t\theta, M})_j V_{mj}^t + (R_m^{\phi\theta, M})_j V_{mj}^{\phi}\} \tag{5.100}$$

$$F_{2M}(\theta_0, \phi_0) = \sum_{m=-M_0}^{M_0} \sum_{j=N_c+N_{d+1}}^{N} \{(R_m^{t\phi, M})_j V_{mj}^t + (R_m^{\phi\phi, M})_j V_{mj}^{\phi}\} \tag{5.101}$$

where N_c and N_d are the number of triangles on the outer conductor surface S_{ce} and the dielectric surface S_d, respectively. The R functions can be expressed as

$$(R_m^{t\theta, J})_j = \sum_{q=1}^{4} jC\{j(J_{m+1} - J_{m-1}) \sin \nu' \cos \theta_0 - 2 \cos \nu' \sin \theta_0 J_m\} \tag{5.102}$$

$$(R_m^{\phi\theta, J})_j = -\sum_{q=1}^{4} C(J_{m+1} + J_{m-1}) \cos \theta_0 \tag{5.103}$$

$$(R_m^{t\phi, J})_j = \sum_{q=1}^{4} C(J_{m+1} + J_{m-1}) \sin \nu' \tag{5.104}$$

$$(R_m^{\phi\phi, J})_j = \sum_{q=1}^{4} jC(J_{m+1} - J_{m-1}) \tag{5.105}$$

with

$$C = \pi j^m \exp(jm\phi_0) \exp(jk_0 z' \cos \theta_0) T_{q'} \tag{5.106}$$

and

$$(R_m^{t\theta, M})_j = \frac{1}{\eta_0} (R_m^{t\phi, J})_j \tag{5.107}$$

$$(R_m^{\phi\theta, M})_j = \frac{1}{\eta_0} (R_m^{\phi\phi, J})_j \tag{5.108}$$

$$(R_m^{t\phi, M})_j = -\frac{1}{\eta_0} (R_m^{t\theta, J})_j \tag{5.109}$$

$$(R_m^{\phi\phi, M})_j = \frac{1}{\eta_0} (R_m^{\phi\theta, J})_j \tag{5.110}$$

The far-field component can finally be put in compact form as

$$\begin{bmatrix} E_\theta \\ E_\phi \end{bmatrix} = \frac{-j\omega\mu_e \exp(-jk_0 r_0)}{4\pi r_0} \sum_{m=-M_0}^{m=M_0} \sum_j \begin{bmatrix} (R_m^{t\theta, J})_j (R_m^{t\theta, M})_j (R_m^{\phi\theta, J})_j (R_m^{\phi\theta, M})_j \\ (R_m^{t\phi, J})_j (R_m^{t\phi, M})_j (R_m^{\phi\phi, J})_j (R_m^{\phi\phi, M})_j \end{bmatrix} \times \begin{bmatrix} I_{mj}^t \\ M_{mj}^t \\ I_{mj}^{\phi} \\ M_{mj}^{\phi} \end{bmatrix} \tag{5.111}$$

This completes the formulation of the problem and determination of the radiated fields from the computed equivalent currents. The external field near the antenna or the field within the dielectric substrate can also be determined, but are omitted here for brevity.

Two different properties of the Y and Z submatrices may be used to reduce the computation time and cost. One is the matrix symmetry. This property is evident from eqns. 5.50–5.57 and gives

$$\left.\begin{aligned}(Z_m^{tt})_{ij}&=(Z_m^{tt})_{ji}\\(Z_m^{t\phi})_{ij}&=-(Z_m^{t\phi})_{ji}\\(Z_m^{\phi\phi})_{ij}&=(Z_m^{\phi\phi})_{ji}\\(Y_m^{tt})_{ij}&=-(Y_m^{\phi\phi})_{ji}\\(Y_m^{\phi t})_{ij}&=-(Y_m^{\phi t})_{ji}\\(Y_m^{t\phi})_{ij}&=-(Y_m^{t\phi})_{ji}\end{aligned}\right\}\qquad(5.112)$$

This means that only one-half of the matrix need be created. The other half can be generated from the above symmetric property. The second property is the mode symmetry. This enables one to generate the Y and Z submatrices of the modes with negative indices from the positive ones, as

$$\left.\begin{aligned}Z_{-m}^{tt}&=Z_m^{tt}\\Z_{-m}^{\phi t}&=-Z_m^{\phi t}\\Z_{-m}^{t\phi}&=-Z_m^{t\phi}\\Z_{-m}^{\phi\phi}&=Z_m^{\phi\phi}\end{aligned}\right\}\qquad(5.113)$$

The same relations are valid for the Y submatrices. Similarly, the elements of the excitation for an electric dipole, generated using eqns. 5.71–5.76, satisfy the relationships

$$\left.\begin{aligned}(V_{-m})^t&=V_m^t\\(V_{-m})^\phi&=-V_m^\phi\\(I_{-m})^t&=-I_m^t\\(I_{-m})^\phi&=I_m^\phi\end{aligned}\right\}\qquad(5.114)$$

Thus the solution of the matrix equations, given by eqns. 5.41, gives the following symmetric relationships satisfied by the current coefficients:

$$\left.\begin{aligned}I_{-n}^t&=I_n^t\\M_{-n}^t&=-M_n^t\\I_{-n}^\phi&=-I_n^\phi\\M_{-n}^\phi&=M_n^\phi\end{aligned}\right\}\qquad(5.115)$$

Accordingly, one only need compute the coefficients of the positive modes, i.e. one-half of the mode coefficients. This results in a major reduction of computation time and cost. The mode symmetry can also be used in calculating the far-field components from eqn. 5.111. The required relationships are

$$\left.\begin{aligned} R^{t\theta,J}_{-m} &= R^{t\theta,J}_{m} \\ R^{\phi\theta,J}_{-m} &= -R^{\phi\theta,J}_{m} \\ R^{t\phi,J}_{-m} &= -R^{t\phi,J}_{m} \\ R^{\phi\phi,J}_{-m} &= R^{\phi\phi,J}_{m} \end{aligned}\right\} \tag{5.116}$$

which can reduce the summation over the modes to one from $m=0$ to $+M_0$.

5.3.5 *Different integral-equation formulations*

Using the auxiliary equivalent problems in Figure 5.2*a* and *b*, one can obtain eight surface-integral equations for the four unknown currents. To solve for these currents only four equations are required, and consequently different sets of equations can be obtained. Details of generating different formulations can be found in Reference 5. However, a simplified form for generation of different formulations will be given here for convenience. One of the possible formulations was given in eqns. 5.15–5.18. This formulation is known as E-PMCHW. One can use a modified form of the surface-integral equations to contain most of the possible formulations in a compact form as

$$+\alpha_{ce}\{+L_e(\boldsymbol{J}_{ce}+\boldsymbol{J}_d)+k_e(\boldsymbol{M}_d)\}_{tan}+\beta_{ce}\left\{-K_e(\boldsymbol{J}_{ce}+\boldsymbol{J}_d)+\frac{1}{\eta_0}L_e(\boldsymbol{M}_d)\right\}_{tan}$$

$$=\alpha_{ce}\boldsymbol{E}^{inc}_e|_{tan}+\beta_{ce}\boldsymbol{H}^{inc}_e|_{tan} \tag{5.117}$$

$$+\alpha_{cd}\{+L_i(\boldsymbol{J}_{cd}+\boldsymbol{J}_d)-k_i(\boldsymbol{M}_d)\}+\beta_{cd}\left\{K_i(\boldsymbol{J}_{cd}+\boldsymbol{J}_d)+\frac{1}{\eta_d}L_i(\boldsymbol{M}_d)\right\}_{tan}$$

$$=-\alpha_{cd}\boldsymbol{E}^{inc}_i|_{tan}-\beta_{cd}\boldsymbol{H}^{inc}_i|_{tan} \tag{5.118}$$

$$\alpha_{ed}[L_e(\boldsymbol{J}_{ce}+\boldsymbol{J}_d)+K_e(\boldsymbol{M}_d)+\alpha_{id}\{L_i(\boldsymbol{J}_{cd}-\boldsymbol{J}_d)+K_i(\boldsymbol{M}_d)\}]_{tan}$$

$$=\alpha_{ed}\boldsymbol{E}^{inc}_e|_{tan}+\alpha_{id}\boldsymbol{E}^{inc}_i|_{tan} \tag{5.119}$$

$$\beta_{ed}\left[-K_e(\boldsymbol{J}_{ce}+\boldsymbol{J}_d)+\frac{1}{\eta_0^2}L_e(\boldsymbol{M}_d)+\beta_{id}\left\{K_i(\boldsymbol{J}_{cd}+\boldsymbol{J}_d)+\frac{1}{\eta_d^2}L_i(\boldsymbol{M}_d)\right\}\right]_{tan}$$

$$=\beta_{ed}\boldsymbol{H}^{inc}_e|_{tan}+\beta_{id}\boldsymbol{H}^{inc}_i|_{tan} \tag{5.120}$$

Here, by a proper selection of the constant parameters α_{ce}, β_{ce}, α_{cd}, β_{cd}, α_d and β_d one can obtain all the formulations given in Table 5.1.

An additional three formulations can also be obtained if we use eqns. 5.117–5.120. The E-field formulation can be obtained by selecting first $\alpha_{ce}=1$, $\beta_{ce}=0$, $\alpha_{cd}=1$, $\beta_{cd}=0$, $\beta_{ed}=0$, $\beta_{id}=0$ ($\alpha_{ed}=1$, $\alpha_{ie}=0$) and then ($\alpha_{ed}=0$, $\alpha_{ie}=1$) to obtain the fourth equation. The H-field formulation can be obtained by selecting first $\alpha_{ce}=0$, $\beta_{ce}=1$, $\alpha_{cd}=0$, $\beta_{cd}=1$, $\alpha_{ed}=0$, $\alpha_{ie}=0$ ($\beta_{cd}=1$, $\beta_{id}=0$) and then ($\beta_{ed}=0$, $\beta_{id}=1$) to obtain the fourth equation. The last formulation is the C-field

Table 5.1 Formulations for integral equations

Formulation type	α_{ce}	β_{ce}	α_{cd}	β_{cd}	α_{id}	β_{id}	α_{ed}	β_{ed}
E-PMCHW	1	0	1	0	1	1	1	1
E-Müller	1	0	1	0	ε_{ri}	μ_{ri}	1	1
H-PMCHW	0	1	0	1	1	1	1	1
H-Müller	0	1	0	1	ε_{ri}	μ_{ri}	1	1

formulation which is the result of combining E-field and H-field formulations. The E-field, H-field and C-field formulations use the electric-field and magnetic-field integral equations, or their combination, respectively, on all antenna surfaces and their name is selected accordingly. The meaning of other formulations in Table 5.1 is that E-, H- or C-integral equations are used on the conductor surfaces and PMCHW or Müller formulation is used on the dielectric surface. Accordingly, if the antenna consists of only a conducting material, then only three formulations can be used, namely the E-field, H-field and C-field formulations. Since only one of the boundary conditions is used, the E-field and H-field formulations have nonunique solutions at the internal resonances of the object. At these particular frequencies one has to use the C-field formulation. Another limitation in using these formulations manifests itself when the antenna has a thin conducting wall, for which the H-field is not recommended since it fails to yield an accurate solution. For antennas that are made of only a dielectric material, five formulations can be used: the E-field, H-field, C-field, PMCHW and Müller formulations. For combinations of dielectric and conductors, all above formulations are possible, namely the E-field, H-field, C-field, E-PMCHW, E-Müller, H-PMCHW, H-Müller, C-PMCHW and C-Müller formulations. These formulations may also suffer from the problem of internal resonances, except the C-field, C-PMCHW and C-Müller formulations. As mentioned previously, the E-field and H-field formulations will be affected by the internal-resonance problem whenever s_{ce}, s_{cd} or s_d encloses a resonant cavity, when s_{ce} and s_d together form a closed resonant cavity, or when s_{cd} and s_d together form a closed resonant cavity. The other formulations that are affected by the internal-resonance problem may also suffer from the internal-resonance problem when either s_{ce} or s_{cd} or both form a closed resonant cavity. Finally, it is not recommended to use the H-PMCHW or H-Müller formulations when s_{ce} or s_{cd} contains thin conductor parts that are not completely attached to the dielectric.

Thus, in using numerical methods for solving an antenna problem, not only must one be familiar with the particular numerical method, but one should also have an understanding of the physical characteristics of the antenna itself. For instance, in waveguide and cavity problems a number of modes may be present which must be considered. Also, when the antenna has a thin edge or corner, the induced currents may become singular, and an adequate number of triangle functions must be used to permit a satisfactory representation of the actual currents. Similar situations may also prevail on parts of two antenna surfaces which interact strongly. Without a clear understanding of the limitations of numerical methods and the antenna physical characteristics one may obtain solutions which are inaccurate or unrealistic.

In using the present method, one may select around 6–10 triangle functions per wavelength of the generating curve, provided that that portion of the antenna is smooth and noninteracting. Otherwise, the number of expansion functions must be larger.

The use of the triangle-expansion functions and their representation by pulses allow one to solve the problem efficiently even for antennas having nonsmooth surfaces, such as open-ended waveguides. The triangle functions overlap on the generating curve as shown in Figure 5.5. They start and end at the points of intersection of the antenna with the axis of symmetry. However, at the points of surface discontinuity or at intersection points of conductors with dielectrics only the tip of a triangle must be positioned. In other words, only halftriangles must be used at these positions, to be combined with the halftriangle of the connecting surface.

5.3.6 Solution verification

One of the difficulties in using numerical methods is the uncertainty of the solution accuracy. Solutions based on moment methods are no different, and it would be desirable to develop an independent verification rule. This is especially important in feed designs, where the crosspolarisation is small and its correct computation demands higher precision from the numerical solutions. Of course, the most appropriate way of verifying the results is an experimental comparison. However, it is costly and should not be used in the design stage. From the designer's point of view it is more convenient to use a numerical experimentation to verify results. To allow this, different integral-equation formulations described in Section 5.3.5 must be used [5], and their solutions must be compared using a moment method. Application of such a multiple-formulation approach to verify and control solution accuracies in antenna design and analysis has recently been investigated [6]. Here, we provide two separate examples to show the process of numerical experimentation and the type of solution accuracy which can be achieved with the proposed solution, and without an excessive numerical search.

The first example we select for verification of the solutions is computation of bistatic scattering of a dielectric-coated sphere. This is not an antenna problem, but is one for which an analytic solution is available using Mei series. Figure 5.6 shows the comparison of solutions for seven different formulations with the analytic one. The formulations are described in Reference 5 and also briefly in Section 5.3.5. The relative permittivity of the dielectric is $\varepsilon_r = 4$ and all seven formulations give satisfactory results for this problem, in predicting peaks and

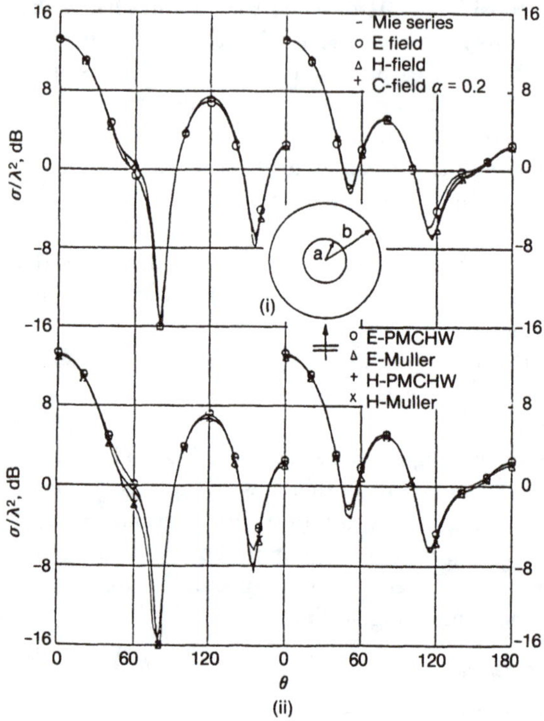

Figure 5.6 *Bistatic-scattering cross-section of a coated conducting sphere* $b/a = 3$, $ka = 1$, $\varepsilon_r = 4$, E-plane left side, H-plane right side

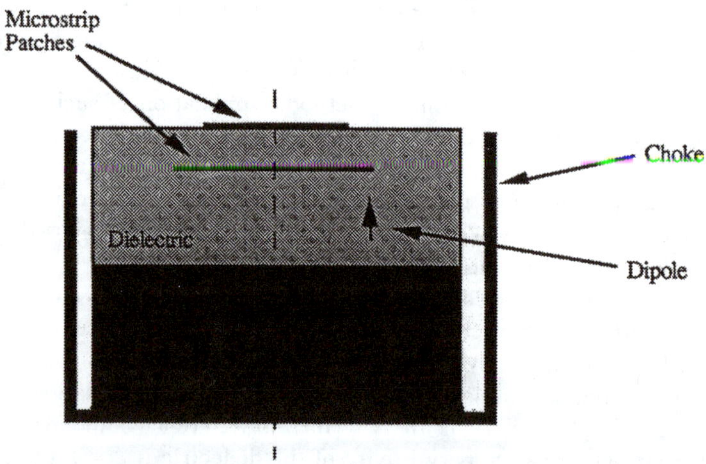

Figure 5.7 *Stacked microstrip antenna*

nulls or their level. Their accuracy is thus acceptable, as far as the scattering by a coated sphere is concerned. The numerical solution of problems involving conducting objects is considerably easier than those of problems involving dielectrics and their combinations, and comparative data are not inlcuded here.

The above comments are also valid for the most complex problems of conductor—dielectric combinations. The most appropriate formulations are those which avoid the application of a single boundary condition, i.e. the E or H, over the entire surface of the object. Eqns. 5.15–5.18 achieved this by applying the electric boundary condition over the conductor and both electric and magnetic boundary conditions over the dielectric. This formulation is the E-PMCHW, as discussed earlier. An alternative formulation with similar applications of boundary conditions is the E-Müller, where the combination parameter between the electric and magnetic currents is weighted by the material constitutive parameters [5]. These two methods were used to generate far-field patterns for a complex feed to show their performance.

The chosen feed is a stacked-microstrip antenna, shown in Figure 5.7. Its performance is discussed in Chapter 14. Here, we only provide a comparison of results for two different solutions. Figures 5.8 and 5.9 show the comparisons for the principal plane patterns and the crosspolarisation, i.e. computed in the 45° plane where it is a maximum. These results show that the accuracy of the H-plane patterns is excellent, as they are coincident down to the −30 dB level. The E-plane data show some differences. In fact the two computed patterns are

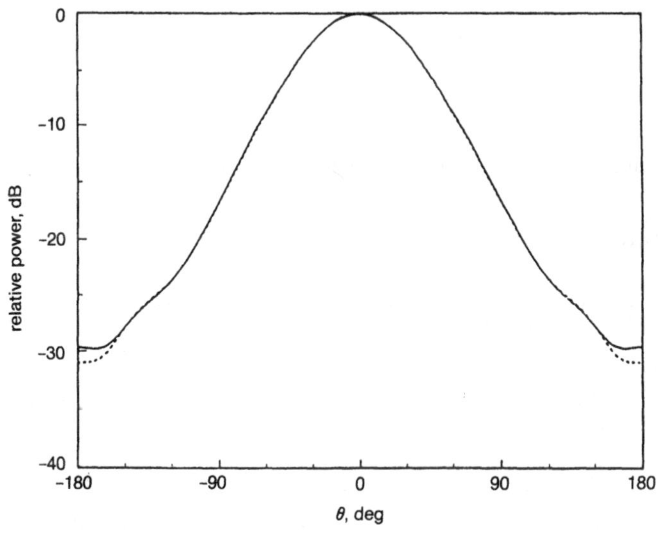

Figure 5.8 H-plane patterns for microstrip feed of Figure 5.7 computed by two formulations

—— EPMCHW
- - - E-Muller

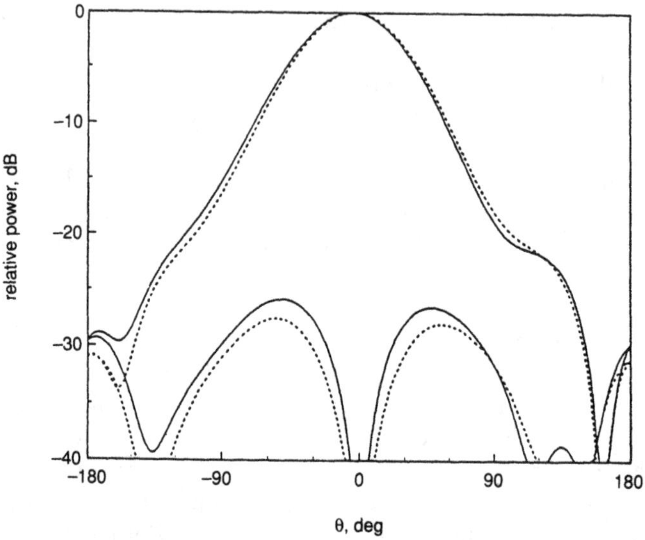

Figure 5.9 Comparison of E-plane polarisation and crosspolarisation of the feed in Figure 5.7 for two formulations

—— PMCHW
····· Muller

again identical in shape down to the −30 dB level, but for two different formulations are just shifted slightly. The crosspolarisation that is around the −28 dB level shows differences of around 2 dB, however. These results were obtained using single-precision computations without making an attempt to improve their solution beyond the recommendations of Section 5.3.5, and show the type of accuracy that one can expect. For higher accuracy levels one can modify surface segmentations and use double precisions in numerical computations. Then the use of multiple solutions can provide a measure of the computational errors which can be used to improve the accuracy further.

5.4 References

1 CLARRICOATS, P.J.B., and OLVER, A.D.: 'Corrugated horns for microwave antennas' (Peter Peregrinus, 1984), IEE Electromagnetic Waves series 18

2 HARRINGTON, R.F.: 'Time harmonic electromagnetic fields' (McGraw–Hill, New York, 1960)

3 MEDGYESI-MITSCHANG, L.N., and PUTNAM, J.M.: 'Electromagnetic scattering from axially inhomogeneous bodies of revolution', *IEEE Trans.*, 1984, **AP-32**, pp. 797–806

4 HARRINGTON, R.F.: 'Field computation by moment methods' (Krieger Publishing Co., 1982)

5 KISHK, A.A., and SHAFAI, L.: 'Different formulations for numerical solution of single and multibodies of revolution with mixed boundary conditions', *IEEE Trans.*, 1986, **AP–34**, (5), pp. 666–673

6 SHAFAI, L., KISHK, A. A., and HOHEB, H.: 'Multiple formulations for solution verification in numerical handling of antenna analysis and design problems', *IEEE Antennas Propag. Mag.*, 1991, **33**, (2), pp. 7–16

Chapter 6
Pure-mode horns

A.D. Olver

6.1 Principles of pure-mode horns

Pure-mode horns radiate from an aperture which contains a single mode and the radiation patterns are almost entirely determined by the transverse fields of the mode and not by the surrounding structure. The most common examples are pyramidal horns and conical horns. Although simple in concept, pure-mode horns do not have good electrical characteristics and for this reason their use as feeds for reflectors is largely restricted to array elements.

The aperture fields of a pure-mode horn are generated by a single mode which is the dominant mode in the waveguide. This excludes horns which use 'hybrid' modes where there is a single mode but it is composed of a hybrid combination of two other modes. Hybrid-mode horns, such as the corrugated horns, radiate much improved radiation characteristics and will be described in other chapters. The analysis of the horns in this chapter are treated by determining the internal modal fields of the horn and then using the aperture-field-prediction methods described in Chapter 3 to compute the radiation fields.

Pure-mode horns were the first type of horn to be developed and have received extensive treatment in textbooks and papers. A series of classic papers was published in the late 1930s and 1940s which described the basic theory and principles of operation. Of particular note are the papers of Barrow and Chu [1], Southworth and King [2], Chu and Barrow [3] and Chu [4]. These and other historic papers are reproduced by Love [5]. The classic books of Silver [6] and Schelkunoff and Friss [7] contain the theory and basic design information and many antenna textbooks since then have included the basic theory [8–10]. Design data have been included in handbooks [11–14].

The applications for pure-mode horns can be divided into small-aperture horns with sizes less than about 1.5 wavelengths and medium-sized apertures with sizes up to about six wavelengths. Small-size horns have radiation patterns which are very dependent on the precise geometry of the horn because the currents on the outside of the horn structure contribute to the radiation fields. A full analysis requires the use of moment-method techniques, as described in Chapter 5, but the general characteristics can be deduced from the information in this chapter. Small-size horns are used as feeds for small-diameter front-fed reflector antennas where the over-riding aim is to reduce the mass and volume to a low level. These are described in Chapter 7. Small horns are also used as elements in arrays of horns where the objective is to control the radiation pattern using array techniques and the requirement from the array element is to be small and easily integrated with neighbour elements. The main use for medium-sized horns is as separate antenna elements rather than as feeds for reflectors. This is because the relatively poor radiation patterns make them unattractive for high-performance feeds. Pyramidal horns, in particular, are widely used as standard gain horns because they can be accurately made and reliably constructed. The gain can be predicted, measured and reproduced over a period of time.

The aim of the material in this chapter is to present information which explains the operation of pure-mode horns and the chapter contains sufficient information for the reader to be able to assess the usefulness of these horns. This is done by considering first the general modal fields in cylindrical waveguides and transverse-field patterns required to give good performance. Then the conical horn is described where the radiation is determined only by the flare angle and aperture size. The pyramidal horn, which is treated next, has two orthogonal planes which act electrically independently of each other but which must be physically designed as a single horn. Arbitrary-shaped aperture horns (i.e. those that are not circular or rectangular) are discussed in Section 6.5. Some shapes of horn can be analysed by modal techniques but the arbitrary shape requires a numerical point-by-point approach using finite-element or finite-difference techniques. Finally Section 6.6 deals with wideband ridged-loaded horns where the aim is to produce a horn which will operate over a broad band of frequency.

6.2 Fields and modes in cylindrical waveguides

The transverse fields in the aperture of a pure-mode horn are generated by the dominant mode in the horn. The horn has a finite flare angle but the aperture field can be considered to be made up of the transverse fields of a waveguide of dimensions equal to that of the aperture, together with a spherical phase cap which is generated by the flare angle. The modes in the waveguide can be either transverse electric (TE) or transverse magnetic (TM).

The field components for TE modes in a cylindrical waveguide are given by [15]

$$\left.\begin{aligned} \boldsymbol{H}_t &= -\gamma\nabla_t\psi_{TE}\exp(-\gamma z) \\ \boldsymbol{H}_z &= K_c^2\psi_{TE}\exp(-\gamma z) \\ \boldsymbol{E}_t &= -Z_{TE}\hat{\boldsymbol{u}}_z\times\boldsymbol{H}_t \end{aligned}\right\} \tag{6.1}$$

where $\boldsymbol{E}_t\boldsymbol{H}_t$ are the transverse electric and magnetic fields, Z_{TE} is the wave impedance and the potential function ψ_{TE} satisfies the 2-dimensional scalar-wave equation

$$\nabla_t^2\psi_{TE}+K_c^2\psi_{TE}=0 \tag{6.2}$$

The propagation coefficient is $\gamma=\alpha+j\beta$, so for propagating modes $\gamma=j\beta$ and for evanescent (or attenuating) modes $\gamma=\alpha$. The wavenumber K_c is given by

$$K_c^2=\omega^2\mu\varepsilon+\gamma^2 \tag{6.3}$$

The boundary condition for TE modes is that $\partial\psi_{TE}/\partial n=0$ so that the tangential electric field and the normal magnetic field vanish on the boundary. This leads to a solution for the cutoff wavenumber K_c which depends on the shape and boundary of the waveguide.

The corresponding fields for TM modes are

$$\left.\begin{aligned}\boldsymbol{E}_t&=-\gamma\nabla_t\psi_{TM}\exp(-\gamma z)\\ \boldsymbol{E}_z&=K_c^2\psi_{TM}\exp(-\gamma z)\\ \boldsymbol{H}_t&=Y_{TM}\hat{\boldsymbol{u}}_z\times\boldsymbol{E}_t\end{aligned}\right\} \tag{6.4}$$

The scalar generating function ψ_{TM} satisfies the same wave equation, eqn. 6.2. The boundary condition for TM modes is $\psi_{TM}=0$. The dominant mode in a cylindrical waveguide is always a TE mode. Hence it is eqn. 6.1 which is expanded using the co-ordinate system appropriate to the geometry.

The solution to the wave equation depends on the geometry of the waveguide and the boundary conditions. This leads to determination of the wavenumber K_c and the transverse fields through eqn. 6.1. The transverse electric fields form the aperture electric fields E_a which can be inserted into the Fourier-transform integrals (eqns. 3.4–3.5) to compute the far-field radiation patterns. The radiation integral yields components in the x and y directions of $f_x(\theta, \phi)$, $f_y(\theta, \phi)$ from which the far-field components $\boldsymbol{E}_\theta$ and $\boldsymbol{E}_\phi$ are determined using

$$\left.\begin{aligned}\boldsymbol{E}_\theta&=f_x(\theta, \phi)\cos\phi+f_y(\theta, \phi)\sin\phi\\ \boldsymbol{E}_\phi&=\cos\theta\{f_y(\theta, \phi)\cos\phi-f_x(\theta, \phi)\sin\phi\}\end{aligned}\right\} \tag{6.5}$$

where the electric field is assumed to be polarised along the y direction and the Green's function $\exp(-jkR)/kR$ is implicit. The far-field copolar electric field E_p and the crosspolar electric field E_q are related to the spherical components using Ludwig's third definition of polarisation [16] by

$$\begin{bmatrix}E_p(\theta, \phi)\\ E_q(\theta, \phi)\end{bmatrix}=\begin{bmatrix}\sin\phi & \cos\phi\\ \cos\phi & \sin\phi\end{bmatrix}\begin{bmatrix}\boldsymbol{E}_\theta\\ \boldsymbol{E}_\phi\end{bmatrix} \tag{6.6}$$

Combining eqns. 6.5 and 6.6 gives

$$\begin{bmatrix}E_p(\theta, \phi)\\ E_q(\theta, \phi)\end{bmatrix}=\cos^2\left(\frac{\theta}{2}\right)\begin{bmatrix}(1-t^2\cos 2\phi) & t^2\sin 2\phi\\ t^2\sin 2\phi & (1+t^2\cos 2\phi)\end{bmatrix}\begin{bmatrix}f_y(\theta, \phi)\\ f_x(\theta, \phi)\end{bmatrix} \tag{6.7}$$

where $t=\tan(\theta/2)$. This equation yields important knowledge about the form that the aperture fields must take to produce low crosspolarisation. The crosspolar component E_q will only vanish if the terms associated with f_x and those associated with f_y cancel over all values of θ and ϕ. Since both f_x and f_y need to be nonzero for this to happen, it is evident that the transvers electric fields in the aperture must have both x-directed and y-directed components.

This means that the field lines must exhibit some degree of curvature and a purely linear aperture electric field will not radiate zero crosspolarisation. This situation arises in a rectangular waveguide where the aperture field has only a linear component; consequenty the crosspolar level is significant. The exception to the above statement is when the aperture is large, when the radiated fields are confined to a small angular region around boresight and $\tan(\theta/2)$ tends to zero. The term containing f_y is reduced to a very low level and the crosspolarisation can be made small by ensuring that $f_x = 0$, which implies that the aperture field is linearly polarised.

6.3 Conical horns

The smooth-wall conical horn (Figure 6.1), is the first horn to be considered in this chapter because it is conceptually easier to understand the advantages and disadvantages of pure-mode fields. The pyramidal horn, which is often treated first, has the additional factor of two intersecting flares which produce two phase factors. Pure-mode conical horns do not, in general, make good feed horns for reflectors because the pattern symmetry is pure and the crosspolar-radiation level is high. The exception is a small-diameter horn where low crosspolarisation is possible. Conical horns have been studied since the early

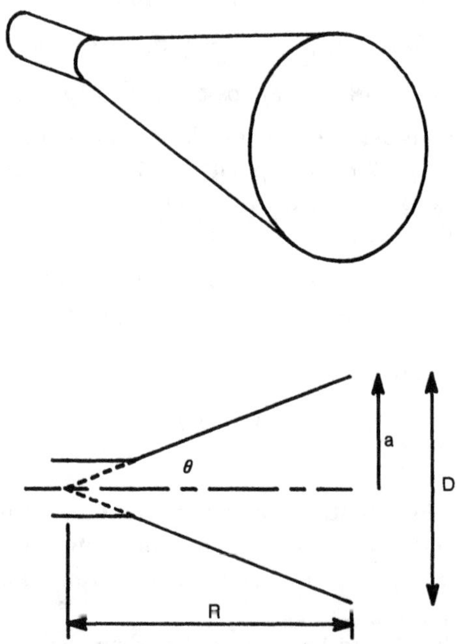

Figure 6.1 Smooth-wall conical horn

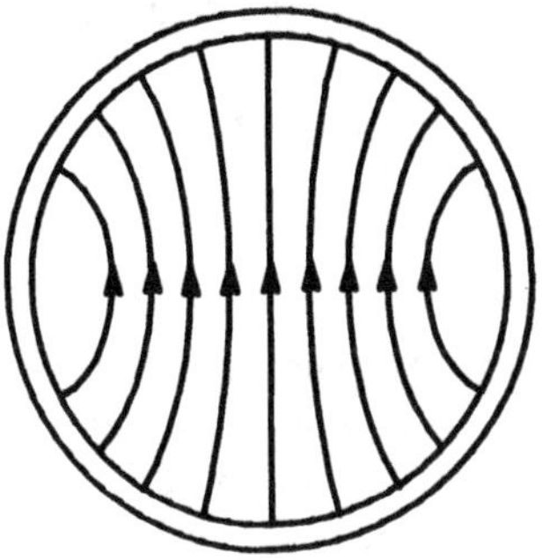

Figure 6.2 Transverse-electric-field pattern of TE_{11} mode

days of microwaves when the basic properties were understood, as described in Section 6.1.

The dominant mode in a circular waveguide or a conical horn is the TE_{11} mode which has the transverse-field pattern shown in Figure 6.2. A conical horn is usually connected to a standard rectangular waveguide by a transition unit (Figure 6.3) which transforms the TE_{11} mode in the circular waveguide into the TE_{10} mode in the rectangular waveguide. This transition is normally designed to be about two wavelengths in length and to taper gradually from the circular geometry to the rectangular geometry. This gives an acceptable bandwidth. It is possible to make a more compact transition by using a series of steps rather than a smooth taper, but the bandwidth is reduced.

The far-field-radiation patterns can be predicted by using a modified form of eqn. 3.26 which includes a factor to account for the spherical phase cap caused by the flare section:

$$\boldsymbol{F}(\theta, \phi) = C \int_0^{2\pi} \int_0^a E_a(r', \phi') \exp\{jkr' \sin\theta \cos(\phi - \phi')\} \times \exp\{-jk(r'^2/2R)\} r' \, dr' \, d\phi' \tag{6.8}$$

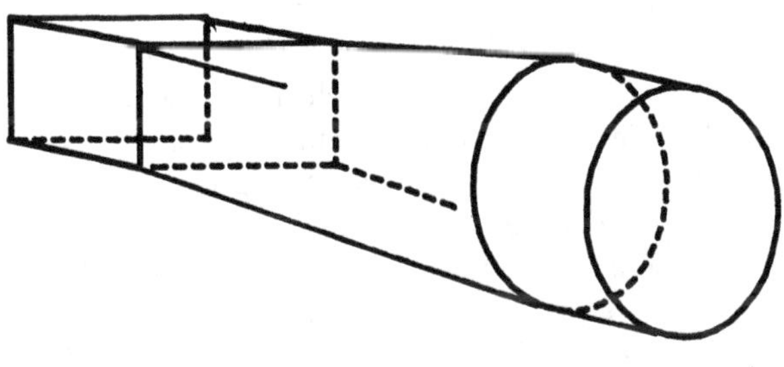

Figure 6.3 Transition from circular to rectangular waveguide

where C is a constant. The spherical phase is contained in the term $\exp\{-jk(r'^2/2R)\} = \exp(-j2\pi s)$.

The quadratic phase factor s is given by

$$s \approx \frac{a^2}{2\lambda R} \approx \frac{a\theta}{2\lambda} \tag{6.9}$$

and a is the radius, R is the axial length, and θ is the semiflare angle (Figure 6.1). This phase factor is very significant in determining the length and flare angle which can be used, as will be apparent below. The aperture electric field in eqn 6.8 is that due to a circular waveguide:

$$E_a(r', \phi') = E_r(r', \phi')\hat{\boldsymbol{u}}_r + E_\phi(r', \phi')\hat{\boldsymbol{u}}_\phi \tag{6.10}$$

where r' and ϕ' are the cylindrical co-ordinates in the aperture plane. The field components are those of the TE_{11} mode in a circular waveguide:

$$\left.\begin{aligned} E_r(r', \phi') &= \frac{j\omega\mu}{K_c^2 r'} AJ_1(K_c r') \sin\phi' \\ E_\phi(r', \phi') &= \frac{j\omega\mu}{K_c} AJ_1'(K_c r') \cos\phi' \end{aligned}\right\} \tag{6.11}$$

where A is a constant and the wavenumber K_c is given by the root of the derivative of the Bessel function $J_1'(K_c a) = 0$. For the lowest-order mode this means $K_c a = 1.8411$.

6.3.1 Radiation patterns

Eqn. 6.8 together with eqns. 6.10 and 6.11 must normally be evaluated numerically. A set of radiation patterns for a conical horn with a semiflare angle of 5° and an aperture diameter of 4λ is shown in Figure 6.4. The patterns show asymmetry between the E-plane and the H-plane and the peak crosspolar level is at -18.7 dB below the peak copolar boresight level. This is a high level and is similar to that predicted for a circular aperture in Chapter 3. In general, it precludes the use of the pure-mode conical horn as a high-performance feed and is one reason why so much effort has been devoted to researching new types of conical horn.

The peak crosspolar level is a function of the diameter of the aperture. This is shown in Figure 6.5. For small-diameter horns, there is a normalised aperture diameter where the crosspolar fields cancel, as described in Section 6.2. The curve in Figure 6.5 was obtained using the electric-field integral equations. The position of the null is dependent on the method of computation, but measurements by Adatia *et al.* [17], have confirmed this prediction method. Figure 6.5 shows that the null occurs around $D/\lambda = 1.15$.

The shape of the copolar radiation patterns is a function of the phase factor s, and it is possible to plot universal patterns independent of the aperture diameter and the flare angle. These are plotted in Figures 6.6 and 6.7 for $s = 0$, 0.1, 0.2, 0.3, 0.4 and 0.6. As s increases, the beamwidth gets broader and

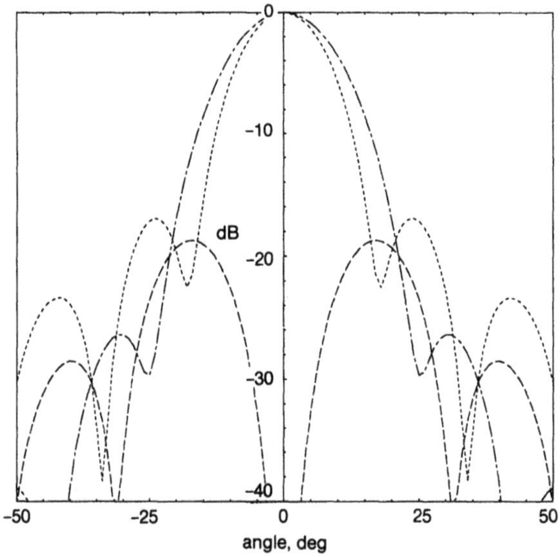

Figure 6.4 *Copolar and crosspolar radiation patterns for a conical horn with 4λ aperture diameter and 5° semiflare angle*

------ E-plane
—·—·— H-plane
———— crosspolarisation

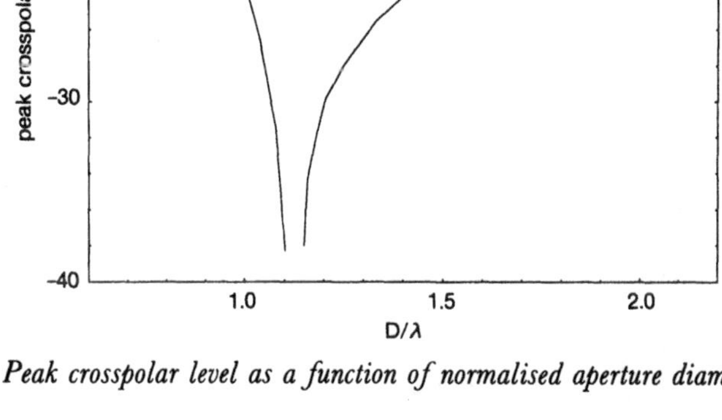

Figure 6.5 *Peak crosspolar level as a function of normalised aperture diameter*

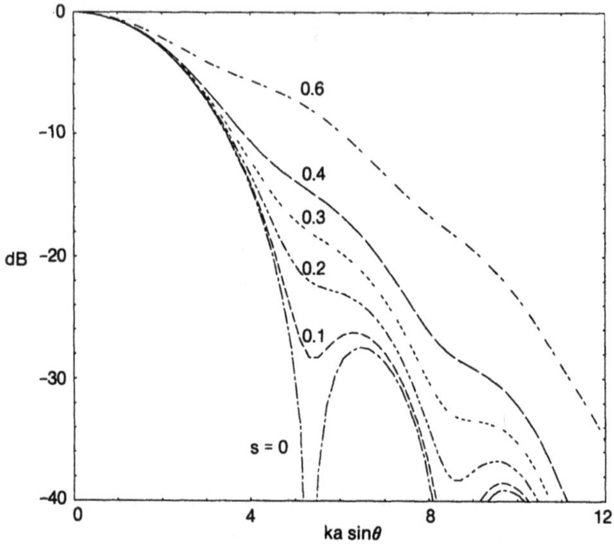

Figure 6.6 H-plane universal patterns for a conical horn

the asymmetry between the E-plane and the H-plane increases. The crosspolar pattern changes only marginally by comparison with the pattern shown in Figure 6.4. The broader beamwidth leads to a loss in gain and hence reduced aperture efficiency.

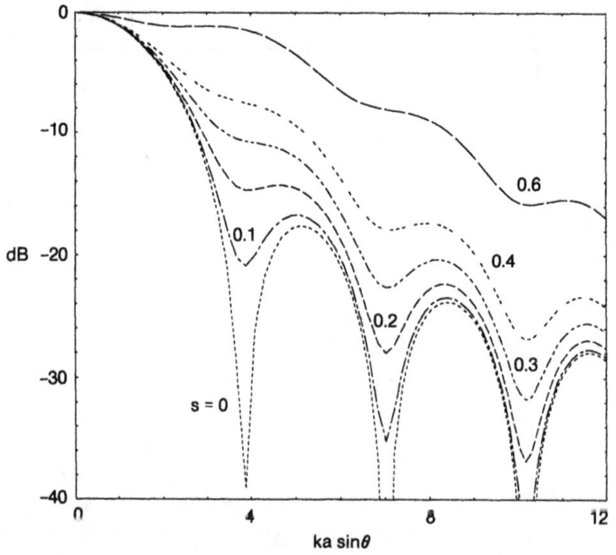

Figure 6.7 E-plane universal patterns for a conical horn

6.3.2 *Gain*

The gain, or directivity, can be computed using eqn. 3.34. It is appropriate to express the total gain as the peak gain due to a uniformly illuminated circular aperture less the gain losses. The peak gain is given by $4\pi A/\lambda^2$ where A is the aperture area. In decibels this is

$$\text{Peak gain} = 20 \times \log_{10}\left(\frac{\pi D}{\lambda}\right) \qquad (6.12)$$

where D is the diameter of the horn aperture.

The gain losses have three components:

(i) the gain loss due to the TE_{11}-mode aperture taper
(ii) the gain loss due to the phase taper across the aperture
(iii) the gain loss due to the crosspolar power.

Even for the high peak crosspolar levels of a conical horn, the crosspolar gain loss is small and can usually be neglected. The other two losses can be combined as a function of the phase factor s and are plotted in Figure 6.8. When $s=0$, the gain loss of 0.77 dB is due to the TE_{11} aperture taper. This corresponds to an aperture efficiency of 84% and is the best which can be achieved from a circular horn. As s increases, the spherical phase error becomes increasingly significant and the losses increase.

6.3.3 *Optimum designs*

The spherical phase factor links the geometric parameters of the horn. Eqn. 6.9 shows that the product of the aperture diameter and the flare angle determines the value of s. Thus high values of aperture diameter are restricted to small flare

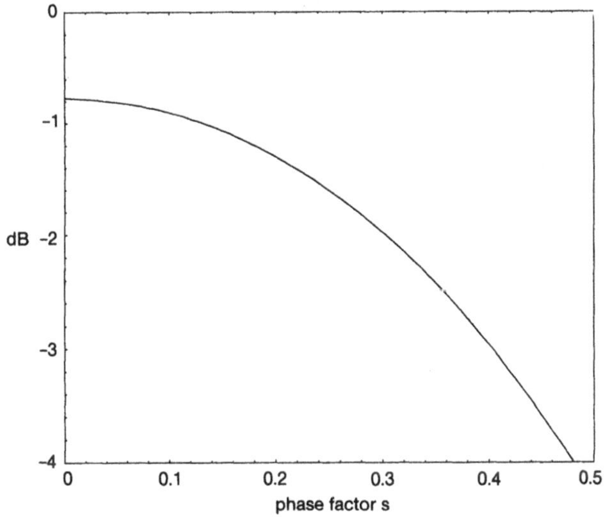

Figure 6.8 Loss in gain due to phase factor in a conical horn

angles in order to produce a given value of s, but small values of aperture diameter can have larger flare angles. The designer of a horn usually has to produce a horn with a specified copolar beamwidth so that the approximate aperture diameter is fixed. The relation between aperture diameter and flare angle is shown in Figure 6.9 for $s=0.1$, 0.2 and 0.4. Observation of the radiation patterns (Figures 6.6 and 6.7) together with the gain loss (Figure 6.8) indicate that $s<0.2$ produces a reasonable pattern with a gain loss of 1.3 dB. Larger values of s give significantly distorted patterns, particularly in the E-plane. This is an important restriction on the maximum size of horn which can be produced. Small flare angles mean physically long horns and for every frequency range there is a length which is just too great to be contemplated. A 6λ diameter horn at 10 GHz satisfying the $s=0.2$ criterion has an axial length of 22.5λ or 680 mm, which is long and bulky. This implies that there is a maximum gain which can realistically be produced by a conical-feed horn of about 25 dB.

As the phase factor increases, the gain loss through increased phase error can be partly compensated by the gain increase due to larger aperture size. There is a value of s, about 0.4, where the gain reaches a maximum. This gives the optimum gain performance, but the copolar-radiation patterns show considerable distortion and may not be suitable for some applications.

The frequency performance of the conical horns can be deduced from the data already presented in the above figures. The main beam does not change much over a reasonable range of frequencies but the gain performance depends on the relation between s and aperture diameter. In practice it is usually the input waveguide which determines the usable bandwidth. The nominal

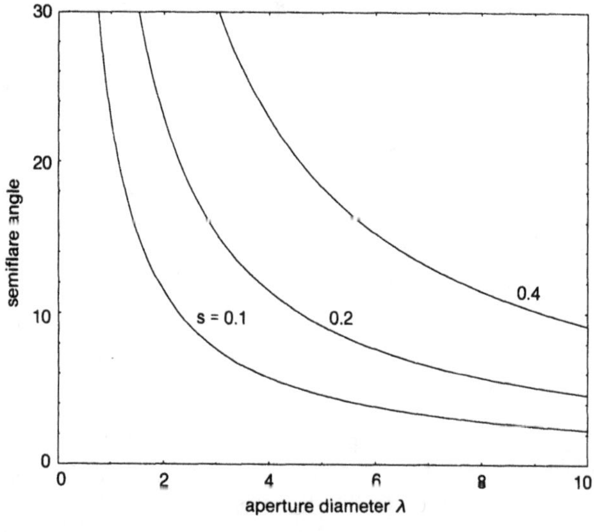

Figure 6.9 Relationship between flare angle and aperture diameter in a conical horn

bandwidth between the cutoff frequency of the TE_{11} mode and the TM_{11} cutoff frequency is 2.1:1 ($K_ca = 1.8411$ to $K_ca = 3.8311$). The practical bandwidth is less owing to the high attenuation near the TE_{11} cutoff. A bandwidth of 1.7:1 is more realistic.

Conical horns which are to be used as elements in an array feed are usually small in diameter to achieve close spacing in the array. They are relatively easy to design and can produce low crosspolarisation, as described above. The aperture diameter is usually less than about 1.5λ, which means a low-flare-angle compact horn with a low phase factor and consequently a high aperture efficiency. The patterns can be obtained from Figures 6.5–6.7.

Conical horns with large flare angles suffer from the additional problem that the junction between the input circular waveguide and the horn causes higher-order modes to be excited. The next-highest-order mode which can easily be excited at a junction between circular waveguide and a conical horn is the TM_{11} mode. The level of excitation can be predicted by using a modal-matching technique (see Chapter 4). Computations indicate that, with a semiflare angle of 5°, 1.5% of the power is converted from TE_{11} mode to the TM_{11} mode. For semiflare angles of 10° it rises to 5.7% and for a semiflare angle of 15° it has risen to 10.5%. The result of this mode conversion is a loss in gain. The shapes of the radiation patterns are only altered to a small extent, because the phase difference between the two modes is approximately 90° which causes only small changes to occur in the TE_{11} patterns, even for large flare angles. Chapter 8 will show how the addition of the TM_{11} mode in the correct phase relationship with the TE_{11} mode can cancel the TE_{11} mode crosspolarisation.

6.3.4 Phase centre

The phase centre of a horn is the point from which the radiation appears to emanate. It is the point on the axis at which spherical waves appear to be centred. The phase centre of a horn should coincide with the focus of the reflector. Usually there is no single phase centre for a horn: each plane of radiation and each set of radiation angles has its own phase centre. This is the case for pure-mode horns. However, an average value for the phase centre can be computed using formulae based on using the predicted phase pattern for the horn [18]. For the conical horn this produces the universal phase-centre curves shown in Figure 6.10, which are plotted in terms of the relative position of the phase centre as a function of the total length of the horn. For no phase factor the phase centre is at the aperture. As the phase factor increases, the phase centre moves back down the horn. Figure 6.10 shows a significant difference between the E-plane phase centre and the H-plane phase centre. For a horn with a diameter of 2λ, the phase centres in the two planes are 0.38λ apart and for a horn of diameter 10λ this rises to 1.9λ apart. This difference means that there is no single position for a conical horn to be positioned when used as a feed for a reflector. Consequently, the efficiency of a conical horn used as a feed in a reflector antenna is reduced.

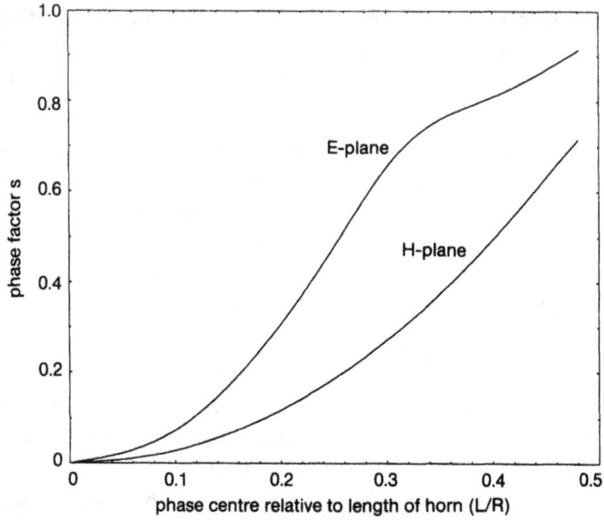

Figure 6.10 Universal phase-centre curves for conical horns

6.4 Pyramidal horns

A pyramidal horn (Figure 6.11) has a rectangular or square aperture and can be considered as being made up of two sectoral horns: an E-plane horn and an H-plane horn. The radiation pattern of the pyramidal horn is the multiplication

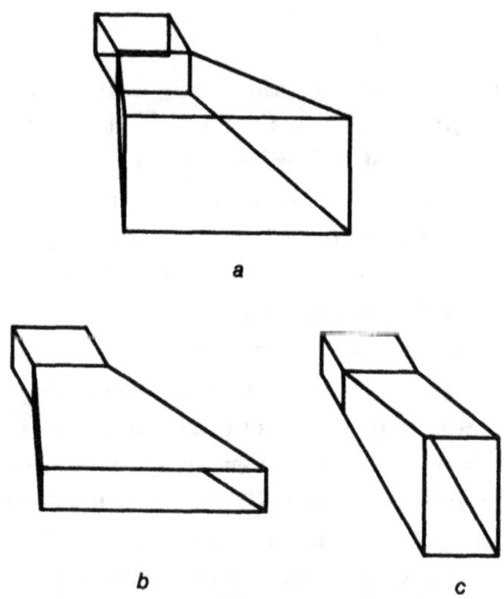

Figure 6.11 Pyramidal and sectoral horns

a Pyramidal
b H-plane sectoral horn
c E-plane sectoral horn

of the radiation patterns of the two sectoral horns. Sectoral horns are rarely used alone as radiators because they have a broad pattern in the narrow plane. A possible application is as a line feed for a cylindrical reflector, but the flare length of the horn is a disadvantage and some type of array feed is preferred.

Pyramidal horns were the first type of horn to be analysed and used. From the 1940s, antenna manufacturers have made standard pyramidal horns. This was done partly by design and partly by trial and error. Once a good horn had been designed, it was continuously reproduced in later years because it was known to produce adequate patterns. Refinements since the 1950s have been confined to better prediction techniques and interest in specific properties such as the crosspolarisation aspects of the radiation characteristics.

The mode function for TE modes in a rectangular waveguide (Figure 6.12) which must be inserted into the general equation, eqns. 6.1 and 6.2 is

$$\Psi_{TE} = \cos\left(\frac{n\pi x}{a}\right)\cos\left(\frac{m\pi y}{b}\right) \tag{6.13}$$

where n and $m = 0, 1, 2 \ldots$ but $n \neq m$ for n or $m = 0$. The wavenumber is given by

$$K_c^2 = \left(\frac{n\pi}{a}\right)^2 + \left(\frac{m\pi}{b}\right)^2 \tag{6.14}$$

so that the cutoff wavelength is

$$\lambda_c = \frac{2ab}{\{(nb)^2 + (ma)^2\}^{1/2}} \tag{6.15}$$

The fundamental mode of a rectangular waveguide is the TE_{10} mode. The next-highest-order mode depends on the ratio of a/b. Almost all standard

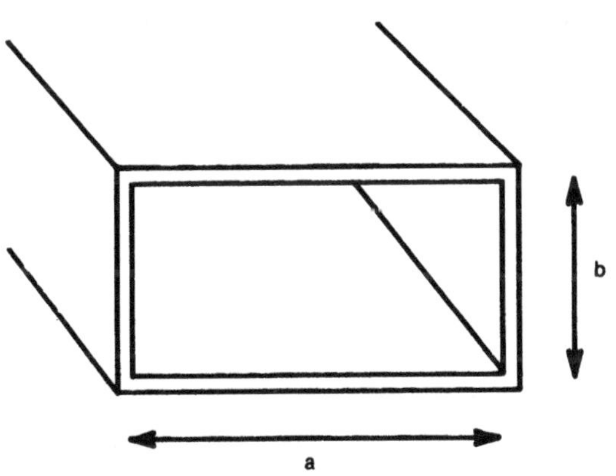

Figure 6.12 Rectangular waveguide

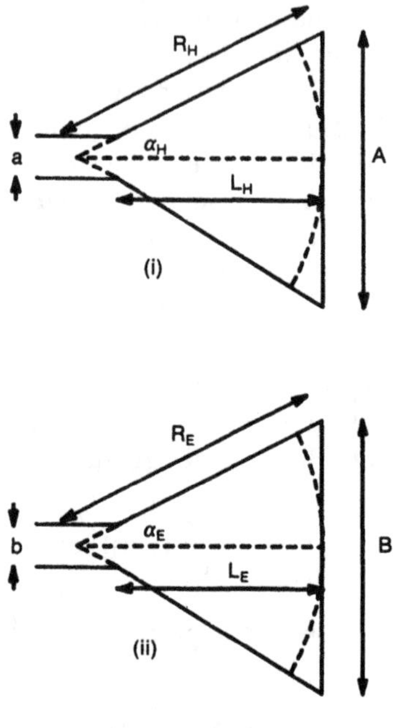

Figure 6.13 Geometry of pyramidal horn

rectangular waveguides have $a = 2b$. The exception is the common X-band waveguide WR90, which has $a = 2.25b$.

The only electric-field component of the TE_{10} mode is the E_y component. This is obtained from eqn. 6.1 as

$$E_y = E_0 \cos\left(\frac{\pi x}{a}\right) \tag{6.16}$$

This is the aperture field of an open-ended rectangular waveguide. The aperture field of a pyramidal horn also needs to account for the influence of the flare angles. For the geometry shown in Figure 6.13, where R_E, α_E and R_H, α_H are the flare lengths and flare angles in the two planes, the aperture field is

$$E_a(x', y') = E_y'(x', y') = E_0 \cos\left(\frac{\pi x}{A}\right) \exp\left[-j\left\{\frac{k}{2}\left(\frac{x'^2}{R_H} + \frac{y'^2}{R_E}\right)\right\}\right] \tag{6.17}$$

6.4.1 *Radiation patterns*

The aperture field can be inserted into the radiation equation (eqn. 3.4) to predict the radiated far fields. It is possible to solve the radiation integral analytically in terms of Fresnel integrals [9, 18]. In a similar manner to the conical horn, the influence of the spherical-phase term can be expressed in terms of two quadratic phase factors so that universal radiation patterns can be produced. The phase factors in the H-plane and E-plane are, respectively,

$$s_H = \frac{A^2}{8\lambda R_H} \qquad s_E = \frac{B^2}{8\lambda R_E} \tag{6.18}$$

The universal patterns for H-plane and E-plane sectoral horns are shown in Figures 6.14 and 6.15, respectively. These are also the patterns for the

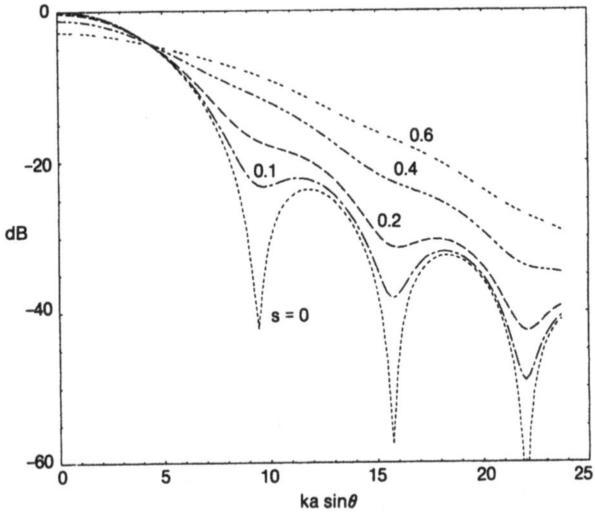

Figure 6.14 Universal patterns for an H-plane sectoral horn

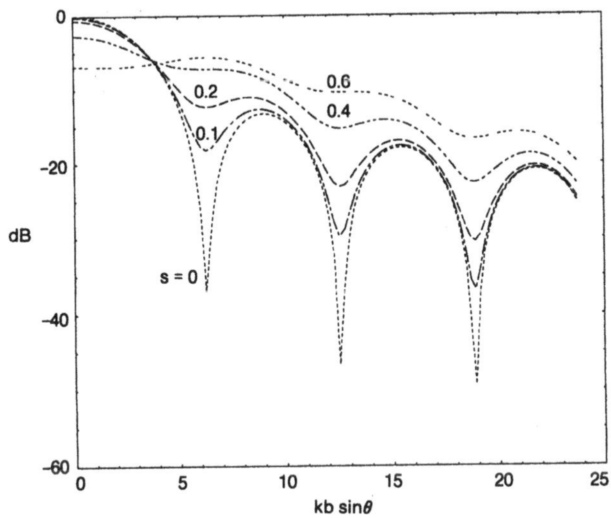

Figure 6.15 Universal patterns for an E-plane sectoral horn

pyramidal horn in the two planes. The patterns have been normalised to the no-phase-error case so that the loss in gain on boresight is evident as the phase factor increases. The distortion in the E-plane patterns is more severe than in

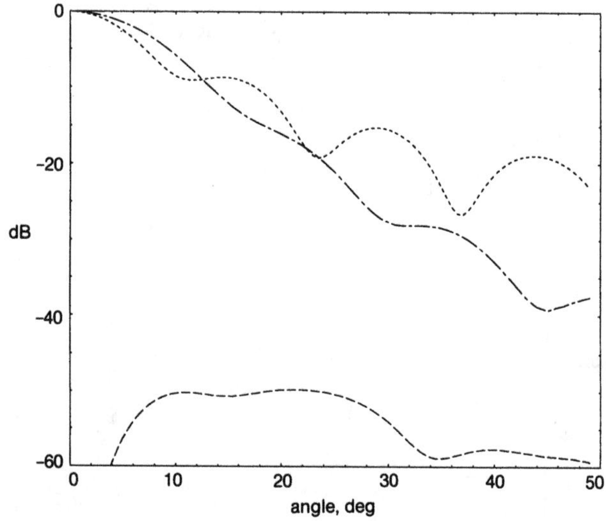

Figure 6.16 Radiation patterns for a 5λ-square-aperture horn with semiflare angle of 12°

------ E-plane
–·–·– H-plane
––– crosspolarisation

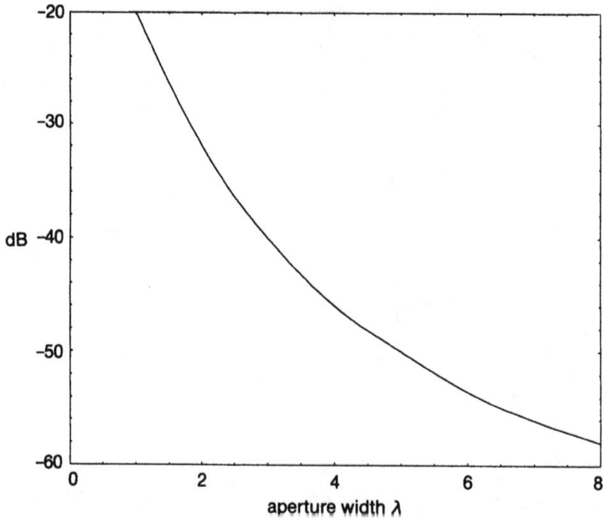

Figure 6.17 Peak crosspolar level as a function of aperture width for a square horn $s = 0.26$

the H-plane. This is particularly clear if the pattern for a horn with a 5λ-square aperture is computed, as in Figure 6.16. However, the two copolar patterns do not interact in the same way as for a conical horn so that the peak crosspolar level is not determined directly by the difference between the two orthogonal copolar patterns. The crosspolar level for this horn in Figure 6.16 is very low at about -50 dB. The peak crosspolar level decreases as the aperture size increases (Figure 6.17). Consequently, the ability to use a pyramidal horn in a given application is usually determined by the asymmetry in the copolar patterns and not by the peak crosspolar level.

6.4.2 Gain

The gain of a pyramidal horn is an important quantity because one of its major uses is as a gain standard. Gain is a measured quantity which includes the losses in the walls of the horn, and the gain of a horn to be used as a standard is always found by measurement using the 3-antenna method or the extrapolation method [19]. The gain, ignoring the losses in the horn, can be calculated by using the standard formula (eqn. 3.34). This can be written in terms of Fresnel integrals and evaluated in closed form. Expressions are given by Schelkunoff and Friss [7] and Balanis [9]. Universal gain curves for a 10λ E-plane or H-plane horn are shown in Figure 6.18. The H-plane curve, showing the normalised gain $G_H\lambda/b$ as a function of a/λ, has a higher gain than that of the E-plane horn of corresponding width. This reflects the beam broadening which

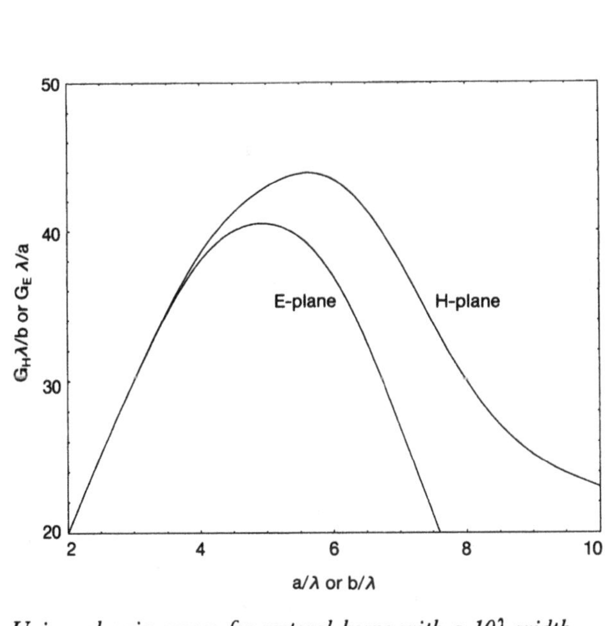

Figure 6.18 Universal gain curves for sectoral horns with a 10λ width

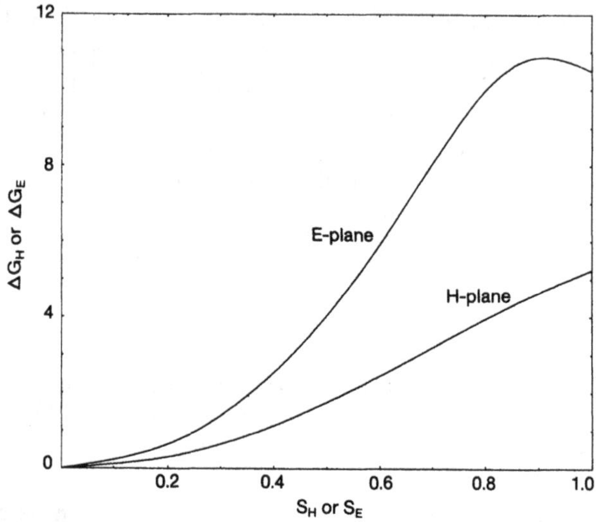

Figure 6.19 Gain-loss factors for pyramidal horns

is evident in Figure 6.15. If the normalised gain curves are plotted for other sizes of aperture, it is found that the peak gains always occur at the same values of phase factor s_E or s_H. These values are approximately

$$s_E \simeq 0.25 \qquad s_H \simeq 0.4 \tag{6.19}$$

The absolute gain can be expressed as

$$G = G_0 \Delta G_H \Delta G_E = \frac{32AB}{\pi\lambda^2} \Delta G_H \Delta G_E \tag{6.20}$$

where G_0 is the peak gain due to a TE_{10} mode in a rectangular aperture and ΔG_H and ΔG_E are the loss in gain from the peak value. These are plotted as a function of the phase factors in Figure 6.19 and, together with eqn. 6.20, can be used to find the gain for any horn.

Equation 6.20 is sometimes called the Schelkunoff gain formula. Because it is computed using physical optics, it includes the influence of the edges of the horn as well as the geometric-optics rays. A more precise analysis needs to add in the effects of higher-order edge-diffracted terms. These can arise either from multiple reflections from the edges or from reflections along the flare which are then diffracted at the edges. A number of authors have addressed these effects [20–22]. The diffracted rays cause a small ripple to appear in the gain-loss curves of Figure 6.19 which decreases as the aperture size increases, being approximately ±0.5 dB at 12 dB, ±0.2 dB at 18 dB and ±0.1 dB at 22 dB [12].

6.4.3 Design of pyramidal horns

The curves given in Figures 6.14 and 6.15 can be used to design pyramidal horns. Detailed information is given by Milligan [13], Love [12] and Balanis [9]. At first sight, it appears as if the E-plane and H-plane can be designed separately because of the orthogonality of the rectangular co-ordinates; however a practical horn must satisfy the constraint that the axial length from the input waveguide to the aperture in the two planes be equal, i.e. the lengths L_H and L_E must be equal, where (Figure 6.13)

$$L_H = \frac{(A-a)}{A}\sqrt{\left(R_H^2 - \frac{A^2}{4}\right)} \qquad L_E = \frac{(B-b)}{B}\sqrt{\left(R_E^2 - \frac{B^2}{4}\right)} \tag{6.21}$$

This inter-relates all the geometric parameters. It is rarely possible to obtain the desired patterns and gain in both planes. The E-plane pattern is more sensitive to the phase factors than the H-plane pattern, so that it is usually better to design for a chosen value of s_E (and hence R_E) and then use the results of eqn. 6.21 to give the value of R_H.

There are several possible goals for the design of a pyramidal horn. The aim can be for a horn to radiate a specified beamwidth, or a horn to radiate a specified gain, or an optimum-gain horn which attempts to produce the most efficient horn for a given size of aperture. An optimum-gain horn should have the values of the phase factors given in eqn. 6.16. According to Love [12], this will then have a gain of

$$G = 8.1 + 10\log_{10}\left(\frac{AB}{\lambda^2}\right) \text{ decibels} \tag{6.22}$$

which is equivalent to an aperture efficiency of about 50%. The full −3 dB beamwidths will be $2\sin^{-1}(0.45\,\lambda/B)$ in the E-plane and $2\sin^{-1}(0.7\,\lambda/A)$ in the H-plane.

Other parameters of interest to the design of pyramidal horns are the impedance and the phase-centre positions. The input impedance of most horns is generally well matched to the input waveguide, unless operation is near the cutoff frequency, or the flare angle is large so that the mismatch at the waveguide–horn junction is significant. Values of VSWR of 1.03 or a return loss of better than −30 dB are readily achievable.

The phase centres in the two planes behave similarly to that of a conical horn discussed in Section 6.3. The position of the phase centre moves back down the horn as the phase factor increases and the movement is greater for the E-plane than for the H-plane. This means that it is not possible to define a single position from which the radiation from the horn is deemed to emerge, and consequently it is never possible to achieve the optimum efficiency if a pyramidal horn is used as a feed for a reflector. This, together with other factors such as efficiency and pattern asymmetry, means that pyramidal horns are rarely used as feeds for reflector antennas. The one exception is for small-aperture horns where the phase factors are very small and the phase centre is effectively at the aperture.

6.5 Other shaped-aperture horns

Most pure-mode horns are either conical horns or pyramidal horns, but there are special applications which require apertures with different shapes. Sometimes the aperture shape produces desirable patterns, for instance in diagonal horns. There are also arrays of horn elements where the requirements of the packing density or the requirements of the array radiation patterns can be best satisfied by a particular shape, for instance a hexagonal-shaped horn.

The techniques for analysis of shaped-aperture horns divide into horns where modal methods can be used and horns where the shapes do not fit a co-ordinate system and numerical methods must be used. In all cases, the objective is to predict the aperture fields formed by the internal electric and magnetic fields. The near-field or far-field radiation patterns can then be computed by standard Fourier-transform techniques. Elliptical-shaped waveguides or horns can be treated by a similar method to that used for conical or pyramical horns. The wave equation (eqn. 6.2) must be solved in terms of Mathieu functions. These functions are difficult and complicated to compute so that prediction is tiresome. Partly for this reason, the pure-mode elliptical horn with a smooth wall has not been analysed or used very much. The other reason is that intuition indicates that the radiation characteristics will be intermediate between conical and pyramidal horns, neither of which have particularly useful patterns.

An alternative modal procedure is to decompose the actual aperture fields into an orthogonal modal set which can be expanded in a numerical manner so as to yield the radiation patterns. Gaussian–Hermite modes [23] have been used successfully to analyse a diagonal horn for millimetre-wave applications. Gaussian modes are an approximate solution to the wave equation, but the error reduces as the spherical phase factor decreases so that a narrow-flare-angle horn or a small horn can give good results. For the diagonal horn, the aperture fields are replaced by a summation of free-space Gaussian–Hermite modes [23]. This method is particularly useful in beam waveguide or quasioptic systems since the Gaussian modes can be reconstructed at any distance in front of the aperture. Patterns for the diagonal horn, which was first used by Love [24], are shown in Figure 6.20. The copolar patterns are symmetric, but the crosspolar level is very high with the peak level at about −16 dB. This is because the transverse electric fields do not have the curvature required to cancel the crosspolar power. Withington and Murphy [23] show that the peak crosspolar level rises as the phase taper across the aperture increases. However, for applications where the crosspolar level is not important, the diagonal horn has good copolar patterns and is relatively easy to make, particularly at millimetre wavelengths.

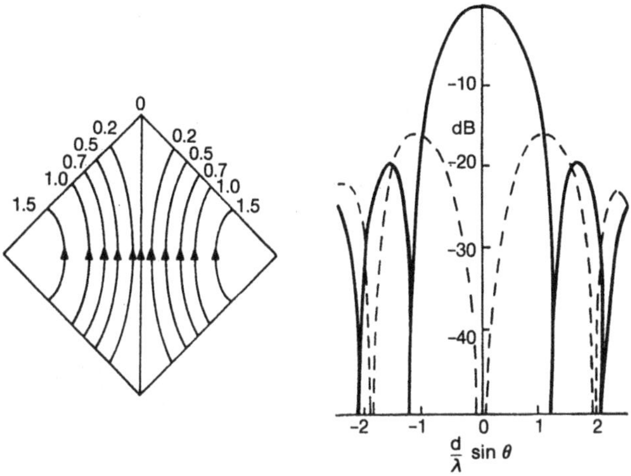

Figure 6.20 Fields and radiation patterns for a diagonal horn

If the shape of the horn aperture cannot be expanded as an orthonormal set of modes, a numerical technique must be used. This then enables an arbitrary shape of aperture to be analysed. Both finite-element [14, 25] and finite-difference [26, 27] techniques have been used. In the finite-element technique the horn aperture is divided into triangular or square subregions. The wave equation is then solved using a variational expression together with a Rayleigh–Ritz procedure. In the finite-difference method, the horn aperture is divided into squares and a linear approximation of the wave equation is applied. Both techniques yield good results, but the finite-difference technique has some advantages in a metal waveguide because the boundary conditions can be satisfied more rigorously.

The finite-difference method will now be outlined. This is based on the work of Beaubien and Wexler [28, 29]. The principle of the technique is to divide the waveguide aperture up into a square mesh (Figure 6.21), and to use a finite-

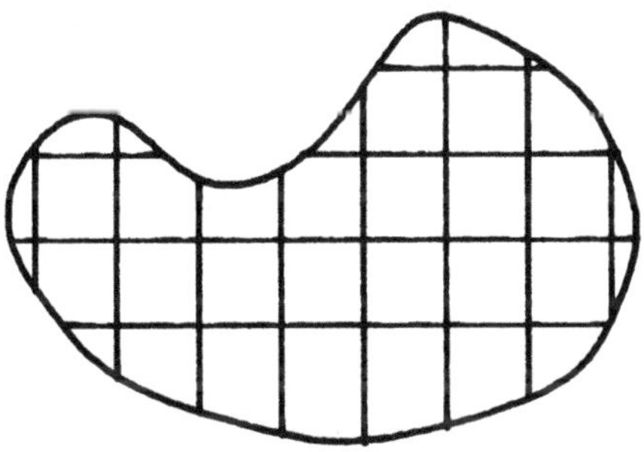

Figure 6.21 Arbitrary-shaped waveguide with square mesh

difference equation to predict the field at points over the grid. The finer the mesh, the more accurate will be the fields, so that a tradeoff is necessary between accuracy and computation time. The wave equation (eqn. 6.2) is replaced by its finite-difference equivalent assuming linear change of field between neighbouring points on the mesh. The boundary conditions are applied at the walls of the waveguide and then the matrix equation for all the mesh points is solved by an iterative method. This then gives a numerical prediction of the aperture fields.

Near the walls, the grid will have unequal-length arms, so that the general case consists of five points whose spacing is nonuniform (Figure 6.22). The arm lengths are *ah*, *bh*, *ch* and *dh* where *h* is the square grid spacing in the middle of the waveguide. The finite-difference equation predicts the field at point 0 from a knowledge of the fields at points 1, 2, 3 and 4. If the distance between the points 1 and 0 is small, the first derivative of the scalar generating function Ψ can be approximated by

$$\left(\frac{\partial \Psi}{\partial x}\right)_{1-0} = \frac{(\Psi_0 - \Psi_1)}{ah} \tag{6.23}$$

By a similar procedure, the second derivative can be approximately represented by linear terms. The 2-dimensional Lagrangian in the wave equation can then be replaced by

$$\nabla_t^2 \Psi = \frac{\partial^2 \Psi}{\partial x^2} + \frac{\partial^2 \Psi}{\partial y^2}$$
$$= \frac{2}{h^2}\left\{\frac{\Psi_3}{c(c+d)} + \frac{\Psi_1}{a(a+b)} + \frac{\Psi_2}{b(a+b)} + \frac{\Psi_4}{d(c+d)} - \left(\frac{1}{ab} + \frac{1}{cd}\right)\Psi_0\right\} \tag{6.24}$$

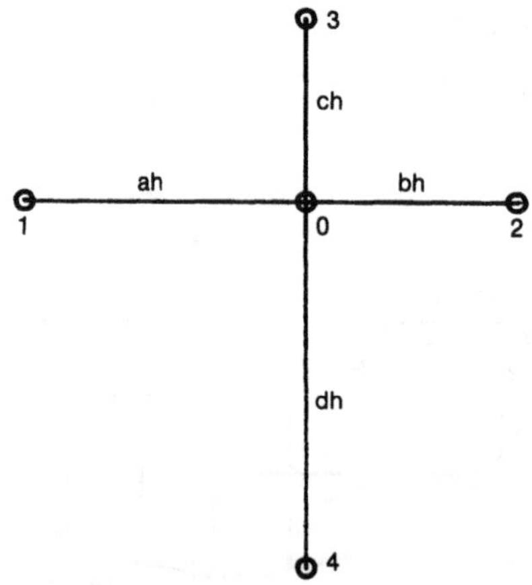

Figure 6.22 Finite-difference technique

This is then inserted into the wave equation (eqn. 6.2) to give the general 5-point finite-difference-operator equation

$$g_{03}\Psi_3 + g_{01}\Psi_1 + (g_{00} - \Lambda)\Psi_0 + g_{02}\Psi_0 + g_{04}\Psi_4 = 0 \qquad (6.25)$$

where $\Lambda = (K_c h)^2$ and g_{ij} are given by the coefficients in eqn. 6.24.

For an internal mesh point which is not adjacent to a boundary, all the arms of the operator are equal and $g_{ij} = -1 (i \neq j)$, $g_{00} = 4$, so that eqn. 6.25 becomes

$$-\Psi_3 - \Psi_1 + (4 - \Lambda)\Psi_0 - \Psi_2 - \Psi_4 = 0 \qquad (6.26)$$

At mesh points next to the boundary, the g coefficients will be modified by the boundary position and the boundary conditions. For TM modes, the condition is that $\Psi = 0$ at the perfectly conducting boundary. However, the main interest is in TE modes and for these the boundary condition is $\partial\Psi/\partial n = 0$. This may be written as

$$\frac{\partial \Psi}{\partial x} \cos\theta + \frac{\partial \Psi}{\partial y} \sin\theta = 0 \qquad (6.27)$$

where θ is the angle which the normal makes with the x axis. The differentials can be expanded as a Taylor series and the equation approximated as a finite series [30].

For a practical waveguide, the mesh produces N mesh points and hence N finite-difference equations similar to eqns. 6.25 or 6.26. These can be combined into a matrix equation

$$[\boldsymbol{G} - \Lambda \boldsymbol{I}][\boldsymbol{\Psi}] = 0 \qquad (6.28)$$

where $\boldsymbol{G}$ is an $N \times N$ square matrix of g coefficients, $\boldsymbol{I}$ is a unit matrix and $[\boldsymbol{\Psi}]$ is a column matrix containing the unknown potential values at the mesh points. Although this can be a large matrix, each row has only a maximum of five nonzero values so that sparse-matrix techniques can be used to evaluate the matrices. Eqn. 6.28 must be solved by iterative techniques to yield the true values for Ψ and the wavenumber K_c. Initial trial values for Ψ can be obtained by the Rayleigh quotient derived from the wave equation.

After the potential values and the wavenumber have been obtained, the transverse aperture fields can be computed from eqn. 6.1 and the radiation characteristics obtained by numerical integration. A starting mesh size of 13×13 has been found to be satisfactory. For most shapes of small waveguide, a final square mesh of 23×23 points is adequate because the fields of the dominant TE mode are well behaved. Larger meshes give potentially more accurate eigenvalues, but at the expense of significantly increased computer time. A typical waveguide takes between 200 and 1000 iterations to converge to a normalised accuracy of within 0.1%. Symmetry in waveguides can be utilised to reduce the computation time, but then mixed boundary conditions must be used.

The finite-difference technique can be verified by comparing the transverse fields and radiation patterns of rectangular and circular horns against the analytical results. Good agreement has been obtained between the numerical

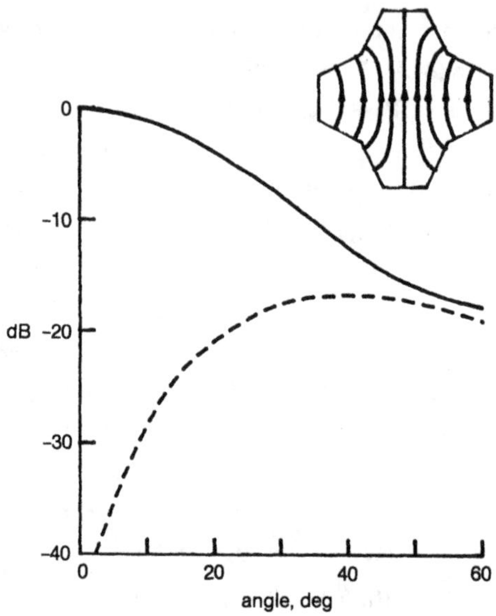

Figure 6.23 *Crossgiude and radiation pattern*

—— 45° copolar pattern
– – – 45° crosspolar pattern

and analytical cases, even for the circular horns where it is not possible to fit the boundary precisely to a square mesh [30]. An example of a horn computed using finite difference is the crossguide shown in Figure 6.23 together with its radiation pattern. This has a field pattern and radiation plots which are similar to those of a circular waveguide. The peak crosspolar level is slightly higher owing to increased field curvature. The radiation pattern from any shape of aperture can be computed using the finite-difference method. It is also possible to extend the idea and use it to synthetise the shape of a horn to produce a specified radiation pattern.

Results for 8-sided horns have been described by Rudge *et al.* [25] (Figure 6.24). Figure 6.25 shows how the peak crosspolar level is altered as the horn changes from a square horn to a diagonal, or diamond, horn. A null appears at one value of aperture size, similar to the case of a circular horn. An 8-sided horn with nearly equal sides has a particularly deep null at a cross-section width of about 1.3 wavelengths, making it suitable for use as an array element. This case has been verified experimentally [25].

6.6 Wideband ridge-loaded horns

The definition of a 'wideband' frequency horn varies according to the field of application. Here a wideband horn will be taken to mean one which spans many octaves. They are used in applications such as surveillance and radar

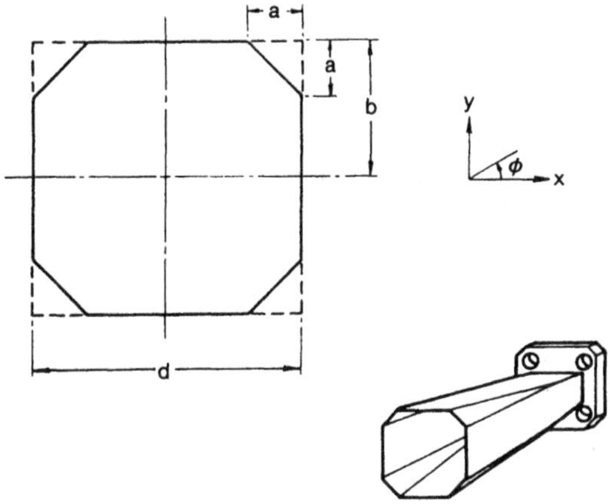

Figure 6.24 Eight-sided horn

where a single antenna is required to transmit or detect over a wide frequency range. There are two aspects to the design of a horn intended to operate over a wideband of frequencies:

(i) The horn should radiate over all the frequency range with pattern characteristics which are as constant as possible.
(ii) The horn should be capable of being fed from the input waveguide over all the frequency range with as constant an input impedance as possible.

These two aspects stem from completely different fundamental constraints. The constant radiation patterns are determined by the size of the horn aperture and

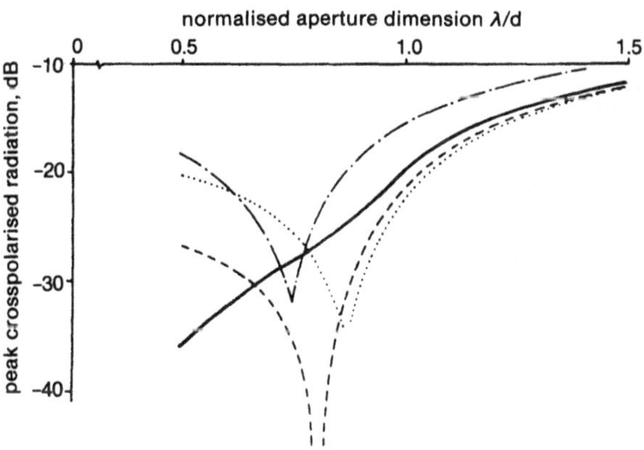

Figure 6.25 Peak crosspolar power in eight-sided horn

—— $a/b=0$ (square)
– – – $a/b=0.4$
· · · · · $a/b=0.6$
—·— $a/b=1.0$ (diamond)

the amount of spherical phase taper across the aperture. The input characteristics are determined by the ability of the throat regions of the horn to propagate signals over the specified frequency range. This latter point is the main constraint on the bandwidth because of the inherent modal properties of waveguides which are essentially bandpass filters. Something must be done to broaden this bandwidth. The most common method is to ridge-load a pyramidal horn or a conical horn, and this method will be described below.

6.6.1 Radiation patterns

The requirement to have constant radiation characteristics might be thought to be impossible owing to the inverse relationship between normalised aperture size and beamwidth. However, examination of the universal radiation patterns for a pyramidal or conical horn (Figures 6.6, 6.7, 6.14 and 6.15) shows that, as the spherical phase taper across the aperture increases, the bandwidth becomes more constant. This means that, as frequency increases, the reduction in beamwidth expected from the increase in the normalized aperture size is partly compensated by the loss in efficiency owing to the increased phase taper. This process only works over a limited bandwidth because eventually the phase taper becomes so large that the monotonically decaying radiation pattern with angle breaks up and a null appears on-axis. The extent to which a constant beamwidth is achievable depends on the type of horn. It is particularly high for a corrugated horn because of the well behaved interaction of the modes and basic symmetry in the radiation patterns. The universal patterns indicate what can be achieved for pyramidal or conical horns. The bandwidth is determined by the value of the phase factor s, which should not be too large at the highest frequency. This can be achieved in practice either with a long horn with a narrow flare angle or a shorter horn with a wider flare angle. Usually the latter is preferred, with the aperture size determined by the lowest frequency of operation, where s is smallest.

6.6.2 Ridge-loaded input section

The bandwidth of the throat region of a horn or the input waveguide of a rectangular or circular waveguide is determined by the frequency difference between the cutoff of the dominant mode (TE_{10} mode for rectangular waveguide, TE_{11} mode for circular waveguide) and the cutoff of the next-highest-order mode which will be excited by the geometry of the input section. For a rectangular waveguide this is a ratio of about 2:1 (depending on the exact ratio of the rectangle). However, it is not possible to utilise all this band because the attenuation is high near to the low-frequency cutoff and it is usual to leave a 'guard' band below the cutoff of the next-highest-order mode. A ratio of about 1.7:1 is more feasible. A method is required to extend the bandwidth and this can be done by introducing ridges into the waveguide (Figure 6.26). The principles will be explained with reference to a rectangular waveguide, but ridges can also be used in circular waveguides. The ridges reduce the cutoff frequency of the TE_{10} mode. Physically, the gap between the ridges begins to

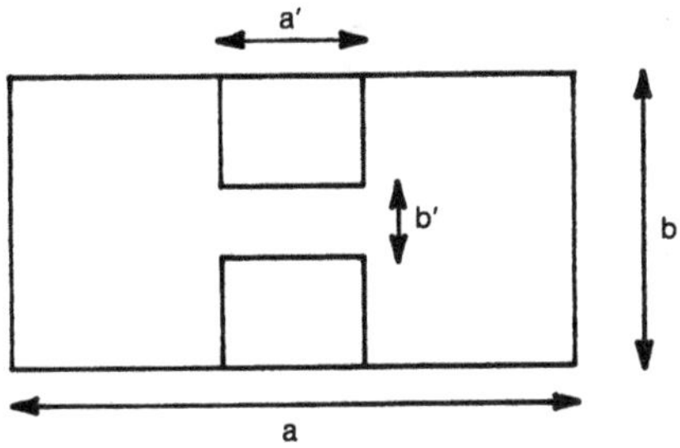

Figure 6.26 Double ridge loading in rectangular waveguide

form a parallel-plate waveguide inside the rectangular waveguide. A parallel-plate waveguide propagates a TEM mode with no low-frequency cutoff so that as the gap is narrowed the cutoff frequency of the TE_{10} mode decreases towards zero frequency.

The potential of ridge-loaded horns and waveguides was realised early in the development of microwave waveguides. The first papers appeared in the 1940s [31, 32] with papers giving design details following in the 1950s [33–35]. These used simple transmission-line theories, as described below, to predict the performance. Very few papers have been published since then, mainly because of the computational difficulty of modelling the geometry and improving on the results of the simple theory. Recently, finite-element and finite-difference techniques have become available, as discussed in Section 6.5, but since these techniques are necessarily approximate, the extra effort involved in implementing them may not be worthwhile.

The double ridge in rectangular waveguide (Figure 6.26) can be approximated by a capacitive susceptance introduced in the middle of the waveguide. The characteristic equation which determines the cutoff frequency is then given for odd TE modes by

$$\frac{b}{b'}\tan\left(\frac{\bar{a}}{2}\frac{a'}{a}\right)+\frac{B}{Y_0}-\cot\left\{\frac{\bar{a}}{2}\left(1-\frac{a'}{a}\right)\right\}=0 \tag{6.29}$$

and for even TE modes by

$$-\frac{b}{b'}\cot\left(\frac{\bar{a}}{2}\frac{a'}{a}\right)+\frac{B}{Y_0}-\cot\left\{\frac{\bar{a}}{2}\left(1-\frac{a'}{a}\right)\right\}=0 \tag{6.30}$$

The geometry is defined in Figure 6.26 and $\bar{a}=2\pi a/\lambda_c=2\pi a f_c/c$. The capacitive susceptance is given by Chen [34] as

$$\frac{B}{Y_0}=\frac{\bar{a}}{\pi}\frac{b}{a}x\left\{\frac{x^2+1}{x}\cosh^{-1}\left(\frac{1+x^2}{1-x^2}\right)-2\ln\left(\frac{4x}{1-x^2}\right)\right\} \tag{6.31}$$

where $x=b'/b$.

The cutoff curves for the first three modes are shown in Figure 6.27 for $a'/a=0.3$. As b'/b decreases so that the width of the gap between the ridges

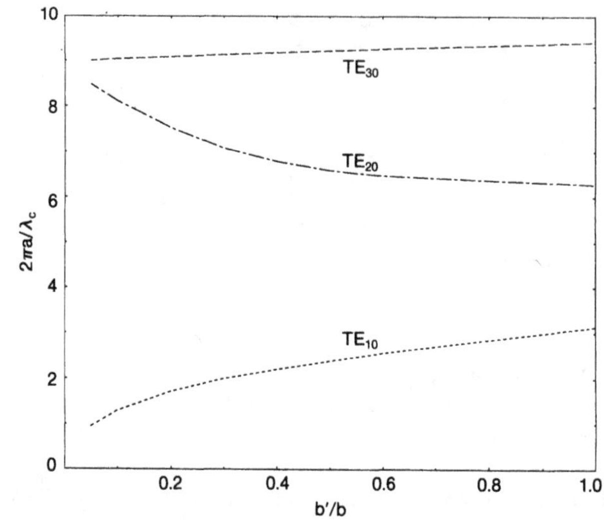

Figure 6.27 Normalised cutoff as a function of b/b' for ridge-loaded waveguide $a'/a = 0.3$

decreases, the cutoff frequency of the TE_{10} mode falls, as expected. At the same time that of the TE_{20} mode increases while that of the TE_{30} mode remains nearly static. This is to be expected since, for the TE_{30} mode, most of the power is at the edges of the waveguide and is unaffected by the ridge. The quantity of interest is the bandwidth between the TE_{10} mode and the next-highest-order mode. If the waveguide and its excitation system is symmetric, then the TE_{20} mode will not be excited so that the relevant bandwidth is that between the TE_{10} mode and the TE_{30} mode. The bandwidth normalised against the TE_{10}-mode cutoff is plotted in Figure 6.28 as a function of the ridge width. A relative bandwidth of 8 is achievable with $b'/b = 0.1$ and $a'/a = 0.5$. Above $a'/a = 0.5$, the approximate theory breaks down. This large bandwidth is in contrast to the value of 2 for a rectangular waveguide without ridges, although it will need to be reduced slightly owing to the high attenuation near the TE_{10}-mode cutoff.

An approximate expression is given for the characteristic impedance by Chen [34] as

$$Z_{0\infty} = \frac{120\pi}{\frac{2}{\bar{a}}\frac{a}{b}\frac{1}{x}\frac{B}{Y_0} + \frac{a'}{a}\frac{1}{x} + \frac{1}{2}\frac{b}{a}\left(1 - \frac{a'}{a}\right)^2} \tag{6.32}$$

This is the impedance at infinite frequency. At any other frequency

$$Z_0 = \frac{Z_{0\infty}}{\sqrt{\left\{1 - \left(\frac{f_c}{f}\right)^2\right\}}} \tag{6.33}$$

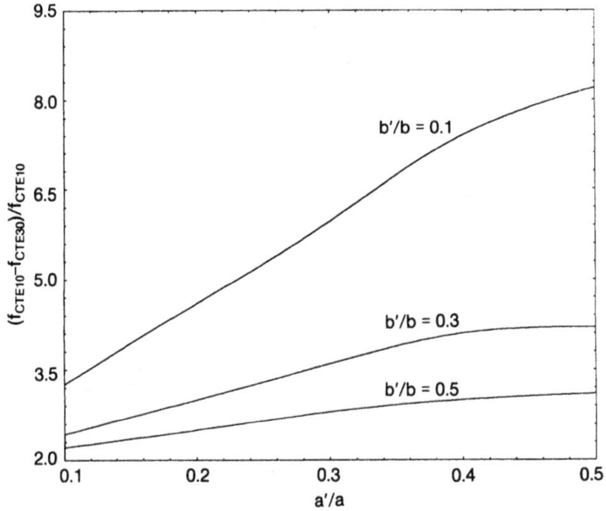

Figure 6.28 Normalised bandwidth for ridge-loaded waveguide as a function of a'/a

The characteristic impedance $Z_{0\infty}$ is plotted against b'/b for $a'/a = 0.3$ in Figure 6.29.

The impedance decreases as the gap between the ridges is reduced. Since the widest bandwidth is achieved with b'/b small, this means that there may need to be a matching network between the ridged waveguide and the coaxial-feed

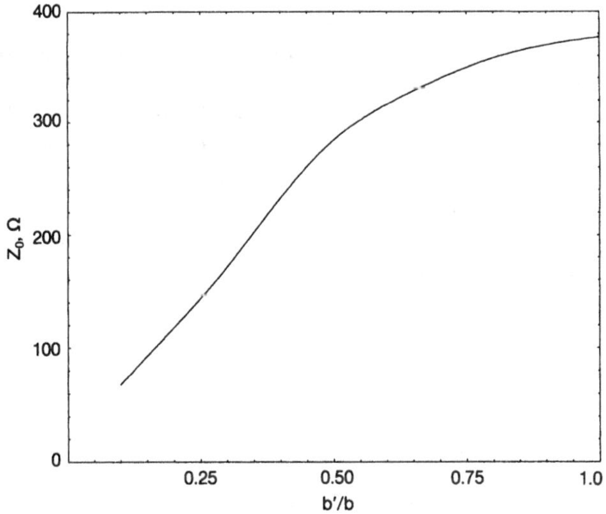

Figure 6.29 Characteristic impedance for ridge-loaded waveguide

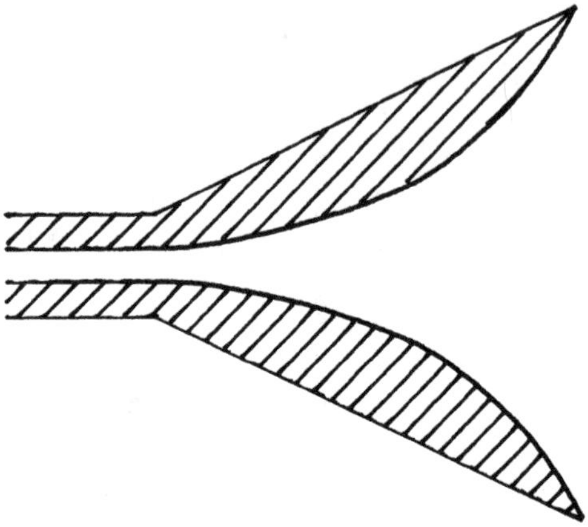

Figure 6.30 Cross-section of ridge-loaded pyramidal horn

line. The characteristic impedance is a function of frequency and may be the limiting factor on the usable bandwidth from a ridged horn. The user must be prepared to accept a poor match at some frequencies if very wide bandwidths are required.

The practical implementation of a ridge-loaded pyramidal horn is shown in Figure 6.30. The ridges start in the input waveguide with the minimum gap and then gradually taper to the aperture where they disappear. The taper is not critical but must be chosen to achieve a smooth change in impedance along the horn. Walton and Sundberg [35] suggest that an exponential taper is satisfactory, and this is known to work in waveguide tapers. The radiation pattern can be predicted to a first order by ignoring the ridges since they are not present at the aperture. In practice, there will be a small distortion to the aperture fields which will modify the radiation patterns.

The above has described the theory and implementation of a double-ridged horn. If dual polarisation, or circular polarisation, is required then a quad-ridged version in square waveguide can be built. To a first order, the double-ridge theory can be used. For more accurate analysis, a finite-difference technique needs to be used. The ridge loading can be applied to a conical horn with similar results, although in this case only a numerical analysis is feasible to study the horn.

6.7 References

1 BARROW, W.L., and CHU, L.J.: 'Theory of the electromagnetic horn', *Proc. Inst. Radio Eng.*, 1939, **27**, pp. 51–64

2 SOUTHWORTH, G.C., and KING, A.P.: 'Metal horns as directive radiators of

ultra-short waves', *Proc. Inst. Radio Eng.*, 1939, **27**, pp. 95–102
3 CHU, L.J., and BARROW, W.L.: 'Electromagnetic horn design', *Trans. Amer. Inst. Electr. Eng.*, 1939, **58**, pp. 333–338
4 CHU, L.J.: 'Calculation of the radiation properties of hollow pipes and horns', *J. Appl. Phys.*, 1940, **11**, pp. 603–610
5 LOVE, A.W.: 'Electromagnetic horn antennas' (IEEE Press, New Jersey, 1976)
6 SILVER, S. (Ed.): 'Microwave antenna theory and design' (McGraw–Hill, 1949; reprinted Peter Peregrinus, 1984), IEE Electromagnetic Waves series 19
7 SCHELKUNOFF, S.A., and FRISS, H.: 'Antenna theory and practice' (Wiley, New York, 1952)
8 STUTZMAN, W.L., and THIELE, G. A.: 'Antenna theory and design' (Wiley, New York, 1981)
9 BALANIS, C.A.: 'Antenna theory, analysis and design' (Harper and Row, New York, 1982)
10 COLLIN, R.E., and ZUCKER, F.J.: 'Antenna theory–Part 1' (McGraw–Hill, New York, 1969)
11 JAKES, W.C.: 'Horn antennas' *in* JASIK, H. (Ed.): 'Antenna engineering handbook' (McGraw–Hill, New York, 1961), Chapter 10
12 LOVE, A.W.: 'Horn antennas' *in* JOHNSON, R., and JASIK, H. (Eds.): 'Antenna engineering handbook' (McGraw–Hill, New York, 1984), Chapter 15
13 MILLIGAN, T.A.: 'Modern antenna design' (McGraw–Hill, New York, 1985)
14 LOVE, A.W., RUDGE, A.W., and OLVER, A.D.: 'Primary feed antennas' *in* RUDGE, A.W., MILNE, K., OLVER, A.D., and KNIGHT, P. (Eds.): 'The handbook of antenna design Vol 1' (Peter Peregrinus, 1982), IEE Electromagnetic Waves series 15, Chapter 4
15 COLLIN, R.E.: 'Field theory of guided waves' (McGraw–Hill, New York, 1960); 2nd ed. IEEE Press, 1991)
16 LUDWIG, A.C.: 'The definition of cross-polarisation', *IEEE Trans.*, 1973, **AP–21**, pp. 116–119
17 ADATIA, N.A., RUDGE, A.W., and PARINI, C.G.: 'Mathematical modelling of the radiation field from microwave primary feeds'. Proceedings of 7th European Microwave Conference, Cophenhagen, 1977, pp. 329–333
18 CALVO, M.: 'Primary feeds for reflectors and lenses' *in* SLETTEN, C. J. (Ed.): 'Reflector and lens antennas' (Artech House, USA, 1988)
19 IEEE Standard Test Procedures for Antennas (IEEE, 1979)
20 JULL, E.V.: 'On the behaviour of electromagnetic horns', *Proc IEEE*, 1968, **56**, pp. 106–108
21 JULL, E.V.: 'Errors in the predicted gain of electromagnetic horns', *IEEE Trans.*, 1973, **AP–21**, pp. 25–31
22 MENTZER, C.A., PETERS, L., and RUDDOCK, R.C.: 'Slope diffraction and its application to horns', *IEEE Trans.*, 1975, **AP–23**, pp. 153–159
23 WITHINGTON, S., and MURPHY, J.A.: 'Analysis of diagonal horns through Gaussian-Hermite modes', *IEEE Trans.*, 1992, **AP–40**, pp. 198–206
24 LOVE, A.W.: 'The diagonal horn antenna', *Microwave J.*, 1962, **5**, pp. 117–122
25 RUDGE, A.W., PHILIPPOU, G.Y., and WILLIAMS, N.: 'Low cross-polar waveguide horns for multiple-feed reflector antennas'. Proceedings of international conference on Antennas and Propagation, London, 1978, pp. 360–363
26 OLVER, A.D., and SANYAL, S.: 'Prediction of radiation from arbitrary cross-section horns using a finite difference technique'. Proceedings of international symposium on Antennas and Propagation, Sapporo, Japan, 1992, pp. 353–356
27 NG, F.L.: 'Tabulation of methods for the numerical solution of the hollow waveguide problem', *IEEE Trans.*, 1974, **MTT–22**, pp. 322–329
28 BEAUBIEN, M.J., and WEXLER, A.: 'An accurate finite difference method for higher order waveguide modes', *IEEE Trans.*, 1968, **MTT–16**, pp. 1007–1017
29 BEAUBIEN, M.J., and WEXLER, A.: 'Unequal arm finite difference operators in the positive definite successive over-relaxation algorithm', *IEEE Trans.*, 1970, **MTT–18**, pp. 1132–1149
30 SANYAL, S., and OLVER, A.D.: 'Propagation and radiation characteristics of arbitrary cross-section waveguide transitions and radiators'. Internal report, Queen Mary and Westfield College, University of London, May 1992

31 COHN, S.B.: 'Properties of ridged waveguide', *Proc. Inst. Radio Eng.*, 1947, **35**, pp. 783–788
32 MARCUVITZ, N.: 'Waveguide handbook' (McGraw–Hill, 1947; Peter Peregrinus, 1986), IEE Electromagnetic Waves series 21
33 HOPFER, W.: 'The design of ridged waveguides', *IRE Trans.*, 1955, **MTT–3**, pp. 20–29
34 CHEN, T-S.: 'Calculation of the parameters of ridge waveguides', *IRE Trans.*, 1957, **MTT–5**, pp. 12–17
35 WALTON, K.L., and SUNDBERG, V.C.: 'Broadband ridged horn design', *Microwave J.*, 1964, **7**, pp. 66–101

Chapter 7

Prime-focus waveguide feeds

L. Shafai and A.A. Kishk

7.1 Introduction

Prime-focus waveguide feeds have seen several stages of development. The early units were simple open-ended waveguides, with small flares to control the radiation patterns [1]. Their performance in reflector efficiency was, however, low. The need to improve the aperture efficiency and polarisation characteristics resulted in the development of corrugated feeds [2, 3]. With a small size and wide flare angles, these horns are designed for an excellent pattern symmetry, low crosspolarisation and low spillover. However, their geometry is complex and difficult to fabricate. These problems stimulated further work for developing other feeds with simpler configurations, resulting in the design of choked-waveguide feeds and dielectric-rod feeds. They are also complex in shape, and the initial development was done experimentally, but both can be investigated by the numerical method provided in Chapter 5. The performance of choked-waveguide feeds will be discussed in this chapter, but the dielectric-rod feeds will be investigated in Chapter 10.

The analysis of reflector focal-region fields by Minnett and Thomas [4] showed that it is a superposition of cylindrical hybrid modes. This led to the development of hybrid-mode feeds using corrugated waveguides and horns [5, 6]. The latter was independently designed by Kay [7]. The focal-region field of paraboloid reflectors can also be realised by coaxial-ring aperture fields, and led to the development of a coaxial feed by Koch [8]. It consisted of a circular waveguide surrounded by multiple coaxial rings, which were individually excited by multiple slots from the waveguide. The individual excitation of coaxial rings made the feed a flexible one, but its geometry was complex to fabricate. The rings, however, can also be excited by the aperture coupling through the open end of the waveguide. In this design the excitation efficiency is not high, but it results in a simpler configuration. Also, the size and location of the rings with respect to the waveguide aperture can be used as design parameters to control the feed-radiation patterns. A feed using this approach was originally designed by Wohlleben *et al.* [9] where, by recessing the corrugated flange behind the waveguide aperture, wide-angle radiation patterns were obtained. Broadening of the radiation pattern by E-plane slots on the waveguide wall was later studied by Shafai [10]. Numerous other investigators have studied these feeds and developed high-performance units [11–13]. The early investigations were primarily based on experimental studies and often did not result in optimised configurations. The development of numerical methods, as described in Chapter 5, simplified the analysis of these feeds and helped in understanding their radiation properties. Since the size of most prime-focus feeds is small, the use of numerical method does not pose computation or memory limitations. Also, as progress in computer technology provides higher-power computers at lower cost, feed design and optimisation become increasingly computer based.

In this chapter, we discuss performance data for small prime-focus feeds which are fed by a circular waveguide. The intention here is not to provide optimum feed designs, but to show adequate data on the effect of different geometrical parameters so that their role and significance can be appreciated. The data presented will concentrate on the feed-radiation patterns and their related parameters such as the phase-centre location or uniqueness, the polarisation characteristics and the reflector gain factor. We have not attempted to survey all feed designs and development; interested readers will find a comprehensive survey in Reference 1.

7.2 Phase-centre determination

In early experimental studies, the amplitude and phase patterns of the feed were usually determined and the phase-centre location was not investigated in detail. The phase-centre location should remain reasonably constant to minimise the phase variation over the radiation sphere of the feed. Experimental work in determining the phase-centre locations for principal planes is tedious and a costly job. However, if the feed-radiation patterns over a radiation sphere are known in both phase and amplitude, the phase-centre location can be determined by a numerical search. This is a useful exercise, since knowledge of the phase-centre location helps in proper focusing of the reflector and an accurate determination of the crosspolarisation.

In addition, in symmetric reflectors, to optimise the gain performance one needs to equalise the E- and H-plane patterns of the feed to provide a symmetric aperture-field distribution which, in turn, minimises the feed and the reflector crosspolarisations. However, equalising the E- and H-plane patterns, i.e. obtaining a symmetric feed-amplitude pattern, is not sufficient to optimise the reflector performance. The aperture phase distribution affects the system performance by a greater extent than the aperture amplitude distribution. Thus, an optimum feed must provide symmetric field patterns for both the field amplitude and phase. This requirement is particularly true in prime-focus reflectors where, neglecting the aperture blockage, the feed dictates the reflector performance. These reflectors require wide-angle feed patterns, which are obtained from circular-waveguide feeds with a wide angle or a 90° corrugated flange. In the design of these feeds, one can equalise the E- and H-plane amplitude patterns by variation of the corrugated-flange geometry and its separation distance from the open end of the waveguide. Such a design process, however, usually causes a separation of the principal plane-phase centres and thus the feed performance deteriorates. Since the phase centre of the feed and its uniqueness strongly affect the reflector-aperture-phase distribution, a knowledge of its dependence on the feed geometry helps in designing high-performance feeds. What is needed is a useful definition for the phase-centre location.

Many definitions are available for the phase centre of a feed. In early studies it was defined as a point halfway between the phase centres of two principal planes [14]. The most useful location for the phase centre was, however, defined by Kildal as a point which maximises the reflector gain [15]. This definition is quite general, but the expressions provided by Kildal enabled one to locate the phase centre within a small distance of a trial point that causes small aperture-phase errors. For wide-angle feeds, estimating the approximate location of the phase centre is generally difficult. The method was generalised by Rao and Shafai [16], for determining the phase centres using any arbitrary initial point. The analysis involves an iteration process, using the Newton–Raphson method, to search for the location of the phase centre from a selected initial point.

For determining the phase centre of primary feeds, three different definitions are used. The first definition is that of Kildal and is determined, using the search method of Davies, Swann and Campey [17], by maximising the reflector phase efficiency. The second definition is due to Rusch and Potter [18], and defines the phase centre as the origin of a sphere with a best nearly constant phase variation in a least-squares sense. Here, a somewhat modified version of this definition is used to determine the phase centre by minimising the phase deviations of the copolar pattern in the $\phi=45°$ plane, for a given angular region. The third definition is simply the average between the principal-plane phase centres, determined by the second definition. Here, we provide the formulation.

The far field of a linearly polarised feed can be expressed in the form (y-polarised)

$$\boldsymbol{E}(R, \theta, \phi)=\{\boldsymbol{A}(\theta) \sin \phi \hat{\boldsymbol{u}}_\theta + C(\theta) \cos \phi \hat{\boldsymbol{u}}_\phi\} \times \frac{\exp(-jkR)}{kR} \tag{7.1}$$

where $\hat{\boldsymbol{u}}_\theta$ and $\hat{\boldsymbol{u}}_\phi$ are unit vectors in the directions of increasing polar angle θ and the azimuth angle ϕ, respectively, the functions $\boldsymbol{A}(\theta)$ and $\boldsymbol{C}(\theta)$ are the principal E- and H-plane patterns of the feed and k is the wavenumber (Figure 7.1). With respect to a co-ordinate origin O$'$, $\boldsymbol{A}(\theta)$, $\boldsymbol{C}(\theta)$ and their sum can be expressed as

$$\boldsymbol{A}(\theta)=|\boldsymbol{A}(\theta)| \exp\{j\Phi_E(\theta)\} \tag{7.2}$$

$$\boldsymbol{C}(\theta)=|\boldsymbol{C}(\theta)| \exp\{j\Phi_H(\theta)\} \tag{7.3}$$

$$\boldsymbol{A}(\theta)+\boldsymbol{C}(\theta)=|\boldsymbol{A}(\theta)+\boldsymbol{C}(\theta)|\exp\{j\Phi(\theta)\} \tag{7.4}$$

where Φ_E and Φ_H are the phase angles of the E- and H-plane patterns, with respect to the origin O′, and Φ is that of their summation. Now, if the phase centre is located at the origin O at the z axis, a distance d from O′, the above functions can be written, with respect to O, as

$$\boldsymbol{A}_d(\theta)=|\boldsymbol{A}(\theta)|\exp\{j\Phi_E(\theta)-kd\cos\theta\} \tag{7.5}$$

$$\boldsymbol{C}_d(\theta)=|\boldsymbol{C}(\theta)|\exp\{j\Phi_H(\theta)-kd\cos\theta\} \tag{7.6}$$

$$\boldsymbol{A}_d(\theta)+\boldsymbol{C}_d(\theta)=|\boldsymbol{A}(\theta)+\boldsymbol{C}(\theta)|\exp\{j\Phi_d(\theta)\} \tag{7.7}$$

with

$$\Phi_d(\theta)=\Phi(\theta)-kd\cos\theta \tag{7.8}$$

For paraboloidal reflectors, the feed must be located at its focus, which means the feed phase centre must coincide with the reflector focal point. According to Kildal's definition, the phase centre of the feed is a point that maximises the reflector phase efficiency. If the reflector aperture angle is ψ_o, the phase efficiency can be defined (with respect to the phase-centre location d) as

$$\eta_p(\psi_0, d)=\frac{\left|\int_0^{\psi_0}\{\boldsymbol{A}_d(\theta)+\boldsymbol{C}_d(\theta)\}\tan(\theta/2)\,\mathrm{d}\theta\right|^2}{\left\{\int_0^{\psi_0}\{|\boldsymbol{A}(\theta)+\boldsymbol{C}(\theta)|\}\tan(\theta/2)\,\mathrm{d}\theta\right\}^2} \tag{7.9}$$

η_p is maximum when Φ_d is constant. In practice, however, Φ depends on d and is never constant. Generally, since the denominator of eqn. 7.9 is independent of d, maximising its numerator maximises η_p. The process provides a location d for the phase centre of the feed, which depends on the reflector aperture angle ψ_0. Denoting this phase-centre location as d_c, the reflector gain factor, excluding the blockage effects, is given by

$$\eta_g=\cot^2\left(\frac{\psi_0}{2}\right)\times\frac{\left[\int_0^{\psi_0}|\boldsymbol{A}(\theta)+\boldsymbol{C}(\theta)|\exp\{j\Phi_{dc}(\theta)\}\tan(\theta/2)\,\mathrm{d}\theta\right]^2}{\int_0^{\psi_0}[|\boldsymbol{A}(\theta)|^2+|\boldsymbol{C}(\theta)|^2]\sin\theta\,\mathrm{d}\theta} \tag{7.10}$$

This equation shows clearly that the gain factor of the reflector depends on the variation of the feed phase over the aperture, and reduces with its increase. Aside from the deterioration of the gain factor, the third definition of Ludwig [19] also shows that the crosspolarisation will increase owing to the nonuniqueness of feed phase centre. According to Reference 19 the crosspolarisation is a maximum in the 45°-plane and is due to $\boldsymbol{A}(\theta)$–$\boldsymbol{C}(\theta)$, the difference of the principal-plane patterns, in complex form.

According to the second definition, the phase centre minimises the phase variation of the copolar pattern in the 45° plane over the reflector aperture angle ψ_0. The objective function is

$$\Delta\Phi(d) = \Phi_{d\,max}(\theta) - \Phi_{d\,min}(\theta) \tag{7.11}$$

where $\Phi_d(\theta)$ is defined in eqn. 7.8. The optimisation of eqn. 7.11 is independent of the reflector, except that it is needed only over the reflector-aperture angle ψ_0. The method gives a value for d, the location of the phase centre, which we denote as d_s.

The third definition utilises eqn. 7.11, for each principal plane, to determine their phase-centre locations d_E and d_H. The average phase-centre location d_{av} is obtained from

$$d_{av} = (d_E + d_H)/2 \tag{7.12}$$

In the following sections we provide the design and performance data for different feeds with increasing degrees of complexity. We start with the circular waveguide. Although it is not an efficient feed for a reflector antenna, the information provided can help in understanding its radiation properties and the degree of improvements due to other feed configurations.

7.3 Circular waveguide feeds

All the configurations discussed are analysed using the method described in Chapter 5. For a waveguide, the geometry of Figure 7.1 was selected and a single electric dipole was used to simulate the excitation. This form of excitation is not a practical one, but provided that the waveguide diameter is small, to attenuate the higher-order modes, the aperture illumination is due to the TE_{11} mode alone. Adding one peripheral choke to the waveguide wall provides one stage of improvement, and its geometry is shown in Figure 7.2.

One of the important parameters of a waveguide is its radius. Thus, the beamwidth, crosspolarisation and backlobe levels of the radiation patterns for

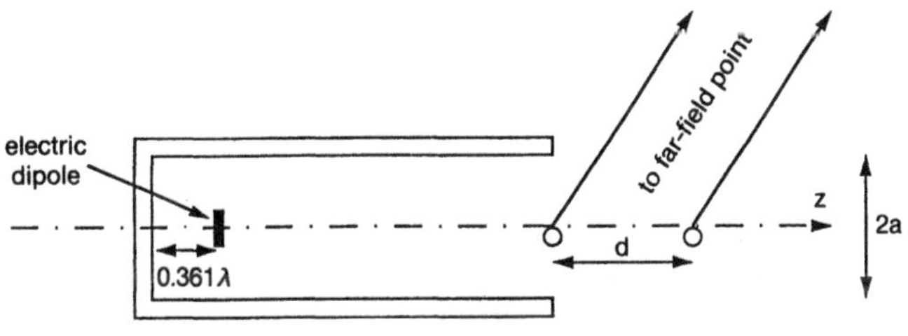

Figure 7.1 Circular waveguide feed

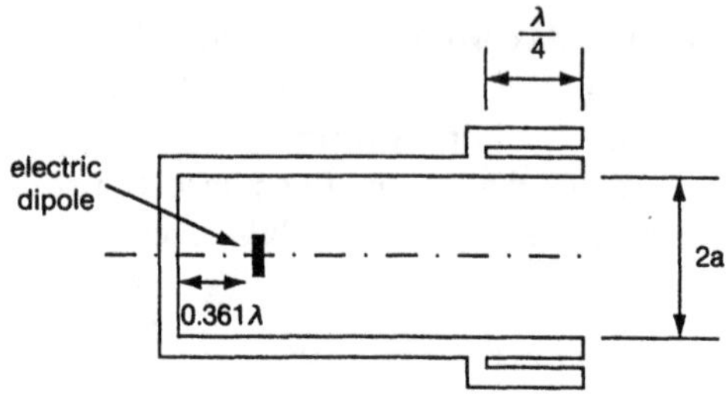

Figure 7.2 *Geometry of circular waveguide with a quarter-wavelength choke*

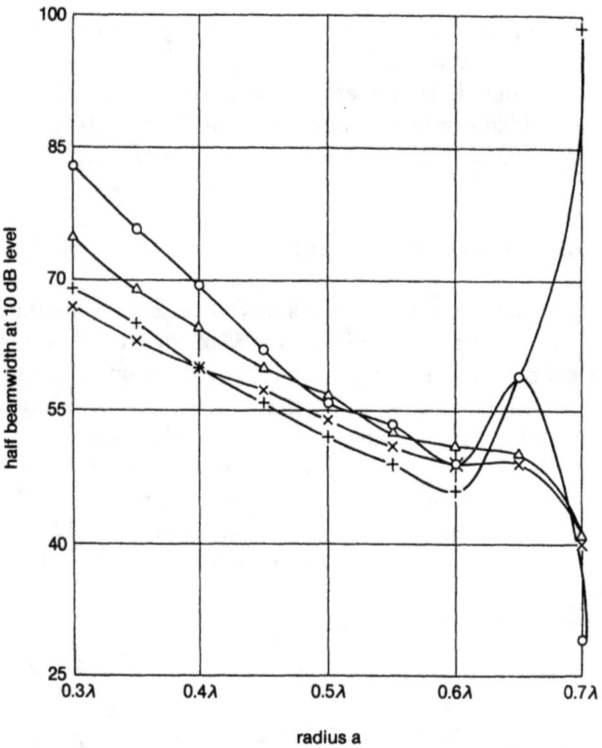

Figure 7.3 *10 dB beam angles for a circular waveguide*

△ H-plane without a choke
○ E-plane without a choke
+ H-plane with a choke
× E-plane with a choke

different radii are determined [20]. The variation of the 10 dB beam angle for copolar patterns is shown in Figure 7.3.

It is commonly known that, for a circular waveguide, the beamwidths at principal planes are different, with the E-plane providing a wider beamwidth. However, Figure 7.3 shows that, even for a circular waveguide, diameters exist which provide equal E- and H-plane beamwidths, which occur between $a = 0.47\ \lambda$ and $0.6\ \lambda$. A sample of copolar and crosspolar patterns at $a = 0.5\ \lambda$ is shown in Figure 7.4, which indicates a good copolar pattern symmetry and a reasonably small peak crosspolarisation. This also indicates a nearly coincidental phase centre of the principal planes. Figure 7.3 also shows that introducing a quarter wavelength choke on the waveguide wall nearly equalises the beamwidths of the principal planes. Figure 7.5 shows a set of computed patterns for a waveguide with a choke, and indicates a negligibly small crosspolarisation.

The complete range of peak crosspolar radiation is shown in Figure 7.6, which compares the results for waveguides with and without a choke. The effect of a single choke on the crosspolarisation is clearly evident. This figure also shows that, for a waveguide without a choke, the crosspolarisation decreases until $a = 0.6\ \lambda$, where the crosspolarisation is a minimum and increases with a further increase of the waveguide size.

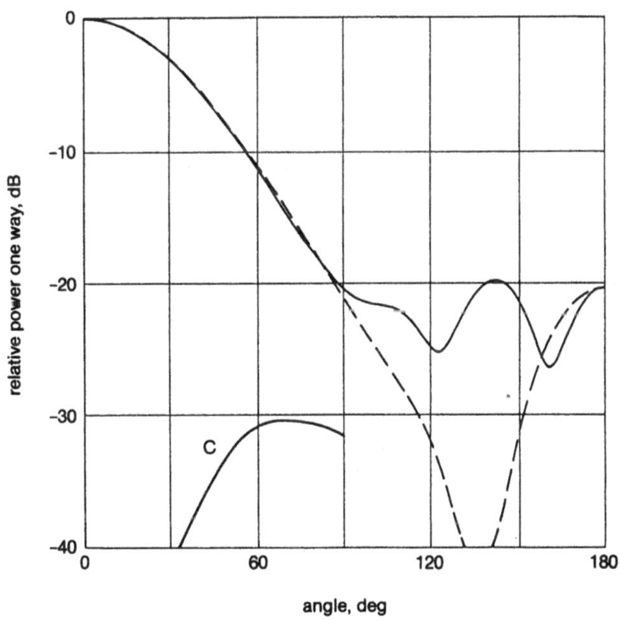

Figure 7.4 Copolar and crosspolar patterns of a waveguide without a choke

$a = 0.5\ \lambda$
——— E-plane
– – – – H-plane
—C— crosspolar

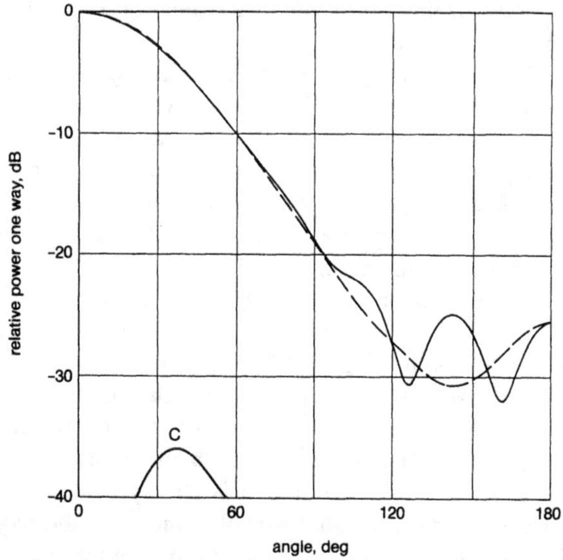

Figure 7.5 *Copolar and crosspolar patterns of a waveguide with a choke*

$a = 0.4\,\lambda$; ——— E-plane; - - - - H-plane; —C— crosspolar

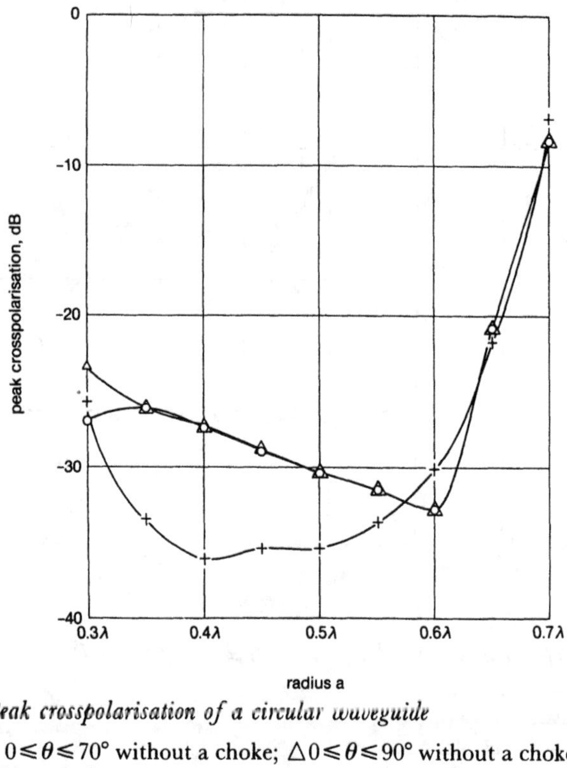

Figure 7.6 *Peak crosspolarisation of a circular waveguide*

○ $0 \leqslant \theta \leqslant 70°$ without a choke; △$0 \leqslant \theta \leqslant 90°$ without a choke; + $0 \leqslant \theta \leqslant 90°$ with a choke

The effect of a quarter-wavelength choke on the backradiation level is shown in Figure 7.7. The improvement is about 10 dB, which can be generalised for most feeds using chokes to reduce backradiation. As an aid in designing feeds for reflector antennas, the gain factor and the spillover efficiency of a paraboloid reflector using a circular waveguide are also computed. The results are shown in Table 7.1. The introduction of the choke improves both gain factor and the spillover efficiencies.

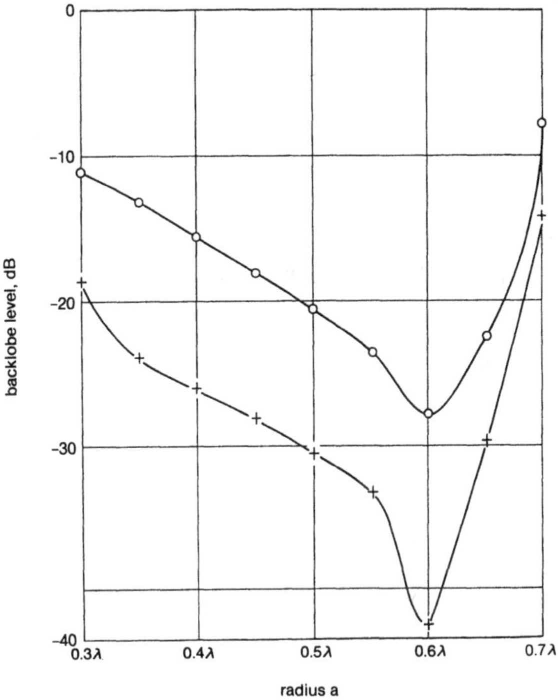

Figure 7.7 *Backradiation level of a circular waveguide*

○ without a choke
+ with a choke

Table 7.1 *Gain factors and spillover efficiencies of circular waveguides*

	Gain factor		Spillover		f/D	
a/λ	Without choke	With choke	Without choke	With choke	Without choke	With choke
	(%)	(%)	(%)	(%)		
0.30	68.72	72.78	81.95	85.10	0.33	0.40
0.35	70.07	74.12	82.42	86.42	0.37	0.42
0.40	71.78	75.04	84.47	86.97	0.40	0.45
0.45	73.54	75.84	84.80	88.47	0.45	0.47
0.50	75.15	76.65	86.07	88.17	0.49	0.51
0.55	76.58	77.35	88.48	89.15	0.51	0.54

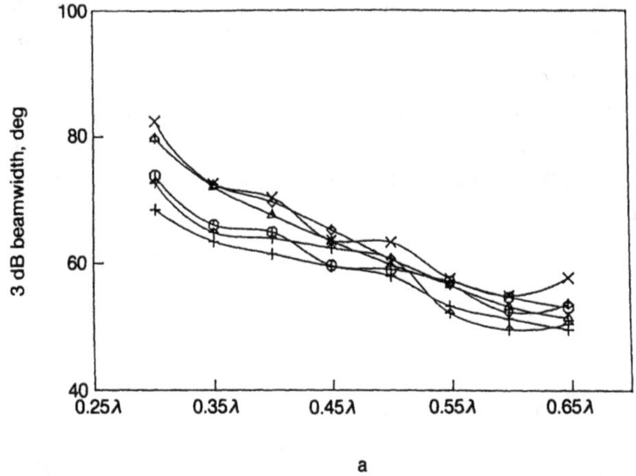

Figure 7.8 3 dB beamwidths for waveguides of different radii and wall thicknesses

H-plane	E-plane
○ $t=0$	× $t=0$
△ $t=0.05\ \lambda$	◇ $t=0.05\ \lambda$
+ $t=0.30\ \lambda$	↑ $t=3.0\ \lambda$

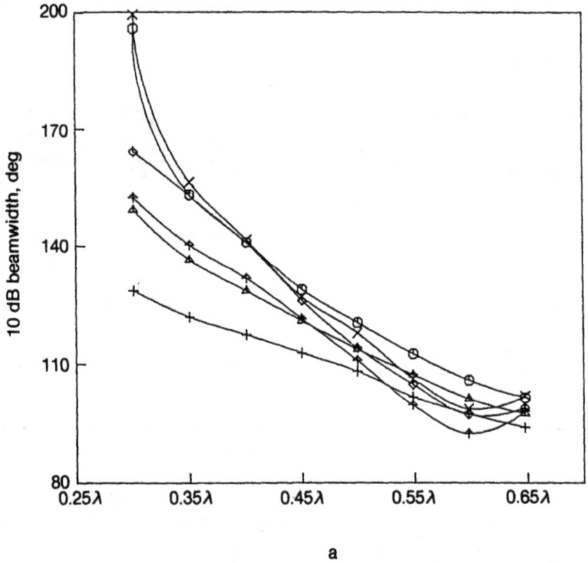

Figure 7.9 10 dB beamwidths for waveguides of different radii and wall thicknesses

H-plane	E-plane
○ $t=0$	× $t=0$
△ $t=0.05\ \lambda$	◇ $t=0.05\ \lambda$
+ $t=0.30\ \lambda$	↑ $t=3.0\ \lambda$

Another parameter of importance is the thickness of the waveguide wall. The effect of this parameter on the waveguide-radiation patterns has not been studied well and to some degree is neglected. However, the current on the end surface of the wall and diffraction effects alter the radiation patterns. For this reason, a set of computed results is provided in this section, for waveguide sizes in the range $0.3\,\lambda \leq a \leq 0.65\,\lambda$, $0 \leq t \leq 1.0\,\lambda$, the inner length of the waveguide being $1.5\,\lambda$ [21].

The beamwidths of the radiation patterns at the 3 dB and 10 dB points are shown in Figures 7.8 and 7.9, each being divided into two groups for clarity of presentation. Near $a = 0.3\,\lambda$, the separation between the beamwidth values in the E- and H-planes is large, since the waveguide is near the cutoff frequency of the radiating TE_{11} mode. Their separation decreases as the radius increases. The effect of wall thickness t on the principal plane beamwidths is also quite different, but shows small effects around $a = 0.5\,\lambda$. The effect of wall thickness on the backlobe level is shown in Figure 7.10. Again, for small radii the effects are oscillatory and tend to decrease with increasing t.

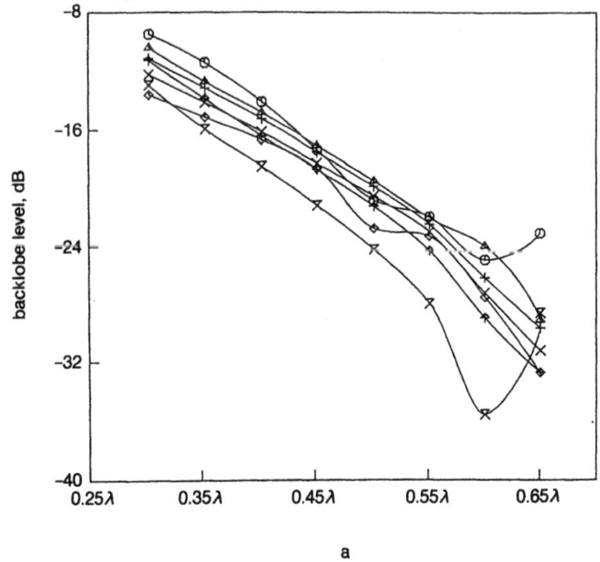

Figure 7.10 Backlobe levels of waveguide-radiation patterns

○ $t=0$
△ $t=0.02\,\lambda$
+ $t=0.05\,\lambda$
× $t=0.10\,\lambda$
◇ $t=0.30\,\lambda$
↑ $t=0.50\,\lambda$
⊼ $t=1.0\,\lambda$

In Figure 7.11, the crosspolarisation decreases with increasing a, up to $a = 0.6\lambda$, after which it increases rapidly. The level of peak crosspolarisation for small radii is quite sensitive to variation of t, and it is difficult to determine an obvious relation for its variation when $0 \leqslant t \leqslant 0.1\ \lambda$. For $0.3 \leqslant t \leqslant 1\ \lambda$ the relation is easy to understand: the crosspolarisation deteriorates for $0.3 \leqslant a \leqslant 0.5\ \lambda$ with increasing t. At $a = 0.5\ \lambda$ the crosspolarisation reaches a minimum at $a = 0.6\ \lambda$. This property has been predicted previously by James [22]. It can be seen that as t is increased the crosspolarisation decreases, according to the results presented by James [23] and James and Greene [24] as a function of the aperture diameter. Here a larger thickness simulates a larger ground plane surrounding the waveguide aperture. For this reason, for very large t one can approach the limit of zero crosspolarisation at $a = 0.6\ \lambda$. In an attempt to understand the reason for the behaviour of the crosspolarisation, the phase-centre location of both E- and H-planes are determined. For brevity, the case of $t = 0.5\ \lambda$ is considered and the results are shown in Figure 7.12. Here, Δd^0 is the

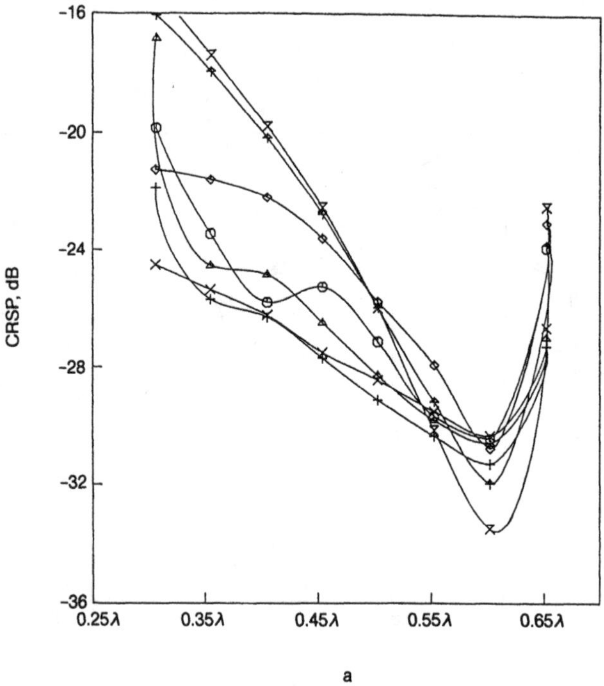

Figure 7.11 Peak crosspolarisation level (CRSP) of circular waveguides

○ $t = 0$
△ $t = 0.02\ \lambda$
+ $t = 0.05\ \lambda$
× $t = 0.10\ \lambda$
◇ $t = 0.30\ \lambda$
↑ $t = 0.50\ \lambda$
⩰ $t = 1.0\ \lambda$

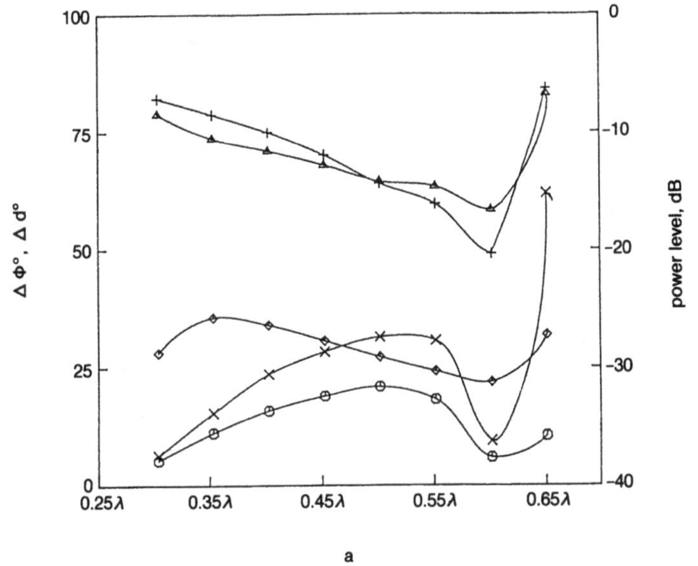

Figure 7.12 Difference between phase centres of the E- and H-planes (Δd), phase difference between E- and H-planes after d_c ($\Delta\phi$) and the power levels of the E- and H-planes at the angle of the crosspolarisation peak for circular waveguides

○ $\Delta\phi$ at θ_c
△ H-plane at θ_c
+ E-plane at θ_c
× Δd°
◇ CRSP at $\theta \leq 70^{\circ}$

axial distance between the phase-centre locations of the two principal planes calculated in degrees, which is determined using the copolar pattern for $0 \leq \theta \leq 70^{\circ}$. Also, $\Delta\phi^{0}$ refers to the phase difference between the phase patterns of the two principal planes. For $0 \leq \theta \leq 70^{\circ}$ the peak crosspolarisation is determined and its angular position is denoted by ϕ_c. The combined phase centre d_c of the waveguide for this angular point ϕ_c is computed and used to determine $\Delta\phi$. The results are also shown in Figure 7.12. It is evident that the peak crosspolarisation depends on $\Delta\phi$, which in turn depends on Δd. $\Delta\phi$ and Δd behave similarly and the minimum peak crosspolarisation occurs for a radius of $a = 0.6\ \lambda$, which has the minimum $\Delta\phi$.

As a check to confirm the numerical results, sample cases were selected to be compared with the measured results in Reference 23. Figure 7.13 shows the computed and measured radiation patterns of a circular waveguide of radius $a = 0.37\ \lambda$ and flange thickness $t = 1\ \lambda$. Excellent agreement between the measured and computed data for both copolar and crosspolar patterns is found. Furthermore, the computed and measured peak crosspolarisations of this waveguide with different thicknesses $0.1\ \lambda \leq t \leq 1\ \lambda$ are compared in Table 7.2.

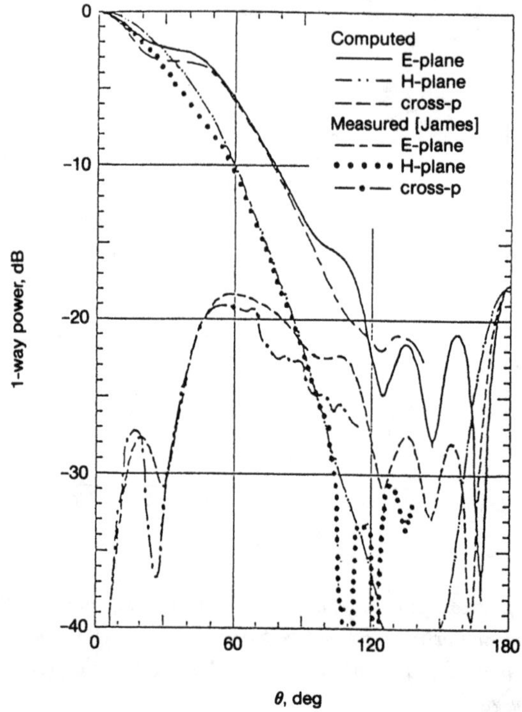

Figure 7.13 Computed and measured radiation patterns of a circular waveguide for $a = 0.37\ \lambda$, $t = 1.0\ \lambda$ [23]

Table 7.2 Calculated and measured maximum crosspolarisation of the TE_{11} mode in a waveguide of radius 0.37 λ and varying flange width

	Maximum crosspolarisation	
Flange width λ	Measured [23] (dB)	Calculated (dB)
0.10	−26.0	−26.01
0.25	−21.5	−22.17
0.50	−18.5	−18.58
0.75	−19.0	−19.17
1.00	−18.5*	−18.36

*See note to Table 7.3.

Again, the same waveguide is tested at different frequencies ranging from 7.5 to 11.5 GHz for two different wall thicknesses. The computed and measured peak values of the crosspolarisation are compared in Table 7.3. The agreement between the measured and computed data is excellent.

To indicate the characteristics of the circular waveguide as a feed for parabolic reflector antennas, its performance with the reflector antenna is considered. Since the phase centre of the feed and the focal point of the reflector must be coincident with each other, the phase-centre location of the feed must be known. The combined optimum phase-centre locations d_c for $0 \leq \theta \leq 70°$ are calculated and shown in Figure 7.14. The phase centre of the waveguide with $t=0$ is again sensitive to the waveguide diameter. In this figure a negative d_c means that the phase centre is in front of the waveguide aperture and positive d_c indicates that it is inside the waveguide aperture. An example of the performance of the waveguide as a feed for a paraboloidal reflector is also investigated for a waveguide with $t=0.1\ \lambda$.

As a reference, the computed phase-centre locations with three definitions are shown in Figure 7.15, and are indicated as d_c, d_s and d_{av}, respectively [25]. For this case, all three definitions give nearly the same location. The variation of the phase-centre location d_c as a function of reflector-aperture angle is indicated in Figure 7.16. It is evident that, for waveguides near cutoff, the location of the phase centre depends strongly on the aperture angle. As the waveguide diameter increases, the phase centre, for aperture angles less than 50°, moves to

Table 7.3 Calculated and measured peak crosspolarisation for the TE_{11} mode in a waveguide of radius 13.41 mm

		Maximum crosspolarisation	
Frequency (GHz)	t (mm)	Measured [23] (dB)	Calculated (dB)
7.5	3.63	−26	−25.66
8.3	3.63	−26	−26.01
9.5	3.63	−27	−27.32
11.5	3.63	−29	−28.45
7.5	36.3	−16.5	−16.57
8.3	36.3	−18*	−18.36
9.5	36.3	−22	−22.56
11.5	36.3	−27	−25.54

*These values in Table 7.2 and Table 7.3 must be the same.

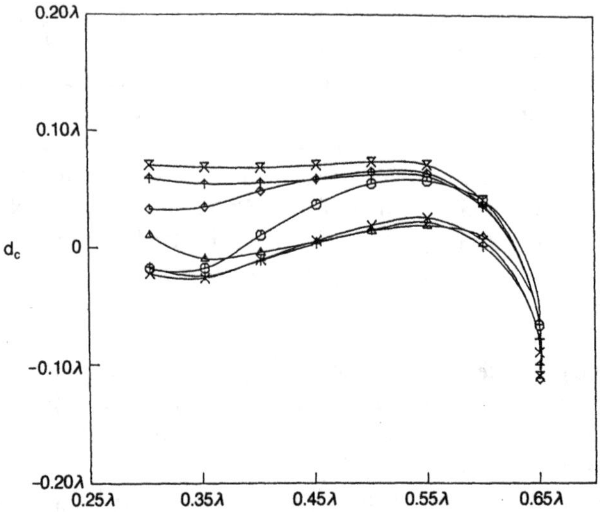

Figure 7.14 *Phase-centre locations from the waveguide aperture*

○ $t=0$
△ $t=0.02\ \lambda$
+ $t=0.05\ \lambda$
× $t=0.10\ \lambda$
◇ $t=0.30\ \lambda$
↑ $t=0.50\ \lambda$
⩙ $t=1.0\ \lambda$

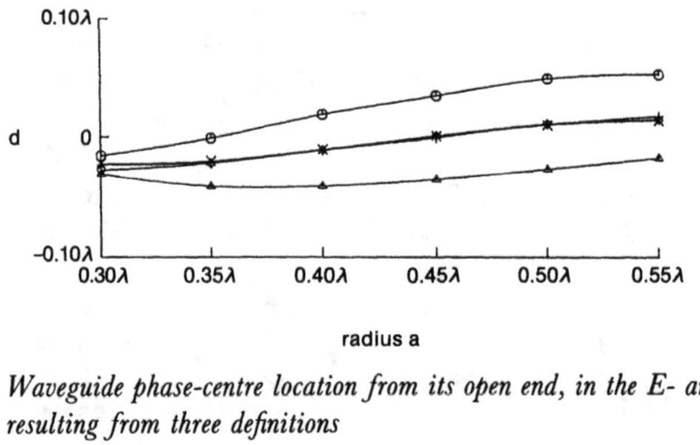

Figure 7.15 *Waveguide phase-centre location from its open end, in the E- and H-planes, resulting from three definitions*

∪ d_E
△ d_H
+ d_{av}
× d_s
◇ d_c

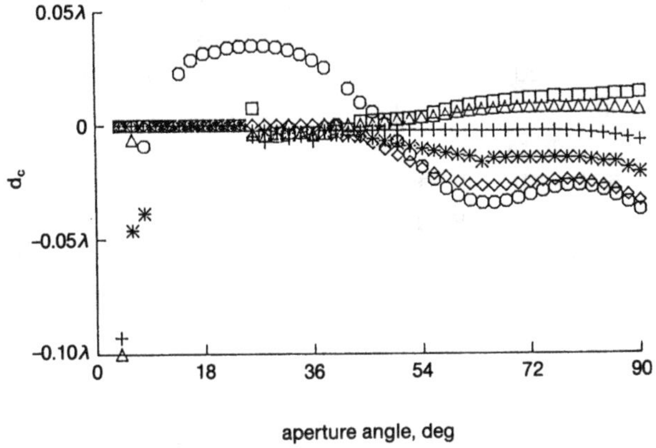

Figure 7.16 Variation of the waveguide phase-centre location d_c with the aperture angle

○ $a=0.3\ \lambda$ + $a=0.45\ \lambda$
◇ $a=0.35\ \lambda$ △ $a=0.5\ \lambda$
* $a=0.4\ \lambda$ □ $a=0.55\ \lambda$

the waveguide aperture. Beyond $\psi_0=50°$, the phase centre is initially in front of the waveguide open end, but moves inside the waveguide for $a\geqslant 0.45\ \lambda$. For $a=0.45\ \lambda$ its location is independent of the aperture angle and is always at the waveguide aperture. Figure 7.17 shows the gain factors of the reflector with a circular-waveguide feed, and for the optimum locations of the phase centre as

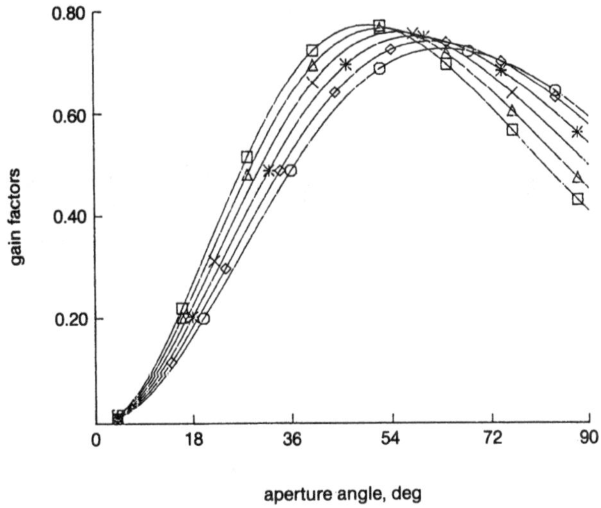

Figure 7.17 Reflector gain factors with a waveguide feed and its phase centre d_c

○ $a=0.3\ \lambda$ × $a=0.45\ \lambda$
◇ $a=0.35\ \lambda$ △ $a=0.5\ \lambda$
* $a=0.4\ \lambda$ □ $a=0.55\ \lambda$

shown in Figure 7.16. These gain factors are computed, using eqns. 7.10 and include the effects of both pattern-phase variations and the spillover power. As expected, the computed gain factor increases with waveguide radius and the peaks move to smaller reflector-aperture angles.

The corresponding results for a waveguide with a choke are shown in Figure 7.18. The reason for improvement of its performance is evident from Figure 7.18 which shows the location of the phase centres. For $a > 0.4\,\lambda$ the phase centres of the principal planes are nearly the same and coincide with all three combined phase centres. For $a < 0.4\,\lambda$ the deviation between d_E and d_H increases as a

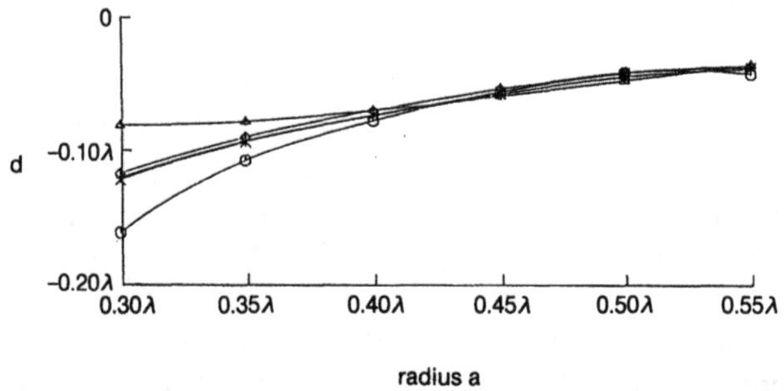

Figure 7.18 Distance of circular-waveguide phase centres from its open end in the E- and H-planes resulting from three definitions

Choke $= \lambda/4$
○ d_E
△ d_H
+ d_{av}
× d_s
◇ d_c

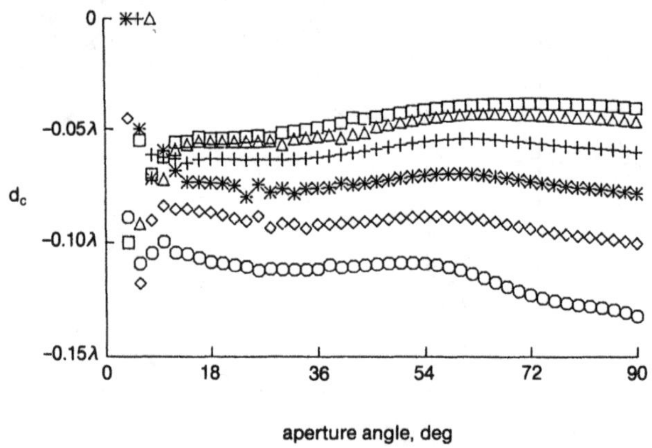

Figure 7.19 Reflector gain factors with a circular-waveguide feed and its phase centre d_c

Choke $= \lambda/4$
○ $a = 0.3\,\lambda$
◇ $a = 0.35\,\lambda$
∗ $a = 0.4\,\lambda$
+ $a = 0.45\,\lambda$
△ $a = 0.5\,\lambda$
□ $a = 0.55\,\lambda$

decreases. In addition, the introduction of the choke moves the phase centre to the front of the waveguide open end. Because of this uniqueness of the phase centre, the phase variation, over the aperture angle $\psi_0 = 70°$ is nearly zero for $a > 0.4\ \lambda$. The variation of the phase-centre location with the aperture angle is shown in Figure 7.19, which shows the similarity of the results for different waveguide diameters. In all cases, the introduction of the choke has made the phase-centre location nearly independent of the aperture angle. However, as the waveguide diameter increases, the phase centre moves towards the aperture (from outside). The expected gain factors are shown in Figure 7.20.

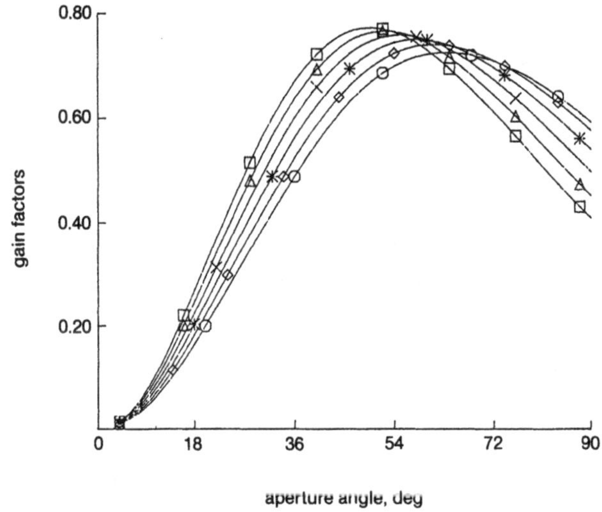

Figure 7.20 *Variation of the circular-waveguide phase-centre location d_c with the aperture angle*

Choke $= \lambda/4$
○ $a = 0.3\ \lambda$
◇ $a = 0.35\ \lambda$
* $a = 0.4\ \lambda$
+ $a = 0.45\ \lambda$
△ $a = 0.5\ \lambda$
□ $a = 0.55\ \lambda$

7.4 Circular waveguide with rectangular corrugated flange

The geometry of the problem is shown in Figure 7.21, where the waveguide has a flange with three corrugations. Since the effect of waveguide radius was studied in previous cases, we study the effect of the parameter L, the separation of the corrugation surface from the waveguide open end. Figure 7.22 shows the results which were obtained again by selecting $\psi_0 = 70°$, and $a = 0.35\ \lambda$. The results indicate that the three definitions no longer give the same phase-centre

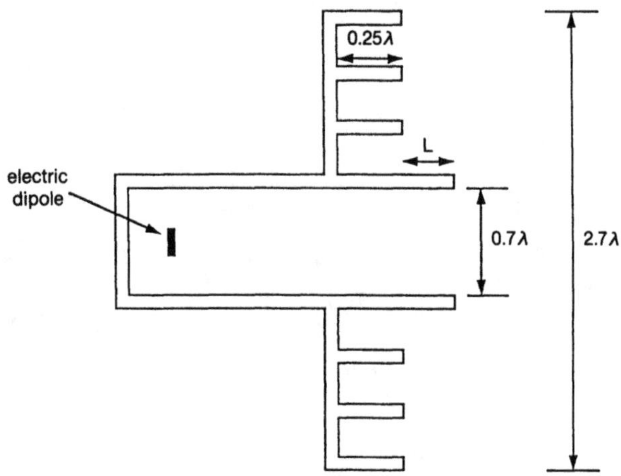

Figure 7.21 Geometry of a circular waveguide with rectangular corrugations on the flange

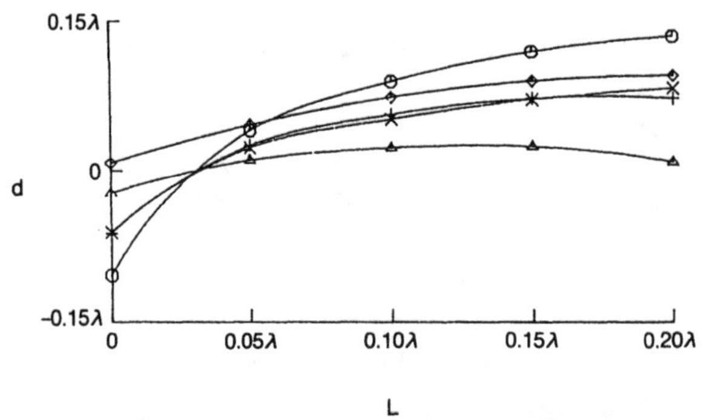

Figure 7.22 Distance of phase centres from waveguide open end for geometry of Figure 7.21

$a = 0.35\ \lambda$ + d_{av}
○ d_E × d_s
△ d_H ◇ d_c

locations. For all values of L, $d_s \simeq d_{av}$, but d_c is somewhat different. In addition, with increasing L, the phase centre moves inside the waveguide. Also, it is interesting that for a certain value of L, here $L=0.06\ \lambda$, the phase centres of the E- and H-plane patterns coincide and thus minimise the feed crosspolarisation. The computed phase deviations, within an angular region of 70°, are increasing with L. These phase variations are larger than for the waveguide. The locations of the phase centre d_c, as a function of the aperture angle, are shown in Figure 7.23. As explained previously, with increasing L the phase centre moves inside the waveguide. In fact, the results of this figure show that, for small values of L the shift in location of the phase centre is nearly $L/2$. As L increases, the interaction between the corrugations and the waveguide aperture decreases and the above relationship fails. Thus, beyond $L=0.1\ \lambda$ the phase-centre location remains relatively stable. The computed gain factors are shown in Figure 7.24, which indicates almost similar variations for all values of L. However, since increasing L tends to broaden the feed patterns, the aperture angle for peak gain factor increases with L. Thus, the parameter L can be used effectively to design feeds for different reflector-aperture angles and to control the feed crosspolarisation without significantly affecting the reflector gain. Typical radiation patterns of such a feed are shown in Figure 7.25.

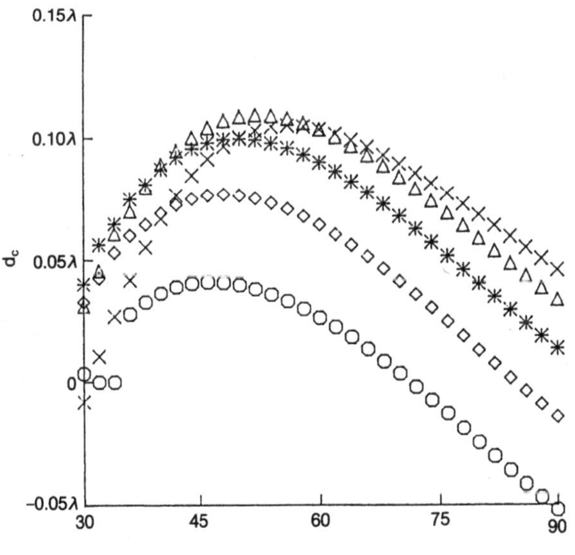

Figure 7.23 *Variation of the waveguide phase-centre location d_c with the aperture angle for the geometry of Figure 7.21*

○ $L=0$ △ $L=0.15\ \lambda$
◇ $L=0.05\ \lambda$ × $L=0.2\ \lambda$
∗ $L=0.1\ \lambda$

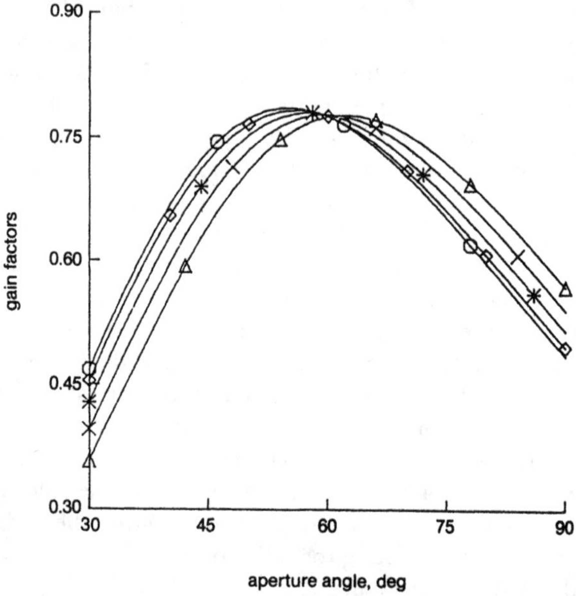

Figure 7.24 *Reflector gain factors with a waveguide feed and its phase centre d_c for the geometry of Figure 7.21*

○ $L=0$ △ $L=0.15\ \lambda$
◇ $L=0.05\ \lambda$ × $L=0.2\ \lambda$
∗ $L=0.1\ \lambda$

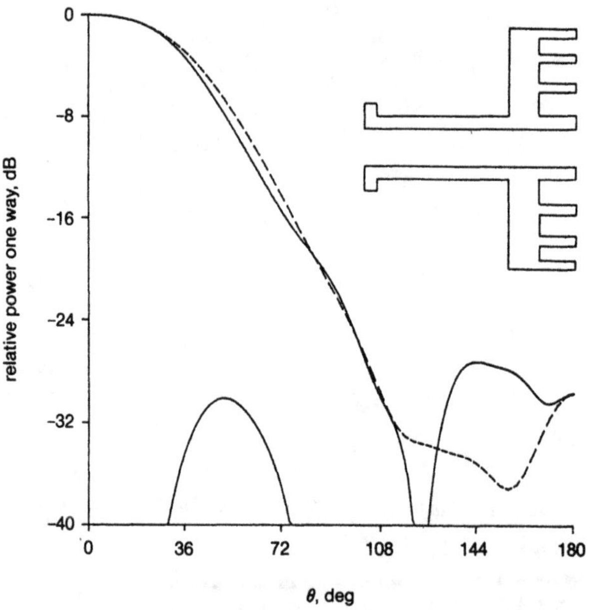

Figure 7.25 *Far-field-radiation patterns for the geometry of Figure 7.21*

7.5 Circular waveguide with trapezoidally corrugated flange

The geometry is shown in Figure 7.26, where the flange has three profiled corrugations of trapezoidal shape [26]. The diameter of the flange is selected to be the same as the previous case for three rectangular corrugations. The computed phase-centre locations are shown in Figure 7.27. The results are similar to the previous case in Figure 7.22, but the separation of the principal-plane phase centres is smaller. In particular, they now coincide for $L=0.1\ \lambda$. The phase variation is also smaller. The most interesting results are shown in

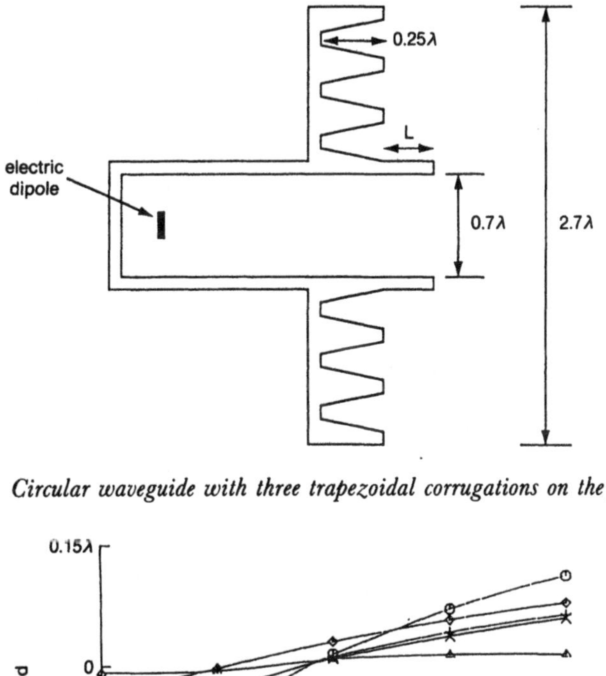

Figure 7.26 Circular waveguide with three trapezoidal corrugations on the flange

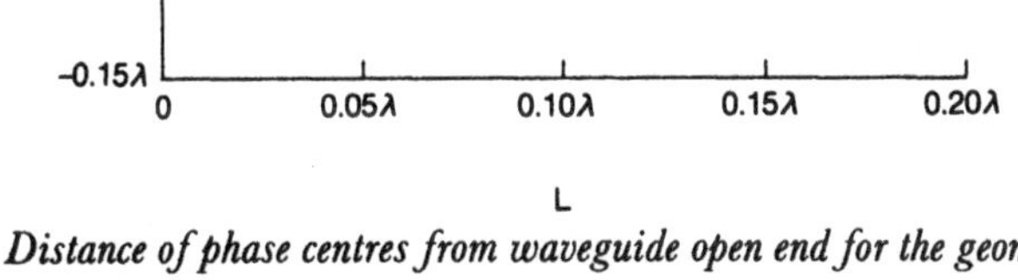

Figure 7.27 Distance of phase centres from waveguide open end for the geometry of Figure 7.26

○ d_E
△ d_H
+ d_{av}
× d_s
◇ d_c

Figure 7.28, which shows the location of the phase centre d_c, as a function of L, where the effect of L is significant for all the values selected. This means the interaction of trapezoidal corrugations with the waveguide aperture is larger, and the phase centre moves inside the waveguide by about $L/2$ for any value of L. This relationship enables one to estimate the phase centre of such feeds from its location at $L = 0$. The gain factors of this feed are shown in Figure 7.29, and are also nearly independent of L. This property also makes feed design much simpler. Radiation patterns for the antenna of Figure 7.26 are shown in Figure 7.30.

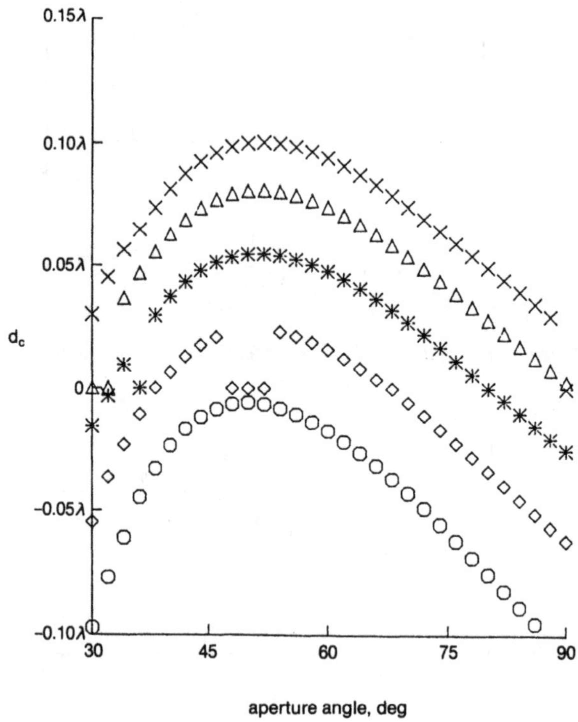

Figure 7.28 *Variation of the waveguide phase-centre location d_c with the aperture angle for the geometry of Figure 7.26*

○ $L = 0$ △ $L = 0.15\lambda$
◇ $L = 0.05\lambda$ × $L = 0.2\lambda$
* $L = 0.1\lambda$

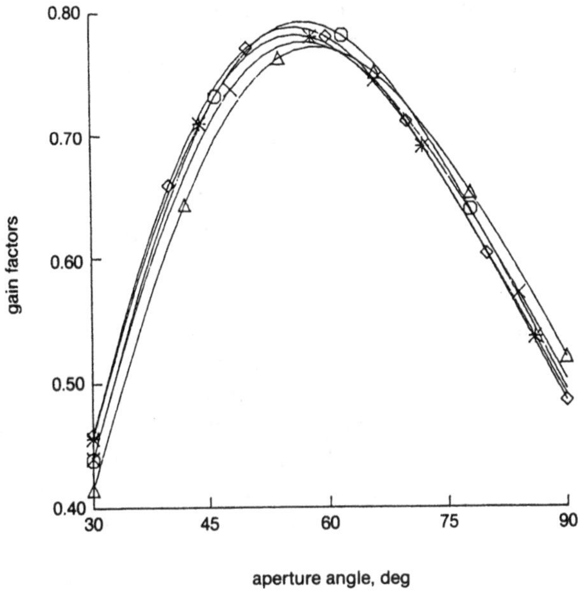

Figure 7.29 Reflector gain factors with a waveguide feed and its phase centre d_c for the geometry of Figure 7.26

Flange with three trapezoidal corrections

○ $L=0$ × $L=0.15\,\lambda$ ∗ $L=0.1\,\lambda$

◇ $L=0.05\,\lambda$ △ $L=0.2\,\lambda$

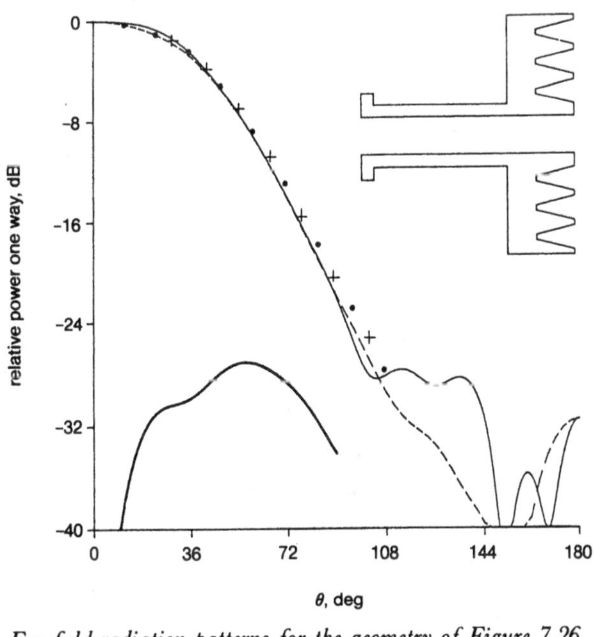

Figure 7.30 Far-field-radiation patterns for the geometry of Figure 7.26

To compare the effect of different corrugations on the phase-centre location and the gain factor, Figures 7.31 and 7.32 are included. In all cases, the corrugation is flush with the waveguide open end. Figure 7.30 shows that, for a single trapezoidal corrugation, the phase-centre location is moved outside the waveguide, but its location has a variation similar to that for the open-ended waveguide. Similarly, for a flange with three trapezoidal corrugations the phase centre moves outside the waveguide, as compared with the rectangular-corrugation example. These results show that the shape of a corrugation has a definite effect on the location of the phase centre. The gain factors are shown in Figure 7.32 which indicates that the aperture angle for peak gain factor is relatively independent of the corrugation shape and number. However, the peak values increase with the number of corrugations.

Table 7.4 compares the performance of different feeds, and shows the phase-centre locations in principal planes, phase deviations within an angular range of 70° and peak crosspolarisations. The radius of the waveguide for all cases is $a=0.35\ \lambda$, and the flange dimensions are the same as in the previous examples. Columns two and three indicate the location of the phase centres in principal

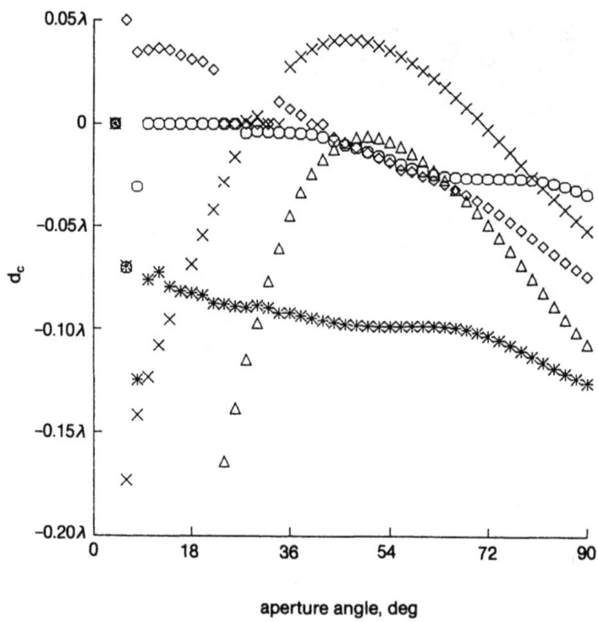

Figure 7.31 Dependence of phase centre d_c on the aperture angle ψ_0 for the geometries of Figures 7.1, 7.2, 7.21 and 7.26 and a waveguide with a single trapezoidal choke

$a=0.35\ \lambda$
$L=0$
○ waveguide
◇ waveguide with one rectangular choke
* waveguide with one trapezoidal choke
× flange with three rectangular corrugations
△ flange with three trapezoidal corrugations

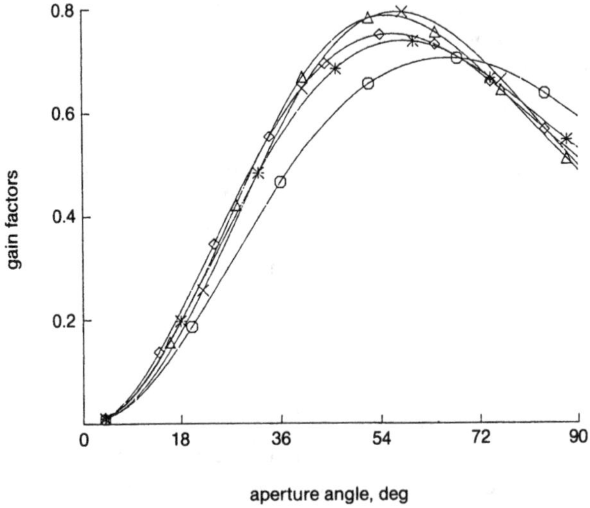

Figure 7.32 Gain factor for cases illustrated in Figure 7.31

- ○ waveguide
- ◇ waveguide with one rectangular choke
- * waveguide with one trapezoidal choke
- × flange with three rectangular corrugations
- △ flange with three trapezoidal corrugations

Table 7.4 Comparison of different feeds in phase-centre location, phase variations and the peak crosspolarisation

Feed type	Characteristics			$\phi_{max}-\phi_{min}$ ($\theta=0$–70)		ϕ_E–ϕ_H		Peak crosspolar
	dE	dH	$\lvert dE-dH\rvert$	E	H	$\theta=0$	θ_c	(dB)
Circular waveguide	0.0	0.0565	0.0565	1.75	2.27	20.34	22.5/73	−25.66
Waveguide with one rectangular choke	0.0544	0.0338	0.0206	2.69	5.64	7.4	11.0/41	−29.72
Waveguide with one trapezoidal choke	0.1447	0.0619	0.0828	2.76	4.18	29.81	36.7/43	−23.17
Flange with three rectangular corrugations	0.0285	0.0109	0.0176	6.96	5.95	6.35	4.8/51	−30.08
Flange with three trapezoidal corrugations	0.1088	0.0269	0.0819	9.73	5.231	29.46	29.76/69	−28.11

d phase-centre distance λ from aperture
ϕ far-field phase, deg
θ_c angle θ at peak crosspolarisation

planes. Column four gives the separations between two principal-plane phase centres, which are large for trapezoidal corrugations and small for rectangular ones. Columns five and six show the maximum phase deviations, in the principal planes and for an angular range of 70°, on a radiation sphere centred at the respective phase centres. Columns seven and eight give the phase deviations between two principal-plane patterns in the directions $\theta = 0$ and θ_c, where the crosspolarisation is a maximum. The final column gives the peak crosspolarisation of each feed. This latter result indicates the strong effect of phase deviations on the crosspolarisation level.

7.6 Phase-error effects

In practice, the phase centre of the feed may not coincide with the focal point of the reflector. It is therefore useful to investigate the sensitivity of the reflector phase and gain efficiencies to their separation distance. For the circular-waveguide feeds, the results are shown in Figures 7.33 and 7.34. Figure 7.34 shows that the open-ended circular waveguide is most sensitive to the phase-

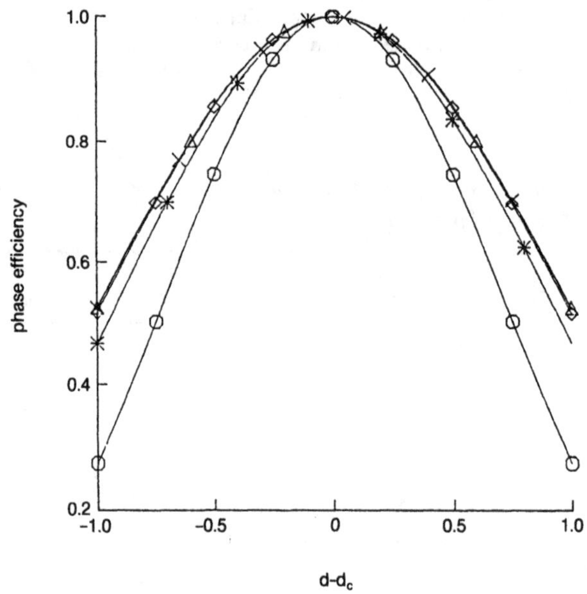

Figure 7.33 *Variation of phase efficiencies with phase-centre displacements corresponding to the five cases illustrated in Figure 7.31*

o waveguide
◇ waveguide with one rectangular choke
* waveguide with one trapezoidal choke
× flange with three rectangular corrugations
△ flange with three trapezoidal corrugations

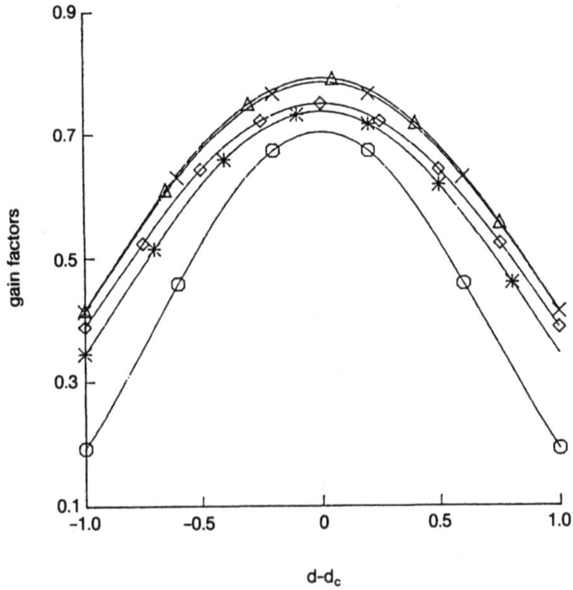

Figure 7.34 *Dependence of gain factors on the phase-centre displacement for the five cases illustrated in Figure 7.31*

○ waveguide, $\psi_0 = 66$
◇ waveguide with one rectangular choke, $\psi_0 = 56$
∗ waveguide with one trapezoidal choke, $\psi_0 = 58$
× flange with three rectangular corrugations, $\psi_0 = 58$
△ flange with three trapezoidal corrugations, $\psi_0 = 58$

centre dislocation, which is partly due to the large reflector-aperture angle ψ for this feed. For other feeds the variations in the phase efficiency are similar. For these feeds and the waveguide, a phase-centre dislocation of 0.5 λ reduces the phase efficiency by about 14% and 25%, respectively. The gain factor of an open-ended waveguide is also reduced, but by a larger amount, with the feed defocusing. Again, for other feeds, although the maximum gain factors are different, their variations are similar. A feed dislocation of 0.5 λ reduces the gain by about 11% and 18%.

7.7 Coaxial waveguide feeds

Kraus and Profera [27] have studied the radiation characteristics of a TE_{11}-mode-excited coaxial waveguide. They have determined the waveguide aperture field using a modal-expansion method and obtained the radiation field by the Kirchhoff–Huygens principle. The analysis indicated that the copolar patterns of a coaxial waveguide lack the desired azimuthal symmetry. To improve the pattern symmetry, they then used a loading circular iris on the waveguide aperture. The aperture field was again determined by the modal-expansion method. While this method is satisfactory for determining the radiation field of large apertures, it is inadequate for coaxial waveguides, since the aperture sizes are generally small. Furthermore, the currents induced on the waveguide outer surface are significant and affect the radiation field considerably.

A more elaborate investigation of coaxial waveguides as feeds was carried out by James [28]. He has treated both corrugated and smooth-wall waveguides using the modal-expansion method. In particular, the radiation-field data for the latter case were obtained using the geometrical theory of diffraction analysis, which provides more accurate results. However, the thickness of the outer wall was assumed to be 0.5 λ. This large wall thickness certainly affects the radiation field and, in particular, the crosspolar radiation. Based on the results obtained, he then concluded that a coaxial waveguide has an inferior performance in both bandwidth and the sidelobe level.

To examine the radiation field of a smooth-wall coaxial waveguide more accurately, we have utilised a numerical method. Initially, the currents on the conducting walls of the waveguide are determined by a numerical solution of certain integral equations. These computed currents are then used to evaluate the radiation fields. It is found that, generally, the numerical method gives more symmetric radiation patterns than the aperture-field method. In addition, the degree of pattern symmetry can be controlled by the waveguide-aperture dimensions. Since a coaxial waveguide has two conductor radii, a variety of radiation patterns can be obtained simply by selecting different aperture dimensions. In particular, a satisfactory pattern symmetry can be found by selecting proper conductor radii and without the need for a corrugated wall or a flange. It should, however, be mentioned that the radiation field of a coaxial waveguide has high sidelobe and backlobe levels. The surface currents on the outer surface also cause oscillations on the radiation patterns. It is found that both problems can be overcome by reducing the surface currents using a quarter-wavelength choke on the waveguide wall. The geometry used for numerical computation is shown in Figure 7.35. A small electric dipole on the waveguide axis is used to excite the structure, by a TE_{11} mode in the coaxial region. Although the configuration used for numerical computations is not identical to practical shapes, the chosen feed-dipole location ensures the aperture illumination by the TE_{11} mode.

It is found that, generally, for any given ratio of the coaxial-waveguide radii, a combination can be found to equalise the E- and H-plane copolar patterns. The computed patterns for one such combination are shown in Figure 7.36. The pattern symmetry is excellent, but both sidelobe and backlobe levels are high, and were found to be of the same order for most waveguides investigated. The crosspolarisation is also excessive, with a peak at about -27 dB level. This high crosspolarisation is not due to the copolar-pattern asymmetry, but is caused by the nonuniqueness of the feed phase centre.

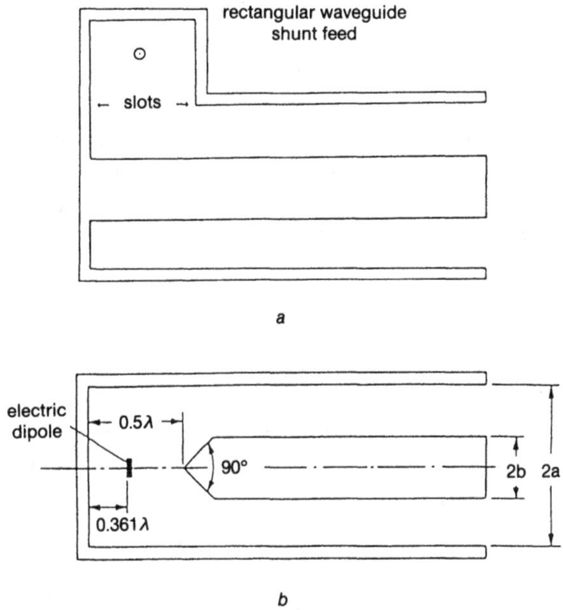

Figure 7.35 Physical and numerical models of a coaxial waveguide
a Physical model; *b* Numerical model

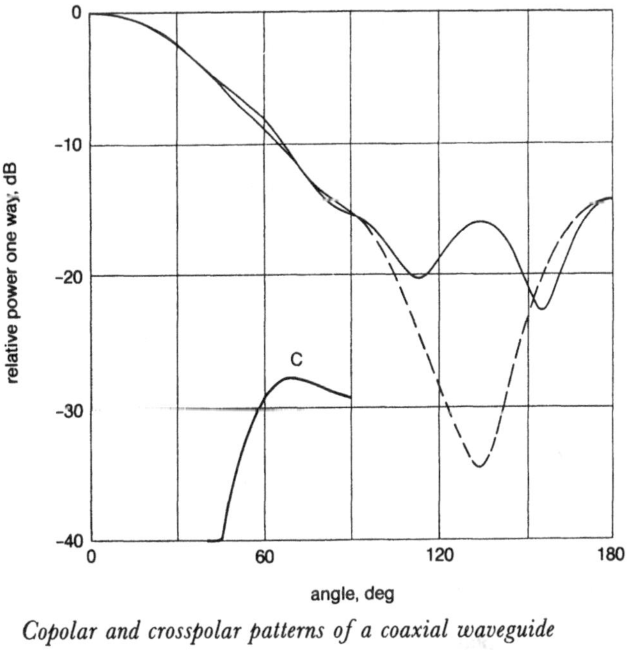

Figure 7.36 Copolar and crosspolar patterns of a coaxial waveguide
$a/b = 4$, $a = 0.4\,\lambda$; ——— E-plane; - - - - H-plane; —C— crosspolar

To improve the sidelobe and backlobe levels of the radiation patterns, a quarter-wavelength choke was introduced on the waveguide outer wall. Figure 7.37 shows the new geometry. Its radiation field was also determined numerically and the computed patterns showed a considerable reduction of the backradiation. A sample of the far-field patterns is shown in Figure 7.38, which can be compared with those of Figure 7.36. The copolar-pattern symmetry is again excellent. The crosspolar pattern also shows considerable improvement with a peak level of about -30 dB.

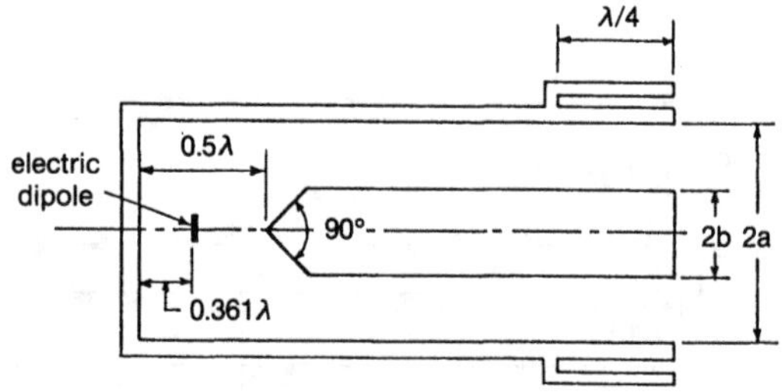

Figure 7.37 Geometry of a coaxial waveguide with a quarter-wavelength choke

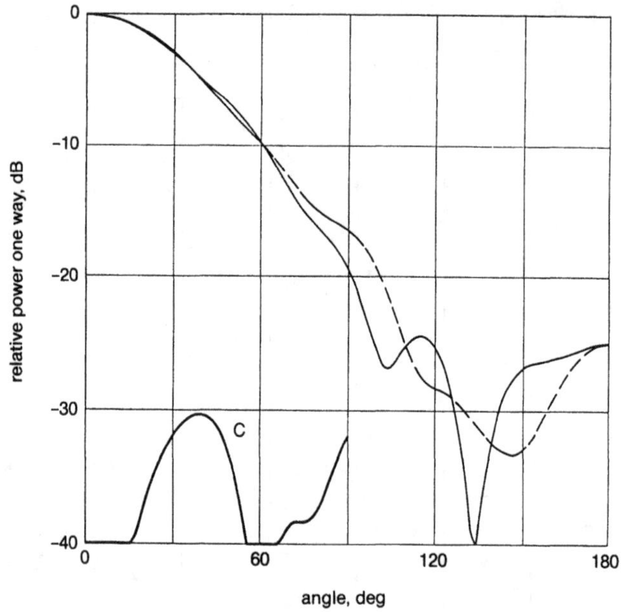

Figure 7.38 Copolar and crosspolar patterns of a coaxial waveguide with a choke

$a/b = 4$, $a = 0.4\ \lambda$
——— E-plane
- - - - H-plane
—C— crosspolar

To provide further design data, the copolar patterns of the waveguide are also computed for different conductor radii. Figure 7.39 shows the results for the backlobe levels. For a waveguide with a quarter-wavelength choke the level of backradiation strongly depends on the outer-conductor radius a, but is relatively independent of a/b. On the other hand, the backradiation of the waveguide without the choke is generally insensitive to the waveguide dimensional parameters and in the region of $a = 0.45\ \lambda$ it is about 10 dB higher. The peak crosspolarisation is also determined for various waveguide parameters. Figure 7.40 shows the results for a waveguide without a choke and within the angular regions $0 \leqslant \theta \leqslant 90°$. For $a \leqslant 0.35\ \lambda$, the crosspolarisation peaks fall within 72° angular range and are minima for $0.4\ \lambda \leqslant a \leqslant 0.45\ \lambda$. The corresponding results for a waveguide with a quarter-wavelength choke are shown in

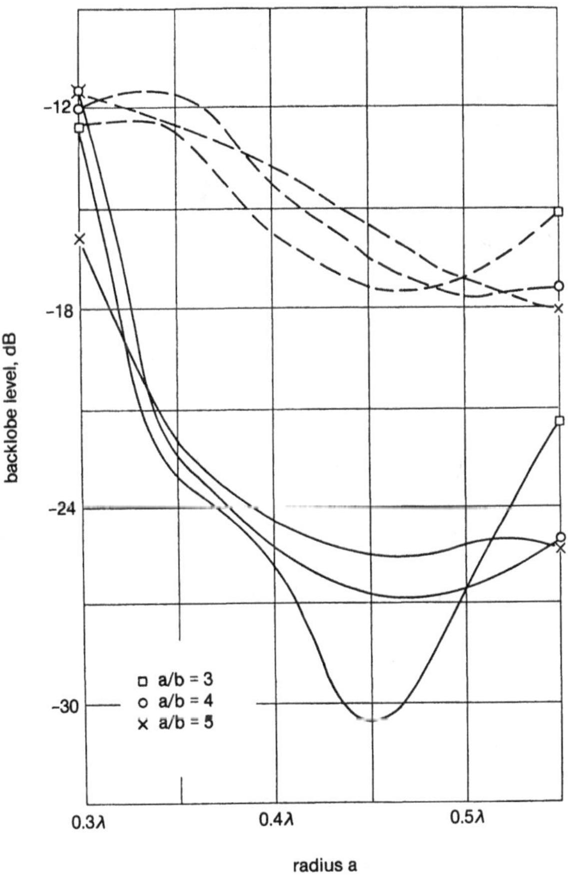

Figure 7.39 Backradiation level of a coaxial waveguide

——— with choke
– – – without choke

Figure 7.41, with the minima for $0.35\,\lambda \leqslant a \leqslant 0.45\,\lambda$. For this case, the minimum peak crosspolarisations are about 3 dB less than those of the waveguide without a choke.

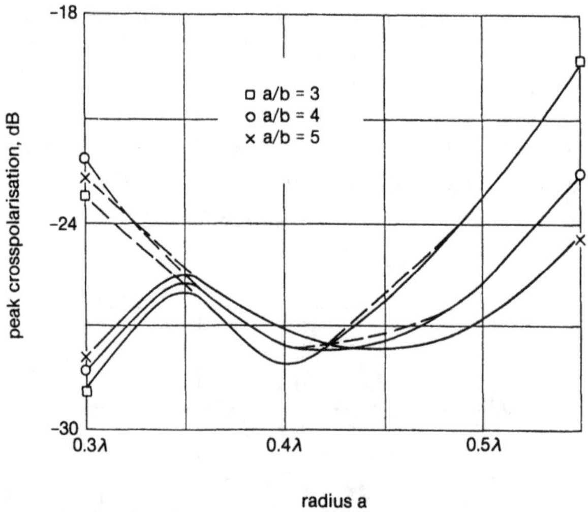

Figure 7.40 Peak crosspolar radiation of a coaxial waveguide

- - - - $\theta \leqslant 90°$

——— $\theta \leqslant 72°$

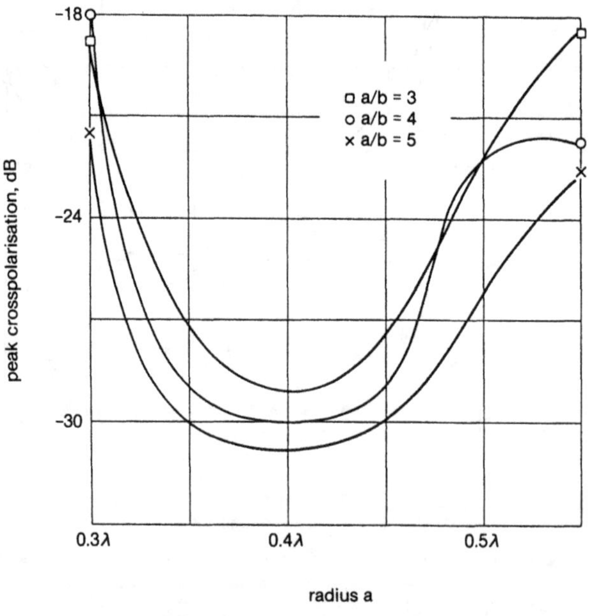

Figure 7.41 Peak crosspolar radiation of a coaxial waveguide with a choke

When the unit is used as a feed for a paraboloid reflector, a parameter of some significance is the 10 dB beamwidth of the copolar patterns. This parameter is therefore also determined for various waveguide dimensions. Figure 7.42 shows the 10 dB beam angles (half beamwidths) for both waveguides, with and without a choke. The results indicate that introducing a quarter-wavelength choke generally reduces both pattern beamwidths and the optimum outer-waveguide diameters. The reductions, however, are small. In addition, the results show that the optimum radius a increases with a/b.

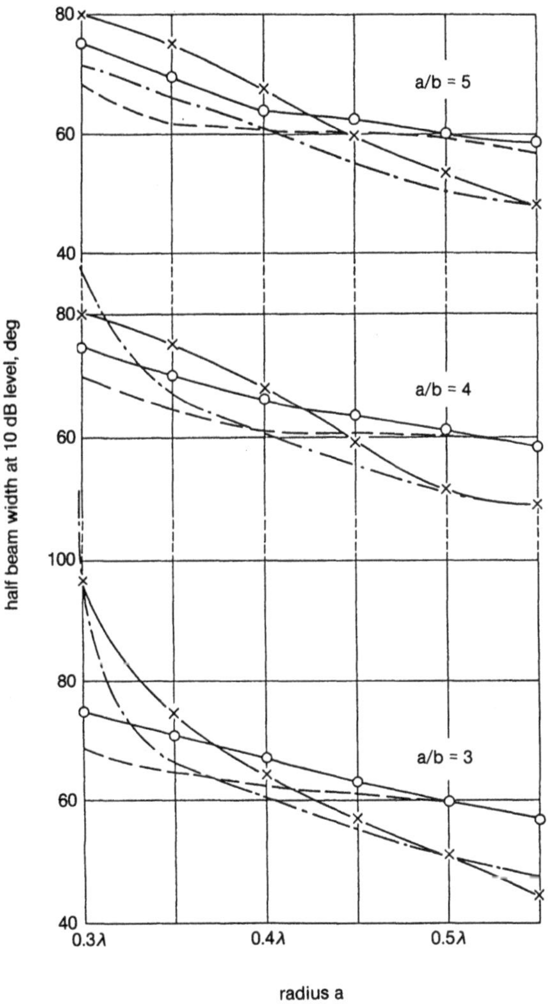

Figure 7.42 10 dB beam angles of a coaxial waveguide

E-plane	H-plane
—×— without choke	—O— without choke
—·—·—with choke	- - - - with choke

The locations of the coaxial-waveguide phase centres computed using the three definitions are shown in Figures 7.43 and 7.44, respectively, for waveguides without and with a quarter-wavelength choke. In Figure 7.43 the phase-centre locations for all three definitions are in front of the aperture and are nearly independent of the radius *a*. The phase deviations are somewhat larger than those of circular-waveguide feeds. For the coaxial waveguide with a choke, the results are shown in Figure 7.44. It is evident that, for $a/b=4$, the introduction of the choke causes a nearly unique phase centre for

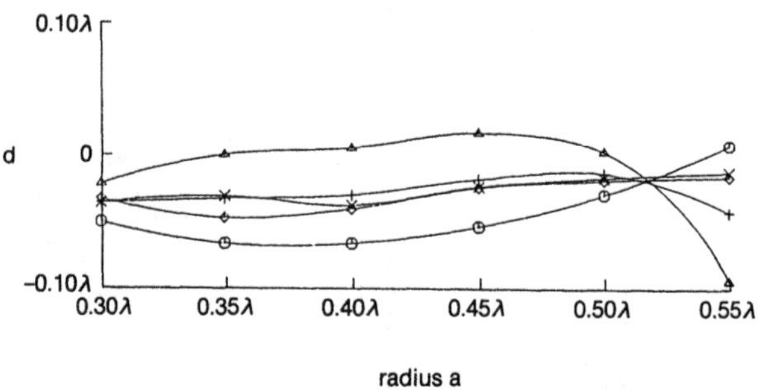

Figure 7.43 Distance of coaxial-waveguide phase centres from its open end in the E and H planes resulting from three definitions ($a/b=4$)

○ d_E × d_s
△ d_H ◇ d_c
+ d_{av}

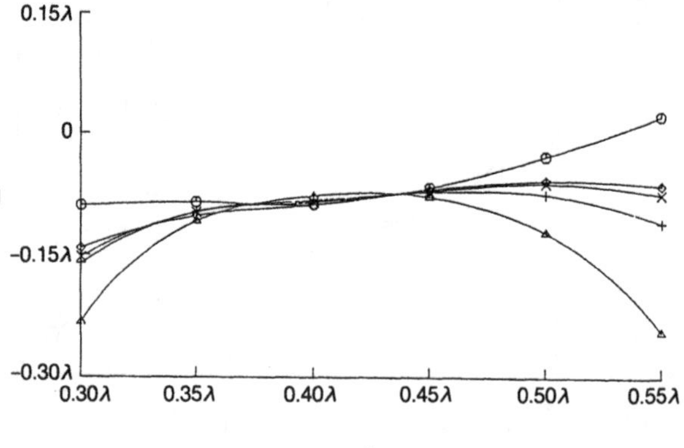

Figure 7.44 Distance of coaxial-waveguide phase centres from its open end in the E- and H-planes resulting from three definitions, $a/b=4$

with a choke + d_{av}
○ d_E × d_s
△ d_H ◇ d_c

$0.35\ \lambda \leqslant a \leqslant 0.45\ \lambda$. Outside this range, the phase centres of the principal planes separate rapidly. The phase deviations within an angular range of 70° are also minima for $0.35\ \lambda \leqslant a \leqslant 0.45\ \lambda$, beyond which they increase rapidly. Finally, the locations of phase centres d_c and the gain factors for coaxial waveguides with and without chokes are shown in Figure 7.45. The introduction of a choke moves the phase centre further out by about $0.05\ \lambda$ and the gain factor increases. Numerical results are shown in Tables 7.5 and 7.6. For a waveguide

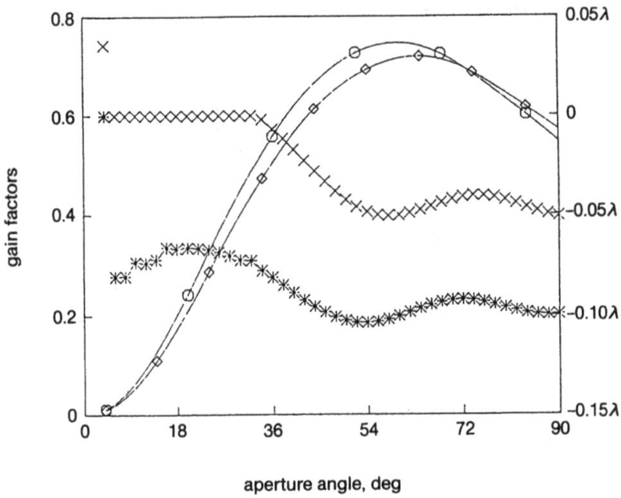

Figure 7.45 Phase-centre location d_c and gain factor for a coaxial waveguide

$a/b = 4$, $a = 0.4\ \lambda$ * d_c with choke
o gain factor with choke   × d_c without choke
◇ gain factor without choke

Table 7.5 Gain factors and spillover efficiencies of coaxial feeds without a choke

a/b	a wavelength	Gain factor (%)	Spillover efficiency (%)	f/D
	0.35	70.3	82.77	0.37
5	0.40	71.6	84.65	0.40
	0.50	75.2	88.14	0.47
	0.35	70.8	83.03	0.37
4	0.40	72.1	85.35	0.40
	0.50	75.61	88.88	0.47
	0.35	71.8	83.69	0.37
3	0.40	73.41	86.33	0.40
	0.50	75.61	86.23	0.51

without a choke, the gain factors peak when the reflector f/D is nearly equal to a/λ, where f and D are the focal length and diameter, respectively. The use of a quarter-wavelength choke to improve the gain factor increases the numerical value of f/D. The equality of the numerical values of f/D and a/λ now holds only when $a \approx 0.5\ \lambda$.

Table 7.6 Gain factors and spillover efficiencies of coaxial feeds with a choke

a/b	a wavelength	Gain factor (%)	Spillover efficiency (%)	f/D
	0.35	73.17	85.16	0.43
5	0.40	74.62	88.03	0.43
	0.50	76.77	87.77	0.51
	0.35	73.05	85.93	0.43
4	0.40	74.76	87.99	0.43
	0.50	76.90	87.74	0.51
	0.35	72.89	87.64	0.40
3	0.40	75.34	88.02	0.43
	0.50	76.84	87.38	0.51

7.8 Feeds for deep reflectors

In low-noise applications it is desirable to use reflectors with small f/D values. The required feed must therefore have a wide-angle pattern. Such feeds are often difficult to design and seldom give high gain factors with low crosspolarisations. With a proper computer optimisation, however, they can be designed to give satisfactory characteristics.

A feed of this type was recently reported by Kim [29] (Figure 7.46). Its computed radiation patterns and crosspolarisation are compared with measured data of Kim in Figure 7.47. The agreement is excellent, and confirms the accuracy of the computation. It also shows high crosspolarisation, computed using the third definition of Ludwig [19]. The computed performance of this feed within a frequency band of 10% is shown in Table 7.7. The gain factors, spillover efficiencies and illumination efficiencies are computed using definitions of Clarricoats and Poulton [30]. The numerical method is used to improve the radiation characteristics of this feed by optimising its dimensions, to achieve symmetric radiation patterns and low crosspolarisation. The new geometry is shown in Figure 7.48, and its far-field patterns in Figure 7.49. The sector-shape radiation patterns show excellent symmetry and low crosspolarisation, which is reduced now to −29 dB. Its performance with frequency is

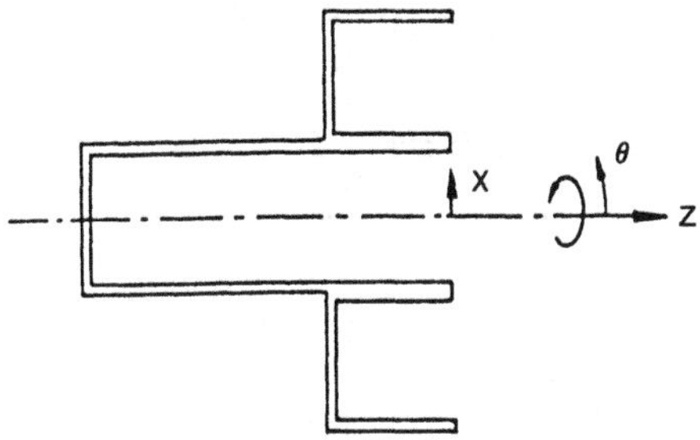

Figure 7.46 *Geometry of a widebeam feed*

Waveguide diameter = 0.794 λ
Coaxial-cavity radii = 0.467, 1.07 λ
Cavity depth = 0.47 λ

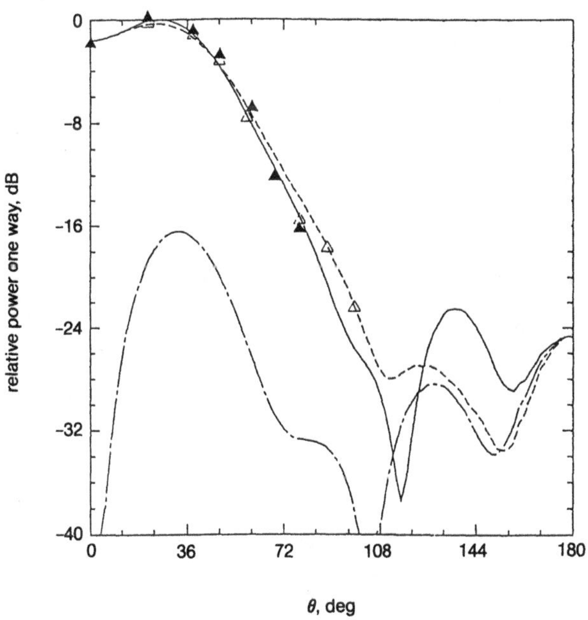

Figure 7.47 *Radiation patterns of the feed illustrated in Figure 7.46*

Computed	Measured [29]
——— E-plane	▲ E-plane
– – – – H-plane	△ H-plane
—·—— crosspolarised	

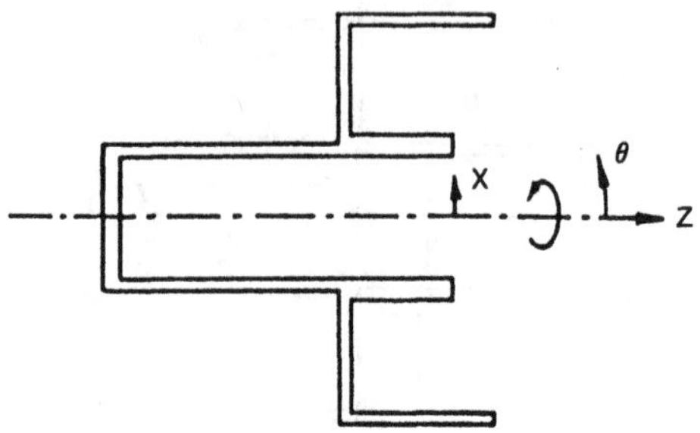

Figure 7.48 Geometry of optimised widebeam feed

Waveguide diameter = 0.7
Coaxial-cavity radii = 0.42, 1.023 λ
Cavity depth = 0.57 λ
External wall extension = 0.1 λ

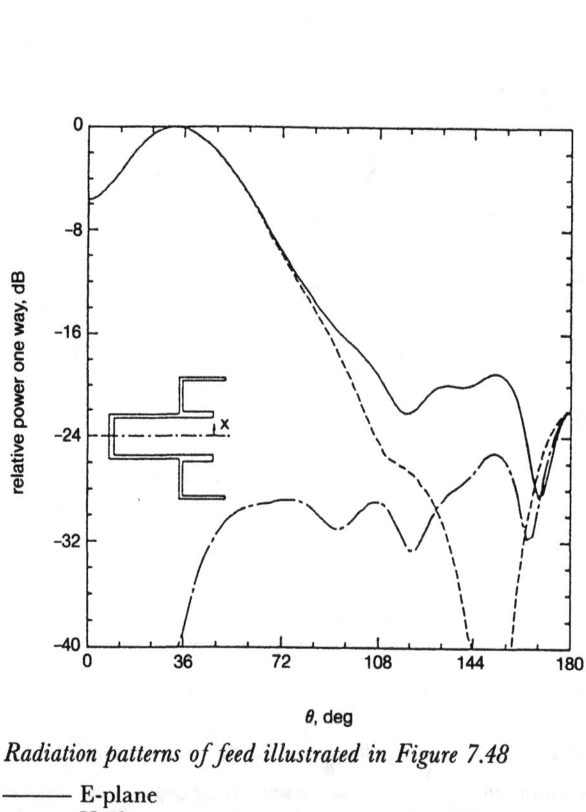

Figure 7.49 Radiation patterns of feed illustrated in Figure 7.48

——— E-plane
- - - - H-plane
—·—·- crosspolarised

Table 7.7 Performance of feed in Figure 7.46

Frequency	$0.95\ f_0$	$0.975\ f_0$	f_0	$1.025\ f_0$	$1.05\ f_0$
ψ_0	62	60	60	58	58
η_g	81.57	81.48	81.40	81.20	81.04
η_s	90.63	89.37	89.93	88.89	89.56
η_i	90.02	91.17	90.51	91.35	90.49
χ_p	−15.87	−16.70	−16.88	−16.92	−17.25
d_c			0.11		

ψ_0 = aperture angle of reflector antenna at which gain factor is maximum
η_g = maximum gain factor at ψ_0
η_s = spillover efficiency at ψ_0
η_i = illumination efficiency at ψ_0
χ_p = maximum crosspolarisation level ($0<\theta<90°$)
d_c = phase-centre location from wavelength aperture

Table 7.8 Performance of feed in Figure 7.48

Frequency	$0.95\ f_0$	$0.975\ f_0$	f_0	$1.025\ f_0$	$1.05\ f_0$
ψ_0	68	66	64	62	60
η_g	82.23	81.52	81.47	81.57	80.77
η_s	89.13	88.99	88.89	88.90	88.49
η_i	92.28	91.61	91.68	91.77	91.29
χ_p	−20.41	−30.87	−28.77	−25.93	−23.94
d_c			−0.04		

shown in Table 7.8. Comparing Tables 7.7 and 7.8, one can observe that the new feed has lower crosspolarisation and achieves the peak gain factor for larger reflector angles.

Neither of the above feeds incorporates chokes and their backradiations are high. The computed data for the optimised feed of Figure 7.48, after adding a single choke were also generated. The radiation patterns are shown in Figure 7.50 and its performance with frequency in Table 7.9. The choke improves the gain factor and spillover efficiency, but the peak crosspolarisation shows no improvement. This is because the feed dimensions were initially optimised for low crosspolarisation without a choke. One expects to improve it further by reoptimising the feed dimensions, after the inclusion of the choke.

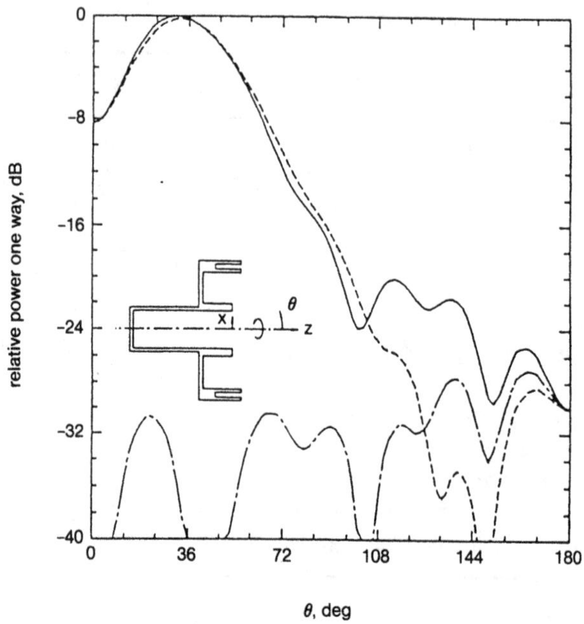

Figure 7.50 Radiation patterns of feed illustrated in Figure 7.48 with λ/4 choke

——— E-plane
– – – – H-plane
—·—- crosspolarised

Table 7.9 Performance of feed in Figure 7.48 with choke

Frequency	$0.95\ f_0$	$0.975\ f_0$	f_0	$1.025\ f_0$	$1.05\ f_0$
ψ_0	70	66	64	62	60
η_g	83.72	82.16	82.62	82.13	82.47
η_s	90.84	90.22	91.08	89.85	90.47
η_i	92.01	91.10	90.71	91.41	91.17
χ_p	−18.53	−26.16	−30.48	−24.41	−23.29
d_c			−0.04		

7.9 Dualband feeds

In some applications the feed is required to operate in multiple frequency bands. When the frequency bands are adjacent, a waveguide feed can be used, where the chosen waveguide diameter must be large enough to yield a useful radiation beam at the lowest-frequency band. The radiation patterns and crosspolarisation can be improved by multiple peripheral chokes, each about a quarter-wavelength at the respective frequency. Numerous designs have been reported in the literature. Here, we provide two examples for dualband operation. A simple feed using a choked waveguide was reported by Cowan [12], in which two separate chokes, one for each frequency band, were used to improve the patterns and crosspolarisation (Figure 7.51). The feed was designed for operation in two adjacent bands of 11.6 GHz and 14.2 GHz. For the second example, we include a horn-type feed. It again has two separate corrugations and was optimised numerically for operation at the above two bands. Its geometry is shown in Figure 7.52 and its radiation patterns in Figure 7.53.

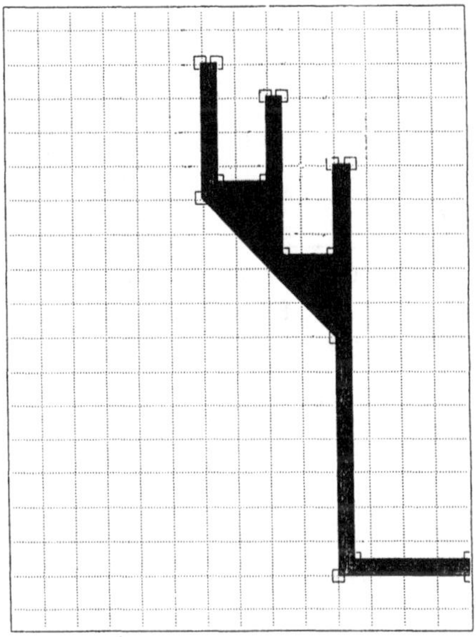

Figure 7.51 Geometry of a numerically optimised dualband feed

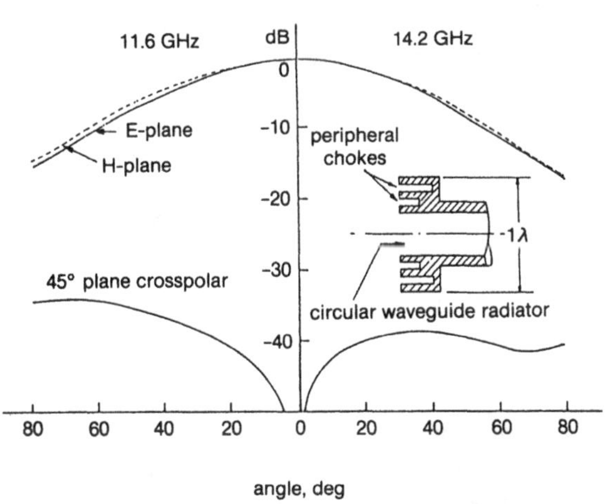

Figure 7.52 Geometry of a dualband-choke feed [12]

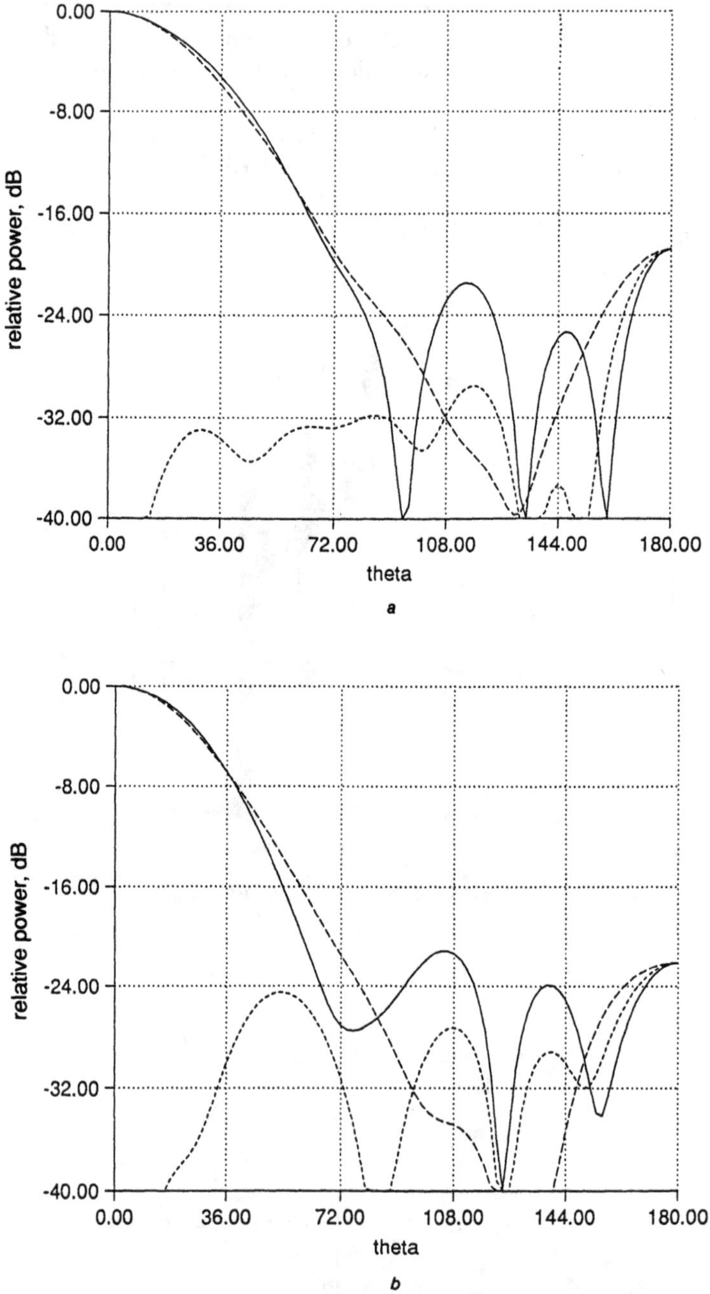

Figure 7.53 Radiation patterns of the dualband feed illustrated in Figure 7.52

——— E-plane
– – – – H-plane
- - - - - crosspolar

a 12 GHz band
b 14 GHz band

When the feed is required to operate at two or more frequency bands which are widely separated, it may not be possible to use the above concept, and alternative approaches may be used. For instance, in dual-mode operation the feed may be designed to operate as a choked-waveguide feed at the low-frequency band, and as a dual-mode feed at the high-frequency band [31, 33]. With this type of feed the performance of each band can be optimised separately, but dual-mode horns or waveguides may be narrowband.

A convenient method for designing multiband feeds, with widely separated frequency bands, is to use coaxial-type waveguides or horns. As indicated earlier, coaxial-waveguide radii, with or without a choke, can be selected appropriately to yield symmetric radiation patterns. This simplifies the multiband-feed design, but care must be taken to ensure that the phase centres of different bands are at nearly the same point. An example of this type of feed was designed for operation at 4 and 6 GHz bands, using a dielectric rod inside the centre conductor to reduce its diameter, which also acted as the upper-band 6 GHz feed. Its geometry and radiation patterns are shown in Chapter 10.

7.10 Small-horn feeds

For high performance a horn should be corrugated and, as a prime-focus feed, must have small size. Numerous different designs are available in the literature which provide performance ranging from narrow angles for shallow reflectors to wide angles for deep reflectors. Their analysis can be handled analytically, as is described in Chapter 4, and designed as a corrugated horn, as discussed in Chapter 9. However, because of their small size they can also be analysed or designed numerically. In this case the shape of horn and corrugations can be made arbitrary to optimise for performance.

Because their analysis and design are covered in Chapter 9, only a simple design example is provided here. The following steps may be used to facilitate the design. Initially, one may determine the horn-aperture size approximately using data available for conical horns (smooth-wall horns), since the H-plane pattern is approximately the same as the pattern of a corresponding corrugated horn, with a slightly narrower beamwidth. Also, the 10 dB beamwidth in the H-plane of a conical horn is approximately twice its 3 dB beamwidth. Thus, one may use the following steps:

(i) select f/D or the aperture angle of the reflector which corresponds approximately to the 10 dB beam angle of the feed, which is also approximately the 3 dB full beamwidth of the conical horn

(ii) from the 3 dB beamwidth compute the horn aperture size, using 55% horn aperture efficiency which is the efficiency of an optimum horn

(iii) select the diameter of the waveguide that feeds the horn; this completes conical horn design

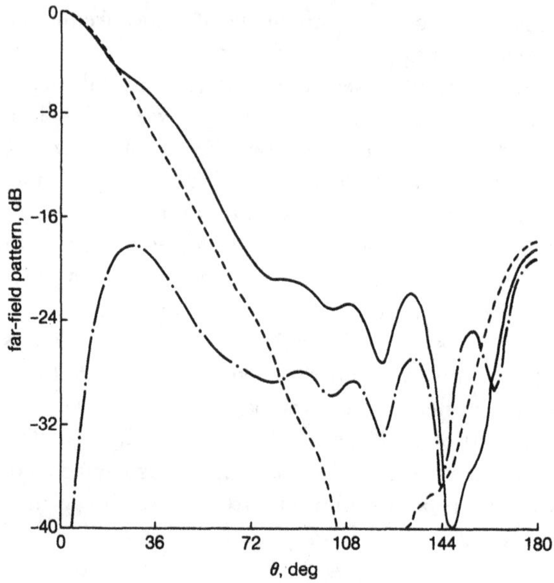

Figure 7.54 *Far-field patterns of the smooth-wall conical horn*

——— E-plane
—·— H-plane
- - - - - crosspolarised

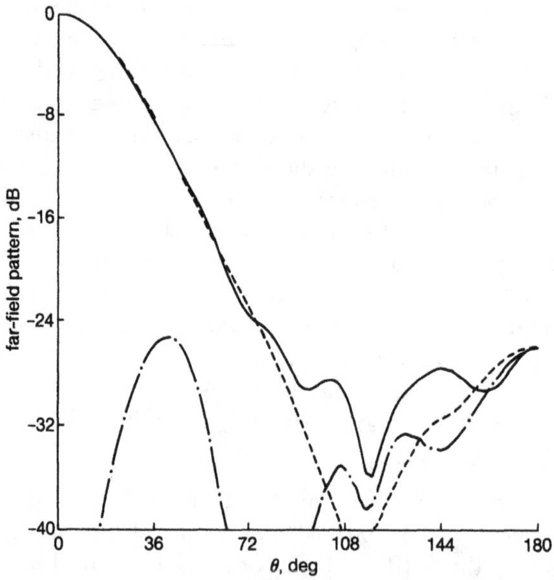

Figure 7.55 *Far-field patterns of the corrugated conical horn corresponding to Figure 7.54*

——— E-plane
—·— H-plane
- - - - - crosspolarised

(iv) select an initial corrugation geometry, i.e. axial, radial or profiled corrugations, etc., and determine the horn radiation patterns
(v) optimise corrugation geometry and/or the horn dimensions to improve its performance requirements.

Figures 7.54 and 7.55 show one such example. The centre frequency is 10 GHz. The conical horn selected has the far-field patterns given in Figure 7.54. Introducing corrugations (not optimised) improves its performance, as shown in Figure 7.55. The beamwidth of the corrugated horn is approximately the average of the conical-horn principal-plane beamwidths.

7.11 References

1 RUDGE, A.W., MILNE, K., OLVER, A.D., and KNIGHT, P. (Eds.): 'The handbook of antenna design' (Peter Peregrinus, 1982), IEE Electromagnetic Waves series 15 and 16
2 LOVE, A. W. (Ed.): 'Electromagnetic horn antennas' (IEEE Press, New York, 1976)
3 CLARRICOATS, P. J. B., and OLVER, A. D.: 'Corrugated horns for microwave antennas' (Peter Peregrinus, 1984), IEE Electromagnetic Waves series 18
4 MINNETT, H. C., and THOMAS, B. MacA.: 'Fields in the image space of symmetrical focusing reflectors', *Proc. IEEE*, 1968, **115**, (10), pp. 1419–1430
5 MINNETT, H. C., and THOMAS, B. MacA.: 'A method for synthesizing radiation patterns with axial symmetry', *IEEE Trans.*, 1966, **AP–14**, pp. 654–656
6 RUMSEY, V. H.: 'Horn antennas with uniform power patterns around their axis', *IEEE Trans.*, 1966, **AP–14**, pp. 656–658
7 KAY, A. F.: 'The scalar feed'. AFCRL, report 64-347, AD601609, March, 1964
8 KOCH, G. F.: 'Coaxial feeds for high aperture efficiency and low spillover of paraboloid reflector antennas', *IEEE Trans.*, 1973, **AP–21**, pp. 164–169
9 WOHLLEBEN, R., MATTES, H., and LOCHNER, O.: 'Simple small primary feed for large angles and high aperture efficiency', *Electron. Lett.*, 1972, **8**, (19), pp. 474–476
10 SHAFAI, L.: 'Broadening of primary feed patterns by small E-plane slots', *Electron. Lett.*, 1977, **13**, (4), pp. 102–103
11 JAMES, G. L.: 'Primary feeds with wide beamwidth', *Electron. Lett.*, 1973, **14**, (14), pp. 447–448
12 COWAN, J. H.: 'Dual-band reflector-feed element for frequency reuse application', *Electron. Lett.*, 1973, **9**, (25), pp. 596–597
13 MIYATA, K., and SUZUKI, M.: 'Effects of choke-load position on radiation properties in double-choked small horn antennas', *Electron. Lett.*, 1983, **19**, (9), pp. 329–330
14 HU, Y. Y.: 'A method of determining phase centers and its application to electromagnetic horns', *J. Franklin Inst.*, 1961, **271**, pp. 31–36
15 KILDAL, P. S.: 'Combined E- and H-plane phase centers of antenna feeds', *IEEE Trans.*, 1983, **AP–31**, (1), pp. 199–202
16 RAO, K. S., and SHAFAI, L.: 'The phase center analysis of reflector antenna feeds', *IEEE Trans.*, 1984, **AP–32**, (7), pp. 740–742
17 BOX, M. J., DAVIES, D., and SWANN, W. H.: 'Non-liner optimization techniques' (Oliver and Boyd, 1969)
18 RUSCH, W. V. T., and POTTER, P. D.: 'Analysis of reflector antennas' (Academic Press, New York, 1970), chap 3, pp. 68–71
19 LUDWIG, A. C.: 'The definition of cross-polarisation', *IEEE Trans.*, 1973, **AP–21**, (1), pp. 116–119
20 SHAFAI, L., and KISHK, A. A.: 'Coaxial waveguides as primary feeds for reflector antennas and their comparison with circular waveguides', *Arch. Elektron. Uekertrag.*, 1984, **39**, (1), pp. 8–14

21 KISHK, A. A., and SHAFAI, L.: 'Accurate design data for circular waveguide antennas', *Int. J. Electron.*, 1989, **67**, (1), pp. 101–110
22 JAMES, G. L.: 'Radiation properties of 90° conical horns', *Electron. Lett.*, 1977, **13**, pp. 293–294
23 JAMES, G. L.: 'Radiation performance of flanged waveguides'. CSIRO Division of Radio-physics, report RPP 2033(L), 1978
24 JAMES, G. L., and GREENE, K. J.: 'Effect of wall thickness on radiation from circular waveguides', *Electron. Lett.*, 1978, **14**, pp. 90–91
25 SHAFAI, L., and KISHK, A. A.: 'Phase centre of small primary feeds and its effects on the feed performance', *IEE Proc., H*, 1985, **132**, pp. 207–214
26 KISHK, A. A., SHAFAI, L., and ITTIPIBOON, A.: 'Performance of a small primary feed with trapezoidal and sinusoidal corrugations', *Electron. Lett.*, 1983, **19**, (8), pp. 297–298
27 KRAUS, L. A., and PROPERA, C. E.: 'A technique for obtaining pattern symmetry and low sidelobe from a TE_{11} mode coaxial radiator', *IEEE Trans.*, 1977, **AP–25**, pp. 365–369
28 JAMES, G. L.: 'Propagation and radiation properties of corrugated cylindrical coaxial waveguides', *IEEE Trans.*, 1983, **AP–31**, pp. 477–483
29 KIM, C., and RHODEN, W.: 'Radiation pattern analysis of coaxial cavity feed horn'. IEEE APS, international symposium, Philadelphia, 1986, vol. 1, pp. 95–98
30 CLARRICOATS, P. J. B., and POULTON, G. T.: 'High efficiency microwave reflector antennas — a review', *Proc. IEEE*, 1977, **65**, pp. 1470–1504
31 GRUNER, R. W.: 'A 4- and 6-GHz prime focus, CP feed with circular pattern symmetry'. IEEE/AP symposium, 1974, pp. 72–74
32 VAN ARDENNE, A., BREGMAN, J. D., and SONDAR, L. H.: 'Compact dual-polarised coaxial feed at 327 MHz', *Electron. Lett.*, 1984, **20**, (18), pp. 723–724
33 COLLIN, R. E., and SCHILLING, H. W.: 'Dual-mode coaxial feed with low cross polarisation', *IEE Proc. H*, 1984, **131**, (6), pp. 405–410

Chapter 8

Multimode horns

A.D. Olver

8.1 Introduction

This chapter is concerned with horns which are designed to radiate or receive more than one mode. The properties of the different modes are used to obtain radiation patterns which either enhance the patterns of the dominant mode or provide extra information at the horn terminals. The aim of this chapter is to explain the fundamental operation of multimode horns and to describe how the modes can be generated. Most of the design effort in multimode horns is concerned with the waveguiding properties of the horn. It is relatively straightforward to understand the radiation patterns needed for a given application, but more difficult to ensure that the multiple modes are generated and propagate correctly along the horn.

There are two broad classes of multimode horns: 'shaped-pattern horns' and 'tracking horns'. Shaped-pattern horns are the largest class of multimode horn. The higher-order modes are added, with appropriate amplitude and phase, to the dominant mode to enhance the radiation performance and improve the efficiency. The designer needs to choose a combination of modes in the aperture of the horn so that the radiation characteristics will give the desired patterns. These modes must be capable of propagating in the horn, and a suitable geometry must be devised to excite the modes in the correct ratio. The basic excitation mechanism of higher-order modes is usually either a sharp change in the cross-section dimensions of the horn or the insertion of a piece of dielectric. In both cases, the abrupt discontinuity is intended to generate one or more higher-order modes. The discontinuity may be followed by a section of horn which is chosen so as to phase the multiple modes in the correct ratio. The abrupt-discontinuity-and-phasing section leads to a fundamental limitation of multimode horns: the bandwidth of operation will be restricted. The excitation mechanism depends on factors which are very frequency-sensitive and shaped-pattern multimode horns are therefore inherently narrowband radiators.

Shaped-pattern horns can be subdivided into three types:

(i) horns designed to reduce the crosspolarisation and improve the pattern symmetry. The best known example in this category is the Potter horn [3] which is a conical horn with TM_{11} mode added to the dominant TE_{11} mode to cancel the crosspolar component of the TE_{11} mode and thereby improve pattern

symmetry
(ii) horns designed to alter the shape of the basic Gaussian copolar pattern. This is done so that the fields illuminating a reflector will be more uniform with angle, and the illumination efficiency of the reflector antenna will increase. It can be accomplished in conical horns by adding a proportion of TE_{12} mode, or in corrugated horns by adding a proportion of HE_{12} mode, or by combining a coaxial and conical horn and adding a proportion of TE_{11} coaxial mode to the TE_{11} circular mode
(iii) Matched-feed horns which are designed to use a combination of modes to match the focal region fields of a reflector antenna, thereby improving the radiation performance of the reflector antenna. An example is the use of TM_{11} and TE_{21} mode to cancel the crosspolarisation in linearly polarised offset-reflector antennas.

Tracking horns are designed to provide information from the horn terminals which enable a target to be tracked. An example is an earth-station antenna where the antenna needs to be kept pointing at a satellite. The dominant mode is used to receive the signal information, whilst one or more higher-order modes are used to provide the tracking information. The higher-order modes perform the tracking function by using radiation patterns which have a null on axis. Tracking horns need to be designed to propagate all the modes from the point at which the tracking and signal information is split up and abstracted. Unlike the shaped-pattern horns, where the input waveguide to the horn need only propagate the dominant mode, tracking horns must have input waveguides capable of propagating both the dominant and higher-order modes. A crucial part of the design of tracking horns is the waveguide beam-forming system which introduces the higher-order modes into the horn.

The nature of the modal-interaction mechanism which is inherent in the design of horns means that there is a very wide range of possible geometries which can achieve the same result. For a given set of system constraints, there will be an optimum horn design. This may be partly determined by mechanical factors such as size and weight in addition to the electrical specifications. For instance, a physically compact horn may be a prime design requirement. The consequence is that there are relatively few general designs and each situation must be considered anew. The scope for innovative design in multimode horns is considerable and the ingenuity of the antenna designer continues to produce new designs. For this reason, this chapter is mainly concerned to explain the fundamentals of operation of multimode horns, using examples from the literature where appropriate.

8.2 Multiple modes in cylindrical waveguides

The electromagnetic behaviour of multimode horns can be divided into two parts: the 'internal' behaviour of the fields inside the horn and the 'radiation' problem. The horn is a truncated piece of waveguide and the behaviour of the internal fields comes from studying the properties of modes in cylindrical or conical waveguides. In multimode horns, the fields inside the waveguide can be decomposed into two or more modes. The orthogonality property of an infinite length of waveguide states that each mode carries power independently of any other mode and that no crosscoupling of power between modes takes place. This enables the properties of each mode to be considered separately.

Changes in the cross-section of the waveguide cause crosscoupling between modes, the amount of crosscoupled power depending on the geometry and electromagnetic properties of the fields. It is the objective of the design of multimode horns to control the amount of crosscoupling. The higher-order modes need to be excited and this will require a controlled discontinuity in the waveguide cross-section. Other changes in cross-section are bound to take place, e.g. the input transition unit, the flare section and the termination of the waveguide into free space, and need to be designed to minimise cross-coupling between modes. Hence the main part of the design of multimode horns consists in the understanding and design of the horn section.

When the multiple modes radiate, each mode can again be considered independently of other modes. This means that the radiation problem consists either of adding together the modes at the aperture and using one of the standard radiation-prediction methods, or using one of the standard radiation methods on each separate mode and adding the radiated fields together in the near field or far field. The main numerical problem is to ensure that the amplitude coefficients for all modes are normalised in a standard manner.

Most horns are basically conical in cross-section so that the diameter at the aperture is greater than the diameter at the input waveguide or throat. Thus the horn is made up of a section of conical waveguide. However, for the purposes of analysis and discussion, it is much easier to treat the horn as being composed of sections of cylindrical waveguide, because the electromagnetic behaviour of cylindrical waveguide is easier to understand. This is valid, as was shown in Chapter 4.

The total transverse field in a cylindrical waveguide is the sum of fields due to N modes:

$$\left.\begin{aligned} \boldsymbol{E} &= \sum_{n=1}^{N} \{A_n \exp(-\gamma_n z) + B_n \exp(\gamma_n z)\}\boldsymbol{e}_n \\ \boldsymbol{H} &= \sum_{n=1}^{N} \{A_n \exp(-\gamma_n z) - B_n \exp(\gamma_n z)\}\boldsymbol{h}_n \end{aligned}\right\} \tag{8.1}$$

where $\boldsymbol{e}_n$ and $\boldsymbol{h}_n$ are the transverse modal electric and magnetic fields, respectively, A_n and B_n are the forward and reflected modal coefficients which specify the relative amplitude and phase of the respective modes, and $\gamma_n = j\beta_n$ for propagating modes and $\gamma_n = \alpha_n$ for evanescent modes.

At the aperture, only the forward-travelling modes contribute towards the radiation. If the dominant, or fundamental, mode has $n=1$, the design of

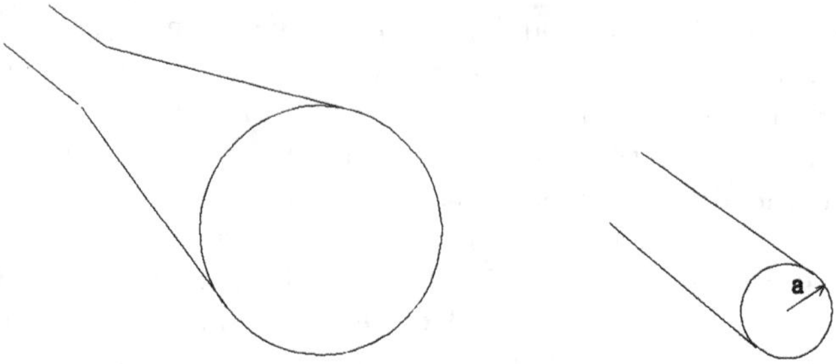

Figure 8.1 Conical horn and circular waveguide

multimode horns consists in choosing the relative proportion of A_n/A_1 for $n>1$.

The fields $\boldsymbol{e}_n$ can be any complete set of modes as long as they satisfy the orthogonality conditions that, for two modes i and j,

$$\left.\begin{aligned} &\iint_s (\boldsymbol{e}_i \times \boldsymbol{e}_i) \cdot \boldsymbol{u}_z \, \mathrm{d}s = 1 \\ &\iint_s (\boldsymbol{e}_i \times \boldsymbol{e}_j) \cdot \boldsymbol{u}_z \, \mathrm{d}s = 0 \qquad i \neq j \end{aligned}\right\} \tag{8.2}$$

and similarly for $\boldsymbol{h}_n$ and for combination of $\boldsymbol{e}_n$ and $\boldsymbol{h}_n$. For rectangular or circular smooth-wall waveguide, sets of TE and TM modes are normally used. For corrugated waveguide the sets are normally hybrid HE and EH modes. The most common types of multimode horns have circular cross-sections so the fields and properties of these modes will be examined in detail to show how multiple modes can be used to alter the radiation characteristics.

8.2.1 Circular waveguides

The TE and TM modes in a multimode circular waveguide of radius a are described by

$$\left.\begin{aligned} E_z &= A_{nm} J_n(K_c r) \cos(n\phi) \qquad \text{TM modes} \\ H_z &= A_{nm} J_n(K_c r) \cos(n\phi) \qquad \text{TE modes} \end{aligned}\right\} \tag{8.3}$$

with the transverse components given by eqns. 6.1 and 6.4. The value of $K_c a$ is given by the roots of either $J_n'(K_c a)=0$ for TE modes or $J_n(K_c a)=0$ for TM modes (Figure 8.1). The amplitude coefficients A_{nm}, for azimuthal order n and mode order m, are determined by changes in the cross-section of the waveguide.

The influence of higher-order modes on the radiation pattern can conveniently be studied by examining the contour patterns for the radiation patterns. These are shown for the relevant modes in Figures 8.2, 8.3, 8.5 and 8.7. The dominant mode is the TE_{11} mode which is the mode naturally excited by coupling a piece of circular waveguide to a section of standard rectangular waveguide propagating the TE_{10} mode. This is therefore the starting point for the discussion of how higher-order modes can alter the radiation characteristics.

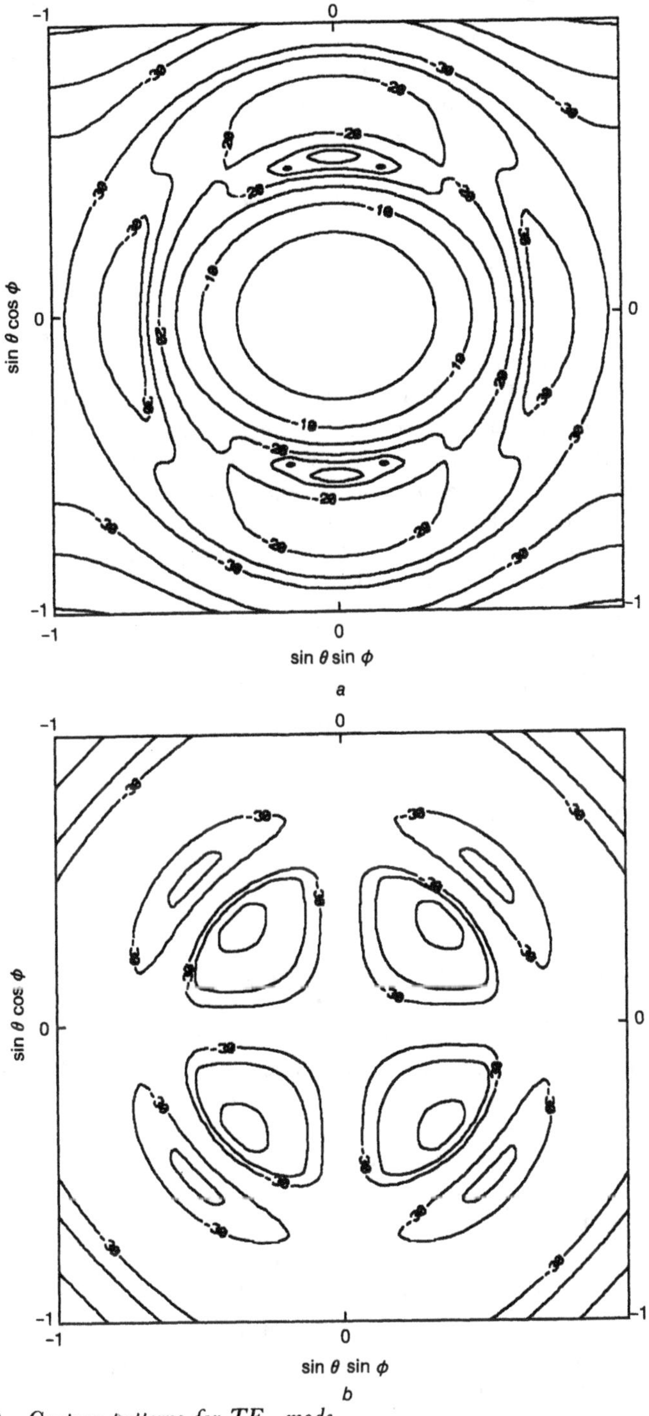

Figure 8.2 *Contour patterns for TE_{11} mode*

a Copolar; *b* Crosspolar

The copolar pattern for the TE_{11} mode (Figure 8.2*a*) shows an elliptical shape. The E-plane, in the vertical direction on the plots, has the narrower beamwidth. The difference in the E-plane and H-plane gives rise to a significant amount of crosspolar radiation, as is shown in the crosspolar contour plot in Figure 8.2*b*. The four lobes have peak levels of -19 dB for an open-ended circular waveguide. The value may change by a few decibels when the waveguide is flared to form a horn owing to the existence of other modes. The asymmetry in the copolar patterns and the high crosspolar levels limit the usefulness of single-mode circular waveguides. What is needed is the addition of a proportion of higher-order mode which will cancel the TE_{11} crosspolarisation. This can be done with the TM_{11} mode, as shown by its patterns in Figure 8.3*a* and *b*. The copolar pattern is of no particular interest with a null along the H-plane. However, the crosspolar pattern shows strong radiation with four lobes having peak levels of about -5 dB relative to the peak copolar levels. Moreover the peaks occur at about the same angles as the peaks of the crosspolar patterns in the TE_{11} mode. Figure 8.4 shows the resultant crosspolar radiation pattern from adding different proportions of TM_{11} mode to TE_{11} mode. The optimum is obtained with a combination of 15% of TM_{11} mode and 85% TE_{11} mode with the modes in phase at the aperture. This will cancel most of the TE_{11}-mode crosspolarisation. In a horn with a finite flare angle, exact cancellation is not possible because of the generation of other higher-order modes (see Section 8.4).

The curves of Figure 8.4 demonstrate an important limitation of multimode horns, namely that the bandwidth is restricted. This follows from the fact that the desired patterns are only achieved by a process of adding modes in the correct amplitude and phase. The actual bandwidth in a conical horn with the optimum combination of TE_{11} and TM_{11} mode is dependent on the flare angle of the horn.

The TM_{11} mode can be excited by changing the horn or waveguide cross-section. However, any mechanism which is designed to excite the TM_{11} mode will also potentially excite the next-higher TE_{1m} and TM_{1m} modes. The contour patterns for the TE_{12} and TM_{12} modes are shown in Figure 8.5. The TE_{12} mode has a narrow main beam with high sidelobes in the H-plane. The result of adding a proportion of TE_{12} mode to TE_{11} mode will be to raise (or lower, depending on the relative phase of the two modes) the H-plane copolar-sidelobe level. The crosspolar pattern for the TE_{12} mode is rich in crosspolarisation (Figure 8.5*b*), but its influence is masked by the high copolar sidelobes. The TM_{12} mode has a null on axis with copolar lobes in the E-plane, so that adding a proportion of TM_{12} mode to TE_{11} mode will raise the E-plane-sidelobe level.

There are two situations to distinguish when considering the desirability of adding some TE_{12} mode and/or TM_{12} mode. First, if the objective is to have a dual-polarisation feed with low crosspolarisation, and secondly where the objective is to shape the main beam so as to achieve higher aperture efficiency. In the first case, neither addition of TE_{12} mode or addition of TM_{12} mode produces desirable results, so the objective of multimode-horn design should be

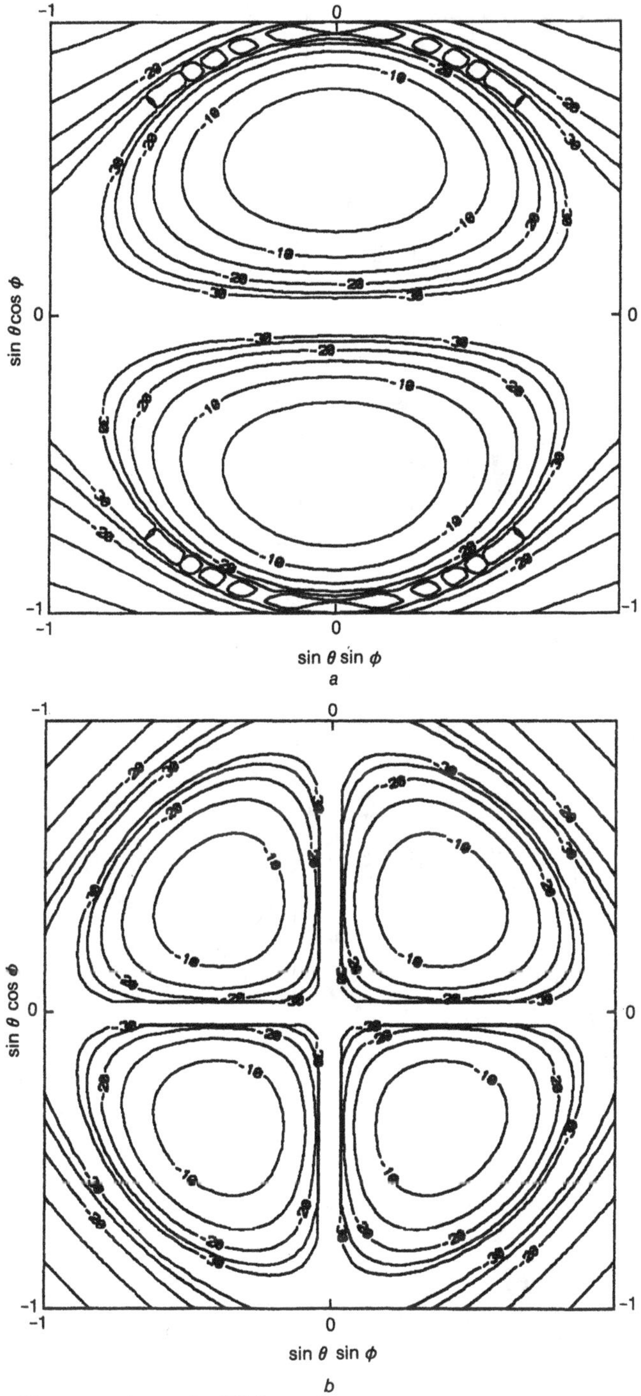

Figure 8.3 Contour patterns for TM_{11} mode
a Copolar; *b* Crosspolar

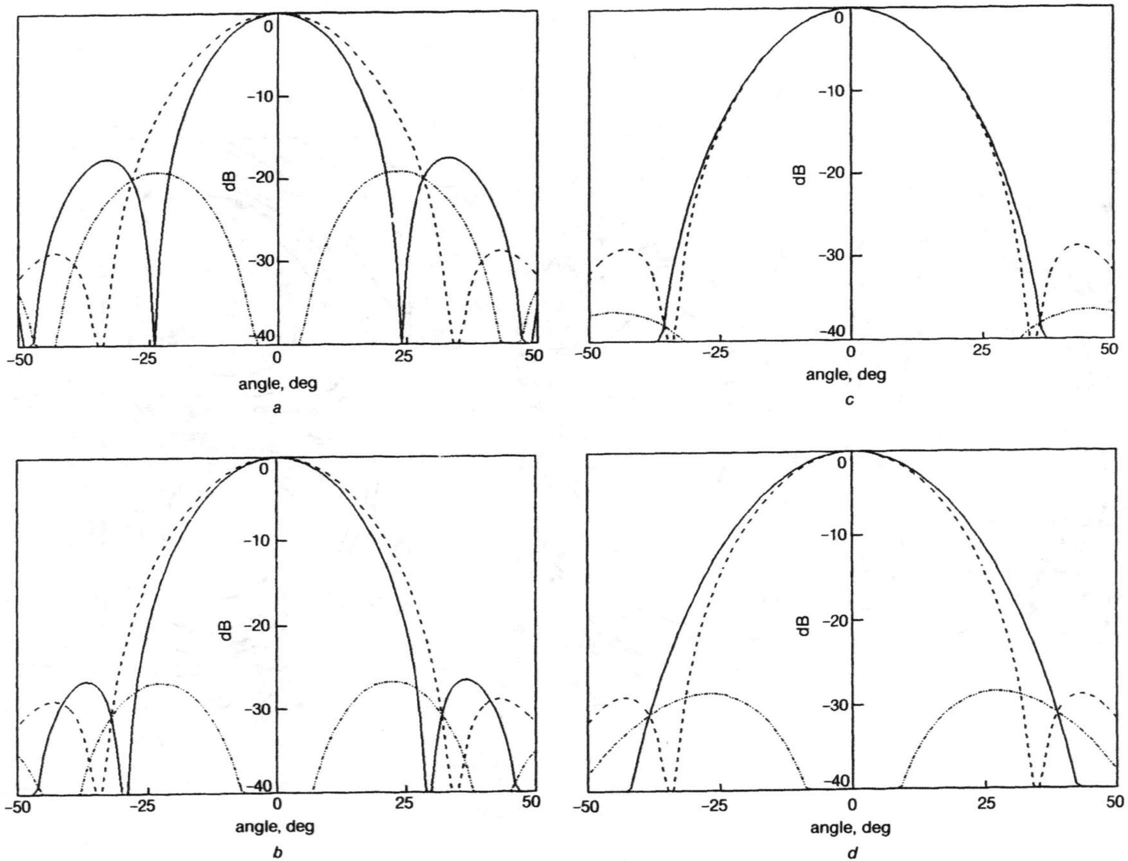

Figure 8.4 Radiation patterns showing addition of TM_{11} mode to TE_{11} mode

——— E-plane
– – – – H-plane
–·–·–·– crosspolarised
a TE_{11} mode alone
b TE_{11} + 5% TM_{11} mode
c TE_{11} + 15% TM_{11} mode
d TE_{11} + 25% TM_{11} mode

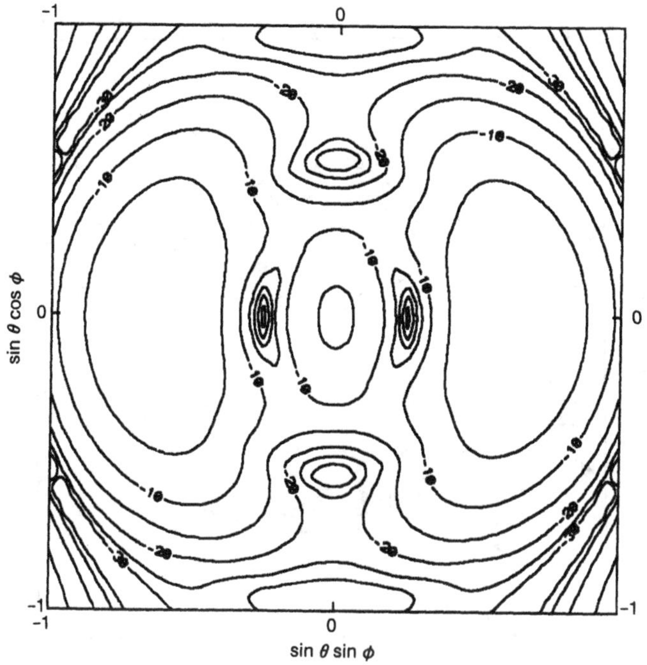

a

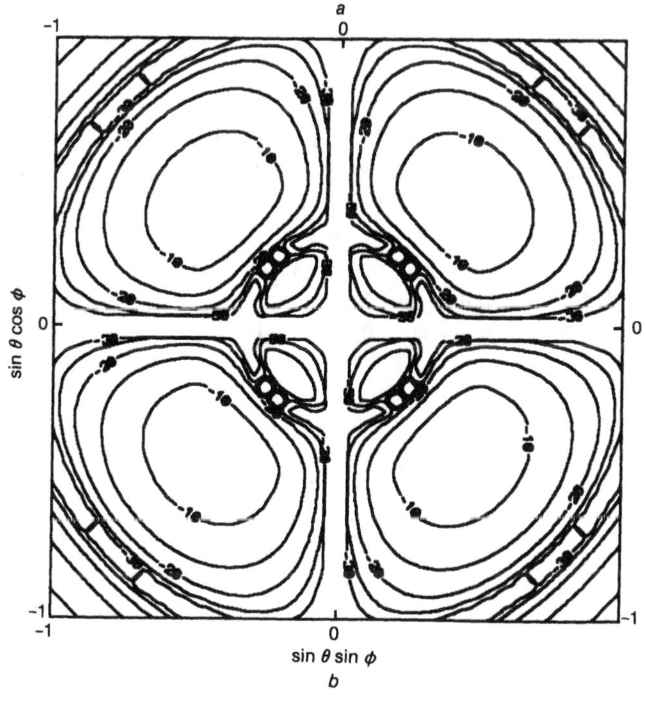

b

Figure 8.5 Contour patterns for TE_{12} and TM_{12} modes

a TE_{12} copolar; *b* TE_{12} crosspolar; *c* TM_{12} copolar

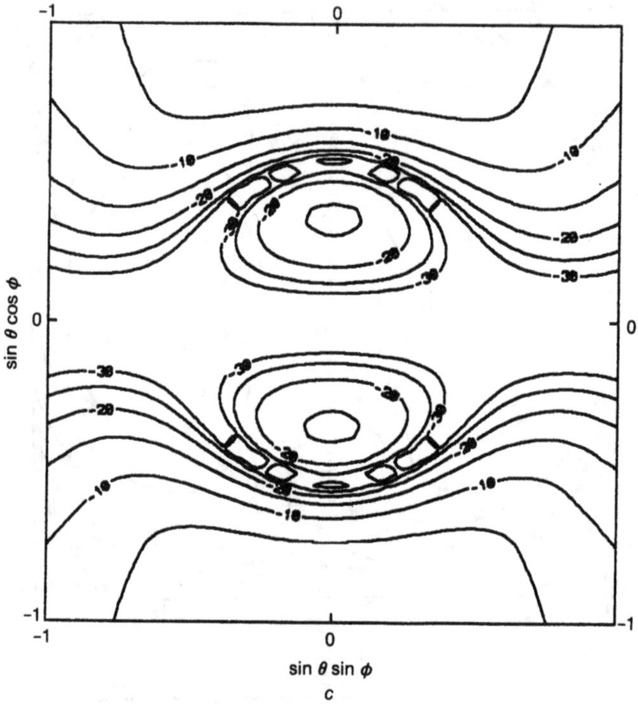

Figure 8.5 continued

to minimise their excitation level. TE_{12} mode is easily excited by increases in flare angle, whilst TM_{12} mode is easily excited by decreases in flare angle. The latter situation arises in profiled horns where the incremental flare angle first increases and then decreases (Figure 8.6). In the second case, the addition of the appropriate amount of TE_{12} mode and/or TM_{12} mode can, in theory, produce a copolar main beam which has a flatter top and a steeper falloff. The crosspolar level may also be raised, but for single-polarisation systems this may not matter.

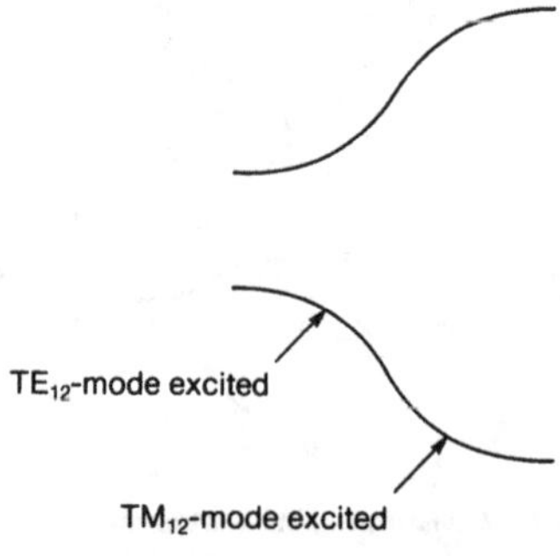

Figure 8.6 Profiled horn

The remaining modes which need to be examined in circular waveguide are modes which can be used in tracking feeds to give monopulse patterns with nulls on axis: the TM_{01}, TE_{01} and TE_{21} modes (Figure 8.7). In these cases the tracking or difference modes will often be used independently of the TE_{11} mode and will be extracted by a suitable beam-forming network behind the horn. The TM_{01} mode has a null along the H-plane and peaks in the E-plane. It can therefore be used to give tracking data in the E-plane. The TE_{01} and TE_{21} modes have very similar patterns with nulls along the E-plane and peaks in the H-plane. Either of these modes can therefore be used to provide tracking data in the H-plane. The TE_{01} mode in circular waveguide was extensively studied in the 1960s and 1970s for millimetre-wave low-loss waveguide and it is still used for high-power plasma diagnostics. High-performance converters have been designed, but they are physically bulky and therefore not well suited to antenna applications. The result is that the TE_{01} mode is not the preferred mode for tracking horns. A combination of TM_{01} and TE_{21} mode will give full tracking data.

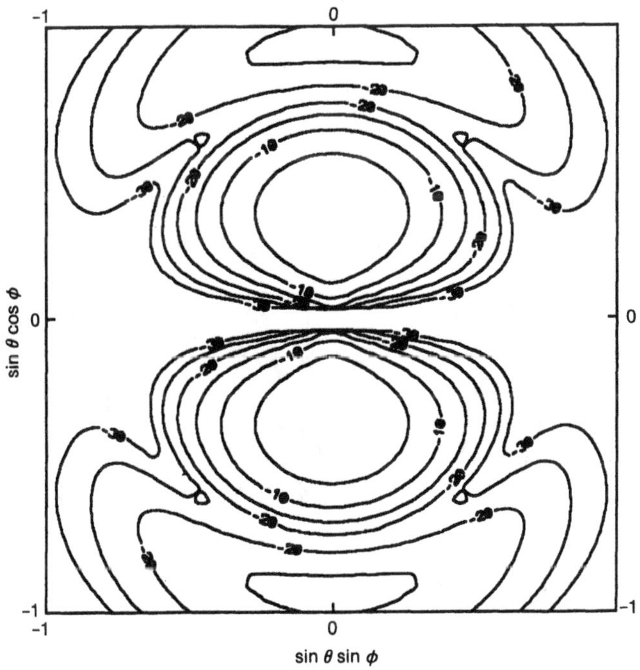

a

Figure 8.7 Contour patterns for various modes

a TM_{01} mode
b TE_{01} mode
c TE_{21} mode

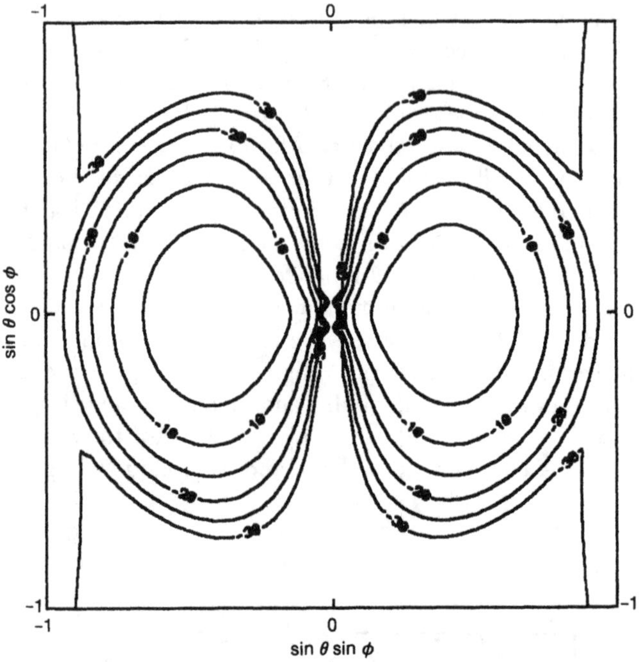

b

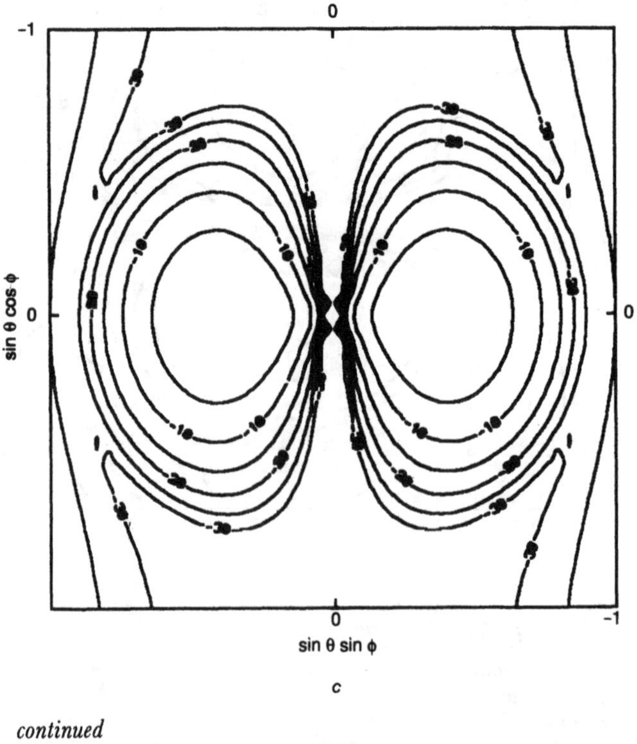

c

Figure 8.7 continued

8.2.2 *Rectangular waveguides*

Rectangular-cross-section horns have found fewer uses than circular horns when used as feeds for reflectors. They have independent E- and H-plane patterns owing to the separability of the two planes in the rectangular-co-ordinate system. This can in some cases be used to advantage, but in general it is a disadvantage. The shapes of horn which produce symmetric patterns are not physically very useful. This is particularly the case when a high-gain aperture is required. The consequence is that the major practical application of rectangular apertures is in arrays when the array pattern is the dominant design factor and the characteristics of the rectangular aperture take second place.

Nevertheless, it is theoretically possible to design rectangular horns so as to produce similar patterns to circular horns. Higher-order modes can be used in the same way to produce symmetric patterns, shape the main beam or give tracking data. The transverse fields in rectangular waveguide for TE_{nm} modes are given by

$$\left.\begin{aligned} E_x &= A_{nm}\frac{j\omega\mu}{K_c^2}\frac{m\pi}{b}\cos\left(n\pi\frac{x}{a}\right)\sin\left(m\pi\frac{y}{b}\right)\\ E_y &= -A_{nm}\frac{j\omega\mu}{K_c^2}\frac{n\pi}{a}\sin\left(n\pi\frac{x}{a}\right)\cos\left(m\pi\frac{y}{b}\right)\end{aligned}\right\}\tag{8.4}$$

and the transverse fields for TM_{nm} modes are given by

$$\left.\begin{aligned} E_x &= -A_{nm}\frac{j\beta}{K_c^2}\frac{n\pi}{a}\cos\left(n\pi\frac{x}{a}\right)\sin\left(m\pi\frac{y}{b}\right)\\ E_y &= -A_{nm}\frac{j\beta}{K_c^2}\frac{m\pi}{b}\sin\left(n\pi\frac{x}{a}\right)\cos\left(m\pi\frac{y}{b}\right)\end{aligned}\right\}\tag{8.5}$$

Note that the TE and TM modes have the same dependence in the x and y planes. The dominant mode is the TE_{10} mode which has $E_x=0$ and no dependence in the y direction. The aperture electric fields are relatively easy to visualise from the above equations. Modes with n or m even have peak fields along the axes. Modes with n or m odd have nulls along the axes and can therefore be used to obtain tracking data. Sketches of the aperture electric fields for the most important modes are shown in Figure 8.8. The TE_{12}/TM_{12} pair can be used in the same way as the TM_{11} mode in circular horns to improve pattern symmetry and lower the crosspolar levels.

8.2.3 *Other waveguide cross-sections*

One of the earliest uses of multiple modes was the diagonal horn proposed by Love [2] (Figure 8.9), which uses a combination of TE_{01} and TE_{10} modes excited with equal amplitude and phase from a section of standard rectangular waveguide. The combination gives symmetric fields in the aperture which radiate a symmetric copolar pattern. Design details are given by Love [2, 37].

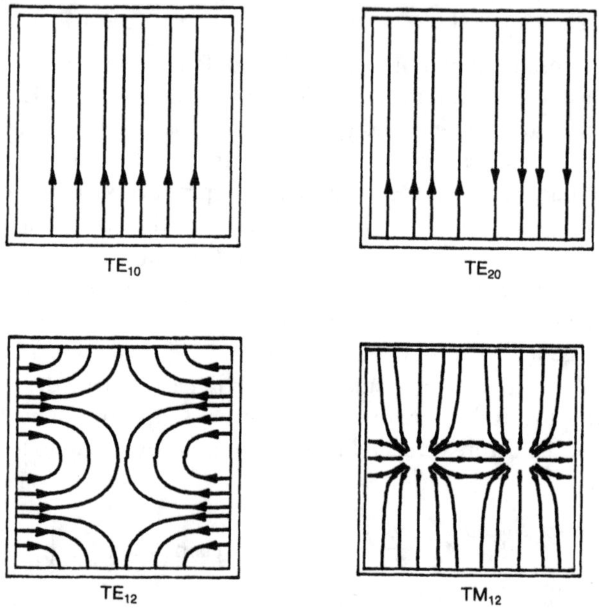

Figure 8.8 *Transverse electric fields for TE_{01}, TE_{20}, TE_{12} and TM_{12} modes in the aperture of a rectangular waveguide*

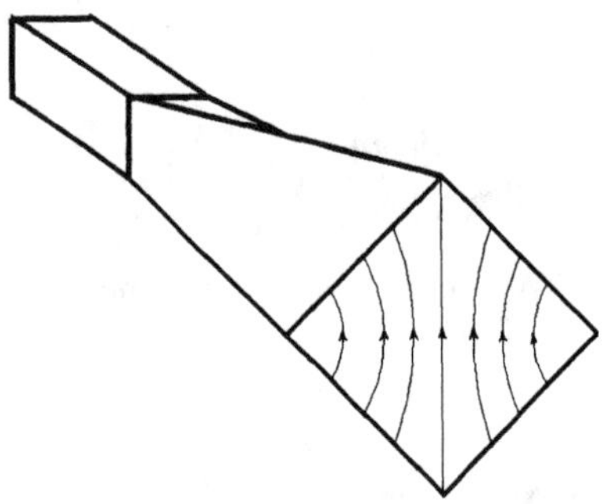

Figure 8.9 *Diagonal horn*

Horns which propagate hybrid modes, such as the corrugated-horn and dielectric-loaded horns, can radiate multiple hybrid modes by an extension of the principles outlined in the previous sections. Most corrugated and dielectric-loaded horns have circular cross-sections and the principles of using combinations of pure TE and TM modes discussed in Section 8.2.2 can be extended directly to circular or conical HE and EH hybrid modes. The corrugated horn with the HE_{11} mode is usually used to achieve low crosspolarisation so that it is not relevant to consider multimode hybrid horns for low crosspolarisation. The

HE_{12} mode can be added to the dominant HE_{11} mode to broaden the copolar main beam and increase the aperture efficiency of prime-focus reflector antennas (Figure 8.10).

The final type of multimode horn to be mentioned is the coaxial waveguide. This is relevant because it can be used in combination with a small-diameter circular waveguide (Figure 8.11), to add a coaxial TE_{11}- and coaxial TE_{12}-mode pattern to the dominant circular-waveguide TE_{11}-mode pattern. The coaxial-radiation patterns by themselves show no useful feature, which is why coaxial cable is never used as a radiator. However, the addition of the coaxial modes to a TE_{11} mode can produce results similar to those sketched in Figure 8.10.

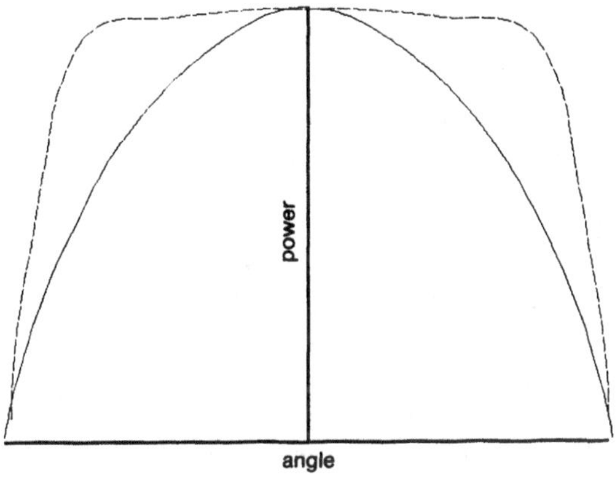

Figure 8.10 Addition of HE_{12} mode to HE_{11} mode in a corrugated horn

——— HE_{11} mode alone
- - - - $HE_{11} + HE_{12}$ modes

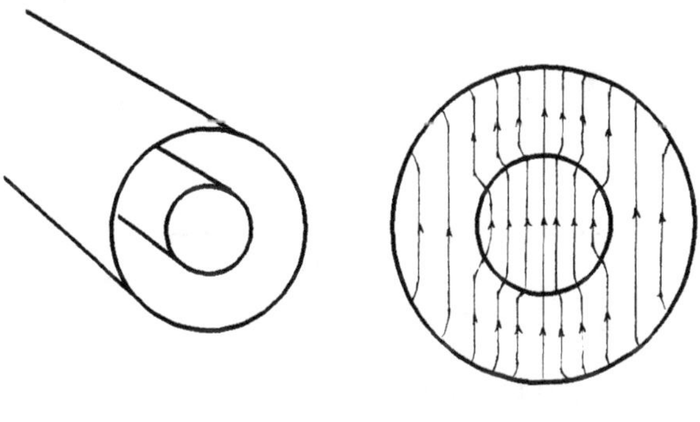

Figure 8.11 Fields in the aperture of a coaxial waveguide

8.3 Excitation of higher-order modes

The principles of the excitation of higher-order modes will be discussed with reference to conical horns which use the dominant TE_{11} mode and one or more higher-order modes with unity azimuthal dependence. The excitation of rectangular higher-order modes follows the same theory. Unity azimuthal modes of the TE_{1m} or TM_{1m} type ($m>1$) can be excited by abrupt or gradual changes in the diameter of the horn or waveguide. This change in diameter will not excite TE_{nm} or TM_{nm} modes ($n \neq 1$), as is evident by examination of the orthogonality condition (eqn. 8.2). To excite nonunity azimuthal-dependent modes, there must be some change in the coaxial symmetry of the waveguide. Possible schemes are offsetting the axis of the input waveguide with respect to the axis of the horn or inserting an asymmetric probe (metal or dielectric) into the waveguide.

The easiest method of exciting higher-order modes with unity azimuthal dependence is to introduce a step change into the diameter of the horn (Figure 8.12*a*). The abrupt junction forces a break in the smooth current pattern and, if the normalised output radius is greater than the cutoff wavenumber for the desired mode (i.e. $2\pi r_2/\lambda > K_c r_2$), some power will be transferred to that mode. The analysis of this type of junction has been studied by Tomiyasu [8], Agarwal and Nagelberg [9] and English [17]. The amount of power will depend on both r_1/λ and r_2/λ. Changes in the flare angle of a horn (Figure 8.12*b*) will also excite higher-order modes, so even the continuous flare angle of all conical horns excites a small amount of higher-order mode. This can be demonstrated by dividing the horn into a series of short steps: a valid procedure as demonstrated by using the modal-matching technique described in Chapter 4. Clearly, each short step can potentially excite higher-order modes although, since the ratio of r_2/r_1 will be very small, only a small amount of power will be excited at each junction.

A common requirement is to excite a proportion of TM_{11} mode to add to the TE_{11} mode, as in the Potter horn [3]. The amount of power converted is shown

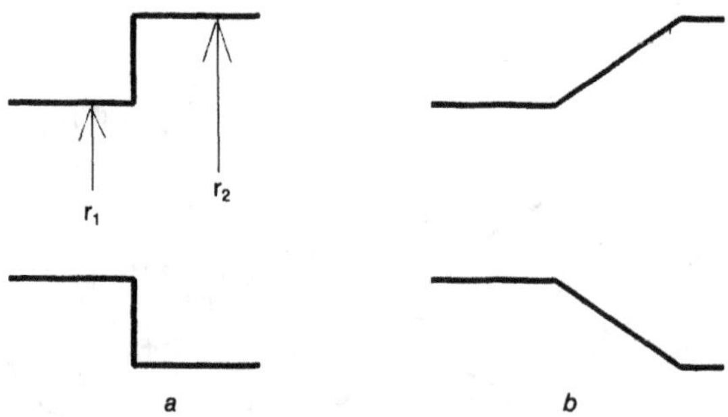

Figure 8.12 Methods of exciting higher-order modes
a Step change; *b* Gradual change

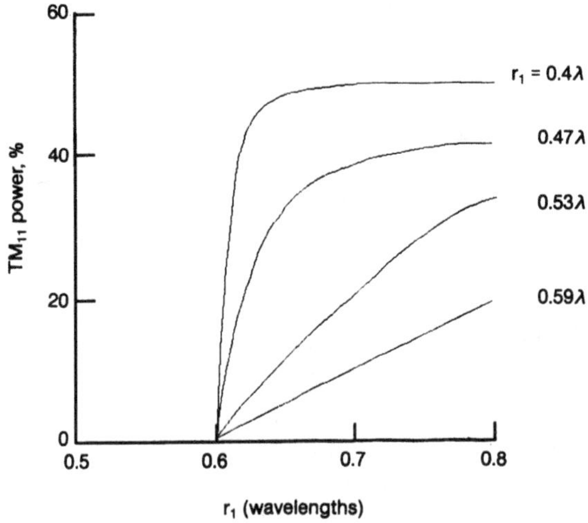

Figure 8.13 Proportion of TM_{11}-mode power converted by step junction illustrated in Figure 8.12a

in Figure 8.13 as a function of the input and output radii. The input circular waveguide is assumed to be propagating the TE_{11} mode. The output radius should be greater than 0.6098 λ in order that the TM_{11} mode can propagate. If it is not greater than this value, the mode will be evanescent. Power will still be converted to the TM_{11} mode but will have a z dependence of $\exp(-\alpha z)$ so that it will decay rapidly. If the junction is followed by a second junction, there may be enough power remaining in the mode to excite more power at the second junction. This situation can be predicted using the modal-matching theory of Chapter 4, but it is complicated and not recommended for a straightforward design.

If the output radius r_2 is greater than 0.8485 λ, the TE_{12} mode will be capable of propagating. It is preferable to have the desired amount of TM_{11} power excited before this radius is reached. Examination of Figure 8.13 leads to the conclusion that more control, and greater bandwidth, is available by operating as far above the TE_{11}-mode cutoff as possible. The input radius is limited to below the cutoff of the TM_{11} mode at $r_1 = 0.6098\ \lambda$. The amount of power which is normally required to be transferred to the TM_{11} mode is between 10 and 20%. A practical design is to use an inner radius greater than 0.53 λ. This value of inner radius is greater than that normally used for single-mode circular waveguides, so two steps are needed. The first junction keeps the power in the TE_{11} mode and the second junction excites the TM_{11} mode. The remainder of the horn is required to:

(*a*) flare the horn so that the aperture diameter will radiate the desired beamwidth (*b*) phase the TM_{11} mode so that it is in phase with the TE_{11} mode at the aperture

(*c*) excite as little extra TM_{11} mode as possible.

Requirement (*c*) implies a very gentle flare, although some power will always be converted. For example, a horn with a semiflare angle of 6.5° and a 3 λ aperture diameter converts 1% of TE_{11} mode to TM_{11} mode along the horn. Requirement (*b*) can be met either by choosing the flare angle appropriately, or by adding a section of constant-diameter waveguide after the exciting step.

An alternative method of exciting the TM_{11} mode is to insert a ring of dielectric into the horn. This was first suggested by Satoh [15]. The dielectric behaves in the same way as the abrupt metal step but, since fields exist inside the dielectric, the process of exciting power into the TM_{11} mode is more complex. Fortunately, this gives both more control and greater bandwidth. A modal-matching-type program is required to design the horn, and simple rules of thumb for the size of the dielectric ring are not available.

8.4 Shaped-pattern horns

The design philosophy for some multimode horns is discussed in this section. This uses the theoretical considerations discussed in previous sections, together with information in the literature.

8.4.1 Horns for pattern symmetry and low crosspolarisation

One of the most successful uses of multiple modes is the cancellation of the high crosspolar power of the TE_{11} mode in narrow-flare-angle smooth-wall conical horns. The principles have been described in Sections 8.2 and 8.3. This type of horn has become known as the 'Potter' horn after the author of the first paper describing such a horn [3]. That paper, published in 1963, was directed towards achieving copolar pattern symmetry and suppressing sidelobes. Dual-polarisation operation was not needed at that time, so the crosspolarisation aspects were not addressed. The horn really came into its own when the need for low crosspolarisation was evident. Together with the corrugated horn, the Potter horn has been the main choice for low-crosspolar feed horns. The limitation of the Potter horn is its narrow bandwidth, which must be weighed against simplicity of construction compared with the corrugated horn.

Potter [3] gave a cross-section sketch of his horn, and this basic design has been followed by most users. Although arrived at largely experimentally, the basic design in Reference 3 is about the best that can be achieved. Figure 8.14 shows the cross-section of a Potter horn with a 3 λ aperture which has been designed using a modal-matching program. The theory of the excitation step has been discussed in Section 8.3. Moving from the left, the first step is needed to move the TE_{11} mode away from cutoff without exciting any TM_{11} mode (which is still cutoff). It also improves the input impedance. The length of the waveguide section L_1 is not critical. The second step excites the TM_{11} mode so

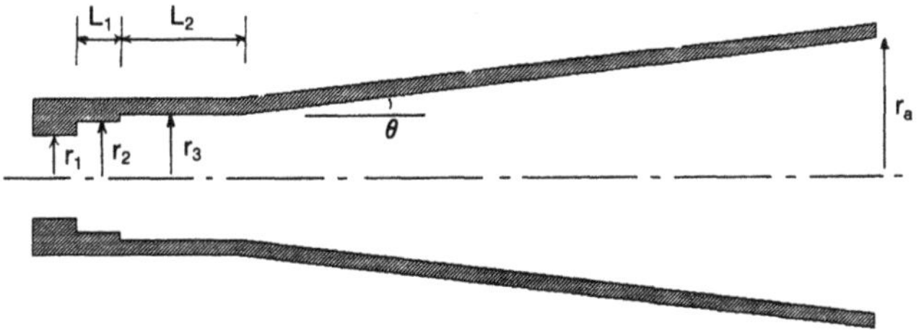

Figure 8.14 *Cross-section of a Potter horn*

that 16% of the power is in the TM_{11} mode. The right-hand-side radius is chosen so that it is below the cutoff of the TE_{12} mode. The waveguide section L_2 is a phasing section whose length is chosen to put the two modes in phase at the aperture. This is 1.25 free-space wavelengths long, but shorter in guide wavelengths. The horn section has a semiflare angle of 6.5°. A small amount of mode conversion occurs at the junction between the phasing section and the conical section. Also a small amount of mode conversion takes place to the TM_{11} mode and other higher-order modes along the conical section. In this design the two components of TM_{11}-mode generation cancel so that the result is still 16% of power in the TM_{11} mode at the aperture. The radiation patterns for the horn are shown in Figure 8.15. The peak crosspolarisation at the design frequency is -30 dB. This is significantly greater than the optimum TE_{11}/TM_{11} mode combination (Figure 8.4) owing to the unavoidable presence of TE_{1m}/TM_{1m} modes. The bandwidth is only a few percent, as observed from Figure 8.16. The flare angle is not particularly critical, except to note that, as the flare angle

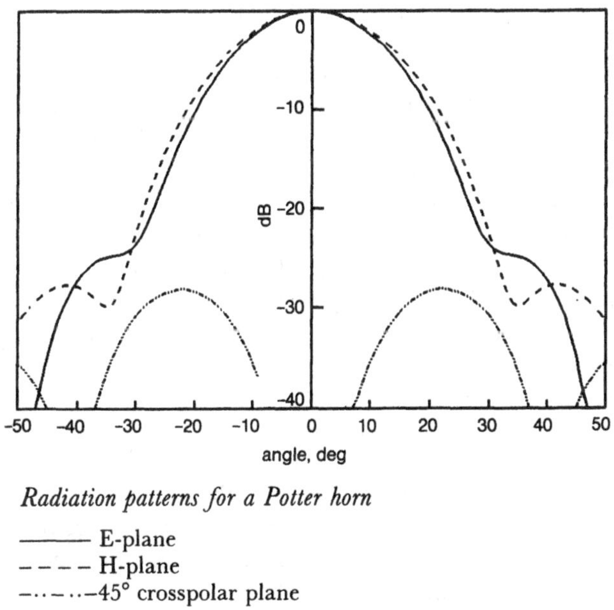

Figure 8.15 *Radiation patterns for a Potter horn*

——— E-plane
– – – – H-plane
–··–··–45° crosspolar plane

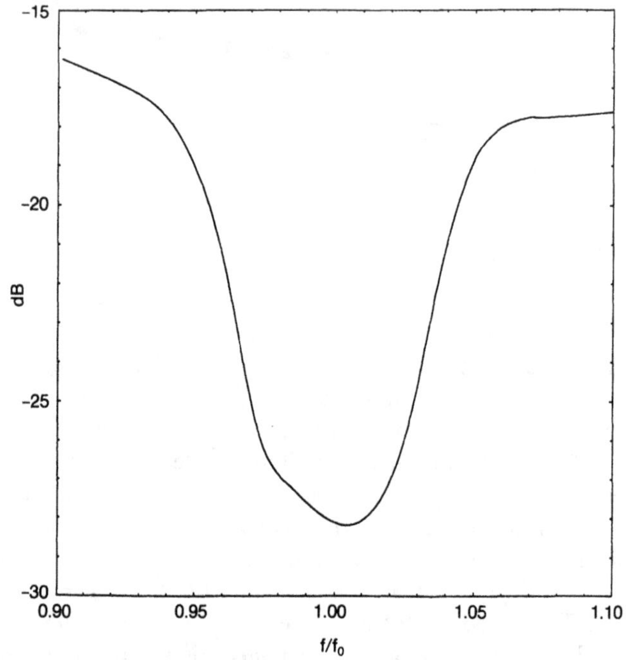

Figure 8.16 Peak crosspolar level against normalised frequency for a Potter horn

increases, the mode conversion along the horn increases. A horn with the same input sections and a semiflare angle of 10° had 21% of power in the TM_{11} mode at the aperture and a peak crosspolar level of −27 dB.

A detailed study of the variation of crosspolarisation with geometry has been carried out by Lier [41]. The TM_{11} mode can also be excited by a sharp change in flare angle (Figure 8.12*a*), as described by Turrin [6] and Cohn [10]. The bandwidth is slightly improved, but not to an extent that makes the greater uncertainty in the design worthwhile. The Potter horn has a role at higher frequencies where it is easier to fabricate at millimetre and submillimetre wavelengths than the corrugated horn, as has been demonstrated by Ediss [40].

Improved bandwidth is possible by using a dielectric ring to excite the TM_{11} mode (Figure 8.17). This was first proposed by Satoh [15,16] and has been demonstrated by Raghavan *et al.* [42] and further analysed by Leong *et al.* [43]. The dielectric section supports hybrid HE_{11} and EH_{11} modes, so that conversion from and to the pure TE_{11} and TM_{11} modes takes place at the input and output to the dielectric section. The horn shown in Figure 8.17*a* is very compact; the resulting improvement in crosspolar bandwidth compared with an equivalent Potter horn is shown in Figure 8.17*b*. For crosspolarisation below −30 dB, the bandwidth has increased from 3% to 9%. The Satoh-type horn could be particularly attractive at millimetre wavelengths.

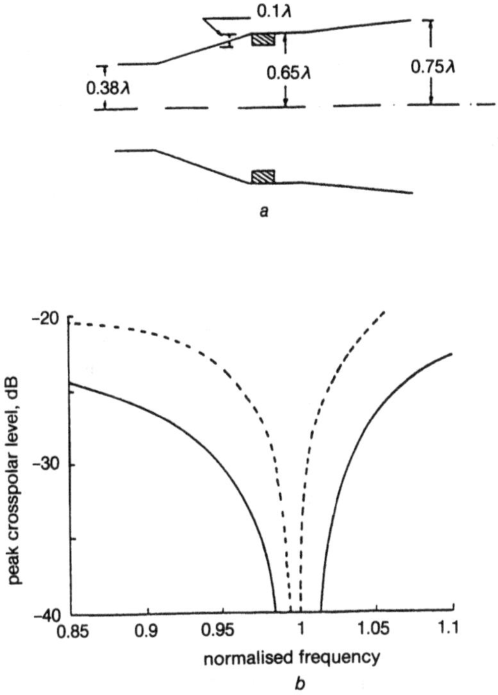

Figure 8.17 *Improvement in bandwidth by use of a dielectric ring*

a Cross-section
b Peak crosspolar level against normalised frequency
——— horn alone
- - - - horn plus dielectric ring

Pyramidal horns can be modified in a similar way to conical horns, by using the TE_{12} and TM_{12} combination. This was first shown by Jensen [4] and studied by others [22,23,29]. The principles of design have been described by Love *et al.* [35]. They follow the same philosophy as for the conical horn. The pyramidal horn has been less used in practice because the horn is more difficult to make accurately, and only produces symmetric patterns for an inconvenient ratio of the two sides of the rectangular aperture. Also the basic crosspolar level of the TE_{10} mode in the pyramidal horn is low, so the need to compensate is reduced.

8.4.2 Horns for increased aperture efficiency

The aperture efficiency of a reflector antenna with $F/D<0.35$ and fed by a prime-focus horn is limited to about 70% by the Gaussian shape of the main beam radiated by a conical or rectangular horn. It can be increased by adding a proportion of a higher-order mode to the dominant mode in the horn. The higher-order mode should have low crosspolarisation and a copolar pattern which, when added to the dominant mode, tends to flatten the top of the main beam. This can be done with TE_{12} mode in a conical horn (see Section 8.2.1) but is not easily implemented for two reasons. First, any scheme which excites the TE_{12} mode will also excite the TM_{11} mode. Secondly, the horn will generally be long, which is undesirable for a prime-focus feed. The principle can, however, be implemented with corrugated waveguide using the HE_{12} mode in addition to the HE_{11} mode. Such a horn was first developed for use with a radiotelescope by the CSIRO group in Australia [12–14]. The principle works with the corrugated horn because the hybrid mode can be thought of as a linear combination of TE and TM modes.

The cross-section of an open-ended corrugated-waveguide horn using the HE_{12} mode is shown in Figure 8.18*a* and the radiation patterns in Figure 8.18*b* for different lengths of waveguide. This shows how the copolar patterns can be changed by the addition of the higher-order mode.

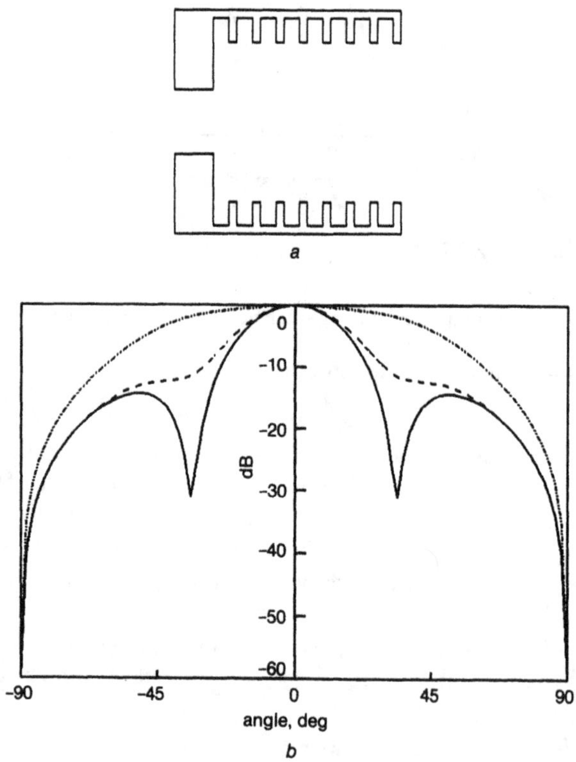

Figure 8.18 High-aperture-efficiency corrugated horn using HE_{11} mode plus HE_{12} mode

a Cross-section
b Copolar radiation pattern
——— four sections
– – – – six sections
–··–··– eight sections

The aperture efficiency can similarly be increased by using the same hybrid modes in a dielectric-loaded horn [44]. This has the added advantage that the dielectric reduces the diameter of the horn. A very compact open-ended waveguide feed with an aperture diameter of 0.8 λ is shown in Figure 8.19. This has a computed and measured efficiency of 75% illuminating a reflector of $F/D = 0.35$.

A number of designs have appeared in the literature for compact circular feeds, meaning horns which are very short in length and designed for prime-focus reflectors. These find ways of exciting the TM_{11} mode in a compact section. They are ingenious empirical solutions for specific applications, and by no means exhaust the possibilities. Koch [18] used multimode coaxial feeds surrounding a circular waveguide, similar to Figure 8.11. The coaxial waveguide was designed to support the TE_{11} and TE_{12} coaxial modes which add together in a similar fashion to the TE_{11}/TM_{11} circular modes to linearise the aperture fields. The correct phase between the inner circular waveguide and the coaxial waveguide was obtained by a dielectric insert in the coaxial waveguide. Efficiencies were measured of between 68% and 75% with a front-fed parabola.

Clavin [25] (Figure 8.20) used a narrow slot at the end of piece of rectangular waveguide and a pair of monopoles to excite the TM_{11} mode in a short length of open-ended circular waveguide. This feed is suitable for reflectors with F/D ratios of about 0.3. The disadvantage is the narrow bandwidth and the high input impedance, although this can be matched at the design frequency. A design by Wolf and Sommer [45] is intended for use as an element in an array (Figure 8.21). This operates on a similar principle to the Koch design and has a

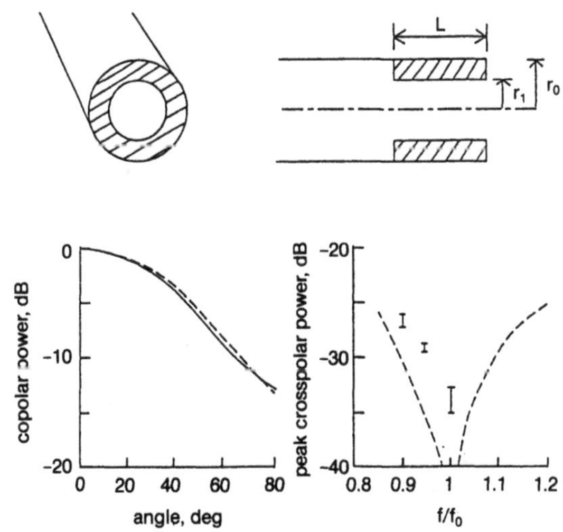

Figure 8.19 High-aperture-efficiency dielectric-loaded horn

Theory and measurement for copolar and crosspolar patterns
——— I measured
– – – – theory

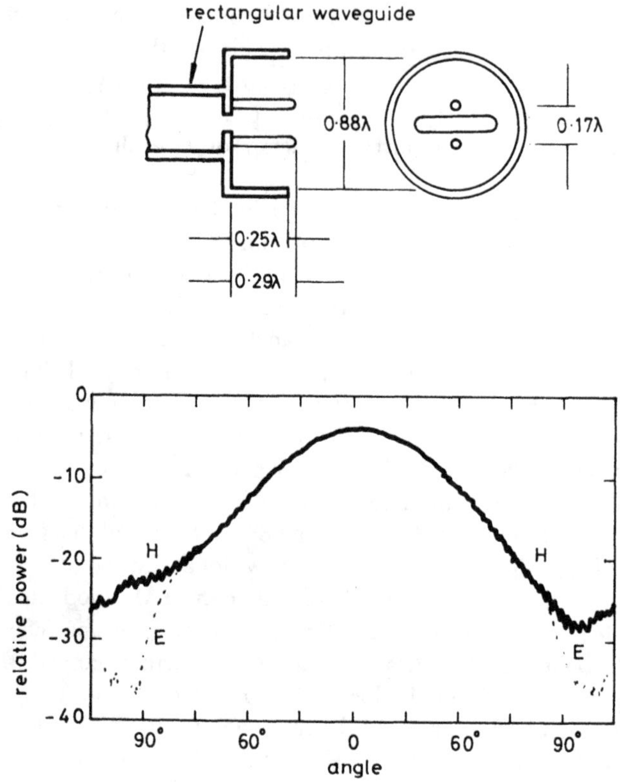

Figure 8.20 Dual-mode prime-focus feed of Clavin [25]

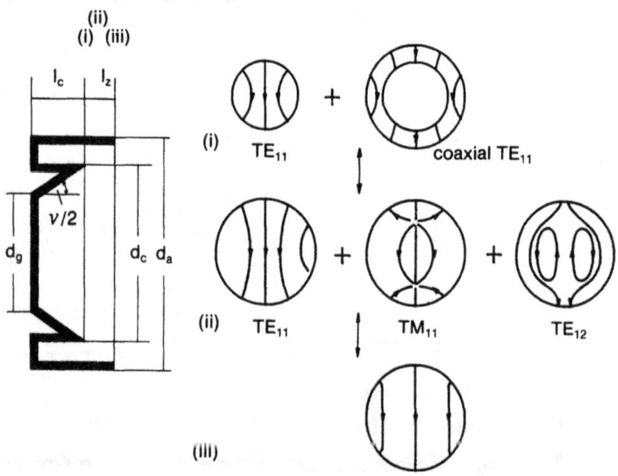

Figure 8.21 Short multimode horn using TE_{11}, TM_{11} and TE_{12} modes

short conical horn section supporting the TE_{11}, TM_{11} and TE_{12} modes and a coaxial section supporting the TE_{11} mode. A version with a 1.2 λ outer diameter has a depth of about one wavelength. The aperture efficiency was 77% with a peak crosspolar level of −28.5 dB.

8.4.3 Matched feed horns for offset reflectors

The focal-region fields of offset-reflector antennas contain asymmetric crosspolar components owing to the offset geometry of a parabola. This results in high crosspolarisation levels for linear polarisation and a beam squint for circular polarisation. Single-mode feeds such as the corrugated horn provide a good match to the copolar fields but not to the crosspolar fields. It was shown by Rudge *et al.* [26–28,30] that a combination of multiple modes could be used to match exactly the focal-region fields of an offset reflector. This is called a 'matched' feed. The theory and implementation is fully described in Reference 35, pp. 399–410. In a conical horn, the TE_{21} mode provides the fields which cancel the asymmetric components and the TM_{11} mode is added as usual to cancel the crosspolarisation. Thus a 3-mode conical horn results. The TE_{21} mode is first excited by an asymmetric step and this is followed by a symmetric step to excite the TM_{11} mode. It is also possible to use a dual-mode corrugated horn using the HE_{11} and HE_{21} modes, or a dual-mode rectangular horn using either the TE_{01} and TE_{11}/TM_{11} pair or the TE_{01} and TE_{20} pair, depending on the plane of polarisation in the offset reflector.

8.5 Tracking horns

Tracking or monopulse horns use multiple modes to obtain tracking data in addition to the signal information. The tracking data need to have a null on axis so that a mode in the horn is required which radiates a difference pattern. The horn must propagate both the normal sum mode (i.e. HE_{11} mode in conical corrugated horns, TE_{11} mode in conical smooth-wall horns, or TE_{10} mode in rectangular horns) and the higher-order modes. Suitable modes in circular waveguides are seen from Section 8.2.1 to be the TM_{01} mode and the TE_{21} mode. These two modes have lobes in orthogonal planes (see Figure 8.7) so that a combination of both modes will give orthogonal tracking data. In tracking horns, the horn section merely needs to be designed to allow all the desired modes to travel along the full length of the horn.

There are two approaches to the design or tracking of monopulse horns. In one approach (Figure 8.22), the higher-order modes are obtained by using two or more rectangular waveguides abruptly joined to large-diameter waveguide

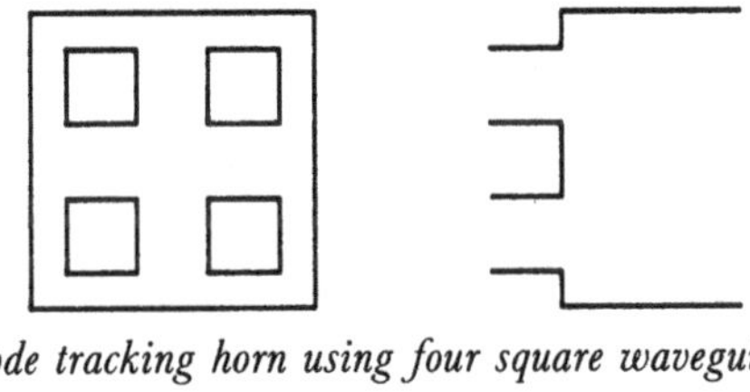

Figure 8.22 Multimode tracking horn using four square waveguides

[31,34,36,46]. The difference and sum patterns are obtained by suitable combinations of the outputs of the rectangular waveguides. In the other approach [33,39,47], the modes are taken into a circular waveguide and a beam-forming network is used to extract the tracking data. The latter method can give a high-quality, high-sensitivity system which does not compromise the performance of the horn for the signal channel. Figure 8.23, from Watson *et al.* [33], shows the beam-forming network to extract the tracking data using the TM_{01}- and TE_{21}-mode combination. Later work by the same author substituted an electronic tracking system which replaces the waveguide beam-forming network [39]. Both these systems are designed for onboard satellite use where high performance is the main design criterion although the cost is relatively high.

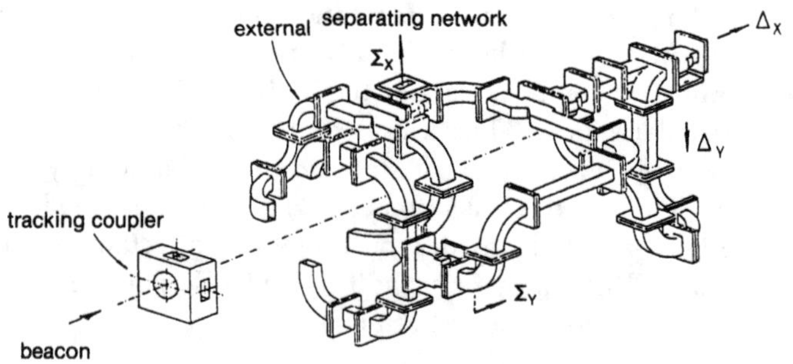

Figure 8.23 Beam-forming network [33]

The monopulse systems using abrupt junctions are lower in cost and more suitable for radar systems to be produced in large numbers, at the penalty of compromised performance in the signal channel. A system using a corrugated waveguide and four square waveguides has been studied by Clarricoats and Elliot [34] (Figure 8.24). Designs were produced which gave good sum and difference patterns with crosspolar levels below −30 dB. The length of the corrugated waveguide is used to phase the sum and difference modes.

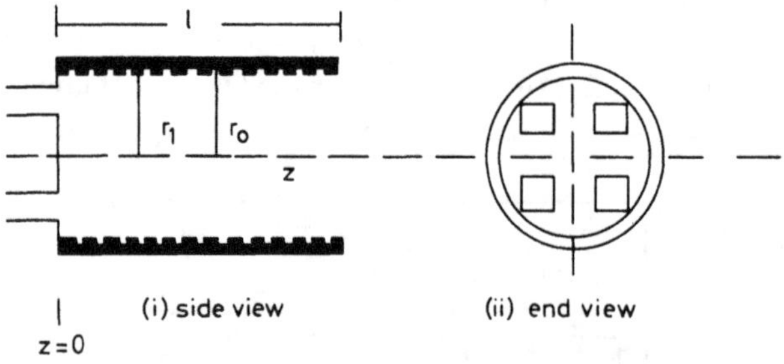

Figure 8.24 Monopulse-horn system [34]

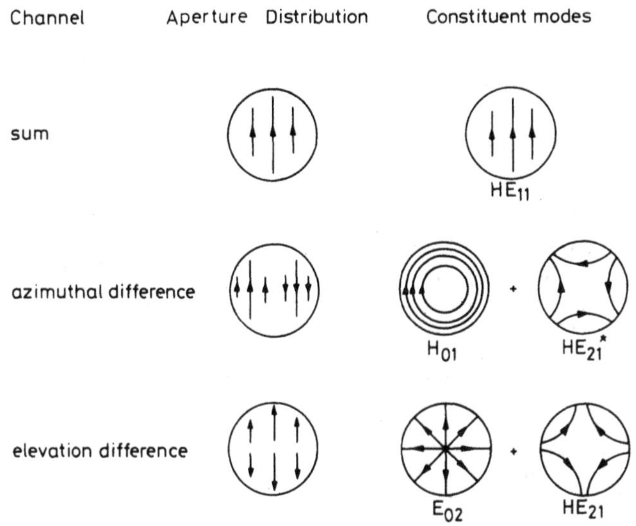

Figure 8.24 continued

8.6 References

1 COLLIN, R.E.: 'Field theory of guided waves' (McGraw–Hill, New York, 1960)
2 LOVE, A.W.: 'The diagonal horn antenna', *Microwave J.*, 1962, **5**, pp. 117–122
3 POTTER, P.D.: 'A new horn antenna with suppressed sidelobes and equal beamwidths', *Microwave J.*, 1963, **6**, pp. 71–78
4 JENSEN, P.A.: 'A low noise multimode Cassegrain monopulse feed with polarization diversity'. IEEE NEREM record, November 1963, pp. 94–95
5 COOPER, D.N.: 'A new circularly polarised monopulse feed system', *Proc. IEEE*, 1965, **53**, pp. 1252–1254
6 TURRIN, R.H.: 'Dual mode small aperture antennas', *IEEE Trans.*, 1967, **AP–15**, pp. 307–308
7 CLARRICOATS, P.J.B., and MASTERMAN, P.H.: 'A multimode monopulse feed for a satellite tracking antenna'. Proceedings of the Microwave Conference, September 1969
8 TOMIYASU, K.: 'Conversion of TE11 mode by a large diameter conical junction', *IEEE Trans.*, 1969, **MTT–17**, pp. 277–279
9 AGARWAL, K.K., and NAGELBERG, E.R.: 'Phase characteristics of a circularly symmetric dual-mode transducer', *IEEE Trans.*, 1970, **MTT–18**, pp. 69–71
10 COHN, S.B.: 'Flare angle changes in a horn as a means of pattern control', *Microwave J.*, 1970, **13**, pp. 41–46
11 AJIOKA, J.S., and HARRY, H.E.: 'Shaped beam antenna for earth coverage from a stabilized satellite', *IEEE Trans.*, 1970, **AP–18**, pp. 323–327
12 VU, T.B., and VU, Q.H.: 'Optimum feed for large radiotelescopes: Experimental results', *Electron. Lett.*, 1970, **6**, pp. 159–160
13 THOMAS, B. MacA.: 'Prime focus one-and-two-hybrid mode feeds', *Electron. Lett.*, 1970, **6**, pp. 460–461
14 COOPER, D.M.: 'Complex propagation coefficients and the step discontinuity in corrugated cylindrical waveguide', *Electron. Lett.*, 1971, **7**, pp. 135–136
15 SATOH, T.: 'Dielectric loaded horn antenna', *IEEE Trans.*, 1972, **AP–20**, pp. 199–201
16 SATOH, T.: 'Dielectric loaded antenna; wideband dual-mode antenna with rotationally symmetric beam and low sidelobe level', *IEEE Trans.*, 1972, **AP–20**, pp. 199–201

17 ENGLISH, W.J.: 'The circular waveguide step-discontinuity mode transducer', *IEEE Trans.*, 1973, **MTT–21**, pp. 633–636

18 KOCH, G.F.: 'Coaxial feeds for high aperture efficiency and low spill-over of paraboloid reflector antennas', *IEEE Trans.*, 1973, **AP–21**, pp. 164–169

19 BALDWIN, R., and McINNES, P.A.: 'Radiation patterns of dielectric loaded rectangular horns', *IEEE Trans.*, 1973, **AP–21**, pp. 375–376

20 VU, T.B.: 'Corrugated horn as high-performance monopulse feed', *Int. J. Electron.*, 1973, **34**, pp. 433–444

21 VU, T.B., and HIEN, N. V.: 'A new type of high performance monopulse feed', *IEEE Trans.*, 1973, **AP–21**, pp. 855–857

22 HARDY, W.M., GRAY, K.W., and LOVE, A.W.: 'An S-band radiometer design with high absolute precision', *IEEE Trans.*, 1974, **MTT–22**, pp. 382–390

23 HAN, C.C., and WICKERT, A.N.: 'Multimode rectangular horn antenna that generates circularly polarized elliptical beam', *IEEE Trans.*, 1974, **AP–22**, pp. 746–751

24 BRAIN, J.R.: 'A simple multi-mode tracking system', *Marconi Rev.*, 1974, **3**, pp. 164–180

25 CLAVIN, A.: 'A multimode antenna having equal E- and H-planes', *IEEE Trans.*, 1975, **AP–23**, pp. 753–757

26 RUDGE, A.W., and ADATIA, N.A.: 'New class of primary feed antennas for use with offset parabolic reflector antennas', *Electron. Lett.*, 1975, **11**, pp. 597–599

27 RUDGE, A.W., and ADATIA, N.A.: 'Matched feeds for offset parabolic reflector antennas'. Proceedings of 6th European Microwave Conference, Rome, 1976, pp. 1–5

28 RUDGE, A.W., and ADATIA, N.A.: 'Primary feeds for boresight jitter compensation of offset reflector radar antennas'. *IEE Conf. Publ. 155*, Oct 1977, pp. 409–413

29 PROFERA, C.E.: 'Dual mode pyramidal horns; complex patterns and fractional power distribution', *IEEE Trans.*, 1977, **AP–25**, pp. 436–438

30 WATSON, B.K., RUDGE, A. W., and ADATIA, N. A.: 'Dual polarised mode generator for cross-polar compensation in offset parabolic reflector antennas'. 8th European Microwave Conference, Paris, 1978

31 SAKR, L., and SURREMEJEM, A.: 'The performance of square and conical corrugated horns for monopulse feeds in millimetric band'. European Microwave Conference, Brighton, UK, 1979

32 LANDSTORFER, F.M., and SACHER, R.R.: 'Optimisation methods applied to horn antennas'. Proceedings of international conference on Antennas and Propagation, York, 1981, p. 319

33 WATSON, B.K., DANG, N.D., and GHOSH, S.: 'A mode extraction network for RF sensing in satellite reflector antennas'. Proceedings of international conference on Antennas and Propagation, York, 1981, p. 323

34 CLARRICOATS, P.J.B., and ELLIOT, R.D.: 'Multimode corrugated waveguide feed for monopulse radar', *IEE Proc. H*, 1981, **128**, pp. 102–110

35 LOVE, A.W., RUDGE, A.W., and OLVER, A.D.: 'Primary feed antennas' *in* RUDGE, A.W., MILNE, K., OLVER, A.D., and KNIGHT, P. (Eds.): 'The handbook of antenna design Vol 1' (Peter Peregrinus, 1982) IEE Electromagnetic Waves series 15, Chapter 4

36 CHOUNG, Y.H.: 'Conical horn excited by circular waveguide; radiation pattern calculation for sum and tracking modes', *IEEE Trans.*, 1984, **AP–32**, pp. 1288–1291

37 LOVE, A.W.: 'Horn antennas' *in* JOHNSON and JASIK, H. (Eds.): 'Antenna engineering handbook' (McGraw–Hill, New York, 1984)

38 PICKETT, H.M., HARDY, J.C., and FARHOOMAND, J.: 'Characterization of a dual-mode horn for submillimeter wavelength', *IEEE Trans.*, 1984, **MTT–32**, pp. 936–937

39 DANG, R., WATSON, B., and DAVID, I.: 'Electronic tracking system for satellite ground stations'. European Microwave Conference, 1985

40 EDISS, L.A.: 'Dual mode horns at millimetre and submillimetre wavelengths', *IEE Proc. H*, 1985, **132**, p. 215

41 LIER, E.: 'Cross-polarization from dual mode horn antennas', *IEEE Trans.*, 1986, **AP–34**, pp. 106–110

42 RAGHAVAN, K., OLVER, A.D., and CLARRICOATS, P.J.B.: 'Compact dual mode dielectric loaded horn', *Electron. Lett.*, 1986, **22**, pp. 1131–1132

43 LEONG, M.S., KOOI, P.S., and YEO, S.P.: 'Modal analysis of dielectric-sleeve loaded circular waveguide for dual mode applications', *IEE Proc. H*, 1987, **134**, p. 369
44 RAGHAVAN, K., OLVER, A.D., and CLARRICOATS, P.J.B.: 'Compact dielectric feeds for prime focus reflector antennas'. Proceedings of IEEE international symposium on Antennas and Propagation, Syracuse, New York, 1988, pp. 350–353
45 WOLF, H., and SOMMER, E.: 'An advanced radiator element for multi-feed antennas'. 18th European Microwave Conference, Stockholm, 1988
46 LEE, K.M., and CHU, R.S.: 'Design and analysis of a multimode feed horn for a monopulse feed', *IEEE Trans.*, 1988, **AP–36**, pp. 171–181
47 PATEL, P.D.: 'Inexpensive multi-mode satellite tracking feed antenna', *IEE Proc. H*, 1988, **135**, p. 381

Chapter 9
Corrugated horns

A.D. Olver

9.1 Introduction

Corrugated horns have become the preferred choice of feed antenna for use in reflector antennas for communications, radar and remote sensing where high performance is required. This is because of their superior radiation performance and in particular their high copolar pattern symmetry and low crosspolarisation. The history, principles, design philosophy, analysis techniques and types of horn are reviewed in this chapter. The intention is not to provide a comprehensive study of corrugated horns, as this is available in Reference [1]. This chapter concentrates on the general principles of design of corrugated horns and then describes developments since that book was published in 1984.

Corrugated horns were first conceived in the 1960s by Kay [2], Simons and Kay [3] and Minnett and Thomas [4, 5] when the interest was in achieving symmetric radiation patterns so that high-efficiency and low-sidelobe reflector antennas could be produced. The first comprehensive analysis of corrugated waveguides and horns was given by Clarricoats and Saha in 1971 [6]. Other significant papers were the first near-field predictions by Clarricoats, Olver and Saha [7]; a review of the design principles by Thomas [8]; an understanding of the propagation and radiation behaviour by Dragone [9]; the design of rectangular corrugated horns by Bryant [10]; and elliptical corrugated horns by Jeuken and Thurlings [11]; and the discovery of the low attenuation properties of corrugated waveguides by Clarricoats, Olver and Chong [12]. In the 1970s it was realised by Parini, Clarricoats and Olver [13] that corrugated horns also radiate very low levels of crosspolarisation, which is essential for dual-polarisation operation or frequency re-use. The low-crosspolarisation properties led to considerable research to accurately analyse corrugated horns and optimise the corrugations [14]. In addition to the papers mentioned above, many other workers have made important contributions. A comprehensive bibliography up to 1984 is given in Reference 1.

At first, the use of corrugated horns was limited. The relatively high mass and volume deterred applications, but by the middle 1980s their electrical advantages strongly outweighed the mechanical disadvantages and they began to be accepted as an ideal feed for a reflector. The development in the early 1980s of the modal-matching technique by Kuhn and Hombach [15] and James [16]

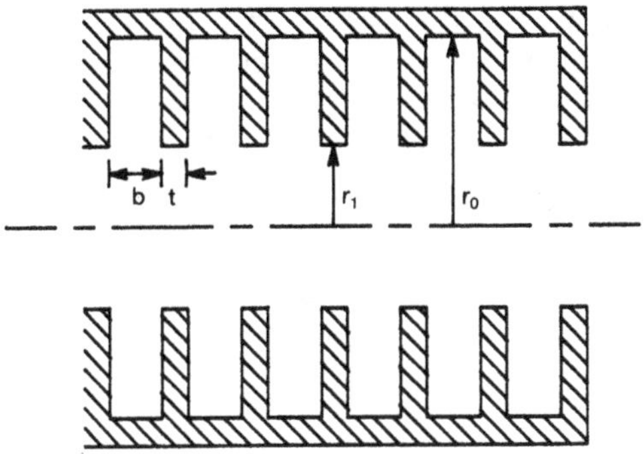

Figure 9.1 Corrugated waveguide

provided a good method of computer prediction of performance so that theoretical predictions can be relied upon to give an accurate indication of practical performance. The opening up of applications at millimetre wavelengths also led to the development of the Gaussian beam-mode method of analysis [17] building on a technique first proposed by Aubry and Bitter [19] which matches neatly with quasioptical design techniques. Corrugated horns may now be considered to be a relatively mature antenna technology, but each communication, radar or remote sensing system is different so the optimum solution cannot always be obtained from past experience. Consequently, the variety of corrugated-horn designs keeps increasing.

This chapter starts by explaining the principles for the good electrical behaviour of the fields in corrugated horns. The procedure for designing horns is outlined. The techniques which are available for analysing the propagation and radiation performance of corrugated horns are discussed. Then the different types of corrugated horns are reviewed. The References at the end of the chapter contain a bibliography of the most important papers which have been published since Reference 1 appeared.

9.2 Principles of operation of corrugated horns

The principle of operation of corrugated horns can be explained by considering the way in which the corrugated walls alter the field pattern in a corrugated waveguide (Figure 9.1). The corrugations change the fields so as to provide the desirable properties of axial beam symmetry, low sidelobes and low crosspolarisation.

The requirement to produce a radiated field which has a symmetrical main beam and a low crosspolar level can only be met by a horn which produces an aperture electric field in which the fields are nearly linear (Figure 9.2). In order to cancel all the crosspolarised components, a small amount of field curvature is required so that the fields must not be precisely linear. This was shown in Chapter 6. However, for the purposes of explanation this small deviation from linearity can be ignored. A linear electric field cannot be produced by waveguides which support pure transverse-electric (TE) or transverse-magnetic (TM) modes since they have aperture electric fields in which the field lines are curved. Only a 'hybrid' mode can produce the desired linear aperture field. This can be seen by examining the aperture fields of an open-ended corrugated waveguide. The dominant mode in a hybrid-mode waveguide produces the aperture electric fields [1]

$$\left.\begin{aligned} E_x &= A_1 J_0(Kr) - \frac{(X-Y)}{kr_1} A_2 J_2(Kr) \cos 2\phi \\ E_y &= \frac{(X-Y)}{kr_1} A_2 J_2(Kr) \sin 2\phi \end{aligned}\right\} \tag{9.1}$$

where $J_0(kr)$ and $J_2(kr)$ are Bessel functions of the first kind, K and k are the transverse and free-space wavenumbers, A_1 and A_2 are amplitude coefficients and X and Y are the impedance and admittance at the boundary $r=r_1$ given by

$$\left.\begin{aligned} X &= -j \frac{E_\phi}{H_z} y_0 \\ Y &= +j \frac{H_\phi}{E_z} y_0 \end{aligned}\right\} \tag{9.2}$$

and y_0 is the admittance of free space.

From eqn. 9.1 it is seen that, if the term $(X-Y)$ vanishes, the aperture field is independent of the angular variable ϕ and $E_y=0$ so that no crosspolarised field exists. For large apertures the radiated fields are the Fourier transform of the aperture fields so that these properties are transferred to the radiation field. The condition $(X-Y)=0$ can be realised either with both X and Y finite and equal or with X and Y both zero. It is this latter condition which is generated by the corrugated walls of the horn. The mode is then described as a 'balanced hybrid' mode because the electric and magnetic fields are exactly balanced to produce

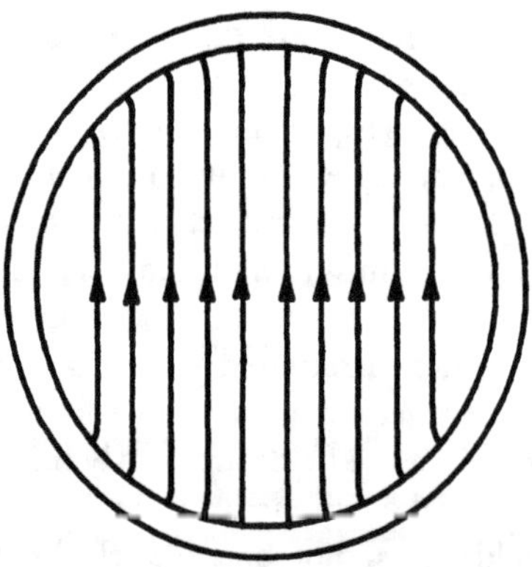

Figure 9.2 Ideal aperture electric field in a corrugated horn

pattern symmetry and low crosspolarisation.

If there are enough corrugations per wavelength, the azimuthal electric field will be zero at the edges of the corrugations, or $E_\phi = 0$ at $r = r_1$, and then $X = 0$. The other condition can be fulfilled by creating the circumstance that no current flows axially along the corrugated ridges. If the corrugations are narrow and quarter of a wavelength deep, they will act as short transmission lines where the short circuit at the end is transferred to an open circuit at the corrugation boundary. This ensures that no axial current will flow, and, since the axial current is generated by H_ϕ, this too will be zero and then $Y = 0$. The quarter-wavelength condition is based on treating the corrugated walls as a flat, plane surface. In fact the walls are annular and this means that the quarter-wavelength depth necessary to produce an open circuit is only correct at large radii. For smaller radii, an open circuit can still be created, but the depth of the slots needs to be increased so that the radial transmission line is open circuit.

The above description is a simple physical explanation as to why the corrugated walls produce an approximately linearly polarised aperture field. The open-circuit condition ($Y = 0$) is wavelength-dependent and hence zero crosspolarisation only occurs at a specific frequency. Away from this frequency, a crosspolar component of the aperture field will exist. However eqn. 9.1 shows that the crosspolar term decreases as $1/kr_1$ so that the larger the diameter of the horn, the wider will be the bandwidth for a given low crosspolar level.

The very low crosspolarisation means that the fields in the aperture are essentially scalar and for this reason, the corrugated horn is sometimes referred to as a 'scalar horn'. This property has most significance for long horns or for wide-flare-angle horns.

The explanation above has described the behaviour for an open-ended waveguide, but almost all horns have a finite flare angle. Then the radiation patterns are partly controlled by the additional factor of the spherical phase taper. This is a function of both the diameter of the horn and the flare angle so that it will increase for both a short horn with a wide flare angle and a long horn with a narrow flare angle. As the factor increases, the radiation pattern becomes more and more determined by this value and less and less by the aperture fields. It then transpires that the copolar-radiation patterns become a function of the flare angle of the horn and not of the aperture diameter. This is a useful property because, to first order, the patterns are no longer a function of frequency and have an approximately constant beamwidth over a wide frequency range. Owing to the scalar nature of the radiated fields, this phase factor influences the pattern in a uniform manner.

9.3 Design procedure for corrugated horns

The design of a corrugated horn starts with a list of specifications and goes through a series of stages to a detailed prediction of the radiation properties and the input impedance. The available techniques for predicting performance are

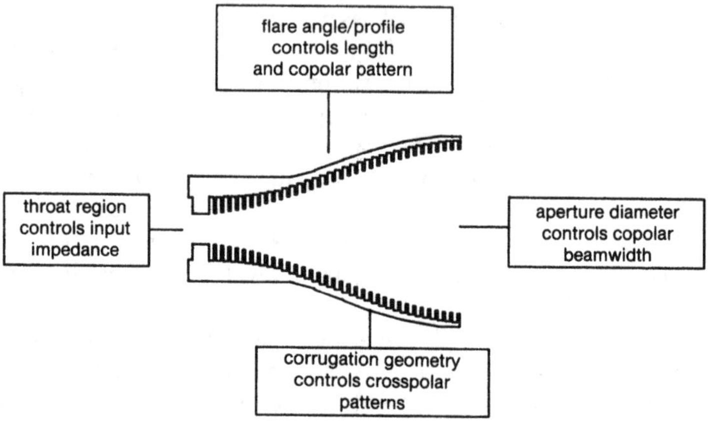

Figure 9.3 Design regions of a corrugated horn

described in Section 9.4. The quality of these means that reliable theoretical predictions are possible. If the measured performance does not agree with the predictions, it can usually be inferred that the practical horn has not been theoretically modelled correctly or that there is some fault in the manufacturing process.

The design procedure for corrugated horn is discussed in detail in Chapter 5 of Reference 1, where step-by-step procedures are given (p. 142). The book was written before the powerful modal-matching technique was available, so the design procedure had to assume traditional methods based on the use of design graphs and the waveguide theories described. Even with the use of modal-matching and other techniques, it is still recommended that a 1st-order design should be done using the procedures discussed in the book. The use of simple formulae, graphs and waveguide theories can lead to a good physical insight which will help in optimising a design. The simple theories are easy to implement and are quick to run on a personal computer, so they can be used to produce a basic design which will satisfy the specifications. If desired, a modal-matching technique (Sections 9.4.5 and 9.4.6) can then be used to confirm and optimise the design. This multiple-stage design procedure is recommended because there are too many geometric parameters which can be varied and the computation process is too slow for a comprehensive parametric study to be warranted using only the modal-matching theory, even though it predicts reliable results. This is particularly true for corrugated horns because a good corrugated horn is one which radiates a single pure mode and so can be analysed using simple methods.

There are four parts of a corrugated horn which need to designed (Figure 9.3): the aperture diameter which controls the copolar beamwidth; the flare angle and profile which also control the radiation pattern and the horn length; the corrugation geometry which controls the crosspolar performance; and finally the throat region which controls the input impedance and the mode conversion from a smooth-wall input waveguide to a corrugated horn. The relative significance of each of these parts will depend on the horn specifications, which in turn will depend on how demanding are the overall antenna-system specifications.

The design process can be summarised in a series of steps:

(i) The aperture diameter should be chosen to give the desired copolar pattern beamwidth. Design curves which relate the halfbeamwidth (i.e the angle from the axis to the pattern level) for the −3 dB pattern level and the −10 dB pattern level are shown in Figures 9.4 and 9.5. These show the tradeoff between aperture diameter and flare angle. Obviously for zero flare angle (which is achievable with a profiled horn, see Section 9.5.8) the beamwidth decreases continuously with flare angle, but, as the flare angle increases, the spherical phase at the aperture compensates for the reduced beamwidth. Consequently for all nonzero flare angles there is a minimum beamwidth which is achievable. The bandwidth over which a nearly constant bandwidth can occur may be quite large, and this is a useful property of corrugated horns.
(ii) The flare angle and/or profile of the horn should be chosen to satisfy length and copolar-pattern specifications. As the flare angle is increased, the influence of the spherical phase cap becomes dominant and the copolar beamwidth becomes a function only of the flare angle and not of the aperture diameter. The

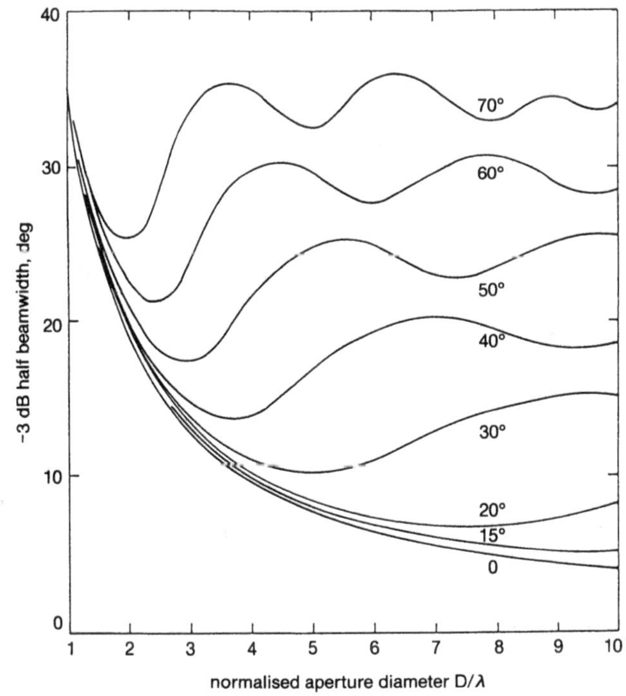

Figure 9.4 −3 dB design curves for corrugated horns

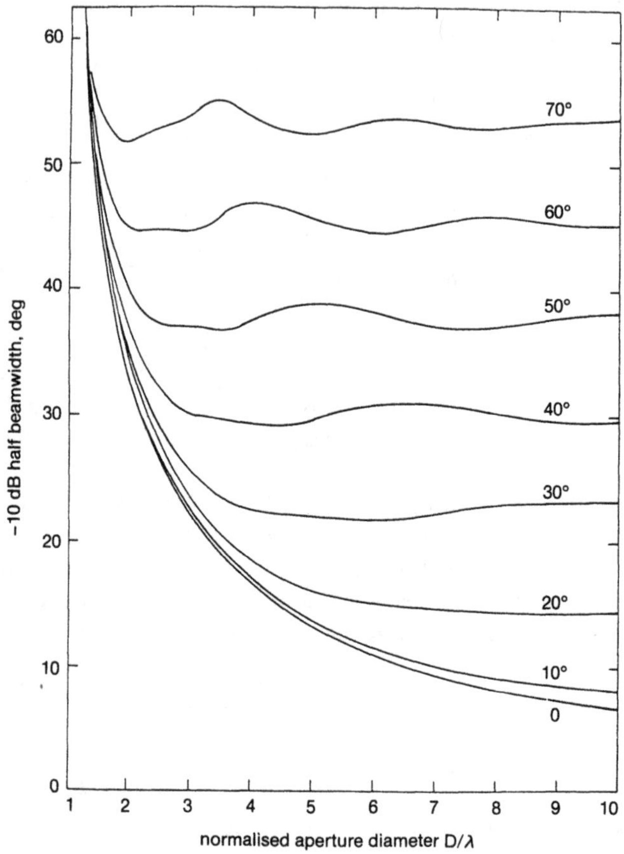

Figure 9.5 — 10 dB design curves for corrugated horns

comments above are relevant. The advantages and disadvantages of wide-flare-angle horns and profiled horns are discussed in Sections 9.5.4 and 9.5.8.

(iii) The geometry of the corrugations should be chosen to give the minimum level of crosspolarisation at the centre frequency. Typical design curves are given in Figure 9.6 and further details are in Section 5.3 of Reference 1. For large aperture diameters this is quarter of a wavelength deep, if there are a large number of corrugations per wavelength. For smaller diameters, the optimum depth increases, as it also does if there are fewer corrugations per wavelength.

(iv) The radiation patterns can be predicted by running programs based on surface impedance or space-harmonic analysis with either cylindrical (for narrow-flare-angle horns) or conical geometries (for wide-flare-angle horns). These will indicate whether pattern specifications are satisfied. The aperture and geometry can be modified as necessary.

(v) The throat region should be designed to give a good impedance match to the smooth-wall horn and to transfer power smoothly from the TE_{11} circular-

waveguide mode to the HE_{11}-mode corrugated-waveguide mode. Normally this means choosing the first slot to be about half a wavelength deep and then tapering the depths of the next few slots down to the operating depth of about quarter of a wavelength. For wide-bandwidth operation, special treatment may be needed using ring-loaded slots; see Section 9.5.6.
(vi) Finally, a modal-matching technique can be used accurately to predict the radiation patterns, mode content and input impedance. The geometry of the horn and the corrugations can be modified if necessary.

9.4 Analytical techniques for predicting performance

A hierarchy of techniques has been developed for predicting the performance of corrugated horns. These range from a single equation which can be solved quickly on a small pocket computer to the sophisticated modal-matching techniques which require long run times on fast computers. As the complexity increases, so does the accuracy of the results. Thus the choice is between quick-and-approximate answers and slow-and-accurate answers. In practice, the best approach is to combine fast initial design methods with slower final predictions of performance.

As with all radiating horns, the analysis is done in two stages: the internal prediction of the aperture fields and the external computation of the radiation from the horn. Most of the effort is in the computation of the internal aperture-field characteristics. The prediction of the radiation pattern uses standard techniques as described in Chapter 3, based on the Fourier-transform integral.

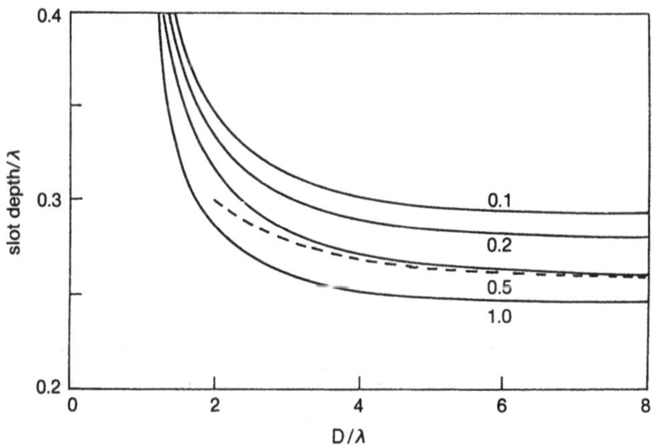

Figure 9.6 Slot depth for minimum crosspolarisation

Slot width $=0.2\lambda$, parameter t/b
— — —surface-impedance result

If the radiation is from an open-ended circular waveguide, or narrow-flare-angle conical horn then the Fourier transform of the aperture electric field E_a is

$$F = \int_0^{2\pi} \int_0^a E_a(r', \phi') \exp\{jkr' \sin\theta \cos(\phi - \phi')\} r' \, dr' \, d\phi' \qquad (9.3)$$

The influence of the flare angle of the horn can be included by adding a spherical phase term so that eqn. 9.3 becomes

$$F = \int_0^{2\pi} \int_0^a E_a(r', \phi') \exp\{jkr' \sin\theta \cos(\phi - \phi')\} \exp(jkr'^2/2r) r' \, dr' \, d\phi \qquad (9.4)$$

An alternative approach is to use spherical waves as explained below.

The detailed treatment of the analytical techniques is given either in Reference 1 or in Chapter 4. The sections below summarise the methods and give the equations necessary to evaluate the performance.

9.4.1 Balanced hybrid model

When balanced hybrid conditions exist in a corrugated horn, eqn. 9.1 shows that the aperture fields in an open-ended corrugated waveguide are

$$\left.\begin{aligned} E_x &= A_1 J_0(kr_1) \\ E_y &= 0 \end{aligned}\right\} \qquad (9.5)$$

Inserting into eqn. 9.4 and evaluating the ϕ integral gives the far-field radiation pattern as

$$F = (1 + \cos\theta) \int_0^1 J_0(kr_1 \sin\theta\, q) J_0(2.405q) \exp\{-j(kr_1^2/2R)q^2\} q \, dq \qquad (9.6)$$

where $q = r/r_1$. This formula is straightforward to evaluate and gives a reasonable estimate for the copolar-radiation patterns. Figure 9.7 shows the results for a 4λ horn with a 9° semiflare angle in comparison with more accurate techniques. Over the main beam to a pattern level of -15 dB, agreement is good. This is the region of interest when feeding a reflector.

9.4.2 Impedance techniques: cylindrical

The balanced hybrid model is only of use when predicting the copolar radiation characteristics near the balanced hybrid frequency. At other frequencies, and to predict the crosspolar patterns, the corrugations must be included in the model. These will change the propagation characteristics in the waveguide so that it is necessary to solve a characteristic equation for the propagation coefficient β. In the impedance techniques the corrugations are represented by an impedance surface so that the waveguide consists of an inner region of radius r_1 bounded by an impedance surface. For this to be valid there must be sufficient corrugations perwavelength for the actual corrugations to behave as a continuous surface. In this case, and assuming that the slots are narrow, only a single nonpropagating

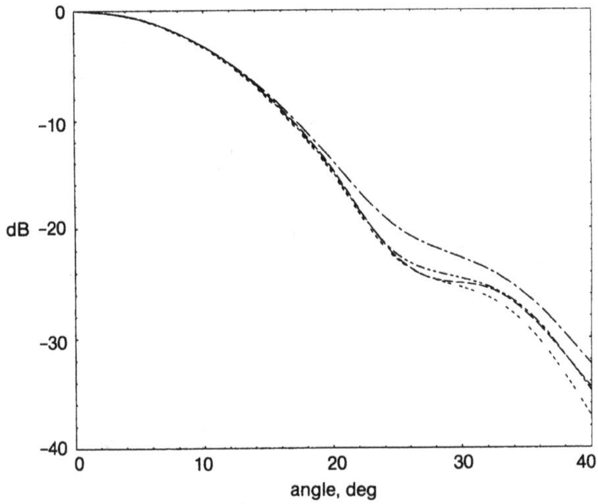

Figure 9.7 Comparison of methods of prediction

— — modal matching
—··— space harmonic
——— spherical modes
—·— balanced hybrid

TM mode will be capable of existing in the slots. Then the surface admittance for an mth-order mode is given (for more details see Section 3.2 of Reference 1) by

$$\bar{Y} = \frac{-jy_0}{kr_1} S_m(kr_1, kr_0) \tag{9.7}$$

where

$$S_m(x, y) = x\frac{J'_m(x)Y_m(y) - J_m(y)Y'_m(x)}{J_m(x)Y_m(y) - J_m(y)Y_m(x)} \tag{9.8}$$

and $J_m(x)$, $Y_m(x)$ are the Bessel functions of the first and second kinds with the prime indicating the derivative.

The electric and magnetic fields in the inner region are

$$E_z = a_m J_m(Kr) \cos m\phi \tag{9.9}$$

$$H_z = a_m y_0 \bar{\Lambda} J_m(Kr) \cos m\phi \tag{9.10}$$

$$E_r = -ja_m \frac{k}{K^2} J_m(Kr)\{\bar{\beta}F_m(x) + m\bar{\Lambda}\} \cos m\phi \tag{9.11}$$

$$H_r = -ja_m y_0 \frac{k}{K^2} J_m(Kr)\{\bar{\beta}\bar{\Lambda}F_m(x) + m\} \sin m\phi \tag{9.12}$$

$$E_\phi = ja_m \frac{k}{K^2} J_m(Kr)\{m\bar{\beta} + \bar{\Lambda}F_m(x)\} \sin m\phi \tag{9.13}$$

$$H_\phi = -ja_m y_0 \frac{k}{K^2} J_m(Kr)\{m\bar{\beta}\bar{\Lambda} + F_m(x)\} \cos m\phi \tag{9.14}$$

where

$$F_m(x) = x\frac{J_m'(x)}{J_m(x)} \tag{9.15}$$

and $\bar{\beta}=\beta/k$, $\bar{\Lambda}$ is the hybrid factor equal to the ratio of H_z/E_z. This factor indicates how nearly balanced are the axial electric and magnetic in the waveguide.

Matching the admittances at the surface $r=r_1$ gives the characteristic equation

$$F_m(Kr_1) - \frac{(m\bar{\beta})^2}{F_m(Kr_1)} = \left(\frac{Kr_1}{kr_1}\right)^2 S_m(kr_1, kr_0) \tag{9.16}$$

which is solved iteratively to find the value of β at a given frequency. Propagation characteristics for corrugated waveguide, obtained from eqn. 9.16, are shown in Figure 9.8. The dominant mode is the HE_{11} mode and this is the mode exclusively used for radiation in a corrugated horn. Higher-order modes can be used to obtain tracking information. A detailed study of the characteristics can be found in Chapter 3 of Reference 1.

The transverse electric fields of the HE_{11} mode can be written in terms of rectangular co-ordinates from eqns. 9.11 and 9.13 as

$$E_x = -j\frac{a_1}{2}\frac{k}{K}\{(\bar{\beta}+\bar{\Lambda})J_0(Kr) + (\bar{\Lambda}-\bar{\beta})J_2(Kr)\cos 2\phi\} \tag{9.17}$$

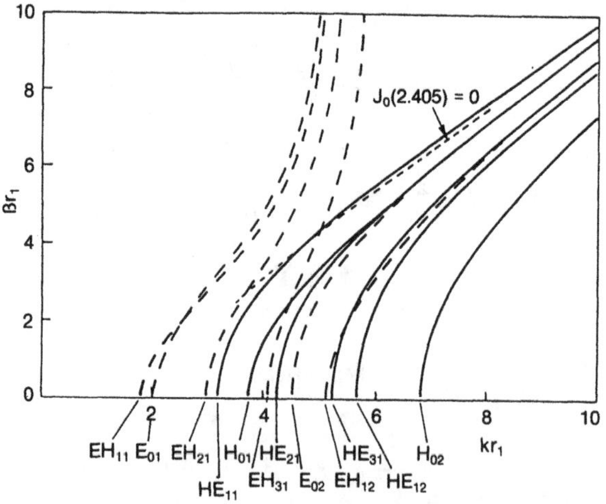

Figure 9.8 Propagation characteristics of a corrugated waveguide
$r_1/r_0 = 0.8$

$$E_y = -j\frac{a_1}{2}\frac{k}{K}(\bar{\Lambda}-\bar{\beta})J_2(Kr)\sin 2\phi \tag{9.18}$$

From these equations it is evident that the balanced hybrid condition which gives a linear aperture field (or $E_y=0$) occurs when $\bar{\Lambda}=\bar{\beta}$. For most applications $\bar{\beta}\rightarrow 1$ so that it is a good approximation to say that the balanced hybrid condition occurs when $\bar{\Lambda}=1$. If more accurate predictions are desired, $\bar{\Lambda}$ can be obtained from

$$\bar{\Lambda} = -\frac{\bar{\beta}}{F_1(Kr_1)} \tag{9.19}$$

For a narrow-flare-angle horn, or an open-ended waveguide, the radiation characteristics can be obtained in closed form by inserting eqns. 9.17 and 9.18 into eqn. 9.3 and evaluating the integrals. This gives the copolar components E_p and crosspolar components E_q in the far field as

$$E_p = F_x\left(\cos^2\frac{\theta}{2}-\sin^2\frac{\theta}{2}\cos 2\phi\right) - F_y\sin^2\frac{\theta}{2}\sin 2\phi \tag{9.20}$$

$$E_q = -F_x\sin^2\frac{\theta}{2}\sin 2\phi + F_y\left(\cos^2\frac{\theta}{2}+\sin^2\frac{\theta}{2}\cos 2\phi\right) \tag{9.21}$$

where

$$F_x = AB\{(\bar{\Lambda}+\bar{\beta})N_0 + (\bar{\Lambda}-\bar{\beta})N_2\cos 2\phi\} \tag{9.22}$$

$$F_y = AB(\bar{\Lambda}-\bar{\beta})N_2\sin 2\phi \tag{9.23}$$

$$N_k = \frac{r_1^2}{\nu^2-(Kr_1)^2}\{\nu J_k(Kr_1)J_1(\nu) - Kr_1J_k(\nu)J_1(Kr_1)\} \tag{9.24}$$

A is a constant, $B=-j(k/K)\pi$, and $\nu=kr_1\sin\theta$. Eqn. 9.21 displays the basic characteristics of crosspolar fields. It is zero in the $\phi=0$ and 90° planes and peaks in the $\phi=45°$ and 135° planes. Figure 9.9 shows typical contour patterns computed with the above equations. The peak value of the crosspolar field lies at about the -9 dB level of the copolar pattern. The optimum value of edge taper, if the horn is used to feed a reflector, is at about -10 dB so that the peak crosspolar level is usually just inside the angle subtended by the feed to the reflector.

The cylindrical-impedance technique described above does not include the influence of the flare angle. However the flare angle has little influence on the crosspolar characteristics so that the study of the crosspolarisation is feasible using eqns. 9.20 and 9.21.

The spherical phase taper on the aperture, created by a small finite flare

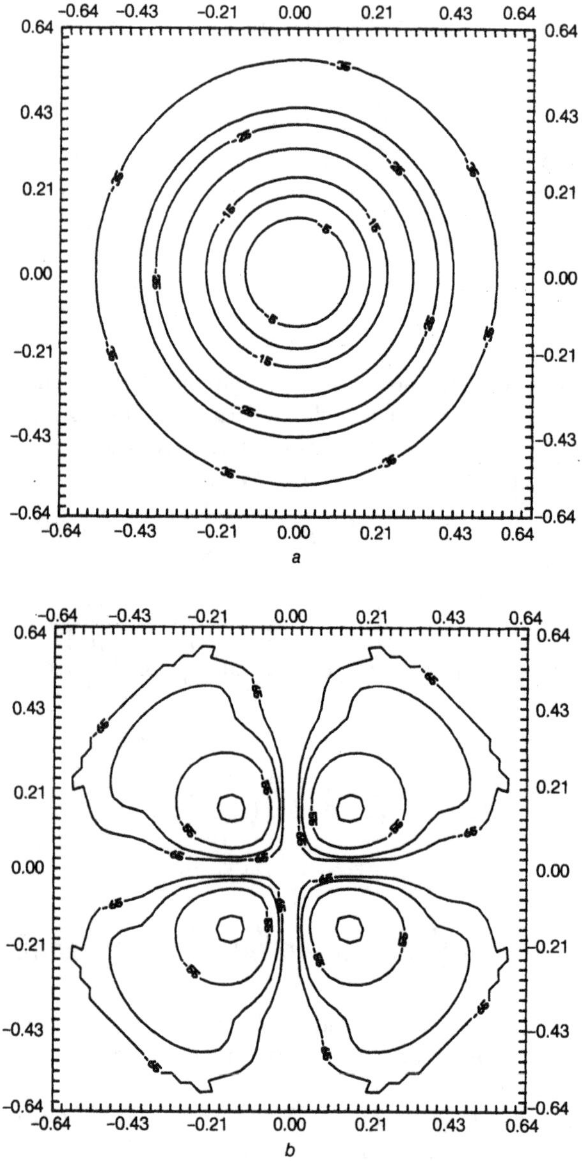

Figure 9.9 Contour patterns for a corrugated horn

a Copolar
b Crosspolar

angle, can be included by using as the radiation formula eqn. 9.4 instead of eqn. 9.3. However, in this case, the integral must be evaluated numerically and a direct analytical result is not feasible.

9.4.3 Impedance techniques: spherical

If the semiflare angle of a corrugated horn is not small (say greater than 20°), the cylindrical-impedance technique breaks down. Instead, a spherical-impedance technique can be used. In this case the modes in the horn (Figure 9.10), are spherical modes. These fit the actual electromagnetic fields in a conical horn much better than a representation using cylindrical modes. The corrugations can still be represented by an impedance surface with the same value as eqn. 9.7.

A detailed study of the properties of spherical modes in conical corrugated horns is contained in Chapter 4 of Reference 1.

The radial electric- and magnetic-field components in spherical co-ordinates are

$$E_R = A\,\frac{\nu(\nu+1)}{jkR^2}\,h_\nu^{(2)}(kR)P_\nu^m(\cos\theta)\cos m\phi \tag{9.25}$$

$$H_R = A\bar{\Lambda}\,\frac{\nu(\nu+1)}{jkR^2 y_0}\,h_\nu^{(2)}(kR)P_\nu^m(\cos\theta)\sin m\phi \tag{9.26}$$

where $P_\nu^m(\cos\theta)$ is an associated Legendre function of order m and degree ν, while $h_\nu^{(2)}(kR)$ is a spherical Hankel function of order ν. The other field components can be obtained from Maxwell's equations and then the admittances matched at the boundary $\theta=\alpha$ together with the condition that $E_\phi=0$ to give the characteristic equation

$$p_\nu^m(\alpha)\{p_\nu^m(\alpha)-\bar{Y}\nu(\nu+1)\}=\frac{m^2\bar{\beta}^2}{\sin^2\alpha} \tag{9.27}$$

where

$$p_\nu^m(\alpha)=\frac{dP_\nu^m(\cos\alpha)/d\theta}{P_\nu^m(\cos\theta)} \tag{9.28}$$

$$\bar{\beta}=j\,\frac{h_\nu^{(2)\prime}(kR)}{h_\nu^{(2)}(kR)} \tag{9.29}$$

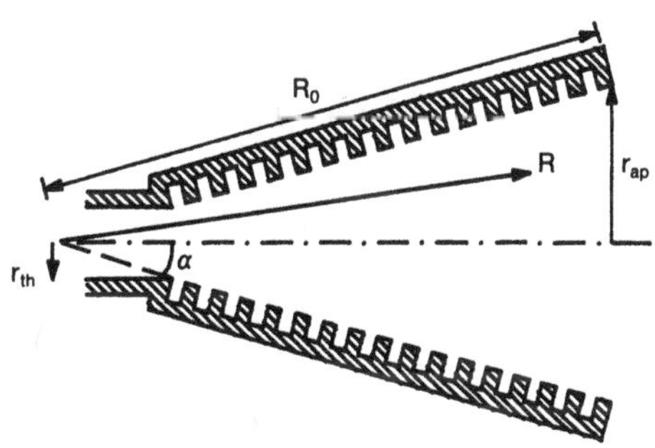

Figure 9.10 Wide-flare-angle horn

$$\bar{Y} = -\frac{jY}{y_0 kR} \tag{9.30}$$

The surface admittance of the corrugated walls can be calculated by the cylindrical model given in eqn. 9.7 or by the good approximation

$$Y = -jy_0\left(\cot kd + \frac{1}{2kR \sin \alpha}\right) \tag{9.31}$$

where d is the depth of the corrugations. This shows that Y is a function of R, which strictly means that the spherical-wave equation cannot be solved due to the interdependence of the spherical co-ordinates. In practice, the radial length R_0 of the horn is normally large and eqn. 9.31 then shows that the dependence on R_0 can be neglected.

Eqn. 9.27 shows that the eigenvalue ν is a function of radius because both Y and β depend on kR. In a conical horn, β is generally complex and has mainly real values for $kR \gg \nu$ and mainly imaginary values for $kR \ll \nu$, with a complex region when $kR \approx \nu$. This behaviour describes the gradual cutoff of spherical modes, in contrast to cylindrical modes which have a sharp cutoff. To avoid this problem, it is usual to assume that $kR \gg \nu$, which is valid except near the throat of a conical horn.

Eqn. 9.27 must be solved iteratively to find the value of ν. Then the aperture fields over a spherical cap will be known. The dominant mode is the spherical HE_{11} mode with $m = 1$. Note that this is not the same as the cylindrical HE_{11} mode and a direct transformation from one to the other cannot be done.

The radiation characteristics can be derived by using a Kirchhoff–Huygens radiation integral [6], but a better method is to use spherical-wave expansions because this ties in directly with the spherical waves of the aperture fields. Assuming that only the HE_{11} mode is of interest and that it is near the balanced hybrid condition where $\bar{\Lambda} \simeq 1$, then the aperture fields can be expressed as

$$E_a(R_0, \theta, \phi) = A_{11} F_\nu^1(\theta) \frac{\exp(-jkR_0)}{R_0} (\hat{\boldsymbol{u}}_\theta \cos \phi + \hat{\boldsymbol{u}}_\phi \sin \phi) \tag{9.32}$$

where

$$F_\nu^1(\theta) = \frac{dP_\nu^1(\cos \theta)}{d\theta} + \frac{P_\nu^1(\cos \theta)}{\sin \theta} \tag{9.33}$$

The electric field can now be expanded in terms of a combination of spherical TE and TM modes. At $R = R_0$,

$$E_a(R_0, \theta) = \sum_{n=1}^{N} C_n h_\nu^{(2)}(kR_0) F_n^1(\theta) \tag{9.34}$$

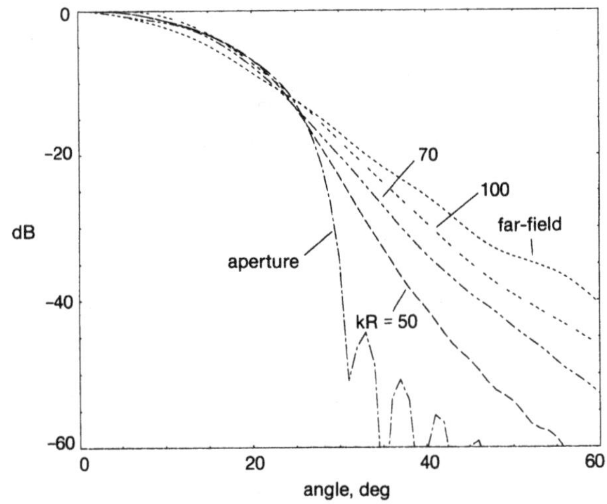

Figure 9.11 *Radiation patterns computed using spherical modes for a 30° semiflare-angle horn of normalised length $kR_0 = 40$*

Parameter: distance kR from apex

The function $F_n^1(\theta)$ is obtained by replacing ν by n in eqn. 9.33. By applying orthogonality relationships, the nth coefficient of C_n is found to be

$$C_n = A\,\frac{\exp(-jkR_0)}{R_0}\,\frac{2n+1}{2n^2(n+1)^2 h_n^{(2)}(kR_0)}\int_0^{\alpha} F_\nu^1(\theta)F_n^1(\theta)\sin\theta\,\mathrm{d}\theta \quad (9.35)$$

The far field is the summation of the coefficients as follows:

$$E(\theta) = \sum_{n=1}^{N} \frac{j^{n+1}C_n F_n^1(\theta)}{h_n^{(2)}(kR_0)} \quad (9.36)$$

The series must be terminated at some value. For $N > kR_0$ the coefficients C_n have negligible values so that it is convenient to choose $N = KR_0$.

One of the major advantages of the spherical-wave-expansion method is that the near fields can be computed with very little extra effort. From eqn. 9.34 the electric field at radius R_1 is given by

$$E_a(R_1, \theta) = \sum_{n=1}^{N} C_n h_\nu^{(2)}(kR_1)F_n^1(\theta) \quad (9.37)$$

Computation is also efficient because the values of C_n and $F_n^1(\theta)$ can be computed for all the pattern angles and stored. Then the patterns at different distances in front of the aperture only require computation of the spherical Hankel functions $h_n^{(2)}(kR_1)$ at each value of R_1. Typical patterns for a corrugated horn with a 30° semiflare angle and a normalised length $kR_0 = 40$ $(D/\lambda = 6.4)$ are shown in Figure 9.11.

9.4.4 Space-harmonic techniques

The periodic nature of the corrugated surface can be taken into account by using a space-harmonic technique to predict the aperture fields. This is a development of the cylindrical-surface-impedance technique which includes the geometric parameters of the corrugations. Then the influence of both the slot and ridge widths and the slot depth can be assessed. The analysis is based on Floquet's theorem which states that the fields in the unit cells of a periodic structure consist of those in one cell together with the phase change between cells. For a cell of length p which is a distance Mp from the base cell, the phase change is $\exp(jM\beta_p)$ where M is an integer and β_p is the phase change per unit cell.

The fields of each mode in the inner region consist of the fundamental field of the mode plus a series of $2N$ space harmonics. Thus eqn. 9.9 is replaced by

$$E_z = \sum_{N=-\infty}^{N=+\infty} a_{Nm} J_m(K_{nr}) \exp(-j\beta_N z) \cos m\phi \tag{9.38}$$

In the slots, the field is expanded as a series of P standing waves of the form

$$E_z = \sum_{P=1}^{P=\infty} (C_P S_m^D \cos \eta_p z - jC'_P S_m^{D'} \sin \eta'_P z) \cos m\phi \tag{9.39}$$

where S^D is the denominator of eqn. 9.8, $\eta_P = 2P\pi/b$ and $\eta'_P = (2P-1)\pi/b$. b is shown in Figure 9.1.

The space-harmonic technique is treated in detail in Reference 1 (Appendix) and it is the main method used in that book to explore the dependence of the crosspolarisation characteristics on the corrugation geometry (Chapters 3 and 5 of Reference 1). The implementation is more complicated than that of the surface-impedance technique but in most cases the number of space harmonics which need to be included is less than ± 3 so that the computation time is short. This enables detailed exploration of the geometry to be undertaken.

When a corrugated horn is to be used as a high-performance feed, the quantity of most interest is the peak level of the crosspolar radiation pattern in the 45° plane. Plotted against frequency this will have the form shown in Figure 9.12 with a minimum at the balanced hybrid frequency. The level of the curve is proportional to the aperture diameter, as shown in eqn. 9.1. The precise value of the balanced hybrid frequency is determined by the geometry of the corrugations. For a large-diameter horn and a large number of slots per wavelength, it will occur at a frequency where the slots are $\lambda/4$ deep. As the aperture diameter is decreased, the slot depth required to maintain balanced hybrid conditions increases owing to the radial impedance in eqn. 9.7. The slot depth also increases if the number of corrugations per wavelength is reduced so that the slot width is increased. The same is true if the ratio of the ridge width to the slot width is increased. The form of these effects is shown in Figure 9.6 and more details can be found in Chapter 5 of Reference 1.

9.4.5 Modal-matching techniques: cylindrical

The modal-matching technique has been found to be a very powerful and accurate method of analysing conical horns because it enables the exact transverse fields at the horn aperture to be computed, including any mode conversion which occurs along the horn. The principle of the technique is to divide a horn longitudinally into short cylindrical waveguide sections in which the propagation characteristics are the roots of Bessel functions and their derivatives. Modal matching is then used at the junction of each section to obtain the power coupled to the next section of the waveguide. Scattering matrices are obtained for each of the waveguide junctions and each of the short lengths of waveguide separating the junctions. An overall scattering matrix for the horn is obtained by cascading the individual matrices progressively along the horn. This gives the complex weighting coefficients for each mode at the horn aperture in relation to the input-waveguide modal fields. The radiated fields of the horn are computed by Fourier transforming these aperture fields.

The detailed theory and method of implementation of the modal-matching technique are explained in Chapter 4. When it is used to analyse corrugated

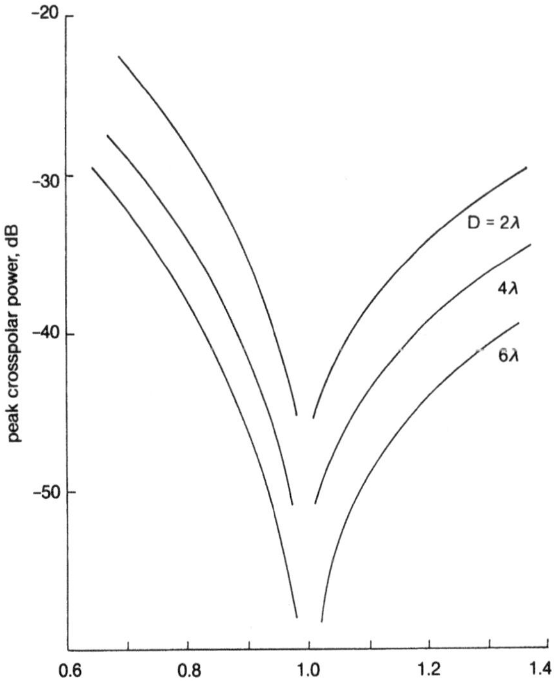

Figure 9.12 Peak crosspolar power as a function of normalised frequency where f_0 is the frequency for minimum crosspolarisation

Parameter: normalised aperture diameter

horns, the sections of waveguide between junctions are lengths of smooth-wall circular waveguide. The propagation coefficients for hollow circular waveguides are given by the roots of Bessel functions $J_n(K_c r_1)=0$ for TM modes and $J'_n(K_c r_1)=0$ for TE modes. This is convenient to use because the roots are known and can be stored in the computer so that the propagation coefficients take the minimum time to evaluate. Most of the implementations of the modal-matching technique assume that all the sections of circular waveguide which form the corrugated horn have a common axis. This means that if a TE_{nm} mode is incident on a junction, the mode conversion at the junction will only excite TE_{nq} or TM_{nq} modes. In other words, modes of azimuthal dependence $\exp(jn\phi)$ will only excite modes of the same azimuthal dependence. The usual situation is that the input mode is a TE_{11} mode, so that the modal matching is concerned with TE_{1q} or TM_{1q} modes. Conversely, it follows that one of the critical manufacturing conditions for a corrugated horn is that it should maintain circular symmetry; otherwise modes of nonunity azimuthal dependence will be excited.

The modal-matching technique can be used to study and optimise the throat region of the corrugated horn. It enables the input impedance of the horn to be evaluated, by computation of the reflection coefficient of all the propagating modes in the throat region. This is particularly useful when the bandwidth properties of a horn are being studied because these are often determined by the design of the throat section. A number of schemes have been investigated to extend the bandwidth. One of the most successful is to use ring-loaded slots in the first few corrugations. These consist of short sections of annular waveguide added to one or both sides of a slot, and are described in Section 9.5.6. The basic modal-matching theory for a standard corrugated horn has been extended to ring-loaded slots by James [16].

Modal matching differs from the analysis techniques described in Section 9.4.4 in that the behaviour of the fields are tracked along the horn from the throat to the aperture. In the other techniques, the horn is modelled as either a single cylindrical corrugated waveguide or a single conical corrugated horn. Only the aperture fields are computed, and there is no information about the way in which the fields change along the horn. Thus the modal-matching technique is inherently a better method of modelling the horn. The penalty lies in the considerable extra computational effort needed to predict the aperture fields. Each junction is a step discontinuity between two smooth-wall circular waveguides. A large number of propagating cylindrical modes are necessary to account for the total power flow. In addition, a large number of evanescent modes must be included in the analysis because adjacent junctions are close together and therefore power which is converted by a junction into an evanescent mode may still have a finite value at the next junction. A useful guideline is that the number of cylindrical modes to be included in the analysis for normal corrugated horns is about ten times the aperture diameter.

The influence of any mismatch between the horn aperture and free space can be included in the analysis by modelling the exact geometry of the flange using a

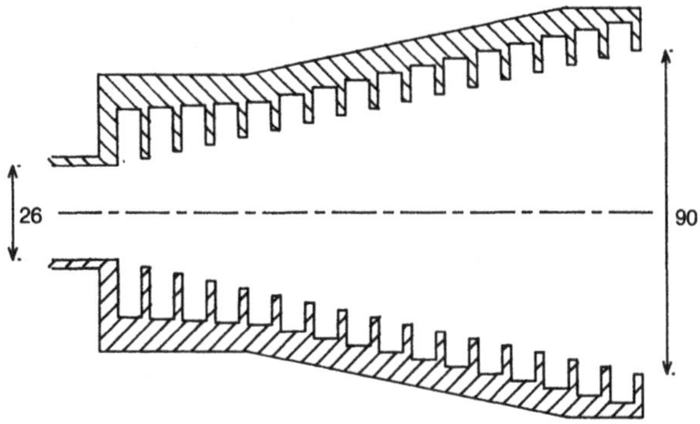

Figure 9.13 *Corrugated horn with 3 λ-aperture diameter at 10 GHz*
Dimensions in mm

method-of-moments approach. This is accurate but needs to be analysed for a particular geometry and involves a significant modification to the theory in Chapter 4. A much simpler approach, which requires no alteration to the modal-matching program, is to add onto the horn a final section of circular waveguide with a large diameter. A large-diameter horn will have negligible mismatch to free space and its characteristic impedance will be almost equal to that of free space, namely 377 Ω. The change in the input impedance can then be observed. Tests using this method show that it works well but also show that aperture diameters have to be small before the influence of the aperture–free-space mismatch becomes significant. For instance, a 3 λ corrugated horn computed with and without a large-diameter output section showed a difference in VSWR of less than 0.01.

The tolerances on the geometry of a corrugated horn can be explored with the modal-matching technique. As an example, consider the horn in Figure 9.13, which has a 3-wavelength aperture diameter at 10 GHz. The copolar and crosspolar radiation patterns computed using modal matching are shown in Figure 9.14. At this design frequency, the peak crosspolar level is low and peaks at -46.9 dB. (The slot depth to achieve this low figure is 0.3λ. This is substantially deeper than the nominal value of $\lambda/4$ owing to a combination of the aperture diameter and the wide slots.) The input return loss for the configuration is -32.9 dB. In practice this means that the return loss of the complete waveguide–feed system will be determined by other components such as tapers and coaxial-to-waveguide transformers and not by the corrugated horn. To see the effect of changes in geometry the depth of the fifth slot in the above horn was increased by 2 mm. This causes mode conversion to occur along the horn and the smooth transfer of power is interrupted. The peak crosspolar level rises by a significant amount to -29.5 dB and demonstrates the need for a smooth change in slot depth along the horn. As another test, the depths of all the slots were assumed to have a random error in depth of 0.5 mm

RMS. This gives a peak crosspolar level of −37.8 dB, which is nearly 9 dB worse than the pure level but for most applications still an acceptable level. In both these tests, the input impedance did not change significantly because the detrimental mode conversion occurs some way down the horn.

9.4.6 Modal-matching techniques: spherical

The modal-matching technique can be implemented using spherical modes (Figure 9.15). This has advantages for wide-flare-angle horns where relatively few spherical modes are needed to represent the internal fields of the corrugated horn. The principles of modal matching using spherical modes are similar to those using cylindrical modes and are described in Chapter 4, Section 4.7. The coefficients in the scattering matrices are more complicated than in the cylindrical case owing to the nonorthogonality of the spherical modes at spherical junctions. Unlike the cylindrical case, there is no common boundary when the radius of curvature of the sphere changes.

Spherical modal matching is inherently less precise than cylindrical modal matching because the corrugations must be represented as an impedance surface. This is because Maxwell's equations become inseparable in a dual conical region. A conical corrugated surface should have corrugations which increase in depth away from the apex. This is not the case in a real horn, so it is better to represent the impedance formed by the corrugations using the same cylindrical model as discussed in Section 9.4.4. Because the throat region is important and of interest, the situation arises that kR is less than or equal to the

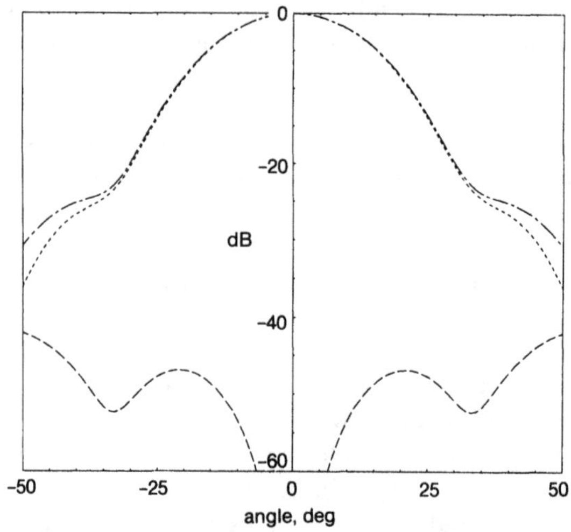

Figure 9.14 Radiation pattern at 10 GHz for horn illustrated in Figure 9.13

— — E-plane
—·— H-plane
- - - 45° plane crosspolar

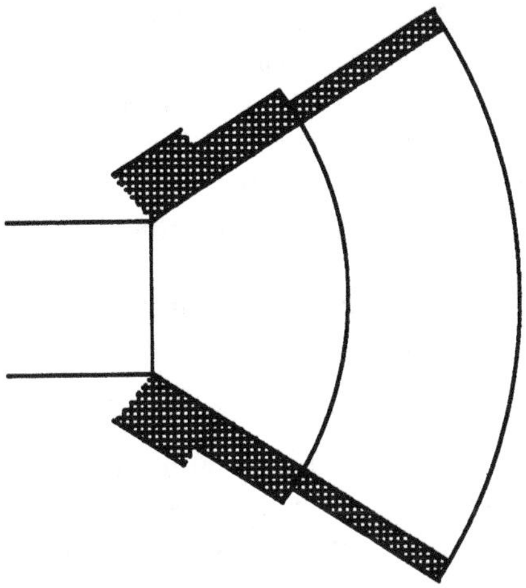

Figure 9.15 Wide-flare-angle horn with sections for spherical modal matching

eigenfunction ν. This means that complex roots of the characteristic equation can occur and complex routines are needed to calculate Hankel functions and associated Legendre functions. The propagation coefficient then has both real and imaginary parts, i.e. $\gamma = \alpha \pm j\beta$. Figure 9.16 shows how the real and imaginary parts of ν vary as kR is increased. The first three modes are shown. The results show that the real part of ν is only weakly dependent on radius and that, for $kR > 15$, the imaginary part of ν can be ignored. Even for small values of kR the imaginary part of ν is much smaller than the real part and ν is a monotonically decreasing function of flare angle. The fact that the dependence of ν on radius is weak confirms that the separation-of-variables assumption in a conical horn is valid and the wave equation can be solved using spherical modes.

A general spherical junction in conical corrugated horn is shown in Figure 9.17. The horn section on the left-hand side may be a smooth-wall circular waveguide, or a smooth-wall conical horn, or a corrugated conical horn. The horn section on the right-hand side is usually a conical corrugated horn but could be a cylindrical corrugated section. The electromagnetic effects of these junctions will now be investigated.

9.4.6.1 Junction of a smooth-wall circular waveguide and a corrugated conical horn

Any mismatch at the junction between the smooth-wall circular waveguide and the corrugated horn will give rise to a return loss and the excitation of higher-order modes. Tests show that five modes on either side of the junction is sufficient for good prediction of the effects of the junction. One of the modes is propagating and four are evanescent. Figure 9.18 shows return loss plotted against normalised slot depth for conical corrugated horns with semiflare angles of 10°, 30° and 60°.

The 10° horn is seen to be matched when the slot depth is half a wavelength, as expected for narrow-flare-angle corrugated horns, but this rule does not work for wide-flare-angle horns. The reflection coefficient then shows less variation with slot depth and is a function of the flare angle. There is therefore no benefit to be obtained in wide-angle corrugated horns by the extra complexity of making the corrugations in the throat region half a wavelength deep. Higher-order modes are excited at the throat junction, but they are usually evanescent for $\alpha < 70°$ and do not contribute very much to the higher-order modes at the aperture.

9.4.6.2 Junction between a smooth-wall conical section and a corrugated conical horn

Addition of a section of smooth-wall conical horn at the throat region (Figure 9.19) has been found experimentally to improve the return loss [8]. This is confirmed theoretically by computing the characteristics of a junction between

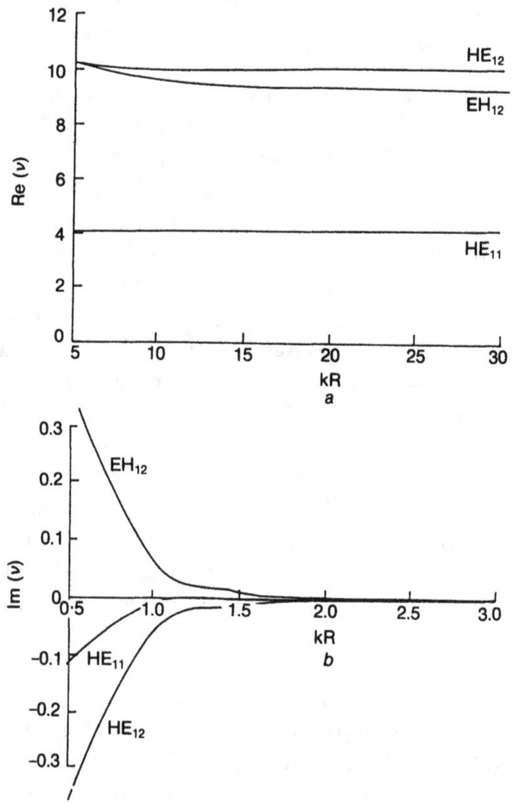

Figure 9.16 Eigenvalue ν as a function of kR

$\alpha = 30°$, $d = \lambda/4$

a Real part

b Imaginary part

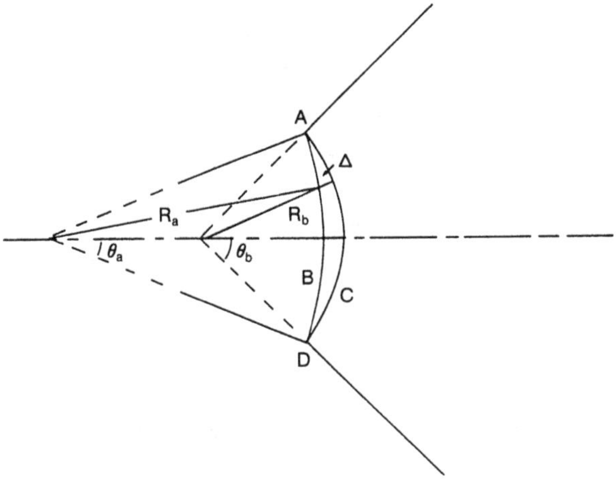

Figure 9.17 Junction between two spherical sections

a smooth-wall conical section and a corrugated conical section. The results are shown in Figure 9.20 and confirm that the horn can be matched with a halfwavelength-deep slot. The amount of higher-order mode excited at the junction is a function of both the length of smooth-wall section and the flare angle of the horn. For the length used in Figure 9.20, the EH_{12} mode is cutoff for the 10° and 30° horns, but for the 60° horn it is above cutoff and power will be converted from the HE_{11} mode to the EH_{12} mode and propagate down the horn to radiate from the aperture.

9.4.6.3 Junctions between two corrugated conical sections

Junctions between two corrugated conical sections occur along the horn when the wall impedance changes. The biggest change in wall impedance occurs

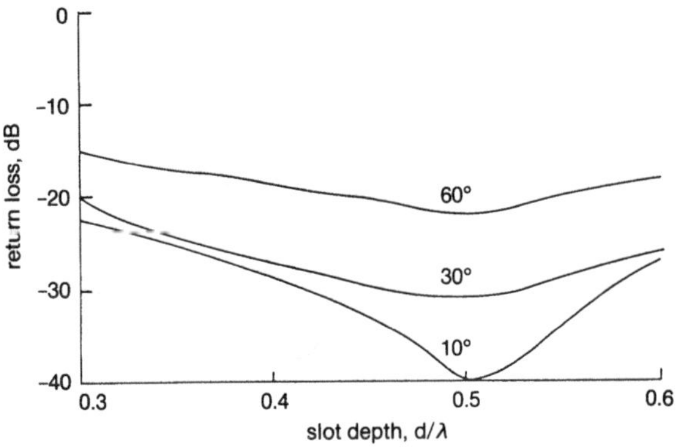

Figure 9.18 Return loss of TE_{11} mode at throat junction against slot depth

Parameter: α. $kr_{th} = 3$

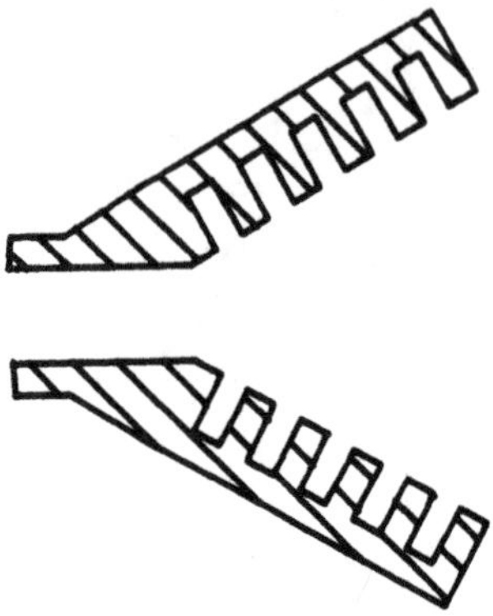

Figure 9.19 Wide-flare-angle horn with smooth conical input section

when the slot depth tapers down from half a wavelength deep to a quarter of a wavelength deep. Since this change takes place at a point in the horn where a number of higher-order modes are above cutoff, the main interest is in the level of excitation of the higher-order modes. Figure 9.21 shows the excitation levels for 12 modes in a 30°-semiflare angle horn with a junction from a 0.4λ-deep slot to a $0.35\ \lambda$-deep slot and $kR = 14$. It is clear from this that the EH modes are more important than the HE modes. The actual levels depicted in Figure 9.21 are particular to those parameters, but the effect is general. The excitation level varies with slot depth but is only weakly dependent on radius kR and almost independent of flare angle.

9.4.7 Gaussian-beam techniques

The number of millimetre-wave systems, particularly in remote sensing, is increasing rapidly. Conventional components become difficult to fabricate and use efficiently at frequencies above about 100 GHz. A powerful alternative to

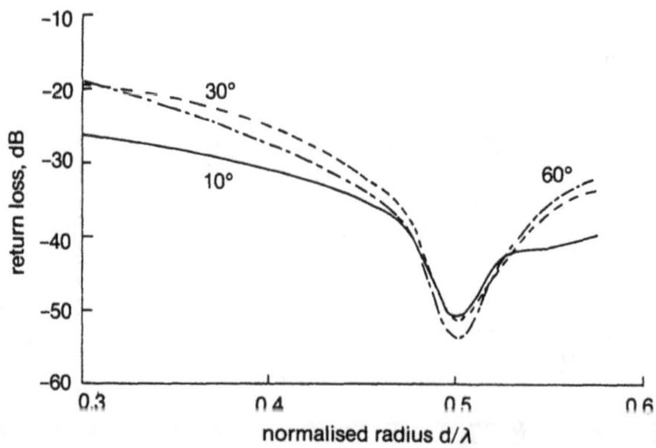

Figure 9.20 Return loss of TE_{11} mode for horn illustrated in Figure 9.19

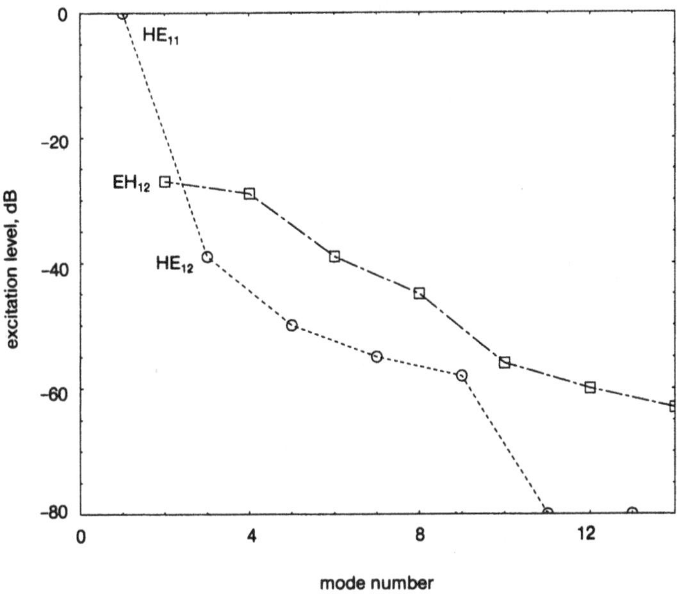

Figure 9.21 *Excitation levels of modes at right-hand side of junction of two corrugated-horn sections*

conventional components is to use quasioptics in which a Gaussian-shaped beam is propagated through a number of processing components such as mirrors, lenses and dichroic surfaces. Quasioptical systems require an antenna which generates a pure Gaussian beam with a low level of crosspolarisation, although the actual level of crosspolarisation is usually not of great interest. The corrugated horn has been used very successfully for this purpose. In this role, the horn is generating, or receiving, a beam which travels through a series of components. A method of analysis is required which enables the characteristics of the quasioptical beam to be computed at any distance from the horn. This can be done using Gaussian-beam modes. The aperture fields of the horn are expanded in a set of Gaussian-beam modes which can then be propagated forward to any distance.

The Gaussian-beam technique is an approximate method, but one which is adequate for the design of quasioptical systems. The approximations are:

(*a*) balanced hybrid conditions are assumed
(*b*) the aperture electric field is assumed to be linearly polarised
(*c*) the aperture is assumed to be large so that the energy is confined to a narrow angle around the axis.

The theory of this technique was given by Wylde and Martin [17,18], and the description below is based on their analysis. The fields in the aperture of the horn are those used for the simplest techniques and will assume that operation is under the balanced hybrid condition, and assume a linearly polarised electric field. Adding a phase factor to eqn. 9.5

$$\left.\begin{aligned} E_x(r) &= AJ_0(K_c r)\exp\left(\frac{-jkr}{2R_0}\right) \\ E_y(r) &= 0 \end{aligned}\right\} \tag{9.40}$$

where R_0 is the radius of the horn (Figure 9.10), and K_c is the root of the zero-order Bessel junction, or $K_c a = 2.405$, where a is the inner radius (i.e. $a = r_1$). The exponential term represents the spherical phase error owing to a finite flare angle on the horn.

The E_x electric field is expanded in terms of an orthogonal set of Gaussian–Laguerre modes, so the Bessel function in eqn. 9.40 is replaced by

$$J_0(K_c r) = \sum_{p=0}^{\infty} A_p L_p^0\left[\frac{2r^2}{W^2}\right]\exp(-r^2/W^2) \tag{9.41}$$

This approach was first used by Aubry and Bitter [19]. $L_p^0(x)$ is a zero-order Laguerre polynomial and W is the $1/e$ diameter of the beam. The values of the coefficients A_n are found to be [18]

$$A_n = \frac{4}{(W/a)^2}\int_0^1 J_0(2.405u) L_N^0\left[\frac{2u^2}{(W/a)^2}\right]\exp\left\{\frac{-u^2}{(W/a)^2}\right\} u\, du \tag{9.42}$$

where $u = r/a$. The choice of W/a is arbitrary. Aubry and Bitter [19] maximised the power contained in the sum of the arbitrary number of modes under consideration, but Wylde [18] proposed to choose W/a so that it maximised the power in the fundamental mode, thus minimising the number of modes in the series. This condition is satisfied by optimising the value of W/a using iteration, and gives

$$T = \frac{W}{a} = 0.6435 \tag{9.43}$$

The values of A_n and the power in each mode are given in Table 9.1 [18] for the first 11 modes. These values apply to all corrugated horns operating under balanced hybrid conditions. 98% of the power is in the fundamental mode so that, in many cases, it is adequate to consider only the fundamental Gaussian mode.

The theory for the propagation of Gaussian-beam modes has been given by Kogelnik and Li [20]. The beam obeys the scalar wave equation

$$\nabla^2\psi + k^2\psi = 0 \tag{9.44}$$

The beam propagates along the z axis and can be represented as a superposition of an orthogonal set of Gaussian–Laguerre modes

$$\psi(r, z) = \sum_{p=0}^{m} A_p\left(\frac{2}{\pi W^2}\right)^{1/2} L_p^0\left[\frac{2r^2}{W^2}\right]\exp\left\{-jkz - j\left(\frac{kr^2}{2q(z)}\right) + \phi_p(z)\right\} \tag{9.45}$$

where it has been assumed that ψ varies slowly with z so that the second derivative can be ignored. The parameters for each mode are the radius of curvature $R(z)$, the beamwidth $W(z)$ and the phase slippage $\phi_p(z)$ associated with the nonplanar phase front. The relationships between these parameters at any point along the axis are given by

$$R(z) = (z - z_0)\left(1 + \frac{1}{\bar{z}^2}\right) \tag{9.46}$$

$$W^2(z) = W_0^2(1 + \bar{z}^2) \tag{9.47}$$

$$\phi_p(z) = (2p + 1)\tan^{-1}\bar{z} + \phi_{p0} \tag{9.48}$$

where $\bar{z} = \{\lambda(z - z_0)\}/(\pi W_0^2)$ and z_0 and W_0 are the values of z and W at the point where the diameter of the beam is a minimum. This is called the 'beamwaist'. ϕ_{p0} is a constant.

The radiation pattern in the far field can be found by applying a paraxial approximation to obtain a simple series result [18]:

$$\psi(\theta) = \sum_{p=0}^{\infty} A_p\left(\frac{Q}{\pi F^2\cos^2\theta}\right)^{1/2} L_p^0(Q\tan^2\theta) \times \exp\left(-Q\tan^2\frac{\theta}{2}\right)\exp[j(2p+1)\{\pi/2 - \tan^{-1}(T^2S)\}] \tag{9.49}$$

where F is the far-field distance and

$$Q = \frac{(ka)^2T^2}{2(1 + T^4)S^2} \tag{9.50}$$

Table 9.1 Normalised power coefficients for Gauss–Laguerre modes

Mode number	Power coefficient
0	0.9792
1	4.90×10^{-9}
2	1.45×10^{-2}
3	1.86×10^{-3}
4	3.81×10^{-4}
5	1.16×10^{-3}
6	3.97×10^{-4}
7	1.50×10^{-8}
8	1.59×10^{-4}
9	2.33×10^{-4}
10	1.12×10^{-4}

$$S=\frac{ka^2}{2R_0} \tag{9.51}$$

is the spherical phase factor at the aperture of the horn.

The design process normally starts with a knowledge of the $1/e$ beam diameter W or the beamwaist W_0. Alternatively, the halfpower beamwidth may be specified. The relationship between the full -3 dB beamwidth and the beamwaist is the same as that for any diffraction-limited aperture [20]:

$$\Delta\theta=\frac{\lambda}{\pi W_0} \tag{9.52}$$

The above equations enable the analysis of corrugated horns to be conducted using Gaussian-beam techniques.

9.5 Types of corrugated horn

The following sections review various different types of corrugated horn with the emphasis on the behaviour of the copolar and crosspolar radiation characteristics.

9.5.1 Standard narrow-flare-angle horns

The standard type of corrugated horn which radiates low levels of crosspolarisation and good pattern symmetry is a horn with a constant flare angle and semiflare angle below about 15°. This can be analysed using any of the techniques described in Section 9.4, but is most appropriately done using the balanced hybrid model, cylindrical surface impedance, space harmonics, cylindrical modal matching or Gaussian-beam techniques. The radiation pattern consists of the radiation from an open-ended corrugated waveguide together with the influence of the spherical phase cap caused by the conical horn. If the flare angle is low, then this spherical phase cap is small and can often be ignored. It has negligible effect on the copolar and crosspolar patterns within the angle subtended to a reflector. The design procedure is described in Section 9.3 and in more detail in Chapter 5 of Reference 1.

The design of the corrugations is one of the main variables. The optimum depth of slots at the aperture depends on the aperture diameter. For large apertures, $\lambda/4$ is the normal value, rising to 0.3λ for aperture diameters of about 2λ. There should be at least three cells per wavelength with a slot-width-to-ridge-width ratio of between 2 and 10. The choice of this value is very flexible and can be determined by manufacturing considerations. Generally, a lower ratio will be desirable at higher frequencies. These recommendations for the corrugation geometry can be implemented without difficulty for operation at frequencies below about 40 GHz, and at higher frequencies by appropriate choice of geometry.

The depth of the slots in the throat region needs to be chosen to provide a match between the input waveguide and the conical horn, as well as act as a mode converter between the circular-waveguide TE_{11} mode and the corrugated-horn HE_{11} mode. It has been found that a good design rule is to choose the first slot to be $\lambda/2$ deep and then to taper down over not less than a wavelength, or up to about ten slots, to the standard depth slots. It is important to carry out the tapering over a short distance because mode conversion along the horn will be increased by changing the slot depth. The $\lambda/2$ depth is chosen so that the short circuit at the base of the slots is transferred to a short circuit at the top of the corrugations which then presents a continuous conducting surface to the fields of the TE_{11} mode in the circular waveguide. The precise design of the throat region can be done either by using a knowledge of the surface impedance of the slots, or by using a modal-matching computer program. The impedance method is described in Chapter 5 of Reference 1 using graphs. The latter is very accurate, but it is not easy to be certain that an optimum design has been achieved without a considerable number of computations.

9.5.2 Small-aperture horns

The radiation and impedance characteristics of corrugated horns with aperture diameters below about 1.5 wavelengths depend greatly on the precise construction of the total horn. The currents flowing on the flange and external parts of the horn influence the radiation patterns. The basic design can be done using the simple theories, but a precise design requires the use of a method which models the whole structure of the horn. The best known of these techniques is the method-of-moments, as described in Chapter 5, and it is recommended that small-aperture corrugated horns be analysed using the methods described there and in Chapter 7.

The horn section can be designed in a similar fashion to a normal horn, but the currents flowing on the flange radiate significant amounts of unwanted power so it has been found desirable to suppress the currents by inserting $\lambda/4$-deep rings around the aperture (Figure 9.22*a*). With a small aperture, the beam is broad and not that much different from that of an open-ended waveguide. For this reason, the internal corrugations can be removed and the horn then consists of a series of rings around a circular aperture (Figure 9.22*b*). This type of horn makes a good feed for prime-focus applications in symmetric reflectors with F/D below about 0.35. They are sometimes referred to as a type of corrugated feed, but this is not the case. There is no hybrid mode in the horn and the radiation patterns depend on the summation of the radiation from the open-ended waveguide and the ringed flange.

9.5.3 Millimetre-wave horns

There are a growing number of applications using millimetre waves which require radiators with good performance for use either as direct radiators or as part of a quasioptic system. The corrugated horn is an ideal horn for this purpose because it radiates a nearly purely Gaussian beam, as described in Section 9.4.7. Horns have been made to operate at up to 250 GHz with success [21]. The design philosophy of a millimetre-wave corrugated horn is similar to

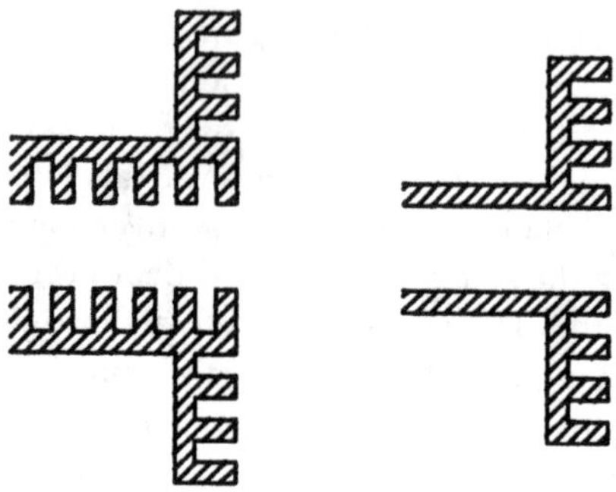

a b

Figure 9.22 Horn with rings in aperture

a Corrugated horn
b Smooth-wall horn

that of a microwave corrugated horn. The horn can have either a linear flare angle or a profile shape to the horn to reduce the length.

The small wavelengths at millimetre wavelengths mean that the major compromise necessary concerns the longitudinal corrugation geometry. The slot width and ridge width are usually chosen to be equal. The maximum number of cells per wavelength should be chosen to be as great as possible but not less than three per wavelength. Thus a horn designed for 200 GHz ($\lambda = 1.5$ mm) should have slots and ridges which are less than 0.25 mm in width. This is feasible with modern production techniques.

Millimetre-wave corrugated horns can be made by electroforming. A mandrel of aluminium is machined with the inverse corrugation pattern and then copper is electroformed onto the mandrel. The mandrel is finally removed with a solvent. The problem is to ensure that the copper is deposited evenly down the corrugations on the mandrel. Horns have been made for frequencies up to 200 GHz, where the corrugation depth is about 0.4 mm and the corrugation width is 0.2 mm. Figure 9.23 illustrates three millimetre-wave horns designed for use at 89 GHz, 150 GHz and 183 GHz. In all cases the aperture diameter is 5.5λ. The accurate mandrel was made by a computer-controlled spark-erosion machine in which a thin brass or molybdenum wire is continuously moved over the surface of a slowly rotating mandrel. A high-frequency pulsed voltage is applied to the wire, causing controlled erosion of both the mandrel and the wire.

9.5.4 Wide-flare-angle horns

Wide-flare-angle horns have copolar patterns which are determined mainly by flare angle and not by aperture diameter. The curves of Figures 9.4 and 9.5 illustrate this point. In this context, wide flare angle means greater than about 20° semiflare angle. The maximum aperture efficiency occurs for relatively small diameters and the optimum size is shown in Figure 9.24. A useful design guide is that the -15 dB-beamwidth level is approximately equal to the semiflare angle of the horn.

The corrugations are normally chosen to be at right angles to the flare and not at right angles to the axis. This may seem to complicate the analysis, since all techniques assume that the corrugations are perpendicular to the axis. In practice it has been found that predictions using corrugations perpendicular to the axis are good, even though the horn may be manufactured with slots perpendicular to the flare.

The best prediction techniques are based on spherical modes because these represent the modal structure of the horn. Spherical modal matching can be used to help design the throat region of a wide-flare-angle horn. This is a crucial design region because the short lengths mean that the principle of slow mode conversion which is effective in narrow-flare-angle horns cannot be used. The principles can be illustrated for a 30°-semiflare-angle horn. The influence of three different types of matching section is studied (Figure 9.25). These have constant quarter-wavelength deep slots; slots decreasing in depth from half a wavelength to quarter of a wavelength at the aperture; and the addition of a short section of smooth-wall conical section. The horns have an input diameter and an aperture diameter of 0.81λ and 4.75λ, respectively. The horns were modelled by dividing them into 10 conical sections. This was found to give good numerical convergence with spherical modal matching. The return loss of the TE_{11} mode in the input smooth-wall circular waveguide, and the level of the first two higher-order modes (EH_{12} and HE_{12}) at the aperture, are shown in Figure 9.26*a*, *b* and *c*, respectively, as a function of normalised frequency. The return loss of the constant-slot-depth horn is quite high because of the mismatch at the throat junction. The level of excitation of higher-order modes along the horn is

Figure 9.23 Millimetre-wave horns

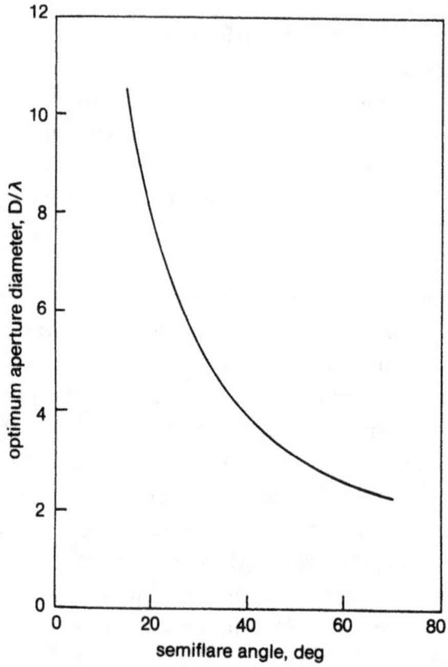

Figure 9.24 Optimum size of wide-flare-angle horns

very low. This is because the wall admittance is only weakly dependent on the radii of curvature kR.

The return loss of the horn with halfwavelength-deep slots shows about a 7 dB improvement over the $\lambda/4$-deep-slot horn. This amount of improvement is significantly less than would be obtained with a narrow-flare-angle horn. For instance, a 12°-semiflare-angle horn shows an equivalent improvement of 18 dB. The need to decrease the slot depth along the horn now means that the surface admittance changes along the horn, and consequently higher-order

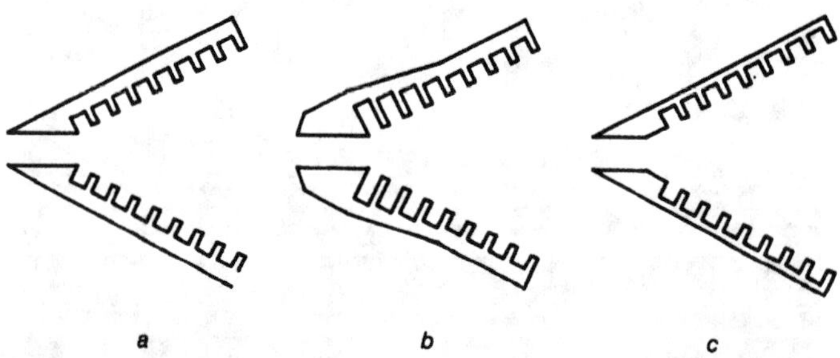

Figure 9.25 Wide-flare-angle corrugated horn with three input sections

a Slot depth constant at $\lambda/4$
b Slot depth decreases from $\lambda/2$ to $\lambda/4$
c Short smooth-wall conical horn at throat

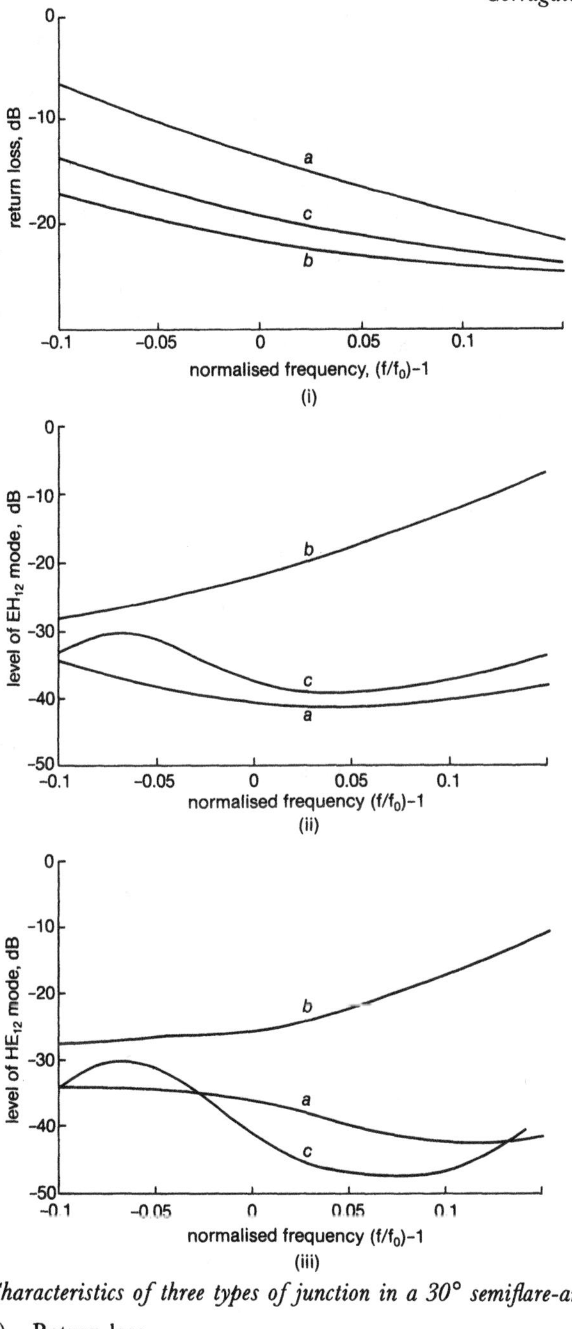

Figure 9.26 *Characteristics of three types of junction in a 30° semiflare-angle horn*

(i) Return loss
(ii) Level of EH_{12} mode
(iii) Level of HE_{12} mode
a Slot depth constant at $\lambda/4$
b Slot depth decreases from $\lambda/2$ to $\lambda/4$
c Short smooth-wall conical horn at throat

modes are excited. At the aperture of the horn there is a significant amount of HE_{12} and EH_{12} mode. The level rises as frequency increases because the cutoff point for the modes moves further down the horn. At 15% above the centre frequency, it reaches -13 dB for the HE_{12} mode and -8 dB for the EH_{12} mode.

The horn shown in Figure 9.25*c* has a smooth-wall conical section of length $L=0.2\lambda$ at the centre frequency. This horn has both a good return loss and a low level of higher-order modes since the mode conversion along the horn is low. The return loss is less than -20 dB over a wide frequency band. Thus this method of matching into a horn is both simpler and more efficient than the deeper slots for wide-flare-angle horns.

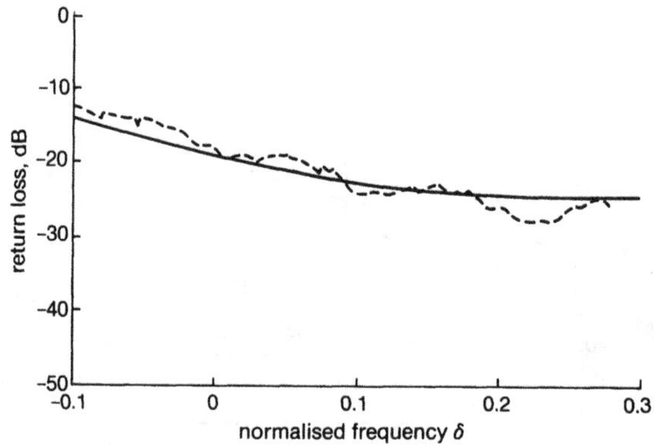

Figure 9.27 Measured and predicted return loss for horn illustrated in Figure 9.25c

$\delta = f/f_0 - 1$

—— predicted

— — measured

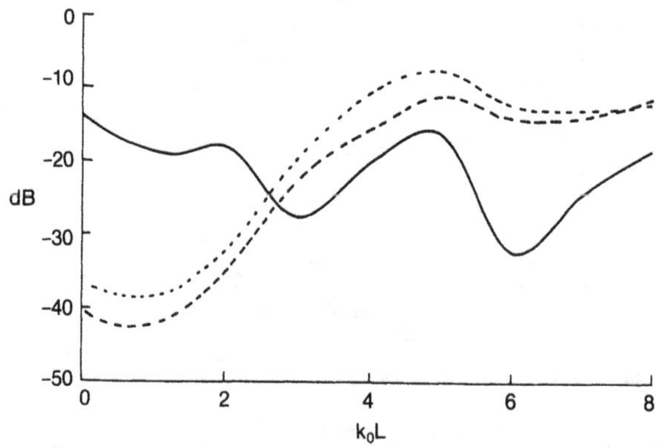

Figure 9.28 Scattering coefficients against normalised length of smooth conical section

—— return loss

- - - level of HE_{12} mode at aperture

—·—·— level of EH_{12} mode at aperture

This conclusion was confirmed by measuring the return loss of a horn with the above parameters. The slots had a width of 0.25λ and there were 3.5 slots per wavelength. Experience has shown that the number of slots along the horn is not critical. The return-loss results are shown in Figure 9.27. These show satisfactory agreement with theory. Most of the discrepancy arises because of reflections at the circular-to-rectangular-waveguide transition.

The use of a smooth-wall section does not always give a low return loss and a low mode conversion, and the length needs to be chosen carefully. Figure 9.28 shows how the scattering parameters are influenced by the length of the smooth-wall section. The levels of the EH_{12} and HE_{12} modes increase as the normalised length kL increases. A longer section means that the junction occurs at a larger diameter and hence further from the cutoff point of the higher-order modes. The return loss depends on both the junctions and oscillates because of phase interaction between the two junctions.

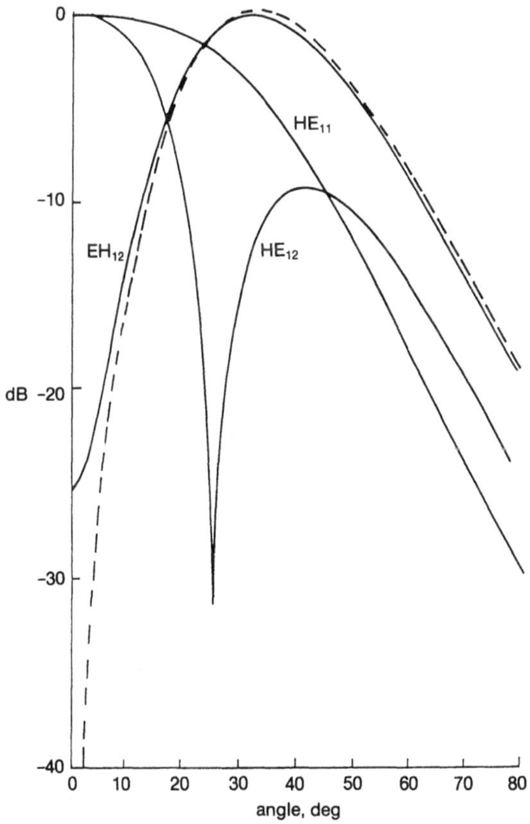

Figure 9.29 Radiation pattern of modes in a conical corrugated horn

——— copolar pattern
– – – – crosspolar pattern

Horns with semiflare angles greater than about 60° do not benefit from the use of a short smooth-wall section because the wider flare angle means that the diameter is larger at the smooth-wall-to-corrugated-horn junction. Higher-order modes are easily excited, with the consequence that mode-conversion levels at the aperture are high.

If higher-order modes are present at the aperture, they will radiate and distort the desirable radiation characteristics of the HE_{11} mode. The patterns of the HE_{11}, EH_{12} and HE_{12} modes are shown in Figure 9.29. The EH_{12} mode has a peak crosspolar level in the 45° plane which is equal in level to the patterns in the principal planes. The addition of this mode to the HE_{11} mode can cause a large increase in the radiated crosspolarisation. The HE_{12} mode has a low level of crosspolarisation but can contribute a high sidelobe to the copolar pattern of the HE_{11} mode.

The total radiation pattern of the horn is obtained by adding together the patterns of the separate modes with the correct amplitude and phase relationship at the aperture. The effects of higher-order modes can be seen for the patterns for two 60°-semiflare-angle horns. These were computed using spheri-

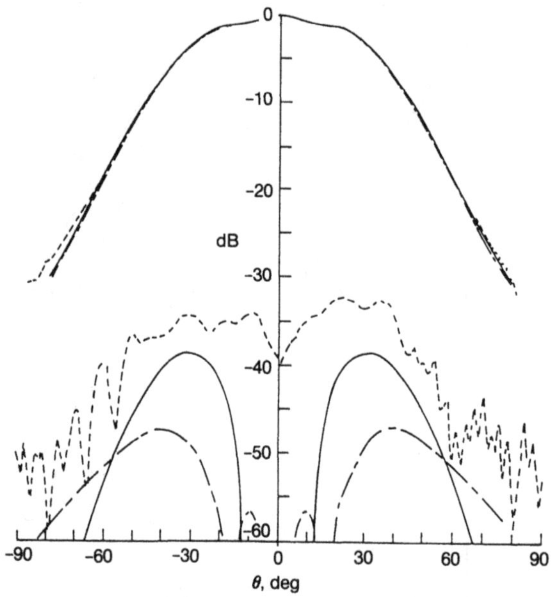

Figure 9.30 Predicted and measured radiation patterns for 60°-semiflare-angle horn at 10.31 GHz

——— predicted with all higher-order modes
—·— predicted with only HE_{11} mode
- - - - measured

cal modal matching. Horns have been made and measured to confirm the theoretical predictions. One of the horns had constant $\lambda/4$-deep slots; the other had a short smooth-wall section at the throat. The horns had an aperture diameter of 5.43λ and were modelled by dividing them into six sections. The flare angle and aperture diameter create a phase error across the aperture which is greater than half a wavelength and so the main beam is broadened.

The results for the constant-slot-depth horn are shown in Figure 9.30. This shows the predicted copolar- and crosspolar-radiation patterns with all modes included, and the predicted patterns with only the HE_{11} mode present. The higher-order modes have negligible influence on the copolar patterns and cause only a small increase in the crosspolar level. The second horn was designed to radiate a significant level of higher-order mode. This is confirmed by the results in Figure 9.31. The smooth-wall section had a length of $L=0.3\lambda$. The EH_{12} mode causes the crosspolar level to be raised by 25 dB and the E-plane copolar pattern to be considerably different from the single-HE_{11}-mode case. The phase of the modes at the aperture has caused the beamwidth to be reduced. The experimental results show good agreement with the theoretical predictions.

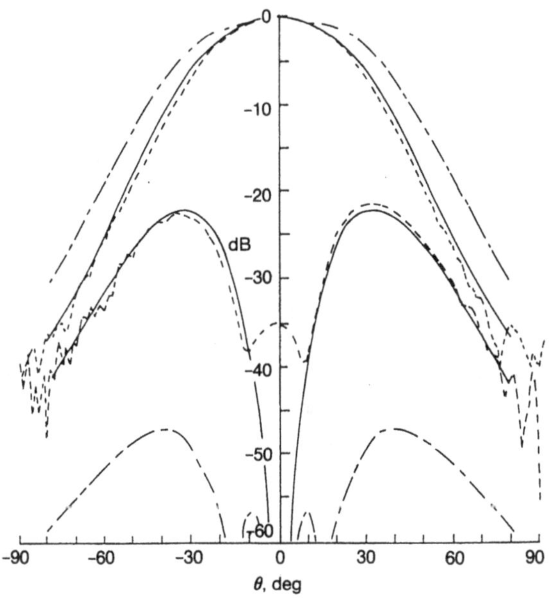

Figure 9.31 *Predicted and measured radiation patterns for 60°-semiflare-angle horn with short smooth section of length L = 8.5 mm*

——— predicted with all higher-order modes
—·— predicted with only HE_{11} modes
- - - - measured

9.5.5 Broadband corrugated horns

The majority of applications require corrugated horns to operate over relatively narrow frequency bands, typically 10% bandwidth but up to 20% bandwidth. There are some applications where operation over a wider bandwidth is required. This may be to radiate simultaneously in two or more discrete bands, and this type is considered in Section 9.5.7. There are also applications where operation is required over a broad band of frequencies. In this context, 'broadband' can be defined as greater than 50% bandwidth. The corrugated horn is particularly useful as a broadband feed because the desirable properties of low crosspolarisation and pattern symmetry are maintained over a broad range of frequencies. This is at first sight surprising, because the principle of operation depends on $\lambda/4$-deep slots. However, it turns out that the rate of change of surface impedance with frequency is relatively slow so that, even at an appreciable frequency distance from $\lambda/4$, the good properties still hold. This characteristic is a function of the aperture diameter, and the bandwidth increases proportionally to aperture diameter.

The corrugated horn has a second potentially useful broadband property. For certain beamwidths it is possible to have a nearly constant bandwidth over a broadband. This is seen by examining Figure 9.4, which shows the -3 dB beamwidth against aperture diameter. For horns with semiflare angles below about 20°, there is a band of frequencies over which the beamwidth is nearly constant. This is due to the compensating effect of the spherical phase cap at the aperture. Over this range, pattern symmetry is maintained. Note that the beamwidth is not variable and in general the designer has to fit the rest of the reflector antenna around the beamwidth which can be provided.

The impedance properties will usually be the factors that limit the usable bandwidth. Even though the slot depths are very tolerant, there is a range outside which operation will be poor. For slot depths under $\lambda/4$ the surface reactance is positive and this means that the EH_{11} mode can propagate. This is a surface-wave mode which has a concentration of power around the radius of the corrugations and then radiates a high level of crosspolarisation. Excitation of the EH_{11} mode can be avoided if there is a very gradual and smooth change from a $\lambda/2$ slot to a $\lambda/4$-deep slot. The throat region is the critical design region. One method of increasing the bandwidth is to use 'ring-loaded' slots. These are described in Section 9.5.6. However, even without ring-loaded slots it is surprising what can be achieved if the pattern specifications are not too severe, as may be the case in many radar-type applications.

To illustrate this point, consider the design of an octave-band horn which is designed to radiate from 8 GHz to 16 GHz with a constant halfpower beamwidth of between 5° and 6°. This is possible with a horn having a semiflare angle of 15° and an aperture diameter of 7λ at 8 GHz, and can be seen by examining Figure 9.4. The slots are chosen to be 7.5 mm deep, making them resonant at 10 GHz. The slot width was selected as 5 mm and the ridge width as 2 mm, giving 2.7 cells per wavelength at 16 GHz. The crucial choice is the depth of the first slot at the throat. With a 24 mm input diameter, the first slot must be at least 10.5 mm deep which puts the cutoff frequency of the HE_{11} mode at 7.7 GHz. This means that the $\lambda/2$ condition occurs at 15.5 GHz. Ideally, the slot ought to be deeper, but this would then restrict the upper frequency. The computed VSWR at 8 GHz is 3.08, which is high but acceptable. At frequencies above 9 GHz it drops to below 1.1. The computed copolar and crosspolar radiation patterns at 8, 12 and 16 GHz are shown in Figure 9.32. The −3 dB

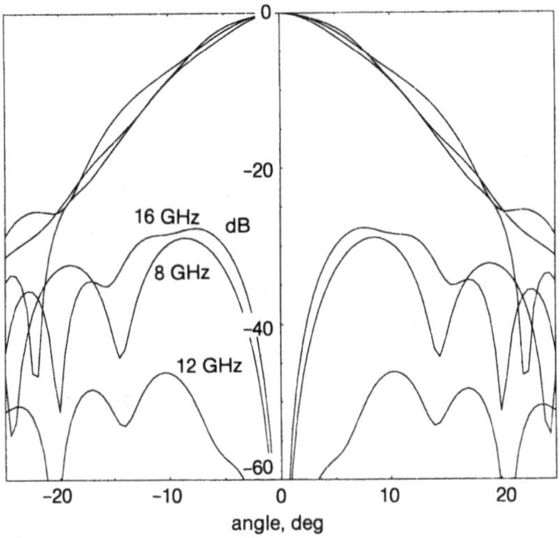

Figure 9.32 E-plane- and crosspolar-radiation patterns for broadband horn at 8, 12 and 16 GHz

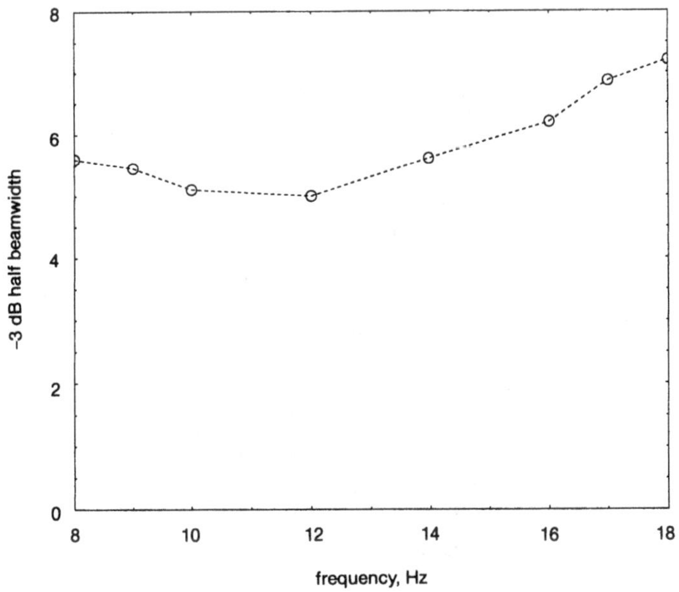

Figure 9.33 −3 dB halfbeamwidth in the E-plane of a broadband horn

halfbeamwidth from 8 GHz to 18 GHz is shown in Figure 9.33. The specification is satisfied up to 15.5 GHz. At higher frequencies the copolar pattern shows a small null on boresight owing to the phase error. The change in beamwidth and peak crosspolar level are monotonic on either side of the balanced hybrid condition, unless higher-order modes are excited, although in this case modal-matching computations show that they are not excited. Hence this example demonstrates the broadband properties of the corrugated horn.

9.5.6 Ring-loaded slots

The potential for ring-loaded slots to increase the bandwidth of the matching section of a corrugated waveguide was first proposed by Takeichi *et al.* [22]. The ring-loaded slots consist of sections of annular circular waveguide which are added into the regular slots of a horn (Figure 9.34). The ring slots may be symmetrical or asymmetrical with respect to the slot in the corrugated wall. To maintain the same pitch as for normal slots, the input slot needs to be much narrower than normal. The theory of ring loading has been studied in detail by James *et al.* in a series of papers [23–25] and also by Kuhn and Hombach [15]. James and Thomas [23] extended the cylindrical modal-matching theory for normal slots to include ring-loaded slots. The basic theory in Chapter 4, Section 4.6, must be modified to include lengths of annular waveguide. The scattering matrices are modified to include terms which contain information about the power-coupling integrals between the annular ring and the circular-waveguide section representing a slot. The details of the theory are contained in Reference 23.

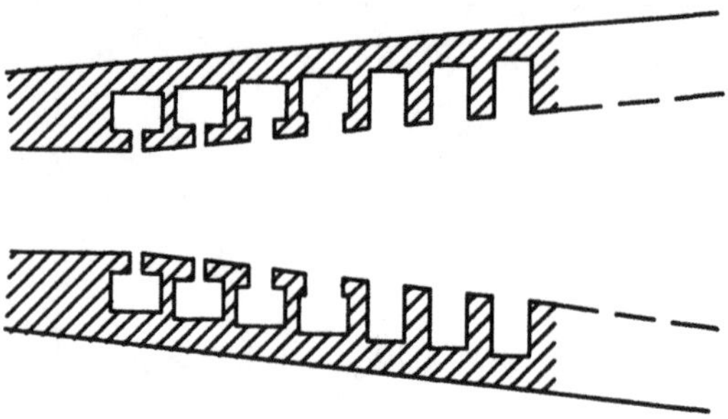

Figure 9.34 Horn with ring-loaded section

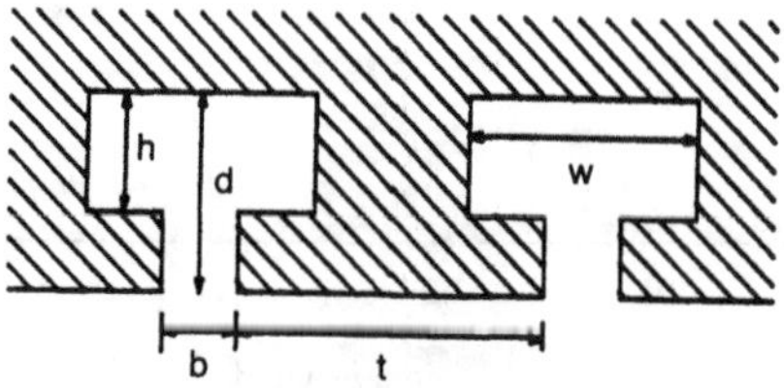

Figure 9.35 Planar 'ring-loaded' slot

A simple demonstration of the broadband properties of a ring-loaded slot can be provided by considering the equivalent case of enlarged slots in a planar corrugated surface (Figure 9.35). This is approximated by a large-diameter ring-loaded waveguide. With a planar corrugated surface, standard transmission-line theory for multiple sections of transmission line can be used. The transmission-line approximation, ignoring any shunt reactance resulting from the discontinuity of the junctions, gives [23]

$$X_z = \frac{b}{p}\left[\frac{\tan(kh) + \frac{b}{w}\tan\{k(d-h)\}}{1 - \frac{w}{b}\tan(kh)\tan\{k(d-h)\}}\right] \tag{9.53}$$

where b is the width of the input slot, p is the pitch of the corrugations, h is the depth of the ring section, d is the total depth and w is the length of the ring section. Normal slots with constant width have an input reactance given by

$$X_z = \frac{b}{p}\tan(kh) \tag{9.54}$$

The frequency performance of the ring-loaded slots, by comparison with the conventional slots, is shown in Figure 9.36 for $d/h = 1.5$. The reactance is infinite for a $\lambda/2$-deep slot (as shown in the Figure) and consequently varies considerably around the resonant frequency. By contrast, the ring-loaded slot has a

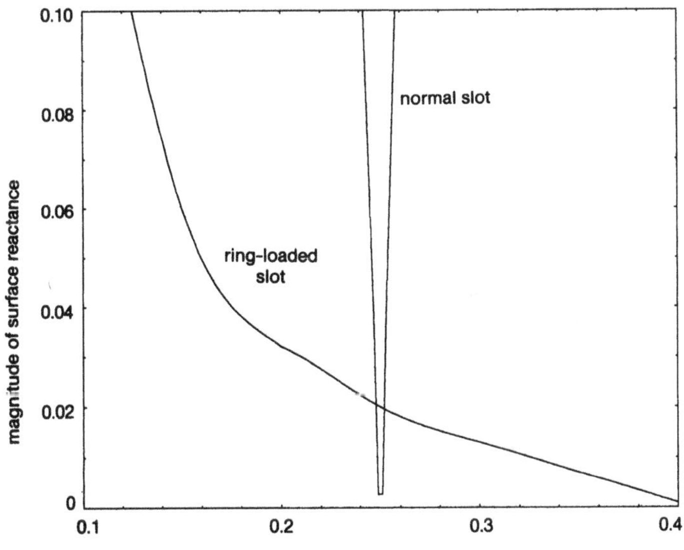

Figure 9.36 Normalised surface reactance of enlarged slots in a plane surface as a function of the normalised slot depth

reactance which is only slowly varying with frequency. Thus the impedance properties of ring-loaded slots are much less frequency sensitive than those of normal slots.

Detailed studies of ring-loaded slots for broadbanding the performance of corrugated horns are given by James [24] and Thomas *et al.* [25]. James proposes some useful design rules. If the length of the horn over which the matching is done is L, then the ratio of ring-loaded-slot depth to total-slot depth is recommended to be $h/d = 3/2$, and the input slot width b_m should be

$$b_m = p\left\{0.1 + \frac{(m-1)}{L}\left(\frac{b}{p} - 0.1\right)\right\} \tag{9.55}$$

where m is the number of the slot from the throat. The slot depth d_m should be chosen to be

$$d_m = \frac{\lambda}{4}\exp\{1/(2.5kr_m)\} \tag{9.56}$$

These values will give a smooth transition from the input TE_{11} mode to the hybrid HE_{11} mode over a 1.5:1 frequency band.

9.5.7 Dual-depth corrugated horns

Dual-depth corrugated horns (Figure 9.37) were conceived [26–29] as a way of obtaining good crosspolar performance in two distinct frequency bands. They differ from broadband horns in that the interest is in obtaining optimum performance in two separate bands rather than the optimum continuous performance over a single wide band of frequency. Dual-depth horns can be analysed either by using the modal-matching technique or by using a modification to the surface-impedance and space-harmonic techniques [28]. The former gives the most accurate method but without the physical understanding which

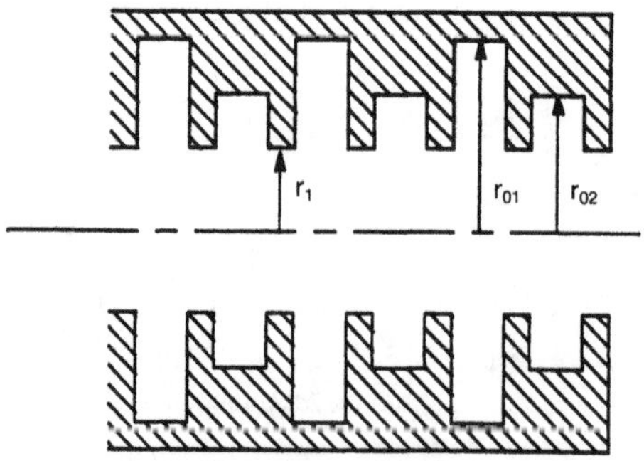

Figure 9.37 Dual-depth corrugated waveguide

comes from the surface-impedance and space-harmonic techniques. The principles will be outlined here with reference to these latter methods.

When there are two interleaved sets of slots (Figure 9.37) with radii r_{01} and r_{02}, the combined surface admittance consists of the series combination of the admittances of the two slots, i.e.

$$Y = \frac{Y_1 Y_2}{Y_1 + Y_2} \tag{9.57}$$

where $Y_q = S_1(kr_1, kr_{0q})$. The propagation equation using the surface-impedance technique is then obtained by replacing eqn. 9.7 by eqn. 9.57. The behaviour of the dual-depth waveguide can then be inferred directly from that of the constant-depth waveguide with appropriate allowance for the dual-depth corrugations. In particular, the condition which gives the balanced hybrid condition and low crosspolarisation occurs when either $Y_1 = 0$ or $Y_2 = 0$. This shows that a dual-depth corrugated waveguide can be expected to operate in two distinct frequency bands, determined by the depths of the slots.

9.5.7.1 Propagation characteristics

Typical propagation characteristics are shown in Figure 9.38. The mode of most interest, the HE_{11} mode, has two branches: first, from cutoff to the point $\beta r_1 = kr_1$ and, secondly, as the high-frequency portion of a mode which has an EH_{12}-type cutoff. The balanced hybrid frequencies are marked. As the ratio of the two slot depths decreases, the turnover point where the modes change from HE to EH becomes sharper and eventually merges together for a single-depth corrugated waveguide. Computed results show that the ratio of the depths of the two slots, and hence the ratio of the two operating bands, can be varied

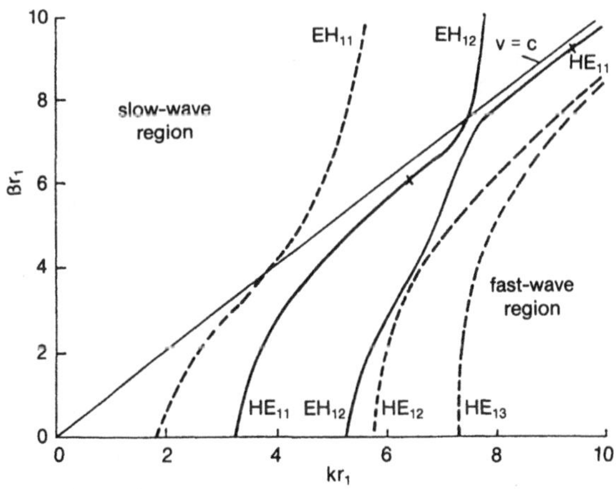

Figure 9.38 Propagation characteristics of dual-depth horn

$r_1 = 40$ *mm*, $S_1/S_2 = 1.67$
× balanced hybrid frequencies
S = slot depth

between 1:1.3 and 1:2. Below 1:1.3 a single-depth horn will suffice, whilst for ratios greater than 1:2 it is possible to use a single-depth horn with the one slot acting simultaneously at two frequencies.

9.5.7.2 Crosspolar-radiation characteristics

The minimum of the peak crosspolarisation level occurs at frequencies where one of the slots is resonant or, in terms of surface admittance, where the admittance for that slot is zero. The value of the resonant frequency is determined principally by the depth of the slot, but also by the number of slots per wavelength and the ratio of the ridge width to the slot width.

One of the most important questions in the dual-depth corrugated horn is how many slots are required per wavelength. Figure 9.39 attempts to answer this question by plotting the peak crosspolarisation against frequency of a dual-depth corrugated horn with aperture diameter 80 mm for four values of the period $p=2b+2t$. In the low-frequency band (Figure 9.39*a*), these values correspond at the crosspolar minimum to $(b+t)/\lambda=7$, 9, 11 and 17 slots per wavelength. The ratio b/t is 10 in all cases. As the number of slots per wavelength is reduced, two consequences are seen. First, the frequency at which minimum crosspolarisation occurs is reduced. This is a direct result of the relative change in the electrical slot width. Secondly, the minimum level of crosspolarisation increases. This is due to the influence of the shallower depth slot which is nonresonant in the low-frequency band. For a large number of slots per wavelength, the nonresonant slot is insignificant in electrical width, but as the slots are increased in electrical width the nonresonant slot acts as a large electrically distorting corrugation which upsets the field coupling between adjacent resonant slots. The behaviour in the high-frequency band (Figure 9.39*b*) is similar except that the minimum crosspolarisation is lower due to the larger aperture. Here the four values of slot width correspond at their minimum crosspolarisation frequency to 4, 5, 7 and 10 slots per wavelength.

Comparison of Figure 9.39*a* and Figure 9.39*b* shows that, for a fixed period, the peak crosspolar level is highest in the low-frequency band. Hence, we can deduce that, to obtain -40 dB crosspolar level, requires a period of 6.6 mm. This corresponds to seven slots per wavelength for an aperture diameter of 3.6λ.

The effect of altering the ratios of the depths of the two slots, and hence the ratio of the centre frequency of the two operating bands, is shown in Figure 9.40. The depth of the deeper slot has been maintained constant so that the lower operating band remains nearly constant except insofar as the nonresonant shallower slot influences the propagation characteristics and aperture fields. The figure shows values of the ratio of the depth of the two slots S_1/S_2 between 1.3 and 2.0. The minimum crosspolarisation in the high-frequency band rises by a relatively small amount as the two depths become more widely separated. This is a consequence of two partially compensating factors. The change in slot depth moves the resonant frequency to a higher value, and thus the number of slots per wavelength decreases, resulting in high crosspolarisation. However, the higher operating frequency means that the electrical aperture diameter

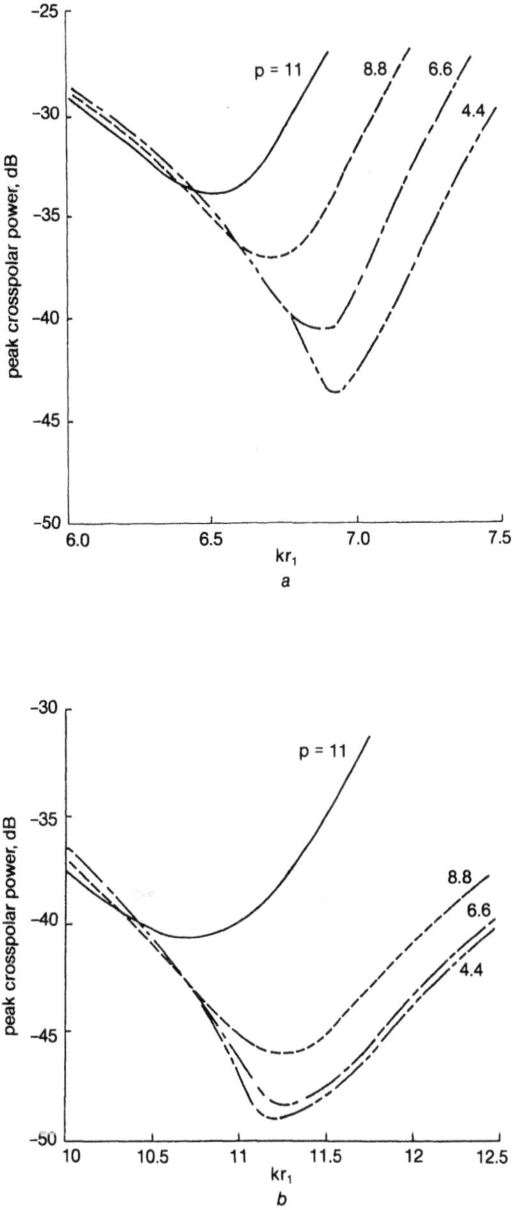

Figure 9.39 Influence of number of slots per wavelength on peak crosspolar level

$r_1 = 40$ *mm*, $S_1/S_2 = 1.67$
Parameter: period, mm
a Low-frequency band
b High-frequency band

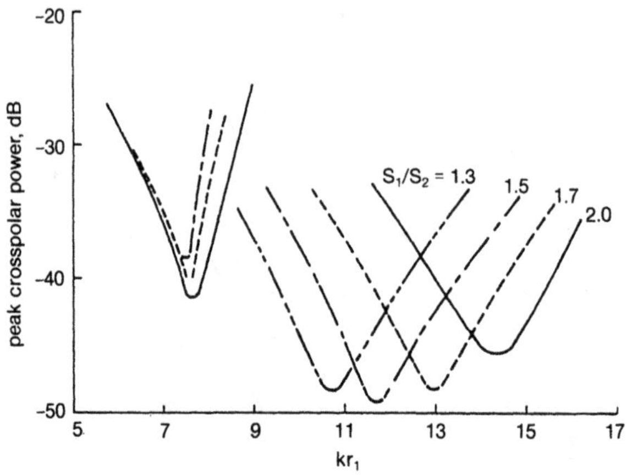

Figure 9.40 Peak crosspolar level against normalised frequency

$r_1 = 41.5$ mm, $r_{01} = 50$ mm, $b = 3$ mm, $t = 0.5$ mm
Parameter: S_1/S_2

increases, and this tends to reduce the crosspolarisation.

The relative bandwidth below which the peak crosspolarisation has a specified value is shown in Figure 9.41. As observed above, the high-frequency bandwidth is greater than the low-frequency bandwidth owing to the larger aperture diameter.

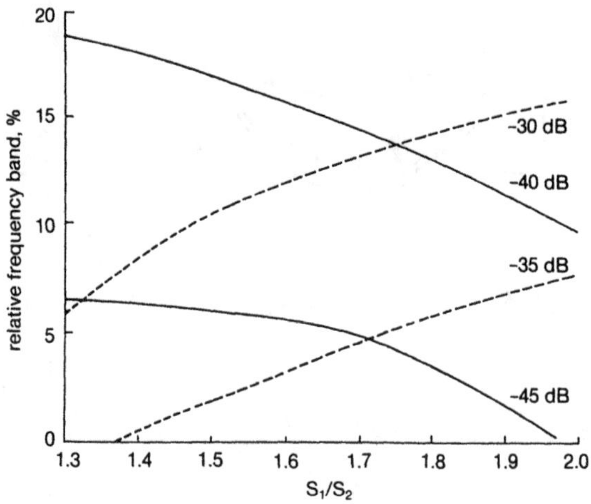

Figure 9.41 Relative-frequency band for which peak crosspolar level is below a specified level

Parameters as Figure 9.40
—— high-frequency band
- - - low-frequency band

9.5.8 Profile corrugated horns

The flare angle for the horn is usually chosen to be linear, but this does not have to be the case. The profile horn (Figure 9.42) was developed as a method of achieving a compact horn which can be down to about two-thirds of the length of an equivalent linear horn and can also have improved electrical properties. The advantages of profiled horns are:

(i) The horn is mechanically shorter and hence the mass of the horn is reduced.
(ii) The flare angle at the aperture is nearly zero so that the horn radiates similarly to an open-ended waveguide with no spherical phase error. This gives the highest efficiency and gain for a given aperture diameter.
(iii) A low level of crosspolarisation continues to be radiated.
(iv) The phase centre is near the aperture of the horn and remains nearly fixed as frequency changes. This can be a useful property when the horn is used as a feed over a wide band.
(v) The small flare angle at the throat is ideal for achieving a good impedance match.

There is one significant disadvantage with a profiled horn. The HE_{12} mode is relatively strongly excited by the change in flare angle near the aperture and then radiates with high copolar sidelobes which coincide in angle with those of the HE_{11} mode. Consequently, the copolar-sidelobe level is raised, as illustrated in Figure 9.43. The amount of increase depends on the maximum flare angle and is the limiting parameter as to the minimum length that a profile horn can be designed. The HE_{12} mode does not radiate significant levels of crosspolar power, so the crosspolar lobes are nearly unaltered compared with those of an open-ended waveguide.

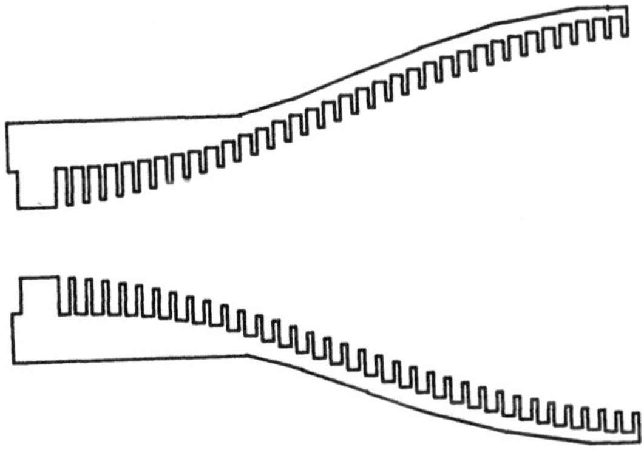

Figure 9.42 Profile corrugated horn

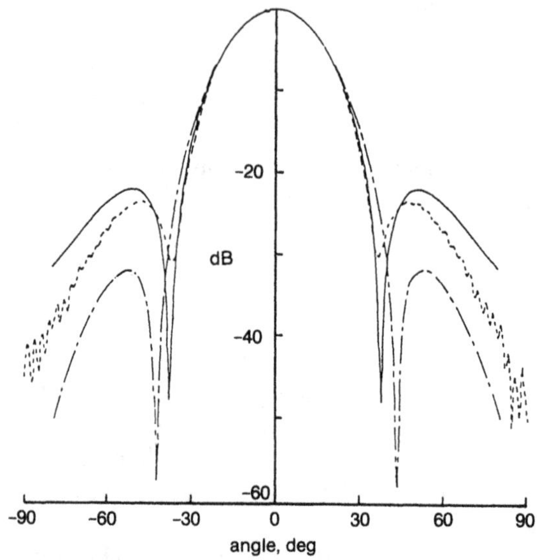

Figure 9.43 Radiation pattern of profile horn with dimensions of horn 2 in Table 9.2

——— predicted pattern
—·— HE_{11} mode only
- - - - measured pattern

The mathematical shape of the profile can be a variety of functions, but a convenient one is to use a sine-squared profile:

$$r(z) = r_{th} + (r_{ap} - r_{th})\left\{(1-A)\frac{z}{L} + A\sin^2\left(\frac{\pi z}{2L}\right)\right\} \tag{9.58}$$

where L is the length of the horn. The parameter A controls the amount of profile which is added to a linear taper. A value of A between 0.7 and 0.9 has been found to give the best results. The maximum flare angle is located half way along the horn and can be found by differentiating the above equation with respect to radius. This gives

$$\theta_{max} = \tan^{-1}\left\{(0.57A+1)\,\frac{(r_{ap}-r_{th})}{L}\right\} \tag{9.59}$$

θ_{max} depends on the length and the difference between the radius at the aperture and the radius at the throat.

It is possible to choose a modified form of eqn. 9.59 in which the point at which the maximum flare angle occurs is not half way along the horn [1]. This gives added flexibility in the design, but in most cases there is little benefit to be achieved from an asymmetrical profile.

Modal matching is ideal for studying the radiation and impedance characteristics of profile corrugated horns. The spherical modal-matching technique

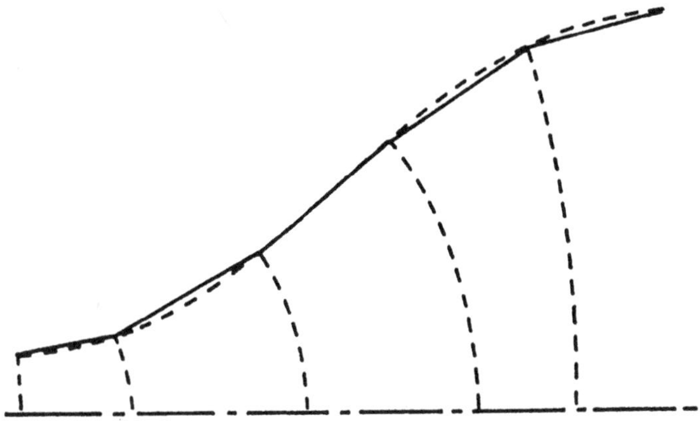

Figure 9.44 Division of a profile horn into conical sections for modal matching

which uses spherical modes to represent the fields in each conical section (Figure 9.44) is better for parametric studies than cylindrical modal matching because fewer modes are needed and there are fewer parameters to vary. In the discussion which follows, spherical modal matching has been used with the slots represented by a surface impedance. This technique is described in Section 9.4.4.

9.5.8.1 Junctions along the horn

The profile horn has three types of junctions along the horn:

(*a*) the junction at the throat between the smooth-wall waveguide and the conical corrugated horn
(*b*) junctions between two conical corrugated sections with the same flare angle but different slot depths
(*c*) junctions between two conical corrugated sections with different flare angles.

Junctions (*a*) and (*b*) occur in wide-flare-angle horns and are discussed in Section 9.5.4. Junctions which are incorrectly designed convert significant amounts of HE_{11}-mode power into the higher-order EH_{12} mode. The EH_{12} mode radiates a high level of crosspolarisation, but it does not significantly influence the HE_{11} copolar pattern. The design of the throat region of a profile horn is similar to that of any other corrugated horn.

Junctions of type (*c*) between two conical corrugated sections with different flare angles mainly excite the HE_{1m} modes. This is in contrast with junctions of type (*a*) and (*b*), which mainly excite the EH_{1m} modes. The HE_{12} mode radiates a strong copolar signal which distorts the HE_{11}-mode copolar pattern. The excitation level in decibels of the HE_{1m} modes is approximately proportional to the magnitude of the difference in flare angles at a junction. It is greater for flare-angle changes near the aperture than for changes near the throat. However, it is almost independent of the actual flare angle.

9.5.8.2 Standard profiled horns

As already mentioned, an advantage of the profiled horn is the ability to design a short, compact horn which reduces the mass, volume and length. The degree of compactness depends on the excitation level of higher-order modes. This is illustrated in Figure 9.45, which shows the return-loss and mode-conversion levels of the EH_{12} and HE_{12} modes in a profile horn. The length of the horn is kept constant at 4.8λ. The slot depth at the throat is $\lambda/2$ and decreases to $\lambda/4$ over a length of 1.4λ. Figure 9.45 shows that the return loss and level of the EH_{12} mode is very low, while the level of the HE_{12} mode is significant and increases as θ_{max} increases. Radiation-pattern calculations show that a level of HE_{12} mode above about -8 dB is unacceptable because the copolar sidelobes will then significantly distort the main beam.

The patterns for two profile horns are shown to demonstrate the design factors. The geometries are given in Table 9.2.

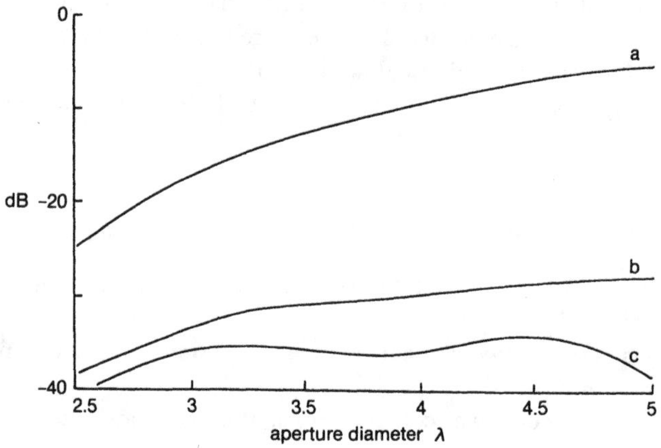

Figure 9.45 *Return loss and levels of higher-order mode at aperture of profile horn with $r_{th}/L = 0.41$ and $L = 4.8\lambda$*

a HE_{12} mode
b Return loss
c EH_{12} mode

Table 9.2 Geometry of profile horns (dimensions in wavelengths)

	L	r_{th}	r_{ap}	d_{th}	d_{ap}	θ_{max}
Horn 1	8.16	0.38	2.60	0.5	0.20	20.0
Horn 2	4.92	0.39	1.34	0.5	0.26	15.1

Both horns have sine-squared profiles given by eqn. 9.58 with $A = 0.7$. Horn 1 is modelled by dividing it up into 17 sections with individual lengths of between 0.3λ and 0.5λ. This is sufficient to model the horn accurately when five modes are included in the computation, further modes having no practical influence. The copolar patterns of horn 1 are shown in Figure 9.46. The predicted level of the HE_{12} mode at the aperture is approximately -10 dB, which causes significant shoulders to appear on the copolar pattern. The crosspolarisation

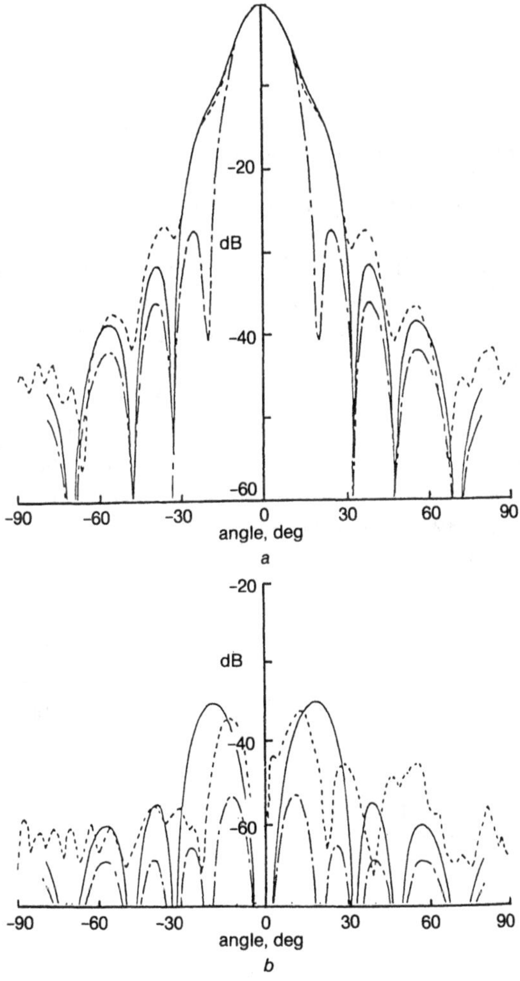

Figure 9.46 Predicted and measured radiation patterns for horn 1

——— predicted pattern
—·— HE_{11} mode only
– – – – measured pattern
a E-plane copolar pattern
b Crosspolar pattern

level is increased by about 12 dB compared with the HE_{11} mode alone, but is still low. An experimental horn was constructed and measured and the results are shown in Figure 9.46. Good agreement is observed between theory and experiment. The HE_{11} and HE_{12} modes are nearly in phase at the aperture. As frequency changes, this phase relationship changes, but not necessarily in a detrimental way. The peak crosspolar level rises to -30 dB over a 20% band of frequencies.

The second horn in Table 9.2 has a more gradual profile than horn 1 and consequently only 12 sections are needed to model the horn theoretically. The HE_{12} mode level is about -22 dB, which leads to less distortion of the copolar pattern. The measured results are shown in Figure 9.43. At the centre frequency, the sidelobe level is raised by 5 dB but this is not sufficient to cause the first sidelobe and main beam to merge. If this horn were used as a feed for a reflector, then the only consequence would be a small reduction in efficiency.

9.5.8.3 Low-mode-conversion discrete-section profile horn

The two horns described above had a smooth sine-squared profile. There is no electromagnetic reason why the profile has to be smooth. The modal-matching model has a number of discrete conical sections and a practical horn could be made with this profile. The mode conversion of the HE_{12} mode will then only occur at the junctions between sections, because the amount of mode conversion along a linear section is very low. The level of HE_{12} mode at the output side of a junction is the sum of the power excited from the HE_{11} at that junction plus the HE_{12}-mode power which was excited at previous junctions. The phase of this latter power will be dependent on the distance between junctions. A synthesis procedure can be devised which seeks to minimise the total level of HE_{12} mode at the aperture by appropriate positioning of the junctions and by adjusting the radius of the horn at the junctions.

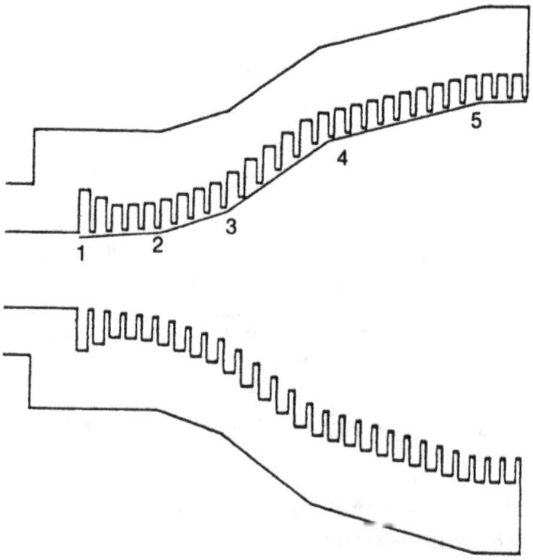

Figure 9.47 Profile horn using discrete sections

The synthesis procedure also needs to take account of the need for low return loss at the throat and a low level of radiated crosspolarised power. To achieve a good return-loss performance, the first and second junctions should be designed so that the first section has a small flare angle and the first slot depth is $\lambda/2$ deep. The depths of successive slots should decrease to the operating depth at the end of the first section. The slot depths in the remaining sections are constant so that the generation of any EH_{12} mode is suppressed. The last junction before the aperture should be chosen to lead to an open-ended waveguide at the aperture. The position of intermediate junctions can then be optimised to minimise the level of HE_{12} mode at the aperture.

This synthesis technique has been numerically demonstrated with a five-section horn (Figure 9.47), which has the same length, throat diameter and aperture diameter as horn 1. Each linear section is divided up into a number of subsections if the axial length is greater than half a wavelength or the slot depth changes significantly. Manual optimisation of the positions of the second, third and fourth junctions led to the values shown in Figure 9.47. The levels of the EH_{12} and HE_{12} modes at the aperture of the horn are shown in Figure 9.48. The level of the HE_{12} mode has been reduced by about 8 dB compared with the continuous profile of horn 1. It is almost independent of frequency. The EH_{12} mode has also been reduced to -40 dB at the centre frequency, but is more frequency sensitive. Figure 9.49 shows the radiation pattern of the synthetised horn. The high main-beam shoulder has been nearly eliminated, and the crosspolar level has been reduced to below -40 dB. The benefit of designing a discrete-section profile horn can be judged from the fact that a sine-squared horn would need to be twice as long as horn 1 to obtain similar copolar patterns.

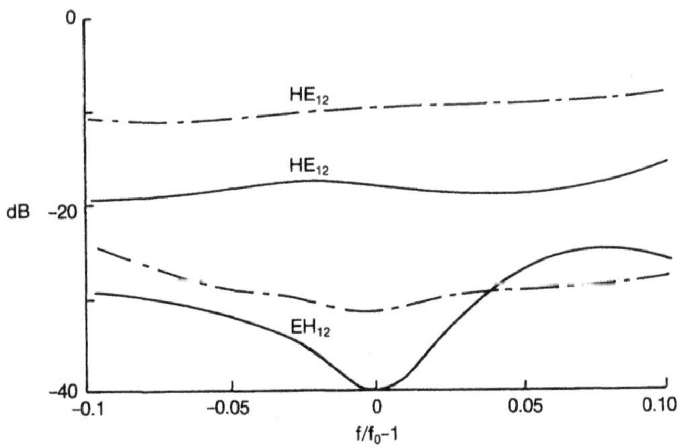

Figure 9.48 Levels of higher-order modes at aperture against normalised frequency

—— synthetised horn
—·— horn 1

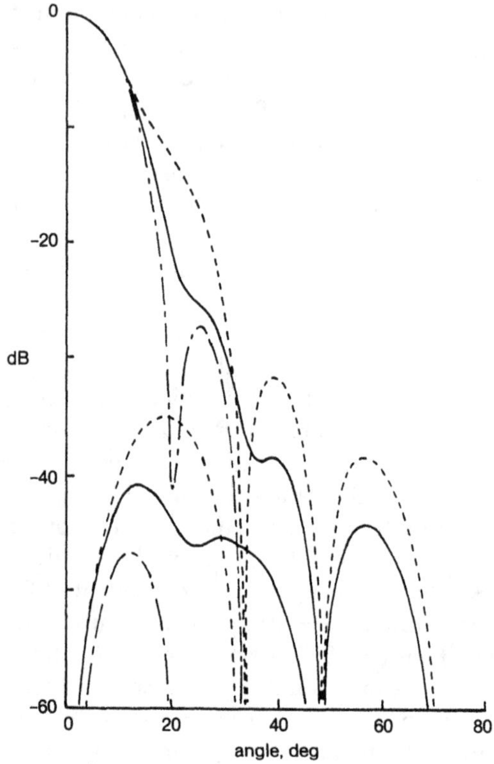

Figure 9.49 Radiation patterns of profile horn with discrete sections

——— synthetised horn
- - - - horn 1
—-— HE_{11} mode

The frequency behaviour is similar to, but slightly less sensitive than, that of horn 2. That is to say, it is more frequency sensitive than a profile horn with a continuous profile. This is to be expected from the interactions which are taking place between the various junctions.

As indicated in the list of advantages at the start of Section 9.5.8, the phase centre of a profiled horn should be near the aperture of the horn. However, the presence of higher-order modes will give a phase pattern which changes with angle. The phase patterns for horn 1 and the synthesised horn are shown in Figure 9.50. The phase of horn 1 changes continuously as the angle increases, but the phase of the synthetised horn stays nearly constant out to about 15°. Any phase error will reduce the phase efficiency of a feed-reflector system. Thus the discrete profile has an additional advantage over a continuous profile.

The above discussion has indicated the potential of profile horns, both with a smooth profile and with a discrete profile. It has also shown how modal-matching analysis techniques can help in designing the optimum geometry.

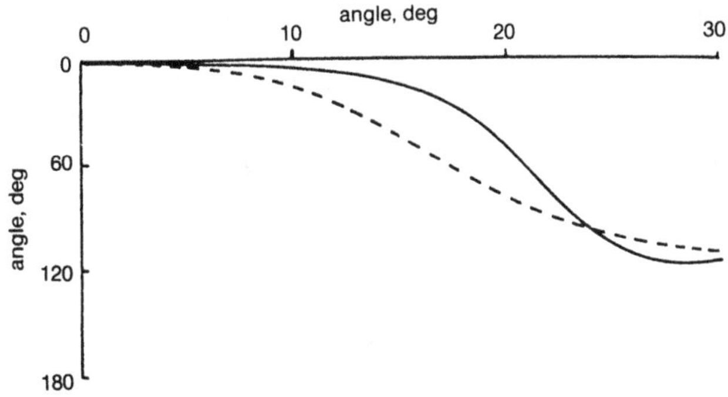

Figure 9.50 *Relative phase patterns*

——— synthetised horn

– – – – horn 1

9.5.9 *Rectangular corrugated horns*

Rectangular corrugated horns have received much less attention in the literature than conical corrugated horns for a number of reasons. The conical corrugated horn has been very successful and the circular geometry gives excellent radiation characteristics. There has been a lack of reliable theoretical techniques for analysis, particularly by comparison with the highly developed and accurate techniques described in the rest of this chapter. The rectangular corrugated horn is difficult to manufacture. Rectangular corrugations do not lend themselves to either machining in one piece or electroforming. Consequently, the horns are expensive to manufacture with much less guarantee that they will perform as predicted.

The advantage of rectangular corrugated horns is their ability, in principle, to radiate elliptically shaped beams with low crosspolarisation. However, an alternative method of producing a shaped beam is to use a conical corrugated horn in conjunction with a shaped reflector. Although the latter is now the preferred method, it is worth understanding the basic characteristics of the rectangular horn because there may be applications where a rectangular horn is a better solution.

The two-wall corrugated horn, with corrugations on opposite walls, can be analysed using impedance techniques. It is studied in Chapter 7 of Reference 1. This enables the basic modal properties to be understood. The mode with balanced hybrid properties which radiates low levels of crosspolarisation is the HE_{12} mode. This mode is symmetric in both the x and y directions and has a single cosine aperture-field distribution in both planes. However, it is not the dominant mode in the waveguide and this means that care has to be taken to excite the correct mode and to avoid mode conversion along the horn. This can be done by a similar method to that used for conical corrugated horns, with deeper slots at the throat. The first slots near the throat should be half a guide wavelength deep (not half a free-space wavelength deep) and taper to one-quarter of a guide-wavelength deep at the aperture.

The four-wall corrugated horn cannot be analysed by impedance techniques because the boundaries cannot be made coincident with the reference directions of a rectangular-co-ordinate system. Most attempts at analysis concentrated on superimposing two orthogonal two-wall structures. Dragone [30] has reported a horn with the corners filled in which showed good agreement with a simple theory based on surface-impedance techniques. The crosspolar level in the principal planes was measured as -35 dB.

When modal-matching techniques became available, a method of analysis existed which permitted study of the horn in the same detail as for conical horns. The first reported use was by Kuhn and Watson [31], who extended the modal-matching technique to a wide range of rectangular geometries. The principles are identical to those set out in Chapter 4 and the scattering-matrix theory is the same. The difference is that rectangular functions are used to describe the modal properties. The mode functions for TE modes in rectangular waveguides with the axes along the centre of the waveguide are

$$\psi = \cos\left(\frac{n\pi x}{a}\right)\cos\left(\frac{m\pi y}{b}\right) \tag{9.60}$$

and for TM modes

$$\phi = \sin\left(\frac{n\pi x}{a}\right)\sin\left(\frac{m\pi y}{b}\right) \tag{9.61}$$

with the wavenumber given by

$$K_c^2 = \left(\frac{n\pi}{a}\right)^2 + \left(\frac{m\pi}{b}\right)^2 \tag{9.62}$$

The mode functions determine the power in the transverse fields which need to be matched across each junction. Kuhn and Watson [31] treated each corrugation as a resonator and used generalised resonator eigenfunctions to represent the fields in each section of the horn. An alternative method is to use standard rectangular functions in a manner exactly analogous to that for circular horns [32]. The selfpower-coupling integrals R_{ij}, Q_{ij} and the mutual power coupling integrals P_{ij} can be evaluated analytically [32]. There are a large number of modes to consider in a rectangular horn and the order of cuton of the modes as frequency increases depends on the ratio of the widths of adjacent sides. If this changes along the horn, then care needs to be taken that the correct set of modes is evaluated. The rectangular-modal-matching technique can be applied to any horn with rectangular geometry. Thus pyramidal and box horns can also be studied.

Kuhn and Watson [31,33] report reasonably good agreement between measurements and predicted levels of crosspolarisation. The rectangular horn is prone to the excitation of higher-order modes along the horn and this means that the peak crosspolar level tends to vary over an operating band. Measurements [31] showed peak levels varying between −26 dB and −40 dB over a 7% band. They devised a form of corrugated horn called a corrugated horn with 'glide' symmetry in which the corrugations on adjacent walls are interleaved. This gave predicted and measured peak crosspolar levels of better than −33 dB [33]. The horn had an aperture size of 3.5λ by 2.5λ and 7°/4.2° semiflare angles tapering to a square throat.

Rectangular corrugated horns may have application at low microwave frequencies where a conical horn becomes difficult to fabricate. Witebsky *et al.* [34] report a horn for the frequency band 1.3–1.7 GHz with a return loss of less than −25 dB and good radiation performance. The corrugations were constructed from bent aluminium sheet which was bolted to another flat aluminium sheet forming the wall. Using a similar technique, it should be feasible to construct a corrugated horn at below 1 GHz.

9.6 References

1 CLARRICOATS, P.J.B., and OLVER, A.D.: 'Corrugated horns for microwave antennas' (Peter Peregrinus, 1984), IEE Electromagnetic Waves series 18

2 KAY, A.F.: 'A wide flare angle horn. A novel feed for low noise broadband and high aperture efficiency antennas'. US Airforce Cambridge Research Laboratories, report 62–757, October 1962

3 SIMONS, A.J., and KAY, A.F.: 'The scalar feed – a high performance feed for large paraboloidal reflectors'. *IEE Conf. Pub. 21*, 1966, pp. 213–217

4 MINNETT, H.C., and THOMAS, MacA.B.: 'A method of synthetising radiation patterns with axial symmetry', *IEEE Trans.*, 1966, **AP–14**, pp. 654–656

5 MINNETT, H.C., and THOMAS, MacA.B.: 'Propagation and radiation behaviour of corrugated feeds', *Proc. IEE*, 1972, **119**, p. 1280

6 CLARRICOATS, P.J.B., and SAHA, P.K.: 'Propagation and radiation behaviour of corrugated feeds Parts 1 and 2', *Proc IEE*, 1971, **118**, pp. 167–187

7 CLARRICOATS, P.J.B., OLVER, A.D., and SAHA, P.K.: 'Near field radiation characteristics of corrugated horns', *Electron. Lett.*, 1971, **7**, pp. 446–448

8 THOMAS, MacA.B.: 'Design of corrugated conical horns', *IEEE Trans.*, 1978, **AP–26**, pp. 367–372

9 DRAGONE, C.: 'Reflection, transmission and mode conversion in a corrugated feed', *Bell Syst. Tech. J.*, 1977, **56**, pp. 835–867

10 BRYANT, G.H.: 'Propagation in corrugated waveguides', *Proc IEE*, 1969, **116**, pp. 203–213

11 JEUKEN, M.E.J., and THURLINGS, L.F.G.: 'The corrugated elliptical horn antenna'. IEEE International symposium on Antennas and Propagation, Illinois, 1975

12 CLARRICOATS, P.J.B., OLVER, A.D., and CHONG, S.L.: 'Attenuation in corrugated circular waveguides. Parts 1 and 2', *Proc IEE*, 1975, **122**, pp. 1173–1186

13 PARINI, C.G., CLARRICOATS, P.J.B., and OLVER, A.D.: 'Cross-polar radiation from open-ended corrugated waveguides', *Electron. Lett.*, 1975, **11**, p. 567

14 CLARRICOATS, P.J.B., OLVER, A.D., and PARINI, C.G.: 'Optimum design of corrugated feeds for low cross-polar radiation'. Proceedings of 6th European Microwave Conference, Rome, 1976, p. 148–152

15 KUHN, E., and HOMBACH, V.: 'Computer aided analysis of corrugated horns with axial and ring-loaded radial slots'. 3rd international conference on Antennas and Propagation, *IEE Conf. Publ. 219*, April 1983, pp. 879–884

16 JAMES, G.L.: 'Analysis and design of TE_{11} to HE_{11} corrugated cylindrical waveguide mode converters', *IEEE Trans.*, 1981, **MTT–29**, pp. 1059–1066.

17 WYLDE, R.J., and MARTIN, D.H.: 'Gaussian beam-mode analysis and phase-centres of corrugated feed horns', *IEEE Trans.*, 1993, **MTT–41**, pp. 1691–1699
18 WYLDE, R.J.: 'Millimetre-wave Gaussian beam-mode optics and corrugated feed horns', *IEE Proc. H*, 1984, **131**, pp. 258–262
19 AUBRY, C., and BITTER, D.: 'Radiation patterns of a corrugated conical horn in terms of Laguerre–Gaussian functions', *Electron. Lett.*, 1975, **11**, pp. 155–156
20 KOGELNIK, H., and LI, T.: 'Laser beams and resonators', *Proc. IEEE*, 1966, **54**, pp. 1313–1329
21 WYLDE, R.J.: 'Corrugated horns at MM wavelengths'. IEE colloquium on Millimetre-Wave and Quasi-Optical Antennas', *IEE Colloquium Digest 105*, 1990, pp. 4/1–4/3
22 TAKEICHI, Y., HASHIMOTO, T., and TAKEDA, F.: 'The ring-loaded corrugated waveguide', *IEEE Trans.*, 1971, **MTT–19**, pp. 947–950
23 JAMES, G.L., and THOMAS, B.M.: 'TE_{11} to HE_{11} cylindrical waveguide mode converters using ring-loaded slots', *IEEE Trans.*, 1982, **MTT–30**, pp. 278–285
24 JAMES, G.L.: 'Design of wide-band compact corrugated horns', *IEEE Trans.*, 1984, **AP–32**, pp. 1134–1138
25 THOMAS, B.MacA., JAMES, G.L., and GREENE, K.J.: 'Design of wide-band corrugated conical horns for Cassegrain antennas', *IEEE Trans.*, 1986, **AP–34**, pp. 750–757
26 ASHTON, R.W.: 'Improvements in or relating to corrugated horns'. British patent 1498 905, 1978
27 GHOSH, S., ADATIA, N., and WATSON, B.K.: 'Hybrid mode feed for multiband applications having dual-depth corrugation boundary', *Electron. Lett.*, 1982, **18**, pp. 860–862
28 OLVER, A.D., YANG, K.K., and CLARRICOATS, P.J.B.: 'Propagation and radiation behaviour of dual-depth corrugated horns', *IEE Proc. H*, 1984, **131**, pp. 179–185
29 GHOSH, S., KUHN, E., and PRATA, A.: 'Simplified high-performance dual-band feed comprising a dual-depth corrugated launcher and a conventional corrugated horn', *Electron. Lett.*, 1984, **20**(13), pp. 532–533
30 DRAGONE, C.: 'A rectangular horn of four corrugated plates', *IEEE Trans.*, 1985, **AP–33**, pp. 160–164
31 KUHN, E., and WATSON, B.K.: 'Rectangular corrugated horns–analysis, design and evaluation'. Proceedings of 14th European Microwave Conference, 1984, pp. 221–227
32 KITCHENER, D.: 'Mutual coupling in finite planar arrays of rectangular apertures'. PhD thesis, Queen Mary and Westfield College, University of London, September 1989
33 KUHN, E., and WATSON, B.: 'Computer-aided design and performance evaluation of fully corrugated rectangular horns'. AP-S International Symposium 1986, Philadelphia, USA, pp. 795–798
34 WITEBSKY, C., SMOOT, G.F., LEVIN, S., and BENSADOUN, M.: 'A large L-band rectangular corrugated horn', *IEEE Trans.*, 1987, **AP–35**, pp. 1310–1313
35 KUHN, E., and PHILIPPOU, G. Y.: 'Fully computer-optimised design of circular corrugated horns'. Proceedings of 14th European Microwave Conference, 1984, pp. 228–233
36 KILDAL, P.-S., JAKOBSEN, K., and SUDHAKAR RAO, K.: 'Meniscus-lens-corrected corrugated horn: a compact feed for a Cassegrain antenna', *IEE Proc. H*, 1984, **131**, pp. 390–394
37 HOMBACH, V., and KUHN, E.: 'Near field of corrugated horns and its influence on dual reflector antenna radiation performance'. Proceedings of fourth international conference on Antennas and Propagation, ICAP 85, Coventry, April 1985, pp. 410–414
38 BRAIN, D. J., and KOL, Y.: 'Performance of a multi-frequency elliptical corrugated horn', fourth international conference on Antennas and Propagation, ICAP-85, Coventry, April 1985, pp. 405–409
39 COOPER, L.J., and AINTABLIAN, H.: 'A multiband corrugated horn feed for satellite communication earth stations'. Proceedings of AP-S international symposium 1985, Vancouver, Canada, June 1985, pp. 307–310

40 SHOGEN, K.: 'An analysis of the elliptical corrugated horn for circularly polarized wave feed'. Proceedings of international symposium on Antennas and Propagation, Japan, 1985, pp. 45–48
41 MORI, I., IWATA, R., and ABE, A.: 'Cross-polarization analysis of dual frequency band corrugated conical horns'. Proceedings of international symposium on Antennas and Propagation, Japan, 1985, pp. 49–52
42 THOMAS, B.MacA.: 'A review of the early developments of circular-aperture hybrid-mode corrugated horns', *IEEE Trans.*, 1986, **AP–34**, pp. 930–935
43 MOLINA, J., CALVO, M., and BESADA, J. L.: 'Design of a meniscus lens corrected corrugated horn as optimum feeder for a Cassegrain antenna'. Proceedings of 5th international conference on Antennas and Propagation, ICAP 87, *IEE Conf. Publ. 274*, York, 1987, pp. 214–217
44 OLVER, A.D., and XIANG, J.: 'Optimum design of wide flare angle corrugated horns'. Proceedings of URSI international symposium on Electromagnetic Theory, Hungary, 1986, pp. 28–30
45 IENCINAR, J.A., and REBOLLAR, J.M.: 'Analysis of corrugated and non corrugated rectangular horns with a general geometry'. Proceedings of URSI international symposium on Electromagnetic Theory, Budapest, Hungary, August 1986, pp. 31–33
46 ABE, A., IWATA, R., and MORI, I.: 'Multi-frequency band corrugated conical horn'. Proceedings of AP-S international symposium, 1987, pp. 688–691
47 OLVER, A.D., and XIANG, J.: 'Wide angle corrugated horns analysed using spherical modal-matching', *IEE Proc. H*, 1988, **135**(1), pp. 34–40
48 BOURGEOIS, A., and LENORMAND, R.: 'Modelling of a nonlinear profile corrugated horn'. Proceedings of JINA '86: Journees Internationales de Nice sur les Antennes, 1986, pp. 135–142
49 KILDAL, P.-S.: 'Gaussian beam model for aperture-controlled and flareangle-controlled corrugated horn antennas', *IEE Proc. H*, 1988, **135**(4), pp. 237–240
50 OLVER, A.D., and XIANG, J.: 'Design of profiled corrugated horns', *IEEE Trans.*, 1988, **AP–36**(7), pp. 936–940
51 ARNOLD, J.M., and DENDANE, A.: 'Intrinsic mode theory of conical corrugated horns', *IEE Proc. H*, 1989, **136**(3), pp. 250–256
52 RAO, K.S.: 'A simple dual-band corrugated horn with low cross polarization', *IEEE Trans.*, 1990, **AP–38**(6), pp. 946–951
53 KILDAL, P.-S.: 'Artificially soft and hard surfaces in electromagnetics', *IEEE Trans.*, 1990, **AP–38**, pp. 1537–1544
54 TUOVINEN, J.: 'Improved accuracy of nonparaxial Gaussian beam (corrugated horn)'. Digest of IEEE Antennas and Propagation symposium, 1991, pp. 1011–1014
55 SHOGEN, K.: 'Design of corrugation depth and velocity dispersion characteristics of elliptical corrugated horn', *Trans. Inst. Electron. Inf. Comm. Eng. B-II*, 1991, **J74**, pp. 309–316
56 SHOGEN, K.: 'Design of corrugation depth and velocity dispersion characteristics of elliptical corrugated horn', *Electron. Commun. Jpn, 1, Commun.*, 1992, **75**, pp. 110–119
57 OLVER, A.D.: 'Corrugated horns', *Electron. Commun. Eng. J.*, 1992, **4**(1), pp. 4–10

Chapter 10
Dielectric-loaded horns

A.D. Olver

10.1 Introduction

This chapter studies conical horns partially loaded with a dielectric cone, as shown in Figure 10.1. These support hybrid modes similar to those for a conical corrugated horn. The consequence is that they can have the same desirable radiation characteristics as corrugated horns. The dielectric-cone-loaded horns are, in principle, simple to manufacture and can be used as high-performance feeds for reflector antennas, particularly at millimetre wavelength where the conventional corrugated horn is difficult to construct. Developments in plastics technology hold the promise of producing high-quality dielectrics, some of which will be space-qualified at reasonable cost.

The dielectric-cone-loaded horn was first discovered by Clarricoats, Olver and Rizk [1] in 1983 as part of an investigation to devise horns with low crosspolarisation over a wide band of frequencies. At that time the theoretical methods available to study the characteristics were limited to the use of surface-impedance models for a cylindrical waveguide. Based on experience with corrugated horns, this method was expected to be adequate to study the performance, and indeed simple theory predicted low levels of crosspolarisation. The desired hybrid modes were created by the inhomogeneously filled metal horn. To verify this proposal a horn, with a 12° semiflare angle was

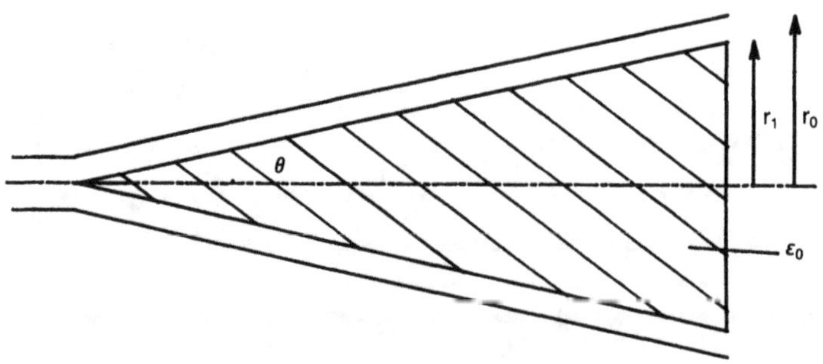

Figure 10.1 Dielectric-cone-loaded horn

constructed and partially loaded with a foam-dielectric cone with a relative permittivity of 1.13. The radiation patterns of the horn were measured and the variation of peak crosspolar level with frequency is shown in Figure 10.2. This demonstrated that the dielectric-cone-loaded horn has a low crosspolarisation over a wider band of frequencies than the equivalent corrugated horn. However, an attempt to predict the performance in more detail was not entirely successful, and it was speculated that this was due to the presence of at least two modes in the aperture. The ability to predict the performance then had to wait until modal-matching techniques (Chapter 4) could be used to predict the performance of the horn accurately including the influence of the throat region and the flare sections.

In 1986, Lier [3] described a similar horn to that shown in Figure 10.1 but using a solid-dielectric cone with a relative permittivity of 2.5, again showing good experimental results. Rizk and Morris [4] reported on measurements with a number of foam-dielectric-filled horns. Tun *et al.* [5–7] also investigated a design using a dielectric with a relative permittivity of 1.4. Both Tun and the present author [14, 15] reported the use of modal-matching analysis to predict the mode conversion along the horn and to report theoretical agreement with measured performance. Knop [8] has also compared the performances theoretically of various cylindrical hybrid structures. A millimetre-wave delectric-core horn giving good performance at 94 GHz was reported by Cahill [9, 26]. Lier extended the work in Reference 3 to elliptical and rectangular horns [10–12]. These demonstrated both experimental and simple theoretical results. The elliptical or rectangular horn enables elliptically shaped radiation patterns to be created but at the expense of a considerably more complicated design,

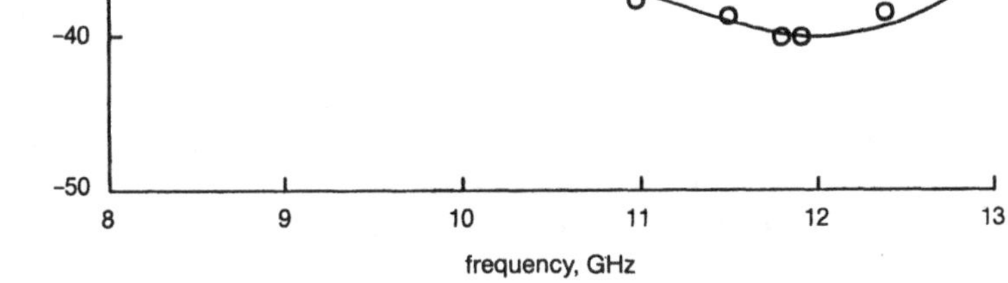

Figure 10.2 Peak level of crosspolarisation against frequency for horn of Figure 10.1

—— predicted using modal matching
– – – predicted using single HE_{11} mode
○○○ measured

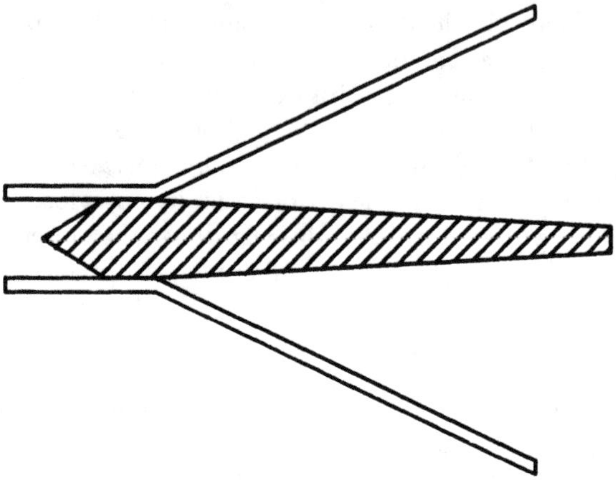

Figure 10.3 Dielectric polyrod in conical horn

particularly in the throat region. Generally, it is easier to produce elliptical or other shaped patterns with a conical horn feeding a shaped reflector.

The dielectric-cone-loaded horn described in detail in this chapter is but one of many types of dielectric horns. The general aim of including dielectric in a horn is to alter the fields in the horn in a beneficial manner so that the transverse fields in the aperture are modified to produce a better pattern than is obtainable without the presence of dielectric. The horns in this chapter do this by creating hybrid modes in the aperture with similar fields to the corrugated horn [2]. Small dielectric feeds are described in Chapter 11. Multimode horns using dielectric to modify the pattern of the dominant TE_{11} mode are described in Chapter 8. There are also medium-aperture horns with rods of dielectric along the axis (Figure 10.3), as described by Newham [13], which produce good characteristics by enhancing the fields near the axis of the horn. The theory described in this chapter can be used without much modification to analyse the dielectric-rod horn as long as the dielectric does not extend beyond the end of the horn.

This chapter starts by explaining the physical principles of operation of dielectric-cone-loaded horns, and then develops the theory necessary to use the modal-matching technique to predict the performance. Parametric studies are reported to show how the geometry and dielectric parameters can be studied to control the performance, including the influence of tolerances and losses along the horn. Examples of horn designs are given, including metallised-foam horns and solid dielectric horns specially suited to millimetre-wave operation. Compact-profile dielectric horns are described which are analogous to the profiled corrugated horn.

10.2 Principle of operation

The basis for the potentially good performance of dielectric-cone-loaded horns can be understood by considering the electric field in the aperture of a hybrid-mode cylindrical waveguide [2] of radius r, i.e.

$$\mathbf{E} = A\left\{J_0(Kr)\hat{\boldsymbol{u}}_x - \frac{(X-Y)U^2}{4kr}J_2(Kr)(\cos 2\phi\hat{\boldsymbol{u}}_x + \sin 2\phi\hat{\boldsymbol{u}}_y)\right\} \tag{10.1}$$

where $J_n(Kr)$ $(n=0,2)$ is a Bessel function of the first kind and of order n, $\hat{\boldsymbol{u}}_{x,y}$ are unit vectors in the x and y directions, U is a normalised transverse wavenumber, $K[=\sqrt{(\omega^2\mu_0\varepsilon_r-\beta^2)}]$ and $k(=2\pi/\lambda_0)$ are transverse and free-space wavenumbers, respectively, and X and Y are the normalised reactance and admittance of the boundary at radius r.

This representation is appropriate to a corrugated waveguide of internal radius r, an unbounded dielectric cylinder with r large in wavelengths, or a dielectric cylinder in a circular waveguide with r large and a small airgap. The first term in eqn. 10.1 is a ϕ-independent copolarised field. The second term is a ϕ-dependent crosspolarised field. Because the radiation field is, to a good approximation, the Fourier transform of the transverse electric field in the aperture, the polarisation properties are transferred to the radiation field. The crosspolarised field can be made to vanish if $(X-Y)=0$, which is possible either with X and Y both zero or with X and Y finite and equal. In the corrugated horn, X is nearly zero owing to the effective zero impedance at the radius of the corrugations, and Y is zero when the corrugation depth is $m\lambda/4$ (m odd). It varies with frequency as the tangent of the corrugation depth. In the large-diameter dielectric-loaded waveguide with a single hybrid mode,

$$\left.\begin{aligned} X &= B_X \tanh(kp) \\ Y &= B_Y \coth(kp) \end{aligned}\right\} \tag{10.2}$$

where $B_{x,y}$ are coefficients, $p=2\sqrt{(\Delta n)}g$, $\Delta n=\sqrt{(\varepsilon_r)}-1$ and g is the width of the airgap.

The function $(X-Y)$ will vanish when

$$kg = \tanh^{-1}\left(\frac{1-\Delta n}{\sqrt{(2\Delta n)}}\right) \tag{10.3}$$

Insertion of typical values of Δn for foam or solid dielectric shows that this condition is satisfied for an airgap of between 0.07 and 0.8 wavelengths. The relationship is frequency-dependent, as with the corrugated horn, but the hyperbolic tangent function is a slow monotonically varying function, so a wide bandwidth of low crosspolarisation can be expected.

The above description is the basis for expecting that a dielectric-loaded conical waveguide will produce low crosspolarisation. However, to predict the performance for a conical horn accurately, a theory is required which accounts for the conical nature of the structure. The first attempt was to treat the horn as

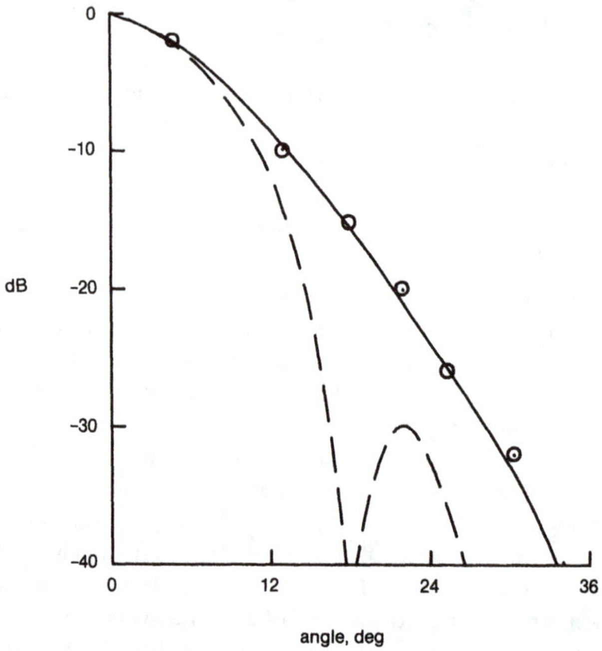

Figure 10.4 Copolar-radiation patterns for horn of Figure 10.1 at 10.5 GHz
Notation as in Figure 10.1

an open-ended waveguide, partially filled with dielectric, propagating the HE_{11} mode. The phase coefficient and fields in the waveguide can easily be computed [17] and the far-field-radiation pattern predicted using Fourier-transform theory. The peak crosspolarisation as a function of frequency obtained in this way is shown by the broken line in Figure 10.2. Although a relatively low level is predicted, there is clearly not good agreement between theory and measurement. The copolar-radiation pattern at 10.5 GHz is shown in Figure 10.4, where the measured pattern (circles) has broadened and merged with the first sidelobe of the pattern predicted from the single-mode theory (broken line). This result, and the prediction of a peak crosspolar level which is higher than the measured results, is an indication that higher-order modes present in the aperture are phased in such a way as to reduce the total crosspolarisation.

10.3 Prediction using modal-matching technique

A detailed study of the operation of the dielectric-cone-loaded horn must be carried out using the modal-matching technique, the principles of which were described in Chapter 4. The horn is divided into a series of circular waveguide sections, each section containing a cylinder of dielectric (Figure 10.5). The modal-matching technique is then applied to calculate the power coupled from

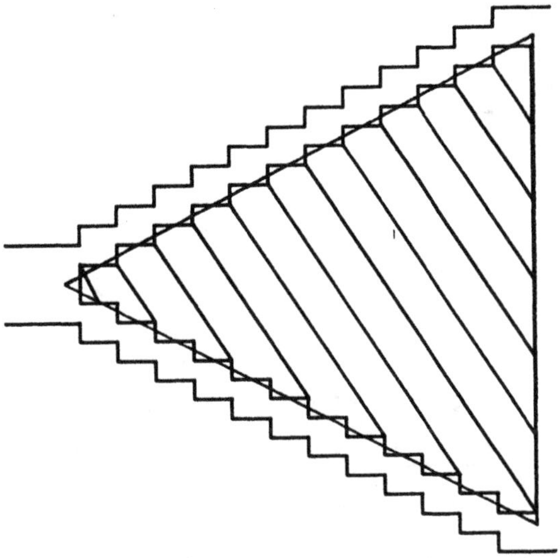

Figure 10.5 Numerical model of dielectric-cone-loaded horn

one section to the next section. The procedures described in Chapter 4 can be applied directly with the important differences that the modes in the inhomogeneous cylindrical-waveguide section are hybrid and the values of the propagation coefficients must be computed using an iterative procedure. The theory and implementation is described in Section 10.3.1. The presence of the inhomogeneous section creates considerable extra computational problems. These stem partly from the uncertainty inherent in an iterative solution process, and partly from the variety of junctions which must be evaluated. The range of possible junctions is shown in Figure 10.6. The power-coupling integrals must be evaluated for each of these cases. There are two options: either to evaluate the integrals numerically, or to find closed-form solutions. The former avoids the need for lengthy analysis and subsequent lengthy code, but leads to very long computation times. The latter means more work initially, but then leads to fast numerical computation and is the preferred method. The total computation time using the analytical approach can be quite reasonable because the step change between sections can be made small so that no more than ten modes are needed in the computation.

10.3.1 Dielectric-filled circular waveguide

The dielectric-filled circular waveguide (Figure 10.7) has been studied extensively [17, 19, 20]. The waveguide supports cylindrical hybrid HE_{1m} and EH_{1m} modes in a similar way to the corrugated waveguide, and under certain conditions can produce linearly polarised transverse fields. This occurs when the relative permittivity of the axial cylindrical dielectric is slightly greater than the relative permittivity of the outer dielectric. (Note the similarity to the

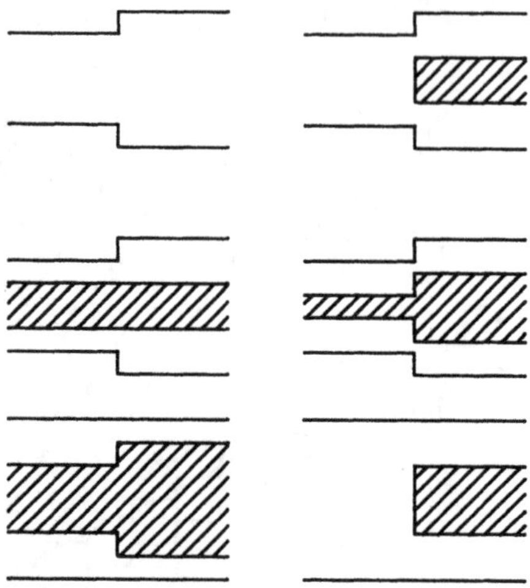

Figure 10.6 Types of junction in dielectric-cone-loaded horn

condition required for balanced hybrid modes in single-mode optical-fibre waveguides.) It can be generated in practice either by using a foam-dielectric cylinder with air as the outer, or by using two dielectrics with slightly different permittivities.

The characteristic equation for a dielectric-loaded circular waveguide is obtained by matching the fields at the radius of the dielectric rod [17]:

$$\left[\frac{\bar{\varepsilon}_0 F_1(Kr_1)}{(Kr_1)^2} - \frac{\bar{\varepsilon}_1 S_1(K_1,r_1,r_0)}{(K_1r_1)^2}\right]\left[\frac{\bar{\varepsilon}_0 F_1(Kr_1)}{(Kr_1)^2} - \frac{\bar{\varepsilon}_1 R_1(K_1,r_1,r_0)}{(K_1r_1)^2}\right] = \bar{\beta}^2\left[\frac{(Kr_1)^2-(K_1r_1)^2}{(Kr_1)^2(K_1r_1)^2}\right]^2 \tag{10.4}$$

where $\bar{\beta}=\beta c/\omega=\beta\lambda/2\pi$ and unity azimuthal dependence has been assumed. This is valid as long as the dielectric cylinder in all sections of the horn is

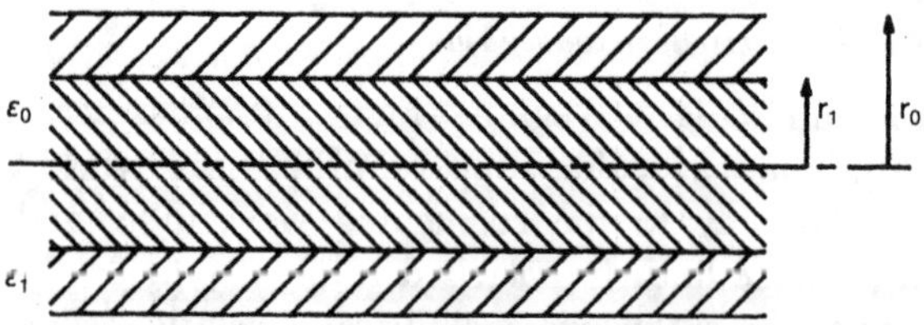

Figure 10.7 Dielectric rod in circular waveguide

coaxial. In this case only modes of type HE_{nm} or EH_{nm} with $n=1$ and m varying are excited. The other parameters in eqn. 10.4 are

$$F_1(Kr_1)=Kr_1\frac{J_1'(Kr_1)}{J_1(Kr_1)} \tag{10.5}$$

$$R_1(K_1,r_1,r_0)=\frac{(K_1r_1)\{J_1'(K_1r_1)Y_1'(K_1r_0)-J_1'(K_1r_0)Y_1'(K_1r_1)\}}{\{J_1(K_1r_1)Y_1'(K_1r_0)-J_1'(K_1r_0)Y_1(K_1r_1)\}} \tag{10.6}$$

$$=\frac{R_N(K_1,r_1,r_0)}{R_D(K_1,r_1,r_0)}$$

$$S_1(K_1,r_1,r_0)=\frac{(K_1r_1)\{J_1'(K_1r_1)Y_1(K_1r_0)-J_1(K_1r_0)Y_1'(K_1r_1)\}}{\{J_1(K_1r_1)Y_1(K_1r_0)-J_1(K_1r_0)Y_1(K_1r_1)\}} \tag{10.7}$$

$$=\frac{S_N(K_1,r_1,r_0)}{S_D(K_1,r_1,r_0)}$$

If the modes are cutoff, so that $K^2<0$ or $K_1^2<0$ then the Bessel functions $J_1(x)$ and $Y_1(x)$ are replaced by the modified Bessel functions $I_1(x)$ and $K_1(x)$, respectively, and the phase coefficient β becomes the attenuation coefficient α.

A useful parameter is the hybrid factor Λ which is defined as the ratio of the axial magnetic field and axial electric fields or

$$\Lambda=\bar{\beta}\frac{\dfrac{1}{(K_1r_1)^2}-\dfrac{1}{(Kr_1)^2}}{\dfrac{F_1(Kr_1)}{(Kr_1)^2}-\dfrac{R_1(K_1,r_1,r_0)}{(K_1r_0)^2}} \tag{10.8}$$

The hybrid factor is positive for HE modes and negative for EH modes. When it is equal to unity the transverse fields are linearly polarised. This is called the 'balanced hybrid condition'. It is the ideal situation for the dominant HE_{11} mode which then radiates a minimum level of crosspolar power.

The characteristic equation (eqn. 10.4) must be solved iteratively to find the relationship between β and ω; this means computing the values of β for all modes at a fixed frequency. The procedure is to start with a trial value which is close to the root of the characteristic equation and then use an interpolation procedure to converge on the answer to within a specified accuracy. This raises the question as to how to choose the trial values. There is no entirely automatic method of doing this, and some knowledge of the behaviour of the propagation coefficients of the modes is required. The best procedure is to start at the throat of the horn in the first section of the waveguide with a large value of β and then track downwards until a root is found. This will be the value of β corresponding to the HE_{11} mode. Then continue tracking with decreasing values of β to find the next mode and so on. The procedure needs to be continued for the nonpropagating modes, i.e. $\gamma = \alpha$, until all the roots have been located. The roots for the next section of the waveguide can be computed using the values of β (or α) for the previous section as the trial values. This sounds simple, but is made more difficult in practice because numerical instabilities, or poles, in the equations can easily lead to a mode being omitted or jumped. This is disastrous for the modal-matching computation and will mean that the sequence of modes is incorrect for the remainder of the horn. The problem can be avoided in most cases by choosing very short lengths of waveguide section so that the propagation coefficient changes only marginally between adjacent sections.

A typical variation of the propagation coefficients along a dielectric loaded horn is illustrated by the results shown in Figure 10.8. This shows the normalised value of the propagation coefficient against position along a horn. At the throat there is only one propagating mode. At the aperture there are four propagating modes. However, the three next cutoff modes also need to be included in the modal-matching calculations because they influence the power crosscoupled at junctions. The oscillatory variation of some of the modes at the start of the horn is due to the apex of the dielectric cone.

The propagation coefficient which has been determined in a waveguide section can then be used to calculate the transverse electric and magnetic fields. These are needed at each junction between waveguide sections in order to compute the power-coupling integrals. The complete fields for a dielectric-loaded circular waveguide are as follows. An $\exp\{j(\omega t - \beta z - \phi)\}$ dependence is assumed.

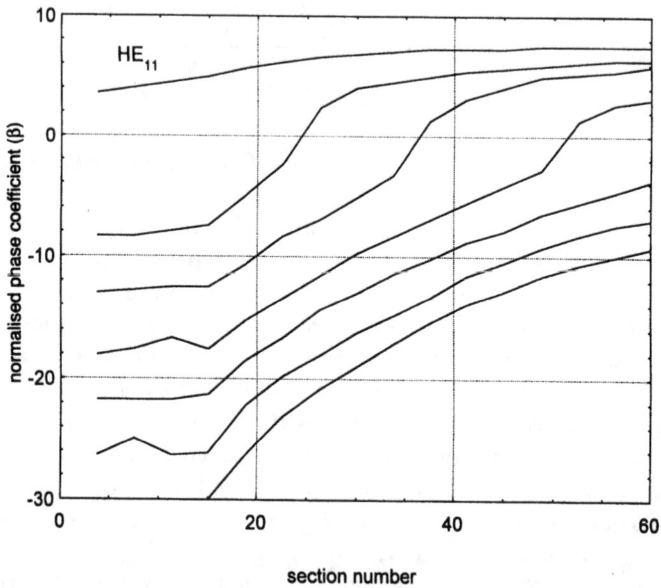

Figure 10.8 Propagation coefficient along a typical horn

In the core dielectric:

$$\left.\begin{aligned} E_z^i &= A_i J_1(Kr) \\ H_z^i &= jA_i \frac{\Lambda}{Z_0} J_1(Kr) \end{aligned}\right\} \tag{10.9}$$

$$\left.\begin{aligned} E_r^i &= -jC_i\{\bar{\beta}F_1(Kr) + \Lambda\} \\ H_r^i &= C_i \frac{1}{Z_0}\{\bar{\beta}\Lambda F_1(Kr) + \bar{\varepsilon}_0\} \end{aligned}\right\} \tag{10.10}$$

$$\left.\begin{aligned} E_\phi^i &= -C_i\{\bar{\beta} + \Lambda F_1(Kr)\} \\ H_\phi^i &= -jC_i \frac{1}{Z_0}\{\bar{\beta}\Lambda + \bar{\varepsilon}_0 F_1(Kr)\} \end{aligned}\right\} \tag{10.11}$$

where

$$C_i = A_i \frac{2\pi r_0^2}{\lambda r(Kr_0)^2} J_1(Kr)$$

and A_i is a constant.

In the outer dielectric (or air):

$$\left.\begin{aligned} E_z^0 &= A_i J_1(Kr_1) \frac{S_D(K_1,r,r_0)}{S_D(K_1,r_1,r_0)} \\ H_z^0 &= jA_i \frac{\Lambda}{Z_0} J_1(Kr_1) \frac{R_D(K_1,r,r_0)}{R_D(K_1,r_1,r_0)} \end{aligned}\right\} \tag{10.12}$$

$$\left.\begin{aligned} E_r^0 &= -jC_0 \left\{ \bar{\beta} \frac{S_N(K_1,r,r_0)}{S_D(K_1,r_1,r_0)} + \Lambda \frac{R_D(K_1,r,r_0)}{R_D(K_1,r_1,r_0)} \right\} \\ H_r^0 &= C_0 \frac{1}{Z_0} \left\{ \bar{\varepsilon}_1 \frac{S_D(K_1,r,r_0)}{S_D(K_1,r_1,r_0)} + \bar{\beta}\Lambda \frac{R_N(K_1,r,r_0)}{R_D(K_1,r_1,r_0)} \right\} \end{aligned}\right\} \tag{10.13}$$

$$\left.\begin{aligned} E_\phi^0 &= -C_0 \left\{ \bar{\beta} \frac{S_D(K_1,r,r_0)}{S_D(K_1,r_1,r_0)} + \Lambda \frac{R_N(K_1,r,r_0)}{R_D(K_1,r_1,r_0)} \right\} \\ H_\phi^0 &= -jC_0 \frac{1}{Z_0} \left\{ \bar{\varepsilon}_1 \frac{S_N(K_1,r,r_0)}{S_D(K_1,r_1,r_0)} + \bar{\beta}\Lambda \frac{R_D(K_1,r,r_0)}{R_D(K_1,r_1,r_0)} \right\} \end{aligned}\right\} \tag{10.14}$$

where

$$C_0 = A_i \frac{2\pi r_0^2}{\lambda r(Kr_0)^2} J_1(Kr_1)$$

10.3.2 Power-coupling integrals

The modal-matching technique involves the matching of the total power across each junction between waveguide sections. The main part of this process is the computation of the power-coupling integrals (see Chapter 4). The integrals to be evaluated are

$$P_{ij} = \int_{s_L} (\boldsymbol{e}_{iL} \times \boldsymbol{h}_{jR}) \cdot \mathrm{d}s_L \tag{10.15}$$

$$Q_{jj} = \int_{s_R} (\boldsymbol{e}_{jR} \times \boldsymbol{h}_{jR}) \cdot \mathrm{d}s_R \tag{10.16}$$

$$R_{ii} = \int_{s_L} (\boldsymbol{e}_{iL} \times \boldsymbol{h}_{iL}) \cdot \mathrm{d}s_L \tag{10.17}$$

where $\boldsymbol{e}_{iL,\,jR}$ and $\boldsymbol{h}_{iL,\,jR}$ are the ith or jth transverse electric and magnetic fields on the left (L) or right (R) of the junction, respectively. The procedure for evaluating these was described in detail in Chapter 4 for hollow circular-waveguide sections. The dielectric-loaded horn is considerably more complicated owing partly to the different types of junction (see Figure 10.6) and partly to the cumbersome field expressions in eqns. 10.9–10.14 which contain combinations of Bessel functions. Analytical evaluation of eqns. 10.15–10.17 have been derived by Raghavan [21].

An example of the results obtained using the modal-matching technique are shown in Figures 10.2 and 10.4. They compare the theoretical results with measured data and show extremely good agreement. This is in line with the level of agreement expected when modal matching is used to predict the performance of hollow cylindrical horns.

10.4 Design considerations

The designer of a dielectric-loaded horn needs to be able to understand the mechanical as well as the electrical properties of the horn. Among these are the length and mass of the horn. For a given aperture size, the mass of a dielectric-loaded horn is directly related to its length. Theoretical computations obtained using the mode-matching technique suggest that, in general, a better crosspolar performance over a wider bandwidth can be achieved using a horn with a smaller semiflare angle. Thus there is a tradeoff between designing a compact horn with adequate crosspolar performance over a narrower band of frequencies and a longer horn which offers a good broadband performance. These aspects will be discussed in Section 10.5. The length of dielectric-loaded horns can be reduced by using a profiled wall geometry. This is described later.

The geometry of the straight-walled linear-taper dielectric-cone-loaded horn is shown in Figure 10.1. The length of the linear horn between the input

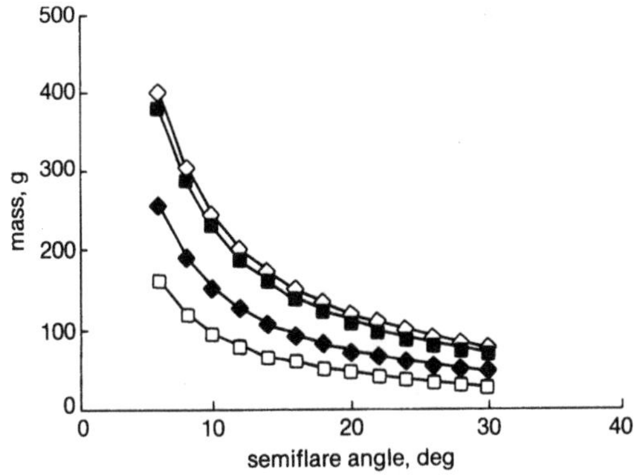

Figure 10.9 *Variation of horn mass with flare angle for a 3-wavelength aperture*

Cone surrounded by air
——■—— $g = 5$ mm
——◆—— $g = 10$ mm
——□—— $g = 15$ mm
Cone density twice that of annular region
——◇—— $g = 10$ mm

waveguide and the aperture can be written in terms of the aperture radius r_0 and the input waveguide r_i as

$$L = (r_0 - r_i) \cot \theta \tag{10.18}$$

The volume of the horn is easily computed and then the mass M of the dielectric loading at which the tip of the cone is in line with the throat of the horn is given by

$$M = \pi\rho_0 \left[(r_0 - g) \left\{ \frac{(r_0 - g)^2}{3} + \frac{\rho_0}{\rho_1} r_0 g \right\} \cot \theta + \frac{\rho_0}{\rho_1} r_i^2 L \right] \tag{10.19}$$

where g is the gap between the inner dielectric cone and the outer metal wall, and ρ_0 and ρ_1 are the densities of the core dielectric and outer dielectric, respectively.

The range of interest is when $0 < \rho_0/\rho_1 < 1$. Figure 10.9 shows the variation in mass for a horn with a cone surrounded by air with three different airgaps of 5, 10 and 15 mm. Also shown is the case where the cone is twice as dense as the surrounding material. It can be clearly seen from eqns. 10.18 and 10.19 that, as the flare angle is increased, both the weight and the length of the horn will be reduced. The copolar- and crosspolar-radiation characteristics and bandwidth specification can be used to determine the maximum acceptable flare angle so as to ensure that the horn is as light and compact as possible. Beyond a semiflare angle of approximately 15°, the change in mass of the horn is fairly small. Below a semiflare of 15° the variation is quite rapid owing to the cot θ

dependence. As the airgap is increased, the cone becomes smaller and the weight of the horn is reduced.

The mass of a dielectric material is proportional to $(\varepsilon_r - 1)$ where ε_r is the relative permittivity. If weight is of prime importance, the relative permittivity should be kept to a minimum. The distribution of the modes in the horn depends on the ratio of the relative permittivities of the cone and surrounding material and on the gap between the horn wall and the inner cone. If a cone with $\varepsilon_r = 1.3$ is placed inside a conical horn, a low-permittivity-dielectric support will be needed. Any support mechanism is likely to degrade the antenna performance, although this might be partially corrected by empirically changing the gap width g. Cahill [9] reports work where the authors constructed and tested several dielectric-loaded millimetre-wave horns using a number of support structures: the authors concluded that a ring support in the waveguide and spokes near the aperture gave the best measured crosspolar of the examples they tested.

10.5 Parametric studies on linear horn

A series of parametric studies is now described in order to explain the properties of dielectric-cone-loaded horns. In this section the horn is assumed to be of the form shown in Figure 10.1 with an airgap between the inner cone and the metal horn. Although this may not be the actual configuration used in practice, it represents the range of possible horns since, if two dielectrics are used, the ratio of the two relative permittivities is the relevant parameter.

10.5.1 Permittivity of dielectric

The results given in Figures 10.2 and 10.4 are for a foam-dielectric cone with $\varepsilon_r = 1.13$, corresponding to an expanded-polystyrene foam with a density of 55 kg/m^3. Other workers have shown that horns with dielectrics of higher permittivity will also operate satisfactorily. The computational and measured results of Tun [6] with $\varepsilon_r = 1.4$ and the measured results of Lier [3] with $\varepsilon_r = 2.5$ (solid polystyrene) can be verified by using the modal-matching technique. Good radiation characteristics are possible with any of these permittivities. The optimum gap found by Tun and Lier agrees approximately with the values calculated using eqn. 10.3. The amount of higher-order mode excited with the solid dielectric is greater than with a foam dielectric, but the peak crosspolar levels are still low. The relative merits of using cones with different relative permittivities depends partly on the application and partly on the aperture diameter. Low-density-foam dielectric material is lightweight, but does not work well with aperture diameters of less than about four wavelengths because of insufficient hybrid-mode guiding action. It is also unsuitable for use with horns of wide flare angle and it needs some form of environmental protection at the aperture, as is usual in all horns used in microwave terrestrial or satellite systems. The solid dielectric material is relatively heavy (typically 1000 kg/m^3

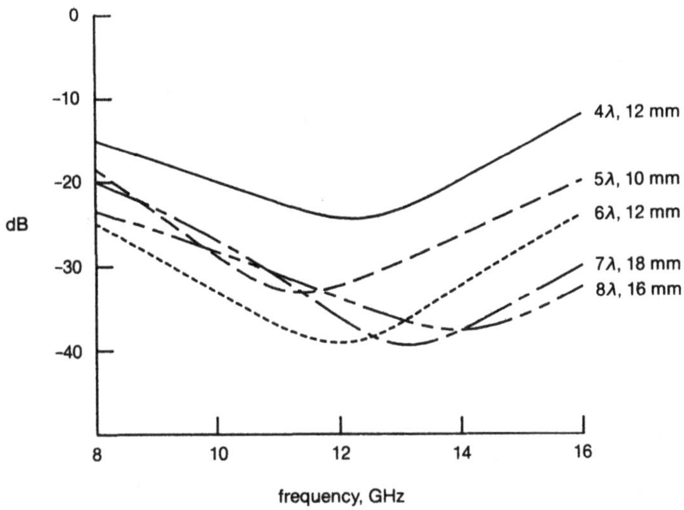

Figure 10.10 Influence of aperture diameter for horn with semiflare angle = 12° and $\varepsilon_r = 1.13$

Parameters: aperture diameter in wavelengths and optimum airgap in millimetres

for polystyrene) but can be used in small-diameter horns. A disadvantage with the solid dielectric is the mismatch at the aperture. For a horn of the dimensions given in Figure 10.1 and with $\varepsilon_r = 2.5$, the input return loss was computed to be about −15 dB over the whole operating band. By comparison, the foam-dielectric cone has a return loss which is mainly due to the conical horn. The empty horn has a return loss of −22 dB over the whole operating band, and the foam-dielectric-cone-loaded horn has a return loss of −20 dB at 7 GHz, falling linearly to −25 dB at 14 GHz.

10.5.2 Variation of aperture diameter

Varying the aperture diameter with constant flare angle is equivalent to altering the length of the horn. The discussion of Section 10.2 indicates that this will mean a change in the phase relationship between the HE_{11} and HE_{12} modes. However, for any given aperture diameter it is possible to adjust the airgap to ensure that low crosspolarisation occurs over the design frequency band. Figure 10.10 shows peak crosspolar levels for a number of aperture diameters of a horn with $\varepsilon_r = 1.13$ and a horn with a 12° semiflare angle. Each horn has an optimum airgap. The four-wavelength horn has relatively poor performance, owing to the lack of guiding action of the foam dielectric. Increasing the relative permittivity restores the guiding action. The peak crosspolar level of a two-wavelength-diameter horn with $\varepsilon_r = 2.5$ and semiflare angle of 20° was computed as −41 dB. The theoretical bandwidth over which the peak crosspolarisation is below a given level, say −30 dB, should increase linearly as the aperture diameter increases. This is indicated in eqn 10.1 and is true for all hybrid-mode structures [2]. In practice the presence of higher-order modes (in this case the HE_{12} mode) limits the high-frequency performance. For these parameters the HE_{12} mode is relatively strongly excited at the upper end of the band.

The dielectric-cone-loaded horn is less frequency-sensitive than the corrugated horn. This is partly because of the resonant nature of the corrugations, which dominates the frequency performance of corrugated horns, and partly because higher-order modes are difficult to suppress at high frequencies in those structures.

10.5.3 Airgap

The airgap should be chosen to satisfy the requirements of low crosspolarisation. Eqn. 10.3 gives the value for very-large-diameter horns. The gap is generally smaller than that predicted by eqn. 10.3, partly because of the need to phase the HE_{12} mode correctly. The airgap is, however, not very sensitive, as is indicated by the broadband performance shown in Figure 10.10. The optimum airgap for smaller-diameter horns must be found either by trial and error or by using an optimisation routine which has the gap as a control parameter.

10.5.4 Change in flare angle

A large flare angle will excite more higher-order modes at the throat junction, but the horn will be shorter. Conversely, a narrow-flare-angle horn will have a lower higher-order-mode content, but will be physically long. These statements are confirmed by Figure 10.11 which shows the peak crosspolar level for a horn with a six-wavelength-aperture diameter and $\varepsilon_r = 1.13$. The horn with a semiflare angle of 8° has a low level of crosspolarisation over a broadband which is nearly monotonically decreasing as frequency increases. However, this may not be realised in practice, owing to the overmoded-input circular waveguide at high frequencies. As the semiflare angle increases, the higher-order modes limit the performance, until at a semiflare angle of 20° the peak crosspolar level only dips below −30 dB at the top end of the frequency band. Figure 10.11 also shows the result for a horn with a semiflare angle of 20° and $\varepsilon_r = 2.5$. The performance has improved considerably, and the minimum level has shifted to a frequency of 8.5 GHz.

10.5.5 Design of throat region

The study above shows that the crucial region of operation of the horn is the throat region, because it is in that region that the higher-order modes are excited and their basic level determined. The design of the throat region is now examined by evaluating the performance for two configurations: type 1, in which the cone extends back into the input waveguide, and type 2, which is a conventional cone-loaded design (Figure 10.12). Results are presented for a two-wavelength-diameter aperture and for three values of semiflare angle.

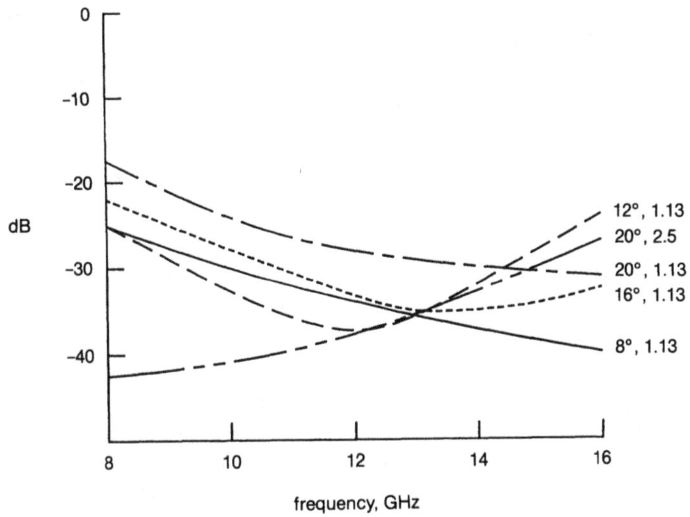

Figure 10.11 *Influence of flare angle*

Aperture diameter = 6λ at 12 GHz
Parameters: semiflare angle in degrees and relative permittivity of the core

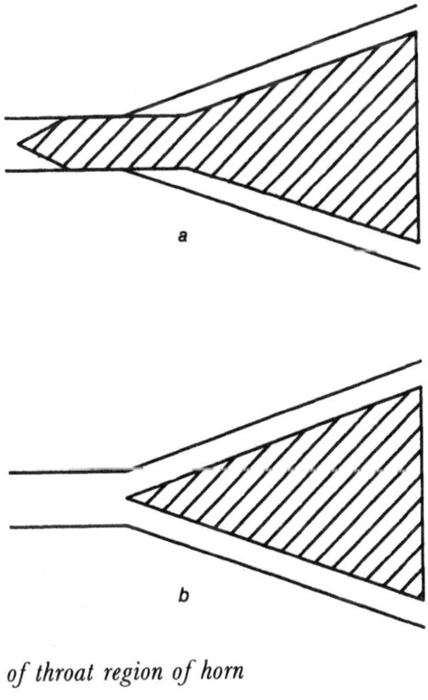

Figure 10.12 *Design of throat region of horn*

a Type 1
b Type 2

Evaluation of the propagation coefficients of the type 2 antenna is more difficult than for the type 1 configuration in certain circumstances. The cutoff EH modes oscillate for the first few sections in the analysis, particularly in the region of the tip of the cone, when the ratio of the permittivity of the cone to that of the annular region is greater than about 1.5:1.0. This is seen in Figure 10.8, which shows the propagation coefficients as a function of position along the horn for a type 2 horn.

For this study, three horns with semiflare angle of 10°, 20° and 30° are used and the dielectric cone is always surrounded by an airgap. Each horn is modelled using seven modes and 60 sections so that four modes are propagating at the aperture. The effect of varying the cone-to-horn-wall gap is investigated, together with varying the cone permittivity, the semiflare angle and the operating frequency. The parameters are altered until a minimum level of peak crosspolar power is radiated. The small aperture size leads to relatively high permittivities of the cone, in the range $\varepsilon_r = 1.5$–2.0. This compares with values of $\varepsilon_r = 1.13$ for a horn with a six-wavelength aperture considered in Section 10.2 and is due to the higher-order modes which must be excited at a smaller horn diameter, and this can only be achieved if the dielectric constant is sufficient to increase the effective diameter of the waveguide, compared with an empty horn.

The peak crosspolar level which is relevant to the feed design is not necessarily the absolute peak level at any angle. This is because, in a parabolic-reflector antenna, the crosspolar pattern is important only over the portion of the pattern which illuminates the reflector, for example down to -10 dB below the copolar peak.

Figure 10.13*a* and *b* summarises some of the properties of the 10°-semiflare-angle horn for a type 1 horn. Figure 10.13*a* shows how the peak crosspolar level and the -3 dB beamwidth vary as the relative permittivity of the core is altered from 1.3 to 1.8. The peak crosspolar level passes through a minimum value of -41 dB at $\varepsilon_r = 1.6$. As the permittivity is increased, the beamwidth steadily rises from 34° to 38°, an increase of more than 10 dB. This is because a larger proportion of higher-order modes is being excited, and they have a broader radiation pattern in the far field. At $\varepsilon_r = 1.6$ the contribution from these higher-order modes is just sufficient to cancel the crosspolarisation to a low level. However, as the power in the modes passes this optimum value, they start to add to the crosspolar level.

Figure 10.13*b* shows the peak crosspolarisation as a function of annular gap in the range 5–10 mm. For a cone relative permittivity of 1.4 there is an insufficient level of higher-order mode present to reduce the crosspolar level below -30 dB for any gap size. With a cone relative permittivity of 1.6 a new minimum level of crosspolarisation, below -42 dB is achieved by changing the gap from 8 mm, to 8.5 mm.

The performance of a 20° horn for both type 1 and type 2 horns over an 8–12 GHz band is shown in Figure 10.14. The crosspolar performance of the type 2 antenna improves to give a minimum value of -43 dB whilst the performance of the type 1 antenna steadily degrades with increasing frequency, being

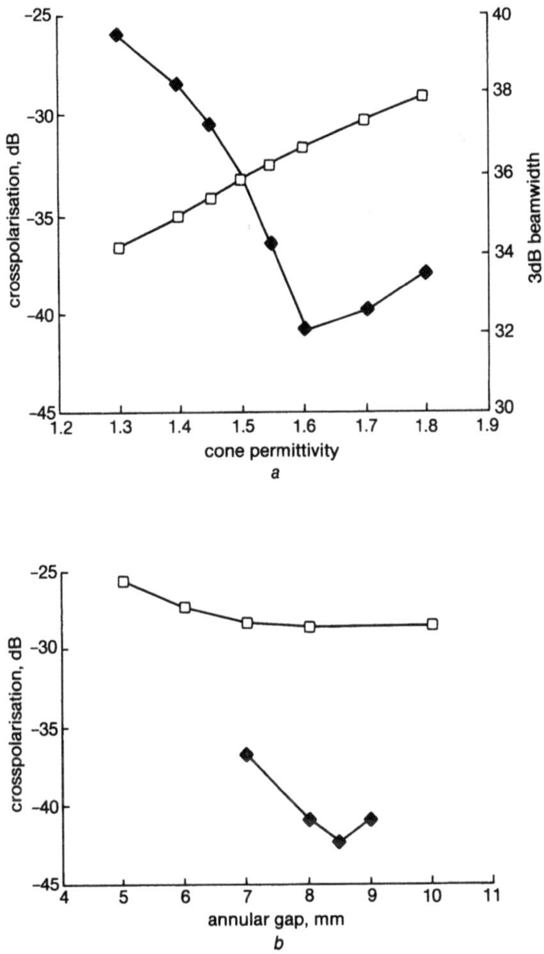

Figure 10.13 *Variation of peak crosspolar level and beamwidth for a type 1 horn at 10 GHz*

a Cone relative permittivity with gap of 8 mm
——◆—— crosspolarisation
——□—— beamwidth
b Annular gap with $\varepsilon_r = 1.4$ and 1.6
——□—— crosspolarisation, $\varepsilon = 1.4$
——◆—— crosspolarisation, $\varepsilon = 1.6$

approximately 10 dB higher than the type 2 configuration over half of the chosen frequency band. This is because the gap and relative permittivity of the cone are the same for each configuration; however, the pointed-cone geometry excites a significantly greater proportion of the higher-order mode than the type 1 configuration. This means that, to improve the performance of the type 1

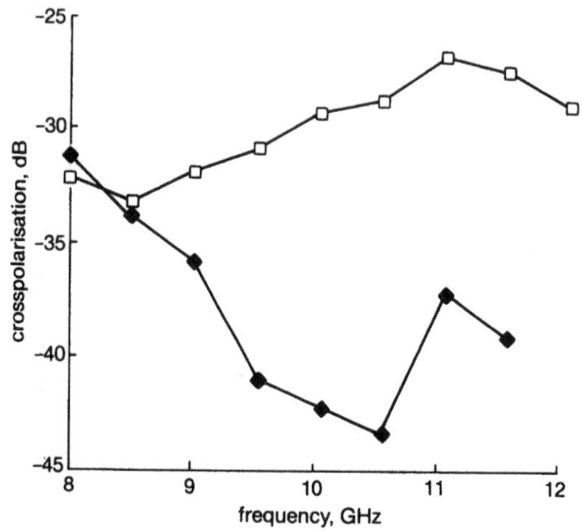

Figure 10.14 *Peak crosspolar level against frequency for a 20° horn with $\varepsilon_r = 1.7$ and gap = 5 mm*

—□— crosspolarisation, type 1
—◆— crosspolarisation, type 2

horn, the permittivity of the cone should be increased.

Figure 10.15 shows the variation in performance of the 30° horn as a function of annular-gap size and cone permittivity for a type 2 horn. The type 2 horn had to be used to obtain a reasonable crosspolar performance, using a dielectric material with a permittivity of 1.7. The optimum crosspolar performance of −37 dB was achieved using an airgap of 4.5 mm whilst the beamwidth remained virtually constant as the gap is altered between 3 and 7 mm. Varying the permittivity of the cone between 1.3 and 1.7, for a 5 mm airgap, produces a peak crosspolar level ranging from −23 dB to −35 dB. Generally, it is more difficult to maintain a good electromagnetic performance as the horn flare angle is increased. It is also more difficult to carry out the root-searching procedure for the sharp cone configuration given large values of flare angle and cone permittivity, owing to oscillations of the evanescent modes.

10.6 Horn-design examples

In this Section a number of dielectric-cone-loaded horns will be described to illustrate the application of the theory described in the previous sections and to show the potential for these types of microwave horns.

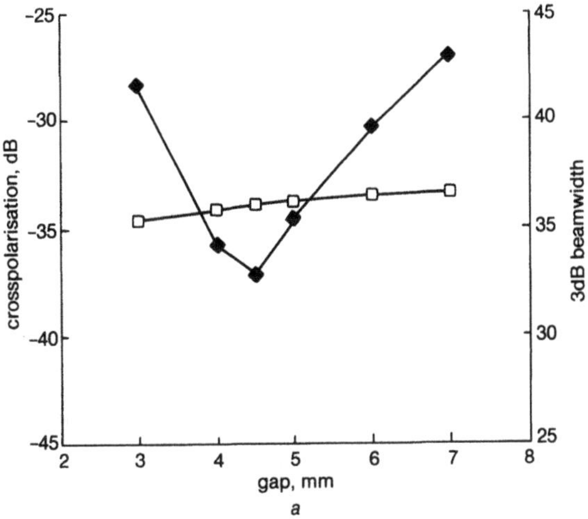

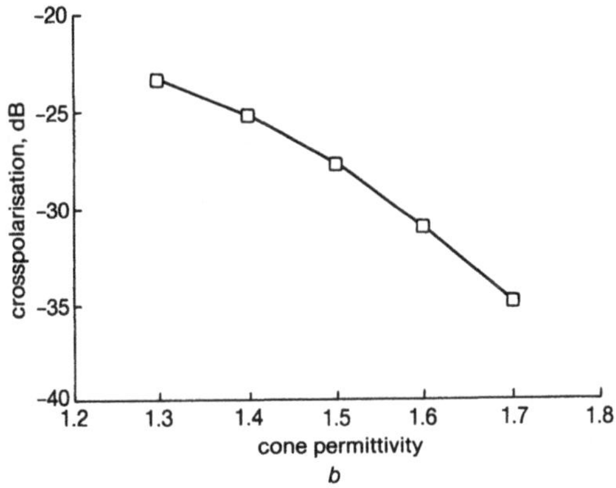

Figure 10.15 *Peak crosspolar levels for a horn with a 30° semiflare angle*

a Against gap
b Against permittivity
$\varepsilon = 1.7$, frequency = 10 GHz
a ——◆—— crosspolarisation ——□—— beamwidth (3 dB)
b ——□—— crosspolarisation (3 dB), gap = 5 mm

10.6.1 Dielectric-foam cone in metal horn

The first horn developed to demonstrate the concept of a dielectric-cone-loaded horn consisted of an X-band conical metal horn with a semiflare angle of 12° loaded with a polystyrene-foam cone which had a relative permittivity of 1.13 (Figure 10.1). The aperture diameter was 154 mm, the input waveguide diameter was 26 mm and the airgap was 12 mm wide. The dielectric was supported by a series of foam rings. The radiation characteristics of this horn are shown in Figures 10.2 and 10.4. The electrical performance is very satisfactory, but the mechanical design was unsatisfactory as a practical horn so alternatives have been considered. Some of the designs which produce practical horns are described in the following sections.

10.6.2 Metallised dual-dielectric cone-loaded horn

An alternative to a conventional metallic horn with a dielectric-cone insert is to surround the inner cone by a low-permittivity-dielectric layer. This ensures that a good electromagnetic performance can be sustained and the cone supported. To maintain the ratio of the relative permittivities between the cone and surrounding medium, the permittivity of the inner cone must be increased. For instance, if the outer material is foam polystyrene, with a relative permittivity of 1.13, and the inner cone has a permittivity of approximately 1.5, then if the outer relative permittivity is increased to 1.3 then the relative permittivity of the cone will need to be approximately 1.7. Raising the permittivity leads to increased problems with dielectric loss and increased reflection at the aperture and input waveguide. However, a significant benefit of surrounding the inner cone with an outer sheath of dielectric is that a layer of metal can be deposited on to the surface, thus completely avoiding the need for a conventional conical horn. The integral structure also implies that the cone can be accurately centred. The ability to mould dielectric foams to the required shapes should mean that batch fabrication is possible.

A dielectric horn was designed and made using two foam dielectrics and a metallised outer layer so as to reduce the weight (Figure 10.16). The horn had a type 1 throat region because this has more mechanical stability than type 2. It was found that a dielectric with a relative permittivity of 1.8 was needed to excite sufficient higher-order modes to cancel the crosspolar lobes to below −30 dB. The outer region then needed to have a relative permittivity of 1.4. The lowest crosspolarisation was achieved using a 10 mm outer region with a

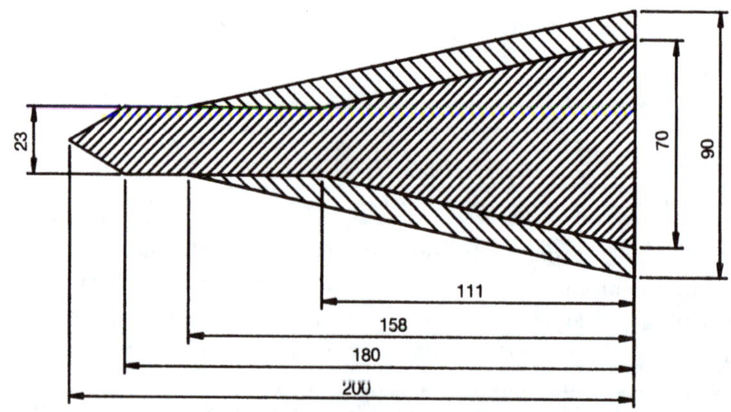

Figure 10.16 Metallised dual-dielectric-loaded horn

semiflare angle of 12°. These values were also confirmed by the constraints of the manufacturing process.

Because of the large change in permittivity between the air-filled waveguide and the dielectric-loaded horn, a matching section was needed to reduce reflection of energy back down the guide. A flared matching section was used at the air-to-dielectric interface within the waveguide. This was modelled using the analysis software. A linear taper was used and this gave a reflection coefficient of -27.4 dB at the design frequency. At the aperture of the horn there will also be a mismatch; in this case a quarter-wave matching layer can be used, although this will restrict the operating bandwidth.

The prototype horn was produced from metallised syntactic-foam material (poly ether sulphone), to which glass spheres were added to adjust the permittivity. The syntactic foams were manufactured by mixing the glass spheres with the base material in exact quantities, thus ensuring that the density and hence the dielectric constant is controlled. The material was produced in the form of a rectangular block, using a metal mould. The constituent thermoplastic resin and glass spheres were heated to the forming temperature of 350 °C and mixed thoroughly. The material was then compressed to a predetermined volume and heated for a period of time. The inner cone and the outer annular region were machined using a lathe. The outer layer was put on to a resin former after the inner material had been removed.

The finished components were mated together and the circular parts of the horn metallised. This was achieved by vacuum deposition, at 4×10^{-4} Torr using a copper target. Several copper coats were applied until a 10 μm layer built up. The resulting copper surface was slightly porous, because it deposited into and around the glass spheres on the surface of the material. DC-conductivity tests confirmed that the metallised surface was continuous.

The prototype metallised dual-dielectric-loaded horn was connected to standard waveguide by means of a brass tube which slid over the cylindrical section of the horn. This proved to be the least satisfactory part of the system. Power leaked from the brass/copper interface. This was stopped by using a conducting paste. In addition, either the tube or the cylindrical section of dielectric was slightly oval which caused problems with the alignment and resulted in some unwanted mode conversion. These unwanted modes had nonunity azimuthal dependence which showed up as asymmetric radiation patterns, particularly for the low-level crosspolar patterns.

Two prototype horns were made. In the first horn it was found that the dielectric had considerable inhomogeneity and had an axial asymmetry. This led to variations in the beamwidth in different planes. A second horn was made to try to overcome the problems. This one was more successful. The radiation patterns at 9 GHz are shown in Figure 10.17. These show reasonable agreement with the theoretical predictions for the copolar patterns but the peak crosspolar level is about 10 dB higher than the theoretical level. The copolar and crosspolar contour plots at 10 GHz are shown in Figure 10.18. These show quite good symmetry in the copolar patterns but the crosspolar contour is

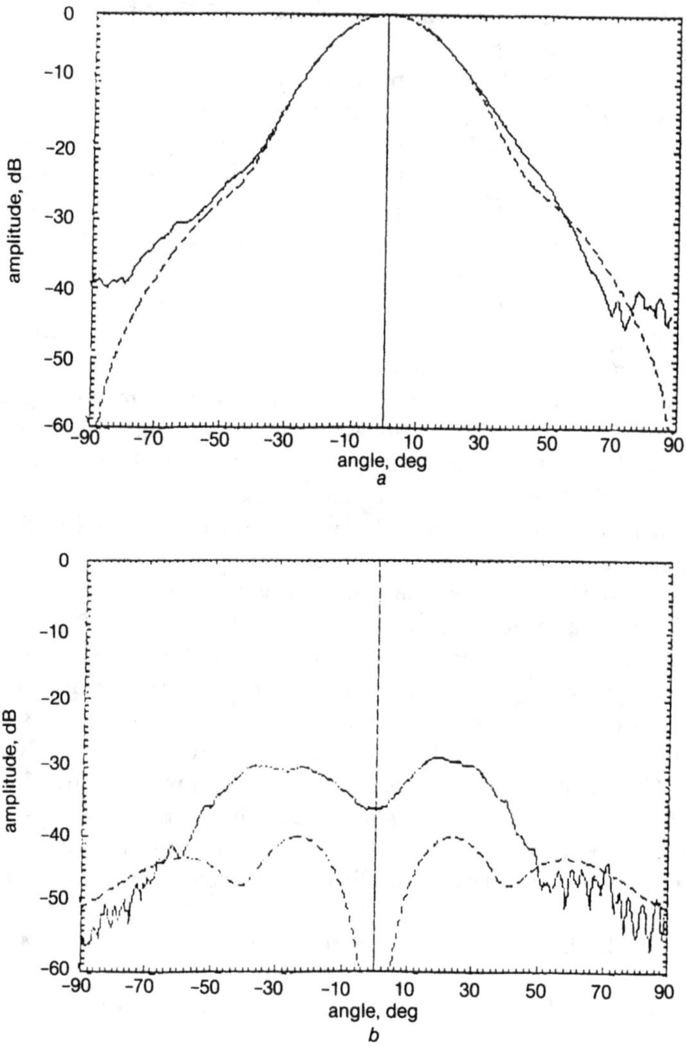

Figure 10.17 *45° patterns for horn illustrated in Figure 10.16*

——— measured
– – – – theory
a Copolar
b Crosspolar

distorted. The high peak crosspolar level and the asymmetrical patterns were investigated in considerable detail, and found to be due to residual nonaxial inhomogeneity in the foam.

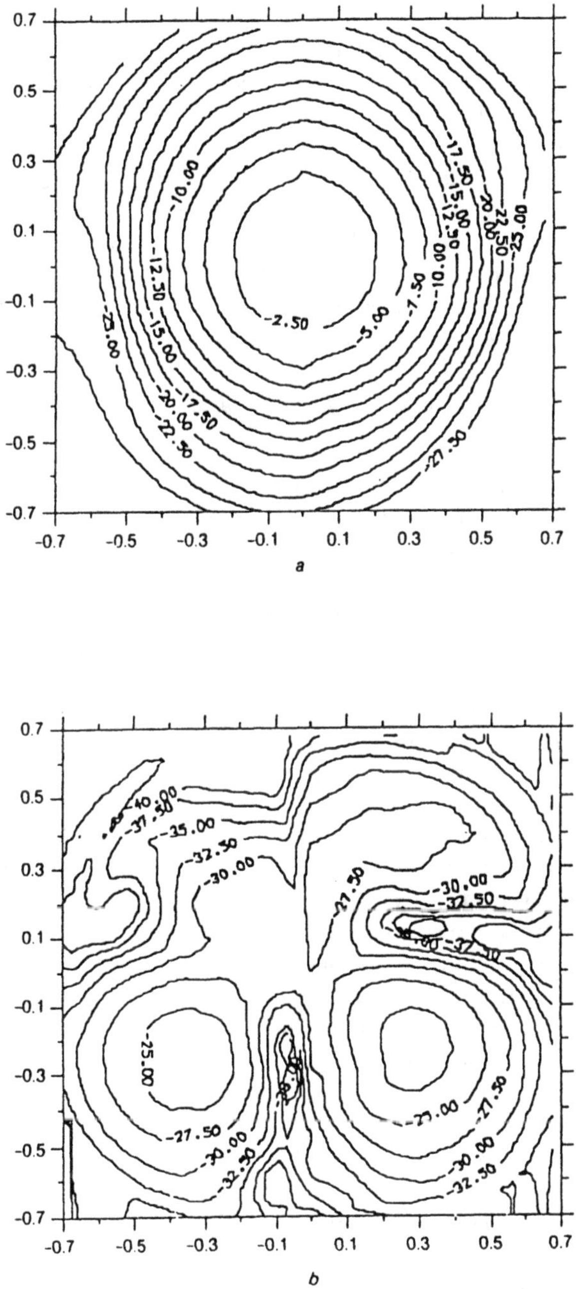

Figure 10.18 *Measured contour patterns for horn illustrated in Figure 10.16*

a Copolar
b Crosspolar

10.6.3 Millimetre-wave horns using a solid-dielectric cone

Dielectric horns can be designed with solid dielectrics, as has been demonstrated by Lier [3]. The primary disadvantage of solid dielectrics is their weight. However, in the millimetre-wave band (30–300 GHz), horns with solid dielectrics will be relatively light because of the small size and offer potential advantages of being easier and cheaper to manufacture than comparable corrugated horns. Cahill [9, 26] has described a horn for 94 GHz which worked well. The author has also made a 94 GHz horn using a cone of polyethylene ($\varepsilon_r = 2.25$) in a conical metal-feed horn, which gave good results. Cahill [27] has also described a design for 183 GHz using PTFE ($\varepsilon_r = 2.1$) which gave measured patterns in close agreement with theory. This horn has a semiflare angle of 9.2°, an aperture diameter of 10 mm and an airgap of 0.24 mm.

10.6.4 Millimetre-wave horns using two solid dielectrics

The procedure described above for making a metallised-foam-dielectric horn can be applied to design a horn with two dielectrics where a metallised layer is deposited on the outer dielectric. This saves the weight of the conical metal horn. Designs using two solid dielectrics will now be described.

For the purpose of design, the relative permittivities of the core and outer material are assumed to be 2.5 and 2.1, respectively. Several polymers have values of relative permittivity in this range. Pure polystyrene has a relative permittivity between 2.49 and 2.55, and a loss tangent below 3×10^{-4} with a typical density of 1050 kg/m^3. Polystyrene thus offers a relatively low-density low-loss material suitable for the core. PTFE has a relative permittivity of 2.1 (density = 2200 kg/m^3) and has negligible losses. As the outer material of the horn, PTFE offers very high volume resistivity, chemical inertness, insignificant water absorption and a surface suitable for electroplating. The molecular bond between PTFE and the electrodeposited material is very strong, thereby eliminating air trapped between the layers.

Two horns are designed using an optimisation routine for different criteria. In the first design the peak crosspolar level is minimised subject to a maximum horn length, with no constraint on the directivity of the horn. This is done with a view of obtaining a low-weight horn. In the second design, there is no constraint on the length of the horn, but the directivity is constrained by a minimum value. Since the optimisation program used has only a limited number of variables, the maxima and minima obtained are strictly local. The horn design is at a frequency of 30 GHz which corresponds to a wavelength of 10 mm. The weights of the horns are computed assuming the core to be made of polystyrene and the outer dielectric to be PTFE with densities equal to the above-quoted values.

The input to the horn must be a pure TE_{11} mode. In a solid-dielectric horn it is necessary to choose the input-guide dimensions in such a way that only the TE_{11} mode propagates while all higher-order modes are cutoff. When the guide is filled with a dielectric material of relative permittivity ε_r, the guide radius at

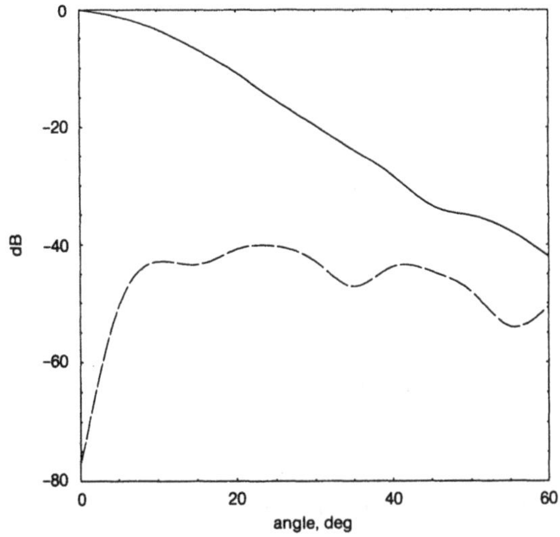

Figure 10.19 45° patterns for millimetre-wave horn 1

——— copolar
- - - - crosspolar

which the next-higher-order mode (TM_{11}) is excited is reduced by a factor of $\sqrt{\varepsilon_r}$. It is important to choose the input-guide radius small enough to cutoff the TM_{11} mode. The waveguide radius r_{th} is, therefore, chosen to satisfy the inequality

$$\frac{1.8411\,\lambda}{2\pi} < r_{th} < \frac{3.832\lambda}{2\pi\sqrt{\varepsilon_r}} \qquad (10.20)$$

If ε_r is taken to be 2.5, then at $\lambda = 10$ mm this requires the input circular guide to have a radius between 2.93 and 3.86 mm. For the horns described here, the radius will be chosen as 3.5 mm.

The first horn is optimised to yield a minimum crosspolar level with a constraint on the maximum length of the horn. The designed horn has a gap between core and horn wall of 1.6 mm and a horn length of 75.0 mm (semiflare angle of about 16°). The computed radiation patterns in the diagonal plane are shown in Figure 10.19 and the performance is summarised in Table 10.1.

Table 10.1 Performance of solid-dielectric horns

	Horn 1	Horn 2
VSWR	1.021	1.009
Peak crosspolar level, dB	−40.4	−50.0
3 dB beamwidth (45° plane), deg	17.9	16.0
Directivity, dB	19.6	21.5
Aperture efficiency, percent	36.5	57.8
Weight, g	67.3	200.4

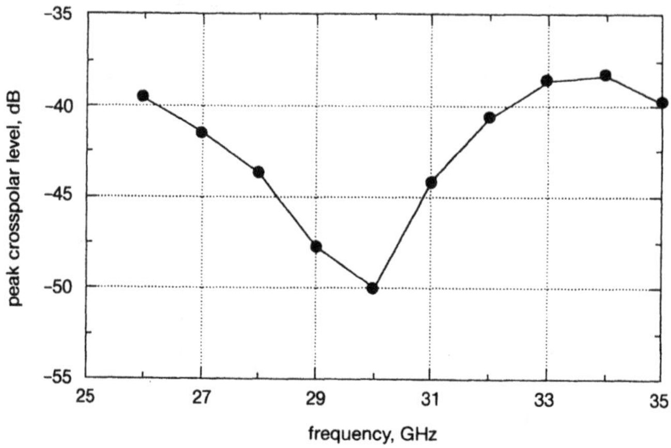

Figure 10.20 Peak crosspolar level against frequency for millimetre-wave horn 2

This horn has low weight due to its short length. The aperture efficiency is, however, low owing to the excessive power coupled to some of the higher-order modes. The reason for this is that higher-order modes are excited and become propagating modes at points close to the throat of the horn owing to the large permittivity of the core. One way of controlling this is to use a small-flare-angle horn so that the points of excitation of higher-order modes are shifted towards the aperture. This increases the horn length, and weight, but yields a lower peak crosspolar level and higher directivity.

The second horn is optimised to yield a minimum crosspolar level with a constraint on the minimum acceptable directivity. There is no constraint on the length of the horn. The designed horn has a type 2 throat, a gap between core and horn wall of 3.5 mm and a length of 180.0 mm. The predicted performance is given in Table 10.1. This horn has a very low crosspolar level. The frequency response is plotted in Figure 10.20. As can be seen from this Figure, the peak crosspolar level remains below -38 dB over a 30% bandwidth.

A comparison of the two designs shows that the second horn gives a much lower peak crosspolar level and a significantly higher aperture efficiency than the first horn, although the former is nearly 3.3 times heavier and more than twice as long. The low directivity of the first horn appears to be a result of beam broadening due to the excessive power coupled to some of the higher-order modes.

An experimental solid-dielectric-loaded horn was manufactured using the metallised-dielectric principles described in Section 10.6.2. The geometry of the horn was a compromise between the two designs described above and is shown in Figure 10.21. It has a type 1 throat for ease of manufacture. Samples of the dielectric were measured to find the actual relative permittivities. This was 2.51 for the crosslinked polystyrene and 2.04 for the PTFE. These values were then used to predict the performance. The measured copolar contour patterns at

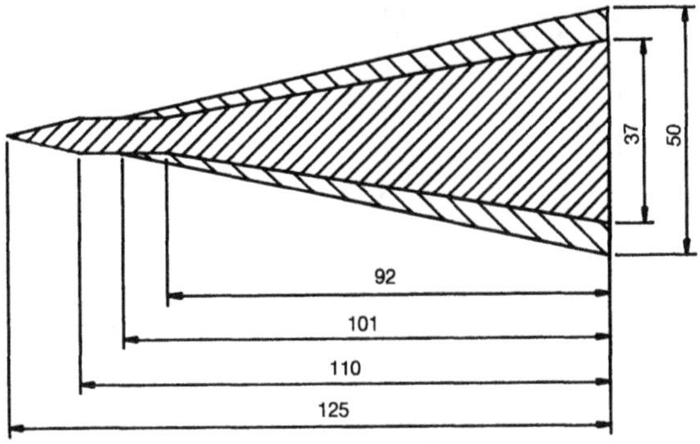

Figure 10.21 *Cross-section of millimetre-wave dielectric horn*

Dimensions in mm

30 GHz are shown in Figure 10.22 and the diagonal measured and theoretical patterns are shown in Figure 10.23. The copolar patterns show good agreement with theory whilst the measured peak crosspolar level is −38 dB, compared with a theoretical prediction of −47 dB. Given the large number of possible sources of error and the relatively low measured level, these results are very promising and show that solid-dielectric-loaded horns are a practical means of making a high-performance horn at millimetre wavelengths.

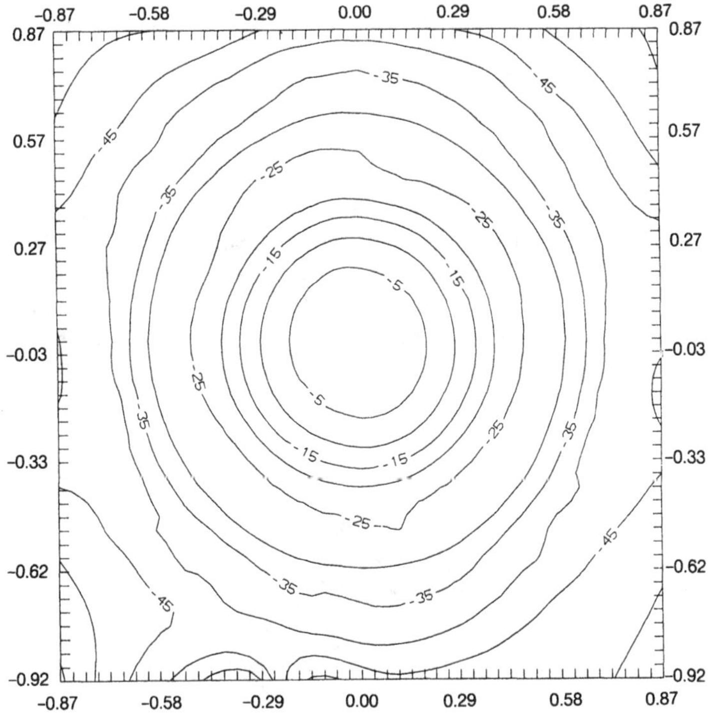

Figure 10.22 *Measured contour pattern for horn illustrated in Figure 10.21*

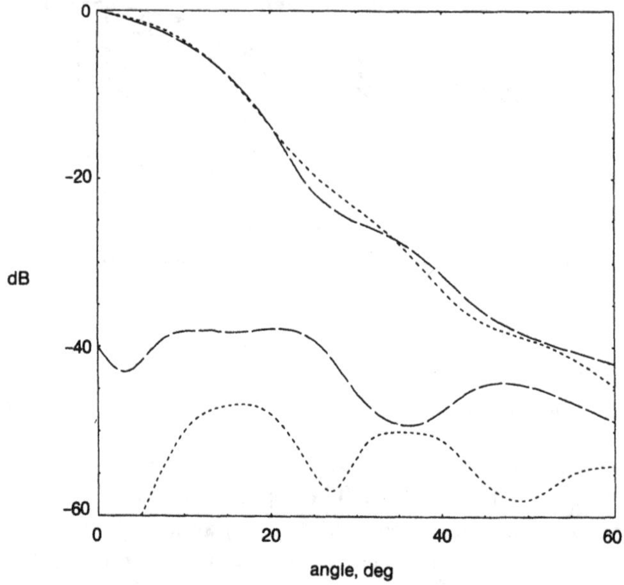

Figure 10.23 45° patterns for horn illustrated in Figure 10.21
30 GHz solid-dielectric horn; — — measured; - - - theory

10.7 Compact profiled dielectric-loaded horns

It has been shown in Chapter 9 that a way of reducing the volume of a corrugated horn is to use a profiled conical corrugated horn. Profiled corrugated horns are compact and have been successfully designed and used. The same philosphy can be applied to dielectric-loaded horns and leads to two main benefits over the horn with a linear flare angle. First, the length and weight of the horn can be controlled and a compact horn can be produced. Secondly, the flare angle at the aperture is reduced to nearly zero so that the phase centre of the horn is near the aperture. Consequently, it remains in this position as frequency changes and, when the horn is used as a feed for a reflector, the efficiency and frequency response are improved. A cross-section of a horn is shown

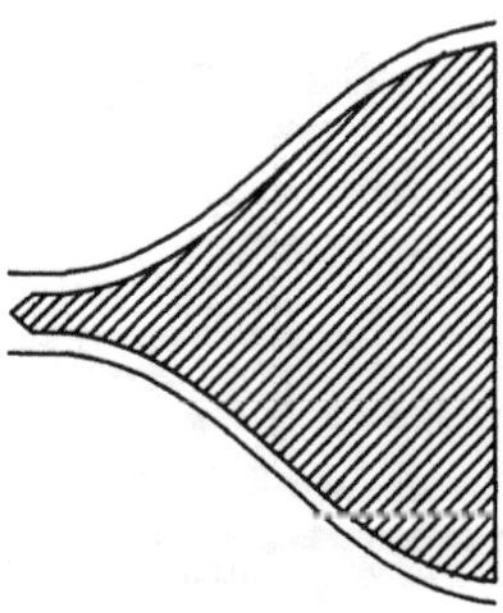

Figure 10.24 Profile dielectric-loaded horn

in Figure 10.24. In principle, the profile for the core dielectric and the outer wall could be different, but this produces a large number of parameters, so for the purposes of this study, a constant gap has been assumed.

10.7.1 Comparison with linear horn

The ability to change the radius of the horn with a profile means that one or more radiation characteristics can be improved. To investigate this a horn with a length of 309 mm, a throat radius of 11.43 mm, an aperture radius of 77.0 mm and a dielectric constant for the core of 1.13 was chosen. The outer dielectric was assumed to be air and the frequency was 10 GHz, giving an aperture diameter of 5.1λ. The optimum radial width of the airgap was found to be 12 mm.

A summary of the predicted radiation characteristics is given in Table 10.2. This compares a linear horn with a profiled horn having the profile

$$r(z) = r_{th} + 3\Delta r\left(1 - \frac{2z}{3L}\right)\left(\frac{z}{L}\right)^2 \tag{10.21}$$

where $\Delta r = r_{ap} - r_{th}$. This was found to be the optimum profile of the types investigated (see later). From Table 10.2, it can be seen that the copolar beamwidth has decreased so that the efficiency has increased and the peak crosspolar level has decreased by nearly 4 dB to -36 dB. The VSWR looking into the horn is similar in both cases, and is low because of the gradual tapering of the dielectric core. For all the cases investigated, the VSWR is of this order so is not referred to further.

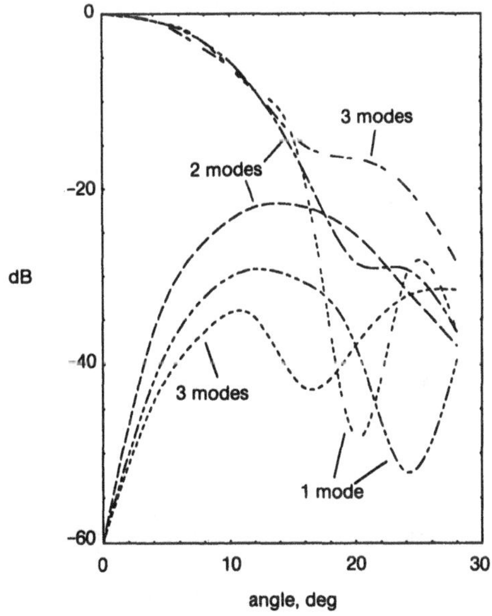

Figure 10.25 Copolar and crosspolar patterns for profile horn

The profiled horn produces a well worthwhile improvement in performance. To see how this is achieved, the copolar and crosspolar radiation patterns are plotted in Figure 10.25 for one mode (the HE_{11} mode), two modes (the HE_{11} and EH_{11} modes), and three modes (the HE_{11}, EH_{11} and HE_{12} modes). Seven modes have been used in the modal-matching analysis, but only the first three modes contain significant amounts of power. This is seen by comparing the curves for three modes in Figure 10.25 with the total radiation pattern in Figure 10.26. Figure 10.25 shows that the EH_{11} mode adds significant amounts of crosspolar power but the HE_{12} mode subtracts this power and reduces the crosspolar level by comparison with just the HE_{11} mode. The HE_{12} mode radiates high copolar lobes at about 20° which fill in the copolar nulls and produce a shoulder on the main beam. This is a feature of profiled horns and is due to the negative rate of change of the flare angle at the aperture.

The frequency performance between 8 GHz and 11 GHz is plotted in Figure 10.27. This shows the peak crosspolar level for both the linear and profiled horns. The level decreases approximately monotonically for the linear horn because the points at which higher-order modes are excited in the horn move linearly along the horn as frequency is increased. In the profiled horn this is not the case, and consequently the smooth behaviour is lost as modes interact in a complicated manner. The ability to optimise the performance at the design frequency also means that the modes are sensitive to frequency and the relative bandwidth is reduced. Figure 10.28 shows the relative power in the first three modes over the frequency band. This makes it clear that the relative powers

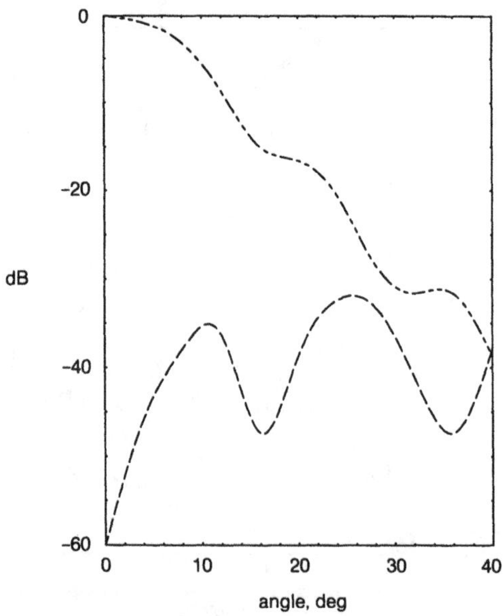

Figure 10.26 Total patterns for profile horn

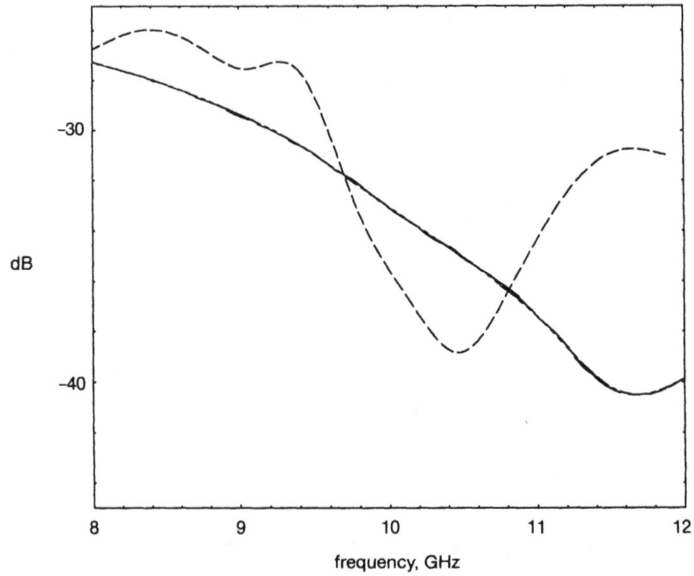

Figure 10.27 Peak crosspolar level against frequency for profile horn

——— linear horn
– – – – profile horn

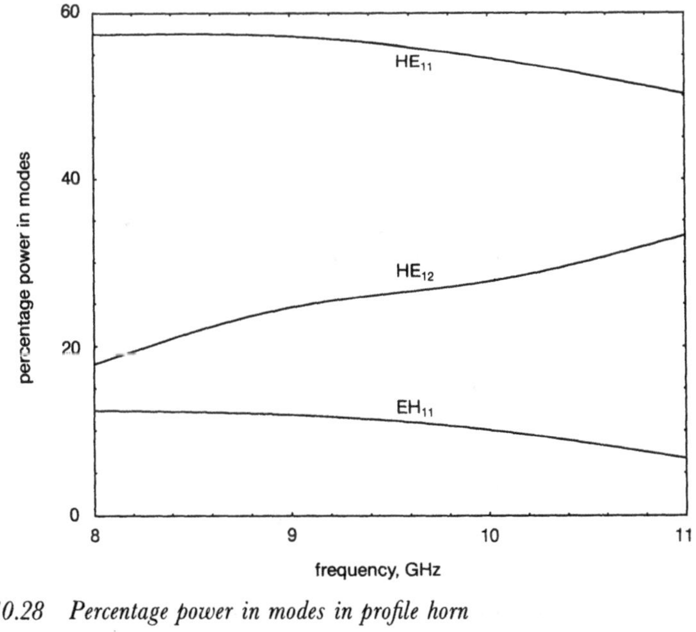

Figure 10.28 Percentage power in modes in profile horn

change in a different way for the three modes. However, it would be possible to take the requirement for wideband performance as an optimisation parameter and obtain the profile which gave a wide bandwidth.

The width of the outer dielectric can be used to optimise the radiation performance. In the above horn, changing the relative permittivity of the inner dielectric to 1.75 meant that the radial width of the airgap needed to be reduced to 2 mm to retain optimum performance. However the lowest peak crosspolar level which could be obtained rose to −32 dB. Generally, foam dielectrics with relative permittivities below 1.5 produce the best performance. It is the ratio of the permittivities of the inner and the outer dielectrics which is relevant to the electrical performance, so that the same performance as Table 10.2 can be produced with relative permittivities of 1.8 and 1.67. There is an increase in the VSWR owing to the increased mismatch with the air at the aperture.

10.7.2 Reduction in length

The profiled horn was used in Section 10.7.1. to improve the radiation characteristics. The profile is now used to reduce the length of the horn using the profiled horn specified above and in Table 10.2. A summary of the results is shown in Table 10.3. Reducing the length by 11% to 275 mm produces excellent performance. Reducing the length by 22% to 240 mm still produces acceptable performance, though with some increase in the peak crosspolar level. For shorter lengths, the profile does not compensate for the effective wide flare angle and the phase taper at the aperture is considerable. This results in increase in both the copolar beamwidth and the peak crosspolar level, as well as significantly reduced efficiency. However, the performance is still better than

Table 10.2 Performance of linear and profile horns

Parameter	*Linear horn*	*Profiled horn*
3 dB beamwidth, deg	14.8	13.7
10 dB beamwidth, deg	26.9	24.8
Directivity, dB	22.1	22.5
Efficiency, per cent	61.8	68.1
Peak crosspolar level, dB	−32.2	−36.0
VSWR	1.04	1.03

Table 10.3 Performance of profile horn as length is reduced

Length (mm)	Peak crosspolar (dB)	3 dB beamwidth (deg)	Efficiency (%)
309	−36.0	13.7	68.1
275	−36.8	13.8	64.4
240	−31.6	14.0	57.1
150	−27.5	16.1	32.0

that of an equivalent linear horn. Figure 10.29 shows the copolar and crosspolar radiation patterns for the horn which is 150 mm long.

The reduction in efficiency is approximately proportional to length, as shown in Figure 10.30. This compares the efficiency of the profiled horn with that of a linear taper horn of length 309 mm. For lengths greater than 258 mm (20% length reduction), the efficiency of the profiled horn is greater than that of the linear horn.

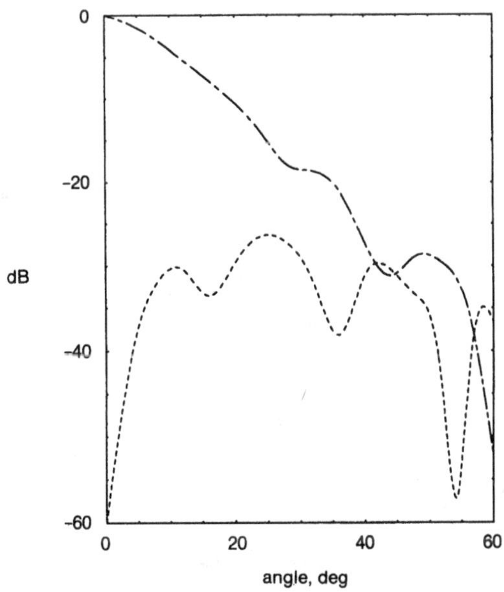

Figure 10.29 Patterns of 150 mm short profile horn

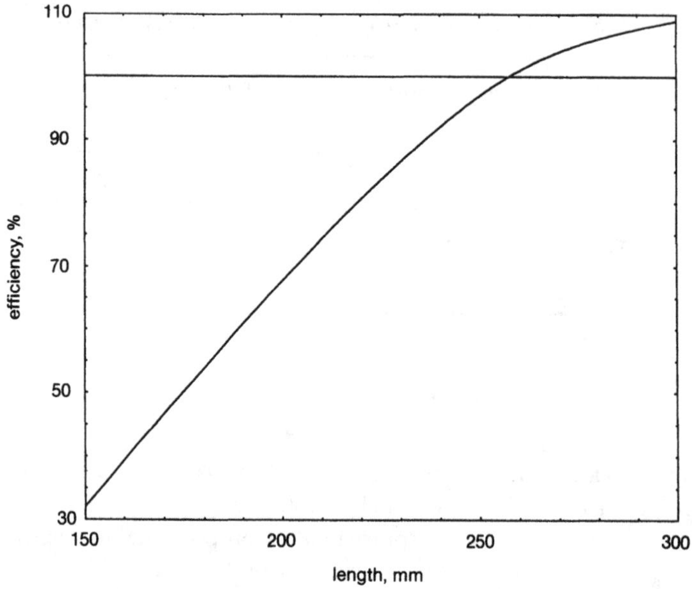

Figure 10.30 Efficiency against length for a profile horn compared with a 309 mm-long linear horn

10.7.3 Influence of shape of profile

The shape of the profile controls both the amplitude and phase of the modes at the aperture. The amplitudes are given by the rate at which the higher-order modes are excited and the phase is given by the position along the horn at which they are excited. In order to see the influence of the shape of the profile, five different types of profile are studied.

The linear horn with a profile

$$r(z) = r_{th} + \Delta r \frac{z}{L} \tag{10.22}$$

has a volume given by

$$V = \frac{1}{3}\pi L(r_{ap}^2 + r_{ap}r_{th} + r_{th}^2) \tag{10.23}$$

where r_{ap} is the radius of the aperture and r_{th} is the radius of the throat.

The first of the profiles is a sine-squared profile (Figure 10.31), given by

$$r(z) = \Delta r \left\{ (1-A)\frac{z}{L} + A \sin^2\left(\frac{\pi z}{2L}\right)\right\} \tag{10.24}$$

where A is a profile parameter ($0<A<1$). $A=0$ gives a straight horn. The best values for A are between 0.7 and 0.9. The volume of this profiled horn is

$$V = \pi L \left\{ A(r_{ap}r_{th} + C\Delta r^2) + \left(\frac{3}{8} - C\right) A^2 \Delta r^2 \right.$$

$$\left. + \frac{1}{3}(1-A)^2(r_{ap}^2 + r_{th}^2) + \frac{1}{3}(1-A)(1+2A)r_{ap}r_{th} \right\} \tag{10.25}$$

where

$$C = \frac{2}{\pi}\left(1 + \frac{\pi^2}{4}\right)$$

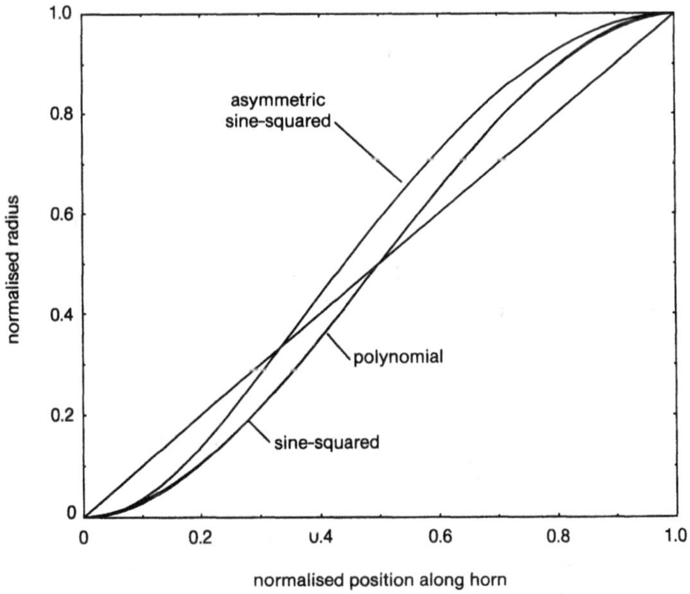

Figure 10.31 Shape of profiles along horn

In the above profile the point of zero change in slope is in the centre of the horn. This does not have to be the case, and the second profile is an asymmetric sine-squared profile given by

$$r(z) = r_{th} + \frac{2\Delta r}{(1+\gamma)} \sin^2\left(\frac{\pi z}{4L_1}\right) \qquad z \leq L_1 \tag{10.26}$$

and

$$r(z) = r_{th} + \frac{2\Delta r}{(1+\gamma)} \left[\frac{(1-\gamma)}{2} + \gamma \sin^2\left\{\frac{\pi(z+L_2-L_1)}{4L_2}\right\}\right] \qquad L_1 \leq z \leq L \tag{10.27}$$

where $\gamma = L_2/L_1$ and $L_1 + L_2 = L$. This profile has been studied for corrugated horns in Reference 2. The profile has zero slope at both ends of the horn.

Another profile which has zero slope at both ends of the horn is a polynomial profile as given in eqn. 10.21. This has a very similar shape to the symmetric sine-squared profile with $A = 0.9$, as can be seen by comparing the profiles in Figure 10.31; however, it does produce slightly improved radiation characteristics. The volume of this horn is given by

$$V = \pi L \left(\frac{13}{35}\Delta r^2 + r_{ap} r_{th}\right) \tag{10.28}$$

Two other profiles are also considered. Both are modifications of a linear flare horn. An exponential profile with

$$r(z) = r_{th} e^{az} \qquad a = \frac{1}{L} \ln\left(\frac{r_{ap}}{r_{th}}\right) \tag{10.29}$$

gives the profile shown in Figure 10.32. This has a volume of

$$V = \frac{1}{2} \pi L \frac{(r_{ap}^2 - r_{th}^2)}{\ln\left(\dfrac{r_{ap}}{r_{th}}\right)} \tag{10.30}$$

The fifth profile has a hyperbolic rather than an exponential profile, given by

$$r(z) = \sqrt{[r_{th}^2 + \{z^2(r_{ap}^2 - r_{th}^2)\}/L^2]} \tag{10.31}$$

The volume of this horn is

$$V = \frac{1}{3} \pi L (r_{ap}^2 + 2R_{th}^2) \tag{10.32}$$

The performances of all the profiles are summarised in Table 10.4. This gives the volume normalised to the linear horn, the copolar beamwidth and the peak crosspolar level. The horn chosen for comparison is the short 150 mm horn of Section 10.7.2. The sine-squared profile has a lower peak crosspolar level than the linear horn, but the volume is 32% larger and the beamwidth has increased so that there is a decrease in efficiency of 15%. The asymmetric sine-squared profile with $L_1 = 50$ mm and $L_2 = 100$ mm ($\gamma = 2.0$) has a smaller volume by

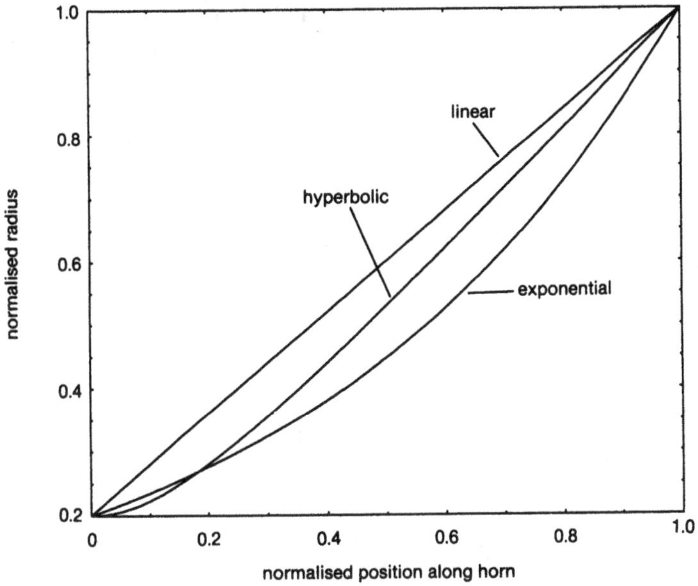

Figure 10.32 Shape of profiles along horn

Table 10.4 Characteristics of different profiles

Type	Volume ratio	Beamwidth	Peak crosspolar
		(deg)	(dB)
Linear	1	16.0	−24.0
Sine-squared	1.32 ($A = 0.9$)	16.6	−26.0
Asymmetric sine-squared	1.19 ($\gamma = 2.0$)	14.4	−25.6
Exponential	0.66	23.3	−23.7
Hyperbolic	0.89	17.4	−25.0
Polynomial	1.07	16.1	−27.5

comparison with the symmetric profile, and a narrower beamwidth. If L_1 and L_2 are reversed, so that $\gamma = 0.5$, then the volume ratio is less than one (0.96) but the beamwidth has increased to nearly 20° and the peak crosspolar level has returned to the same level as the linear horn. The polynomial profile has the best all-round performance of the profiles, with the lowest crosspolar level and a beamwidth similar to that of the linear horn. The volume is 7% greater than that of a linear horn, but the beamwidth is nearly the same and the peak crosspolar level has decreased by 3.5 dB. Examination of Figure 10.31 shows the similarity of this profile to the sine-squared profile, but indicates the sensitivity of the radiation parameters to the exact combination of modes at the aperture.

The exponential and hyperbolic profiles effectively increase the flare angle at the aperture, but lower it at the throat. This means that the volume is substantially reduced, especially for the exponential profile. However, in both cases the radiation performance is reduced. The efficiency of the exponential profile is 20% less than for the linear horn. The equivalent value for the hyperbolic profile is 10% less. Note that all the horns used in this comparison are very short, only five wavelengths in length, so that the efficiency is necessarily low owing to the large spherical-phase taper at the aperture. However, the results achieved are representative of those obtained for a longer horn.

The study has shown that the profiled dielectric horn gives useful control over the performance of the horn. It can be designed in two distinct ways. Either the length of the horn can be kept the same as a linear horn, which will give improved electrical performance; or the length (and weight) of the horn can be reduced by comparison with the linear horn, whilst maintaining similar low levels of crosspolarisation.

10.8 Dielectric losses in a dielectric-loaded conical horn

The loss in the dielectric materials used to make a dielectric-cone-loaded horn must be taken into account in evaluating the performance of a horn. The insertion loss introduced by the dielectric materials has to be kept to a minimum, preferably below about 0.1 dB, for the dielectric horn to be a viable alternative to the corrugated horn. Use of low-loss materials will minimise losses, but it is necessary to have a theoretical estimate of the amount of loss in the horn.

An exact prediction of the loss in the dielectric can be computed by replacing the real relative permittivity ε_r by the complex permittivity $\varepsilon_r(1-j\tan\delta)$, where $\tan\delta$ is the loss tangent of the dielectric. This involves using complex arguments for the cylindrical Bessel functions and the roots of the characteristic equation will have to be tracked in the complex plane. This is the only way of estimating the loss when $\tan\delta$ is large, say greater than 0.1. But in this case the dielectric loss in the horn will be too high to be of any practical use.

When the loss tangent is small, a perturbation technique can be used to evaluate the attenuation constant α in terms of the loss tangent of the material. This technique is accurate for low values of $\tan\delta$ and, once α is obtained, the complex propagation constant can be included in the scattering matrix of the modal-matching horn-analysis program.

10.8.1 Perturbation expression for the attenuation coefficient

A perturbation expression for the attenuation coefficient α for unity azimuthal-dependent modes has been derived by Clarricoats [17] for a circular guide concentrically filled by a lossy dielectric rod (Figure 10.7). The outer dielectric is assumed to be lossless. A more general expression valid for a multilayer annular structure in which all layers are assumed to be lossy, but restricted to circularly symmetric modes, has been given by Olver [20]. The expression given by Clarricoats [17] for a two-layer structure has been generalised to all hybrid modes by McBride [19].

Here the expression given by Clarricoats [17] is used to compute the loss in the dielectric-loaded horn. This assumes a lossy dielectric core surrounded by a lossless outer dielectric. If the outer dielectric is air, this assumption is completely valid. If the outer dielectric is lossy but thin, the power flow is confined almost entirely to the core, and the loss in the outer dielectric introduces little error.

A circular waveguide of radius r_0 is concentrically filled by a lossy dielectric rod of radius r_1 and complex relative permittivity $\varepsilon_r = \varepsilon' - j\varepsilon'' = \varepsilon'(1 - j\tan\delta)$. The dielectric rod is surrounded by a lossless material of relative permittivity ε_1'. Then the attenuation coefficient α for a propagating hybrid mode in the inhomogeneously filled circular guide is given by

$$\alpha = \frac{\pi\varepsilon''\{(\bar{\beta}^2 + \Delta^2)G + 2\Delta\bar{\beta} + (\lambda_0/2\pi)^2 K^2 (G - F(Kr_1)\}}{\lambda_o\bar{\beta}\left[\Delta\left(\bar{\beta} + \dfrac{\varepsilon'}{\bar{\beta}}\right) + (\varepsilon' + \Delta^2)G + \left(\dfrac{K}{K_1}\right)^4 \left\{\varepsilon_1' H + \Delta^2 L - \Delta\left(\bar{\beta} + \dfrac{e_1'}{\bar{\beta}}\right)\right\}\right]} \tag{10.33}$$

where $\bar{\beta} = \beta\lambda_0/2\pi$ is the normalised phase coefficient given by the roots of eqn. 10.4 and

$$G = F(Kr_1) + 0.5\{(Kr_1)^2 - 1 + F^2(Kr_1)\} \tag{10.34}$$

$$H = \frac{2}{\pi^2 S_D^2} - S_1(K_1, r_1, r_0) - 0.5\{(Kr_1)^2 - 1 + S_1^2(K_1, r_1, r_0)\} \tag{10.35}$$

$$L = \frac{2\left\{1 - \dfrac{1}{(K_1 r_0)^2}\right\}}{\pi^2 R_D^2} - R_1(K_1, r_1, r_0) - 0.5\{(Kr_1)^2 - 1 + R_1^2(K_1, r_1, r_0)\} \tag{10.36}$$

The remaining symbols are given in eqns. 10.5–10.7.

The attenuation coefficient α, given by eqn. 10.33, is a function of the angular frequency ω, the ratio r_1/r_0, the relative permittivity of the core ε', and its loss tangent $\tan\delta$. For fixed ε', α is directly proportional to $\tan\delta$. For low values of $\tan\delta$, and $r_1/r_0 \simeq 1$, α increases as $\sqrt{\varepsilon'}$.

If the dielectric completely fills the guide, it is possible to give simpler expressions for the attenuation coefficient α and the phase coefficient β, cf. Collin [18]:

$$\alpha = k_0 \left(\sqrt{\frac{\varepsilon'}{2}}\right)\left(\{(\omega_c/\omega)^2 - 1\} + \sqrt{[\{1-(\omega_c/\omega)^2\}^2 + \tan^2\delta]}\right)^{1/2} \qquad (10.37)$$

and

$$\beta = k_0 \left(\sqrt{\frac{\varepsilon'}{2}}\right)\left(\{1-(\omega_c/\omega)^2\} + \sqrt{\{1-(\omega_c/\omega)^2\}^2 + \tan^2\delta]}\right)^{1/2} \qquad (10.38)$$

If ω is well above the no-loss cutoff frequency ω_c, and $\tan\delta$ is sufficiently small, a binomial expansion can be used to give the well known results

$$\alpha = k_0 \frac{\sqrt{\varepsilon'}}{2} \frac{\tan\delta}{\sqrt{\{1-(\omega_c/\omega)^2\}}} \qquad (10.39)$$

and

$$\beta = k_0 \sqrt{\varepsilon'} \sqrt{\{1-(\omega_c/\omega)^2\}} \qquad (10.40)$$

10.8.2 Losses in dielectric-loaded horn

The theory given above can be used to compute the losses in typical dielectric-loaded horns in order to be able to specify the amount of loss which is tolerable in the dielectric. As a first example, consider the following horn:

aperture radius = 45 mm
gap between core and horn wall = 10 mm
length of horn = 180 mm
frequency = 10 GHz
relative permittivity of core = 1.8 $(1 - j\tan\delta)$
relative permittivity of the outer material = 1.4.

For $\tan\delta = 5\times10^{-3}$, the loss in the horn is computed as 0.95 dB and for $\tan\delta = 5\times10^{-4}$, it is 0.095 dB. For these practical values of $\tan\delta$, the loss in the horn in decibels is directly proportional to the loss tangent. This is consistent with the observation that, for fixed ε' and low values of tan δ, the loss is directly proportional to the latter. The change in the phase coefficient is small enough to be neglected and so there are no changes to the peak crosspolar level or beamwidth. If a loss of 0.1 dB is acceptable, then this corresponds to a tan δ of 5.3×10^{-4}. Such a value of loss tangent is within the range which is to be expected from good-quality commercial foam dielectrics.

As a second example, consider a horn using solid dielectrics for use at 30 GHz:

aperture radius = 25 mm
gap between core and horn wall = 1.6 mm
length of horn = 75 mm
frequency = 30 GHz
relative permittivity of core = 2.5 $(1-j \tan \delta)$
relative permittivity of outer materal = 2.1.

In this case a tan $\delta = 5 \times 10^{-3}$ gives a loss of 1.55 dB and the loss is again proportional to loss tangent. Figure 10.33*a* and *b* plots the transmission coefficient of this horn in decibels along the axis for the first three modes (HE_{11}, EH_{11} and HE_{12}) for the cases of no loss, and for tan $\delta = 0.001$ and 0.005. These figures show that, in the no-loss case, as much as 28% of the total power appears in the EH_{11} and HE_{12} modes. At points on the horn axis where these modes are just above cutoff, the attenuation is much higher for these modes than for the dominant HE_{11} mode. This results in a comparatively higher insertion loss for this horn than would have been the case in a similarly sized horn with less power coupled to higher-order modes.

Profiling the horn reduces the length of the horn, as shown in Section 10.7. This can be seen by comparing the performance of a linear horn with that of a profiled horn with the following parameters:

aperture radius = 45 mm
gap between core and horn wall = 11.5 mm
length for the linear horn = 180 mm
relative permittivity of the core = 2.2 $(1-j0.001)$
relative permitivity of outer dielectric = 1.8.

Using a suitable profile, and adjusting the gap, it was possible to reduce the length of this horn to 135 mm, a reduction of 25%. The performance of the two horns is compared in Table 10.5. Although the crosspolar level of the profiled horn is somewhat higher than that of the straight horn, it has lower insertion loss, higher aperture efficiency and weighs about 25% less. The overall horn loss depends partly on $(\sqrt{\varepsilon'})$ tan δ and partly on the total length of the horn, as well as on the amount of power coupled to various modes. Hence, reducing the horn length by profiling reduces losses as well as decreasing the size and mass of the horn.

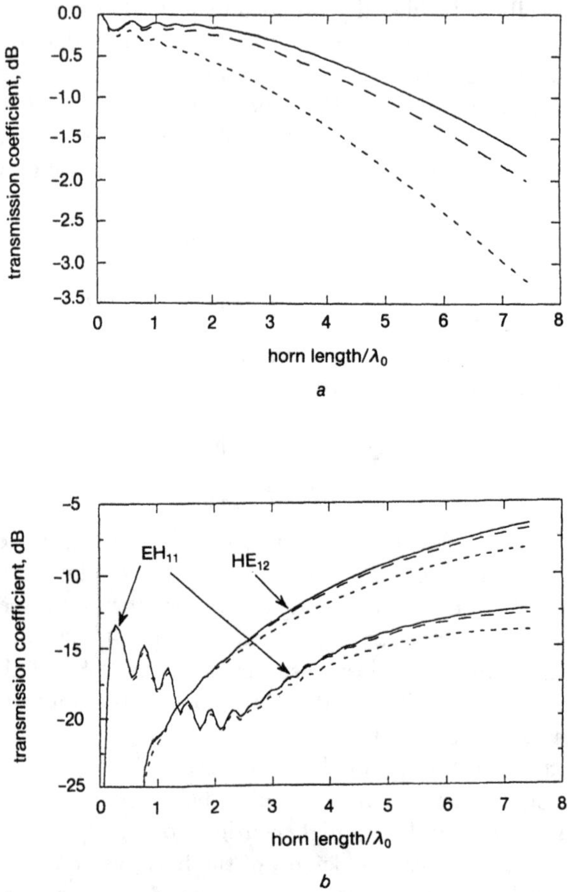

Figure 10.33 Transmission coefficient along horn axis

a HE_{11} mode
b HE_{12} and EH_{11} modes
——— no loss; — — tan $\delta = 0.001$; - - - - tan $\delta = 0.005$

Table 10.5 Comparison of loss and other parameters in linear and profiled horns

Parameter	*Linear horn*	*Profile horn*
Length, mm	180.0	135.0
Insertion loss, dB	0.205	0.156
Peak crosspolar level, dB	−39	−37
−3 dB beamwidth, deg	26.6	23.7
Directivity, dB	17.1	17.7
Aperture efficiency, percent	58	67

10.9 Tolerance on relative permittivity of dielectric

The radiation characteristics of a horn (such as its directivity, beamwidth, crosspolar level and far-field patterns) are calculated by assuming that all mechanical and electrical parameters pertaining to the horn have fixed constant values. For a manufactured horn these parameters can only be within prescribed tolerances, and it becomes necessary to account for these variations in horn design. For a dielectric-loaded conical horn, two major factors which often lead to a significant deterioration of its far-field characteristics are circular asymmetry about the horn axis and inhomogeneity in the dielectric materials. Circular asymmetry is not easy to model theoretically, but tolerances in the dielectric material and longitudinal inhomogeneity can be assessed using the modal-matching program to model a horn with variations in dielectric parameters along the horn.

The tolerance in the dielectric is best specified in terms of the electric susceptibility, $\kappa = \varepsilon_r - 1$ rather than on the relative permittivity ε_r. Thus a 5% tolerance in susceptibility at $\varepsilon_r = 1.40$ means a range of ε_r from 1.38 to 1.42. A horn with the following parameters is used to demonstrate the acceptable tolerances:

throat radius = 11.50 mm
aperture radius = 45.00 mm
gap between core and horn wall = 10.00 mm
length $L = 180.00$ mm
frequency of operation = 10 GHz
relative permittivity of core = 1.80
relative permittivity of outer material = 1.40.

The optimised performance of the horn yields a peak crosspolar level of −46.1 dB and a directivity of 17.27 dB, which is equivalent to an aperture efficiency of 60%.

The variation in the peak crosspolar level and the directivity for up to a ±17.5% variation in susceptibility of the outer dielectric is shown in Figure 10.34. The results do not show a linear change of radiation characteristics with permittivity changes because of the complex interaction between modes which cancel to produce the low crosspolar levels. Similar results are shown in Figure 10.35 for variations in the outer dielectric. The change in peak crosspolar level appears to be large, but it must be remembered that this is due to the difference between the E- and H-plane patterns so that very small changes in the copolar patterns can lead to large changes in low levels of crosspolar radiated power.

In the above example, a ±5% variation in the ratio $\varepsilon_0/\varepsilon_1$ (where ε_0 is the relative permittivity of the core and ε_1 is the relative permittivity of the outer dielectric) raises the crosspolar level by about 5 dB. The increase would be slightly less if the initial crosspolar peak was higher. This conclusion seems to be generally valid for a range of horn designs. The absolute tolerance on the

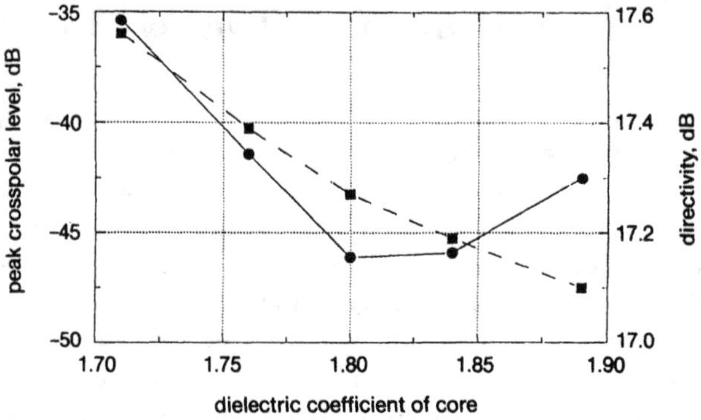

Figure 10.34 *Peak crosspolar level and directivity as a function of the relative permittivity of the core*

See text for geometry
——●—— peak crosspolar level
——■—— directivity

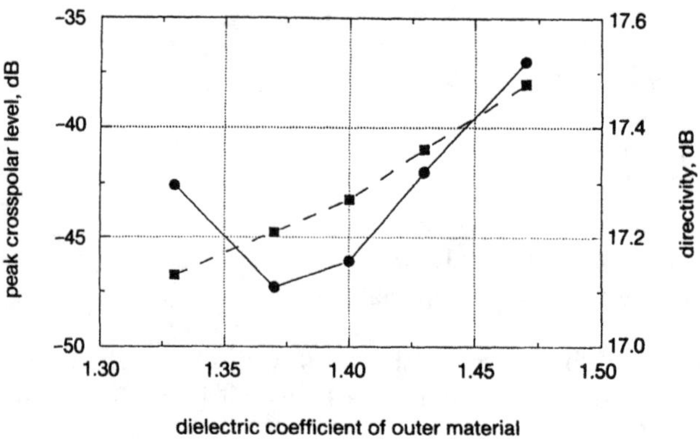

Figure 10.35 *Peak crosspolar level and directivity as a function of the relative permittivity of the outer dielectric*

See text for geometry
——●—— peak crosspolar level
——■—— directivity

relative permittivity of low-density foam dielectrics is more sensitive than for higher-density materials because the susceptibility changes rapidly for small changes in relative permittivity.

The effect of inhomogeneities in the dielectric materials can be of two types: either continuous change in the relative permittivity from the throat of the horn to the aperture or random variations in the relative permittivity. A continuous variation of the relative permittivity within the bulk of a material is possible in several instances. For example, in the manufacture of syntactic foams by mixing tiny glass spheres in a base material (such as a thermoplastic resin), the glass spheres could sink under the action of gravity resulting in slightly higher density towards the bottom. Since motion under gravity of a sphere in a viscous fluid is uniform, this would result in an approximately linear change in the relative permittivity along a vertical axis. If the core and outer dielectric of the horn has a similar variation of the relative permittivity along the horn axis, the effect of this on the performance of a horn can be theoretically studied. Using the horn specified above, a $\pm 5\%$ linear variation in the ratio $\varepsilon_0/\varepsilon_1$ led to the crosspolar level remaining below -43 dB and directivity almost unaltered if the relative permittivities of core and outer material both increase or decrease linearly from the throat to the aperture. As expected, the worst crosspolar level of -31 dB is obtained when the core relative permittivity is decreasing and that of the outer material is increasing. The change in the relative permittivity along the horn alters the distribution of power in the higher-order modes by slightly modifying the position along the horn at which the modes are excited.

Random variations in the relative permittivity are likely to be small, possibly within 0.5% of the RMS value. It is difficult to quantify these variations in a systematic way. Physically, no electrical parameter can vary discontinuously at the microscopic level within the bulk of the dielectric material. Random variations are therefore likely to be unpredictable fluctuations in the value of the relative permittivity at the macroscopic level. These can be theoretically modelled by generating two sets of pseudorandom numbers. The first set of numbers is scaled to obtain a series of random points along the horn axis. The second set of numbers is used to define random values of ε_0 and ε_1 within $\pm 0.5\%$ tolerance. These values of relative permittivity are assigned at the points defined by the first set of numbers. The relative-permittivity values obtained are then connected by a smooth curve, using cubic splines, to define the value of relative permittivity at intermediate points.

Using the horn specified above, a $\pm 0.1\%$ tolerance led to the far-field characteristics remaining nearly unaltered. At $\pm 0.5\%$ tolerance, the crosspolar level increases by about 3 dB. At $\pm 1.0\%$ tolerance, the crosspolar peak rises from -46 dB to -38 dB. These figures are only broadly indicative, since the above analysis assumes axial symmetry which is not valid for random variations. It would, however, seem to suggest that random fluctuations in relative permittivity must be within about $\pm 0.5\%$ for satisfactory crosspolar performance. This is a more stringent specification than for the linear variations in relative permittivity.

10.10 Control of crosspolarisation and efficiency with a lens

The dielectric-loaded horns studied in this chapter have had a plane surface at the end of the dielectric. It is possible to shape the aperture end of the dielectric so as to form a lens out of the central dielectric core (Figure 10.36). A suitably designed lens can be used to improve the horn's radiation characteristics and serves several purposes. It can be used to produce a flatter phase front at the aperture, which will both increase the efficiency and increase the bandwidth of the horn. The profile of the lens can be designed to control one of the radiation parameters, say the crosspolarisation. It is possible to have a horn–lens combination of this type machined into a single integrated feed. As shown in Figure 10.36, the outer wall of the horn is extended in the form of an oversized circular waveguide to constitute an outer boundary for the lens section. The main theoretical reason for the presence of the waveguide wall is to enable the same modal-matching technique to be extended to the horn–lens design. The lens can be analysed in the same manner as the horn by dividing it into a number of uniform waveguide sections. Here, the dielectric-loaded horn with a lens is studied in terms of its crosspolar performance, directivity and other radiation characteristics. Section 10.10.1 deals with the profile of the lens, followed by examples which demonstrate the potential for the dielectric lens horn.

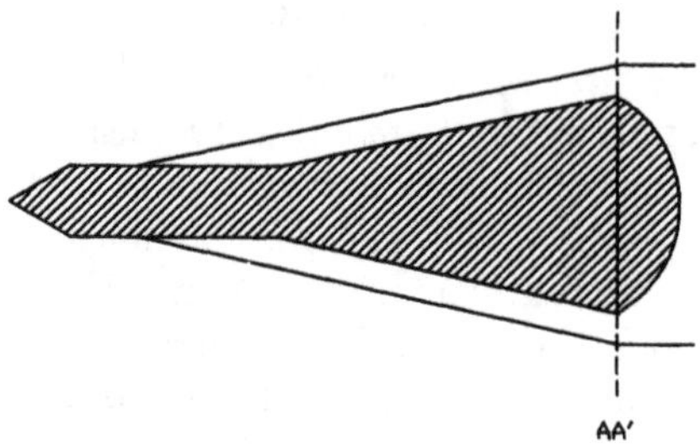

Figure 10.36 Dielectric-loaded horn with lens

10.10.1 Lens profile

The theory of lens design based on geometrical optics is discussed in a number of books [22, 23] and described for lens-corrected horns in Chapter 11. The use of a dielectric lens to improve the radiation characteristics of conical horns has been discussed by Olver and Saleeb [25].

The principle of the design of a lens by geometric optics is to track the optical-ray path through the dielectric horn and lens. If a constant phase front is desired at the aperture, the optical-path length of all rays from the apex of the horn to the aperture must be constant. In practice, the optical-path lengths through the dielectric-loaded horn are not known so that the procedure is to use a program to find the aperture field assuming no lens. This aperture field incorporates the amplitude and phase of the wavefront at the aperture, which can then be used as the input to the geometric-optics design of the lens. There are two possible routes in the design. Either the phase at the output-aperture plane can be made constant so as to yield maximum directivity and efficiency, or the profile can be part of an optimisation loop which attempts to minimise a specific radiation parameter, for instance the crosspolarisation.

If the phase at the aperture plane AA' in Figure 10.36 is spherical (which is always approximately the case), the lens profile can be written in the form

$$r = \left[\frac{1}{n^2} \left\{ F(n-1) + z \right\}^2 - z^2 \right]^{1/2} \tag{10.41}$$

where n is the refractive index of the core with respect to the outer dielectric and

$$F(n-1) = \frac{1}{n-1} \left[n \sqrt{\{(r_{ap} - g)^2 + L^2\}} - L \right] \tag{10.42}$$

g is the gap between the dielectric core and the outer wall and L is the length of the horn.

Alternatively, if $f(r)$ is the computed phase in the aperture plane AA′, then the profile of the lens may be described by the equation

$$z - L = \frac{L(n_0 - n_1) + f(r)}{n_0 \sec \theta - n_1} \tag{10.43}$$

where $\tan \theta = r/L$ and n_0 and n_1 are the refractive indices of the core and outer dielectric respectively.

10.10.2 Design of lens to improve efficiency

The lens can be designed to produce a uniform plane wave of constant phase at the end of the lens. This is demonstrated using the following parameters:

throat radius $r_{th} = 11.50$ mm
aperture radius $r_{ap} = 60.00$ mm
gap g between core and horn wall $= 10.00$ mm
length (without the lens) $L = 240.00$ mm
Frequency of operation $= 10$ GHz
Relative permittivity of core $\varepsilon_0 = 1.75$
Relative permittivity of outer material $\varepsilon_1 = 1.50$.

The performance of this horn–lens combination is given in Table 10.6. Examination of the values in the Table reveals that the horn with the lens is 20.4% longer than that without the lens. The aperture efficiency has increased by 27% with the lens. This is a significant improvement in efficiency. However,

Table 10.6 Characteristics of dielectric-loaded horn with lens to improve efficiency

Parameter	Without lens	With lens
Length, mm	240	289
Peak crosspolar level, dB	−36.4	−31.2
−3 dB beamwidth, deg	19.5	17.7
Directivity, dB	19.7	20.8
Aperture efficiency, percent	60	76

it has been achieved at the expense of an increase in the crosspolar level of 5.2 dB. This is to be expected, because the low crosspolarisation in the dielectric-loaded horn is achieved by cancellation of higher-order modes. The lens has been designed only to improve the HE_{11} phase front and therefore alters the phase relationship for the higher-order modes. Consequently, the cancellation is not quite correct and increased crosspolar radiation occurs.

The transverse aperture-phase variations in the E and H planes are shown in Figure 10.37. The aperture phase is not quite constant. This is because the lens profile is designed using geometric optics. The horn is only 4.5 wavelengths in diameter so geometric-optics design is approximate.

10.10.3 Design of lens to reduce crosspolarisation

It is possible to design a lens which reduces the crosspolarisation by suitably modifying the aperture phase distribution. To design a lens of this type it is necessary to know the aperture phase which will yield low crosspolarisation. At the radiation aperture of the lens, the radius of the core is zero and hence the fields in the aperture can be expressed as a combination of TE and TM modes. From these fields, the aperture fields which produce low crosspolar radiation can be synthetised, since the far field can be expressed in closed form.

Using an empirical approach it is found that a lens profile can be found which produces an aperture distribution that significantly reduces the peak crosspolar level. The profile is given by

$$r = (r_{ap} - g)\cos^m\left\{\frac{\pi(z-L)}{2L_L}\right\} \tag{10.44}$$

where m is an arbitrary positive integer and L_L is the axial length of the lens.

Applying this design to the above horn–lens combination reduces the peak crosspolar level from −36.4 dB to −41 dB. However, the aperture efficiency also reduced from 63% to 56%. Measurements on a large number of trial designs seem to show that a low crosspolar level always reduces the efficiency, although it might be possible using an optimisation loop to find a combination where both high efficiency and low crosspolarisation are satisfied.

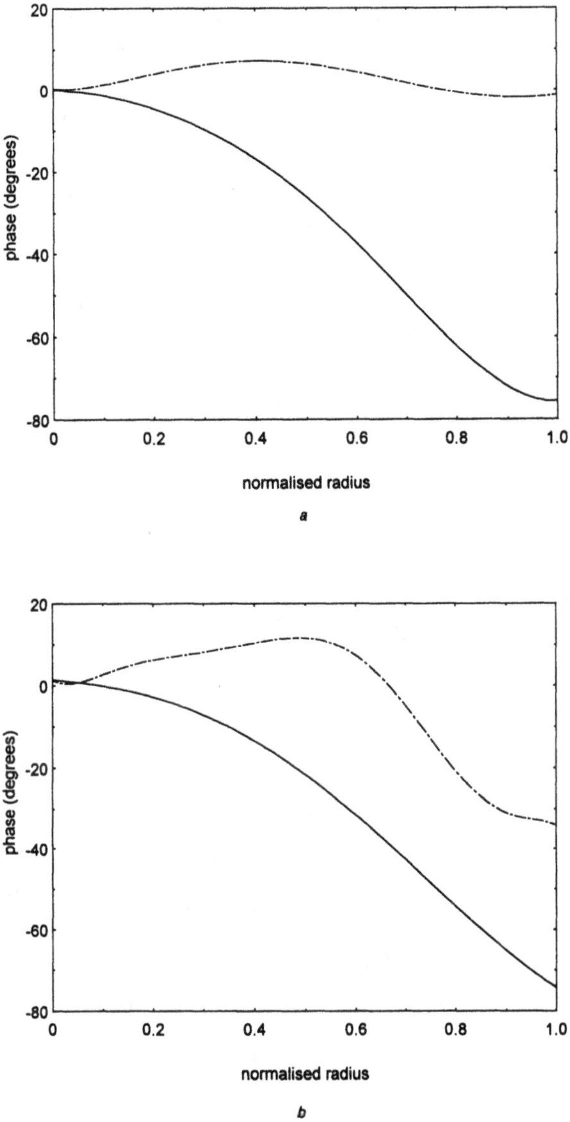

Figure 10.37 *Transverse phase across aperture for horn with lens*

a E-plane
b H-plane
——— no lens
– – – – lens

10.11 References

1 CLARRICOATS P.J.B., OLVER, A.D. and RIZK, M.S.A.S.: 'A dielectric loaded conical feed with low cross-polar radiation'. Proceedings of URSI symposium on Electromagnetic Theory, Spain, August 1983, pp. 351–354

2 CLARRICOATS, P.J.B., and OLVER, A.D.: 'Corrugated horns for microwave antennas' (Peter Peregrinus, 1984), IEE Electromagnetic Waves series 18

3 LIER, E.: 1986, 'A dielectric hybrid mode antenna feed: a simple alternative to the corrugated horn', *IEEE Trans.* 1986, **AP-34**, pp. 21–29

4 RIZK, M., and MORRIS, G.: 'Broadband dielectric loaded hybrid mode horns'. Proceedings of Military Microwaves-86 Conference, Brighton, 1986, pp. 267–272

5 TUN, S.M., PHILIPPOU, G.Y., and ADATIA, N.A.: 'Analysis of rotationally symmetric dielectric-loaded waveguides and horns', *Electron. Lett.*, 1986, **22**(13), pp. 685–686

6 TUN, S.M., PHILIPPOU, G.Y., and ADATIA, N.: 'Analysis and design of rotationally symmetric dielectric loaded waveguides and horns'. Proceedings of 5th international conference on Antennas and Propagation, York, 1987, pp. 438–442

7 TUN, S.M., BUSTAMANTE, R., and WILLIAMS, N.: 'Dielectrically loaded horns' Proceedings of Military Microwaves Conference, 1988, pp. 71–76

8 KNOP, C.M., CHENG, Y.B., and OSTERTAG, E.L.: 'Performance of corrugated and dielectric-loaded metallic-conical horns', *Electron. Lett.*, 1986, **22**, pp. 1253–1254

9 CAHILL, R: 'Effect of edge region on cross polar performance of mm-wave dielectric core horn', *Electron. Lett.*, 1989, **25**, pp. 131–133

10 LIER, E., and RENGARAJAN, S.: 'Radiation from elliptical hybrid-mode waveguides', *IEE Proc. H*, 1990, **137**, pp. 417–419

11 LIER, E.: 'Broadband elliptical beamshape horns with low cross-polarisation', *IEEE Trans.*, 1990, **AP–38**, pp. 800–805

12 LIER, E., and STOFFELS, C.: 'Propagation and radiation characteristics of rectangular dielectric-loaded hybrid mode horn', *IEE Proc. H*, 1991, **138**(5), pp. 407–411

13 NEWHAM, P.: 'A broadband polyrod fed conical horn'. Proceedings of 6th international conference on Antennas and Propagation, Warwick 1989, *IEE Conf. Publ.* 301, pp. 158–162

14 RAGHAVAN, K., OLVER, A.D., and CLARRICOATS, P.J.B.: 'Computer-aided design of dielectric loaded horns'. Proceedings of 5th international conference on Antennas and Propagation, York, April 1987, pp. 434–437

15 OLVER, A.D., CLARRICOATS, P.J.B., and RAGHAVAN, K.: 'Dielectric cone loaded horn antennas', *IEE Proc. H*, 1988, **135**, pp. 158–162

16 OLVER A.D., PEARSON, R.A., PEAKE, G.E.J., and PHILIPS, B.: 'Metallised dual dielectric loaded horn'. Proceedings of international conference on Antennas and Propagation, York, April 1991, pp. 121–124

17 CLARRICOATS, P.J.B.: 'Propagation along unbounded and bounded dielectric rods', *Proc. IEE*, 1961, **108C**, p 177

18 COLLIN, R.E.: 'Field theory of guided waves' (McGraw–Hill, New York, 1960), pp. 182–184.

19 McBRIDE, J.M.W.: 'Properties of dielectric rod junctions in circular waveguide'. PhD thesis, Queen's University, Belfast, 1964

20 OLVER, A.D.: 'Guided electromagnetic wave propagation in radially stratified media'. PhD thesis, University of Leeds, 1966

21 RAGHAVAN, K.: 'Modal matching technique applied to dielectric loaded and smooth-walled conical horns'. Queen Mary College, University of London, internal report, April 1987

22 COLLIN, R.E., and ZUCKER, F.J.: 'Antenna theory—Part 2' (McGraw–Hill, New York, 1969), pp. 104–150

23 SILVER, S. (Ed.): 'Microwave antenna theory and design' (McGraw–Hill, 1949; reprinted Peter Peregrinus, 1984), IEE Electromagnetic Waves series 19
24. OLVER, A.D., and XIANG, J.: 'Design of profiled corrugated horns', *IEEE Trans.*, 1988, **AP–36**, pp. 936–940
25 OLVER, A.D., and SALEEB, A.A.: 'Improved radiation characteristics of conical horns with plastics-foam lenses', *IEE Proc. H*, 1983, **130**, pp. 197–202
26 CAHILL, R.: 'Design of core support mechanism for mm-wave dielectrically loaded horn', *Electron. Lett.*, 1989, **25**, pp. 1248–1249
27 CAHILL, R., and PRIOR, C.J.: 'G-band dielectric core horn', *Electron. Lett.*, 1993, **29**, pp. 130–131

Chapter 11

Lens-corrected horns

A.D. Olver

11.1 Introduction

Horn antennas suffer from the disadvantage of poor aperture efficiency and an inability to control the copolar-radiation characteristics except by altering the diameter of the aperture. This is the case for most of the horns described in the preceding chapters, particularly with short wide-angle horns. In addition to the poor aperture efficiency, the phase centre of wide-angle conical horns changes with frequency of operation. To overcome this problem, the horn must either be made with a narrow flare angle which makes it long, or it must be profiled. An alternative is to place a lens over the aperture of the horn (Figure 11.1). The surface of the lens refracts the radiated energy and collimates it into a narrower beam. In principle, the refraction can be done with either a dielectric lens which has a refractive index greater than unity or a metal-plate lens which has a refractive index less than unity. However the use of metal-plate lenses is rare because the metal plates introduce a polarising effect and their edges cause diffraction. Most lens-corrected horns are therefore composed of dielectric lenses.

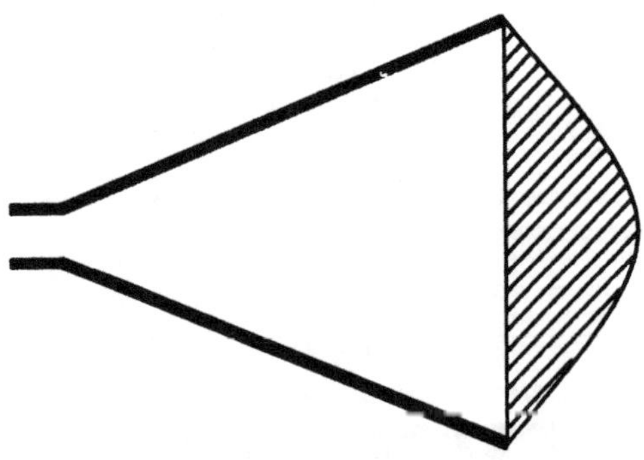

Figure 11.1 Lens-corrected horn

This chapter shows how a lens can be added to the aperture of a conical horn to control the radiation pattern and achieve a high aperture efficiency, potentially good crosspolarisation, good voltage standing-wave ratio (VSWR), and constant phase centre.

The lenses can be made either from solid-plastics dielectrics or from foam-plastics dielectrics. The potentially improved characteristics can be obtained from solid-plastics lenses, but the use of a low-density foam gives an extra degree of freedom because of the ability to alter density and hence the relative permittivity of the foam.

The shape of a lens is chosen to produce a constant phase front across the aperture of the lens. This is achieved using the theory of geometric optics. The use of geometric optics is an approximation, since the lens is placed at the aperture of a horn and therefore in the near field of the antenna. However, alternative design techniques are not yet available and experience shows that the GO design produces acceptable results. The lens will also alter the amplitude distribution, because refraction changes the relative distribution of radial energy. For a solid lens, the amount of change in the amplitude distribution is small. For a plastics-foam lens some control is possible by choosing a value for the relative permittivity of the foam to give a specified amplitude distribution, as is explained below.

Theoretical and experimental studies of lens-corrected horns have been made in the past. An attempt was made by Rust [1] to correct the phase across the aperture of a sectoral horn by using a metal-plate lens. Triangular shapes of Styrofoam were used by Oh *et al.* [2] to alter the phase and power distribution across the aperture of E-plane sectoral horns. Cylindrical dielectric lenses were designed to correct the aperture phase of a sectoral horn by Collin and Zucker [3]. Gunderson and Holmes [4] showed that the aperture and phase of a conical horn could be corrected with solid-dielectric lenses. Clarricoats and Saha [5] computed the radiation pattern of a 40°-semiflare-angle corrugated conical horn when corrected by two types of solid-plastics lenses.

11.2 Theory of lens design

The normal design procedure for lenses is to assume that the principles of geometric optics can be used and to treat the power radiating from the horn apex as rays (Figure 11.2). As has been noted above, this is clearly an approximation since the lens is very much in the near field of the horn. It is however justified to a first order in the main-beam region of the radiation characteristics and the resulting design equations are simple. Experimental results show that the geometric-optics assumptions are surprisingly good. The assumption is also made that the horn has a circularly symmetric radiation pattern resulting in a circular cross-section to the lens. Lenses could, in principle, be designed to compensate for pattern asymmetry. This would lead to lenses with elliptical cross-sections. These have not been used because of the complications of manufacture.

The first principle of geometric optics is that power flows along ray paths and the total power flowing through any given closed surface constitutes a tube of rays. This means that the total power flowing across any cross-section of a ray tube must be constant. This condition can be used to track the strength of the rays. The second principle is Fermat's principle, which states that energy travelling between two points takes the path that makes the optical-path length minimum. Snell's law, which describes the behaviour of optical rays at surfaces, follows from Fermat's principle.

The lens equation will be derived with reference to the one-surface-refracting lens in Figure 11.2. In this lens, all the refraction takes place at the first surface and the ray passes normally through the second surface. It is assumed that the lens is rotationally symmetric so that the design is essentially 2-dimensional. The equation for the lens surface is derived by equating the optical-path length through an arbitrary point P on the surface to an aperture plane with the optical-path length along the axis. Referring to Figure 11.2, this becomes

$$\rho + nl = F + nT \tag{11.1}$$

where F is the focal length, n is the refractive index of the dielectric ($\sqrt{\varepsilon_r}$), and T is the thickness of the lens. Using the co-ordinates (ρ, θ), this equation can be written in polar form as

$$\rho = \frac{(n-1)F}{n \cos \theta - 1} \tag{11.2}$$

This is an equation for a hyperbola of eccentricity n with the origin at one focus. The profiles of other types of lenses can be derived by a similar procedure.

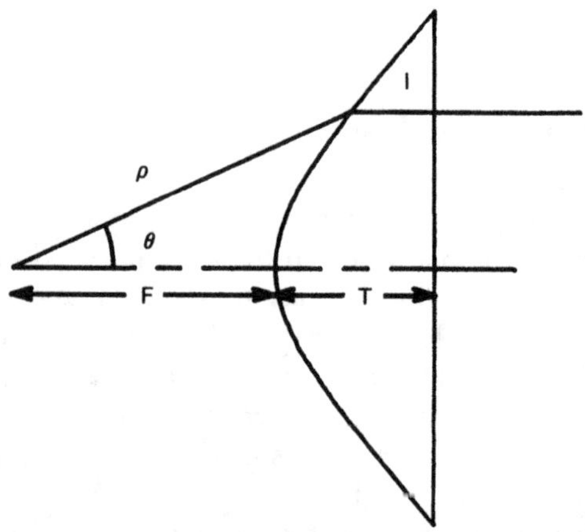

Figure 11.2 One-surface-refracting lens

11.3 Plastics-foam lenses

Lenses are normally constructed from solid plastics. Common plastics are polystyrene with a relative permittivity ε_r of 2.56 and a loss tangent of about 0.0001, and polyurethane with a relative permittivity of 2.4 and a loss tangent about ten times larger than for polystyrene. The use of plastics-foam lenses is much less common, but they offer the antenna designer the important ability to choose the relative permittivity to be any value between unity and the value appropriate to the solid material. This is done by altering the proportions of air and plastics, and hence altering the density of the foam plastics. A linear relationship exists between the density and the relative permittivity. In practice, there is a lower limit to the available density of about 16 kg/m^3, giving a relative permittivity of about 1.02, the exact value depending on the plastics material.

Another factor of importance is the crosspolarisation of a lens-corrected horn. A lens placed in front of a pure-hybrid-mode horn will inherently generate crosspolarisation because of the different transmission properties through the lens of the two planes of polarisation. A lens made of low-density foam reduces this effect to a low level.

The advantages of using foam lenses are not obtained without some potential disadvantages. The foam lens must be homogeneous, requiring care during manufacture. It may be physically large if the relative permittivity is low, although this could be an advantage at millimetre and submillimetre wavelengths. Some environmental protection will normally be needed if the lens is exposed to the air. This can be provided by a thin coating over the plastics-foam surface.

11.4 Shape of lens

The shape of the lens is designed so as to produce a constant phase front at the output aperture. To achieve this aim, four types of lens can be used, as shown in Figure 11.3. Type 1 and type 2 lenses are usually designated single-surface lenses, because all the refraction of energy takes place at one surface of the lens. For type 1 lenses this is at the surface facing the horn. In type 2 lenses, the first surface is spherical, thus matching the wavefront of the horn, and the refraction occurs at the second surface. Type 3 and type 4 lenses are usually designated 'dual-surface lenses' because the refraction takes place at both surfaces. Type 3 lenses are a special case of type 4 lenses but are worth distinguishing because the design equations are much simpler and they are the preferred shape for lens-corrected horns.

Two equations are of interest for each type of lens: an equation describing the profile of the surface; and an equation giving the axial thickness of the lens. The principle of the derivation of the profile of the lenses has been described above

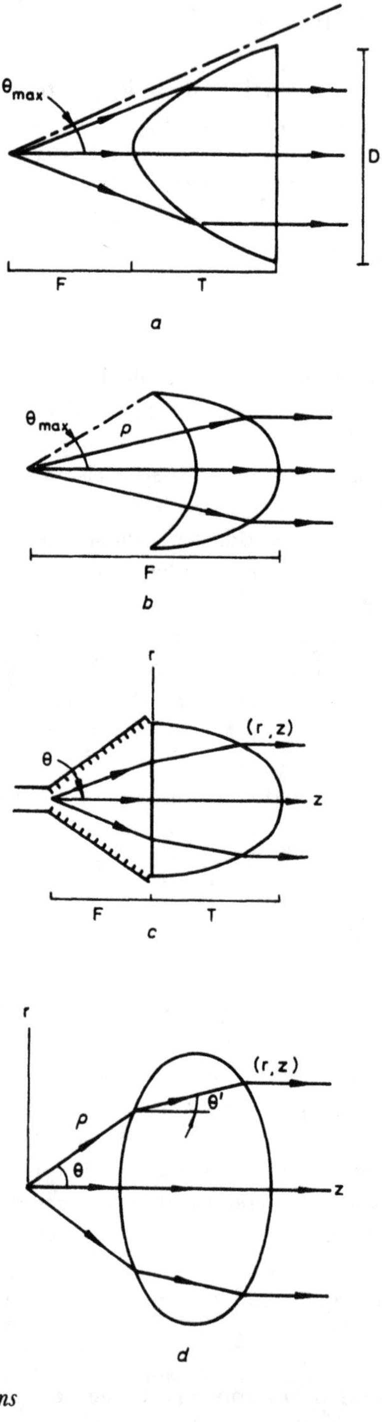

Figure 11.3 Types of lens

a Type 1; *b* Type 2; *c* Type 3; *d* Type 4

and eqn. 11.2 gives the surface contour for type 1 lenses. The ratio T/D of the axial length to the diameter for a type 1 lens is

$$\frac{T}{D}=\sqrt{\left\{\frac{1}{n^2}\left(\frac{F}{D}\right)^2+\frac{1}{4n(n+1)}\right\}}-\frac{1}{n}\frac{F}{D} \tag{11.3}$$

For a type 2 lens the polar equation for the surface profile is

$$\rho=\frac{F(n-1)}{\cos\theta+n} \tag{11.4}$$

and the thickness is

$$\frac{T}{D}=\frac{1}{2\sqrt{(n^2-1)}} \tag{11.5}$$

It is easier to express the shape of type 3 lenses using the transverse radius r of the lens and the longitudinal co-ordinate z. The profile of a type 3 lens is given by

$$r=\frac{F+(n-1)T+(n^2-1)F\sec\theta-FS\tan\theta}{\left(\dfrac{n^2}{\sin\theta}-S\right)} \tag{11.6}$$

and

$$z=(r-F\tan\theta)S \tag{11.7}$$

where

$$S=\sqrt{\left\{\left(\frac{n}{\sin\theta}\right)^2-1\right\}} \tag{11.8}$$

The axial length of a type 3 lens is given by

$$\frac{T}{D}=\left[\sqrt{\left\{1+\frac{1}{(2F/D)^2}\right\}}-1\right]\frac{F/D}{n-1} \tag{11.9}$$

The geometric equations for type 4 lenses cannot be written explicitly; it is necessary to solve a set of simultaneous differential equations, as follows:

$$\frac{d\rho}{d\theta} = \frac{n\,\rho\,\sin(\theta - \theta')}{n\,\cos(\theta - \theta') - 1} \tag{11.10}$$

$$\frac{dr}{d\theta'} = \frac{P(\theta')\sin\theta'\,r_{max}^2}{2r\int_0^{\theta'_{max}} \rho(\theta')\,\sin\theta'\,d\theta'} \tag{11.11}$$

$$\frac{dz}{dr} = -\tan\left\{\theta' + \tan^{-1}\left(\frac{\sin\theta'}{n\cos\theta'}\right)\right\} \tag{11.12}$$

where $P(\theta')$ is the power radiated per unit solid angle in the direction θ' and

$$\theta' = \theta - \frac{\pi}{2} + \tan^{-1}\left(\rho\,\frac{d\theta}{d\rho}\right) + \sin^{-1}\frac{1}{n\sqrt{\left\{1 + \left(\rho\,\frac{d\theta}{d\rho}\right)^2\right\}}} \tag{11.13}$$

The axial length of a type 4 lens is given by

$$\frac{T}{D} = \sqrt{\left\{\frac{1}{n^2}\left(\frac{F}{D}\right)^2 + \frac{1}{4n(n-1)}\right\}} - \frac{1}{n}\frac{F}{D} \tag{11.14}$$

An important limitation of type 1 and type 2 lenses is the relation between the angle θ_{max} and the horn semiflare angle θ_0. The angle θ_{max} is the angle of the asymptote to the type 1 lens and the angle between the axis and the line connecting the focus and the edge of the type 2 lens when the lens is a complete half of an ellipse. This angle must be greater than the semiflare angle of the horn, otherwise the lens can never be in contact with the horn at its aperture. Although it is possible to position the lens in front of the aperture, this is undesirable because of the physical problems of mounting, the increased overall length and the loss of energy due to radiation directly from the horn aperture. The angle θ_{max} is given by

$$\theta_{max} = \cos^{-1}\left(\frac{1}{n}\right) \tag{11.15}$$

For a given horn, there is a lower limit for the angle θ_{max}; hence for the relative permittivity of the material. Figure 11.4 shows the minimum value required for the relative permittivity against the horn semiflare angle. Low-density-foam dielectrics are limited to use with horns having small flare angles, where the benefits of a lens are less obvious. For this reason, the use of type 1 and type 2 lenses for pattern shaping is not considered further, although they can be used to improve efficiency.

Type 4 lenses are more difficult to design and construct than type 3 lenses, and have no particular advantages. Thus a type 3 lens seems the best choice from the aperture-phase requirements. For a type 3 lens, the ratio of focal length to diameter is related to the horn semiflare angle by

$$\frac{F}{D}=\frac{1}{2}\cot\theta_0 \tag{11.16}$$

so it can be seen from eqn. 11.9 that the axial length of the lens is a function only of the dielectric constant of the lens material. This is plotted in Figure 11.5 for four values of θ_0. As the horn semiflare angle increases, a longer lens is required to turn the spherical phase cap into a linear phase front. The practical limit to the length will depend on mechanical constraints and particularly on the frequency of operation. At millimetere wavelengths, short F/D ratios (i.e. long lenses) may be an advantage and ease the manufacture.

The plane phase front at the output aperture of the lens is independent of frequency, as long as the power at the horn aperture can be considered to possess a spherical phase front. This means that the phase centre of the lens plus horn does not move with frequency, an important additional advantage in some feed/reflector combinations.

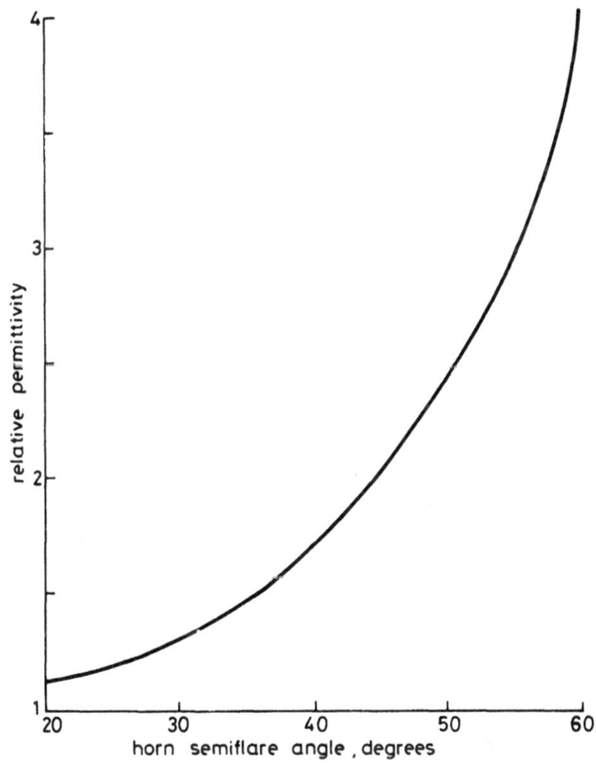

Figure 11.4 Relative permittivity of type 1 and type 2 lens materials against semiflare angle of horn

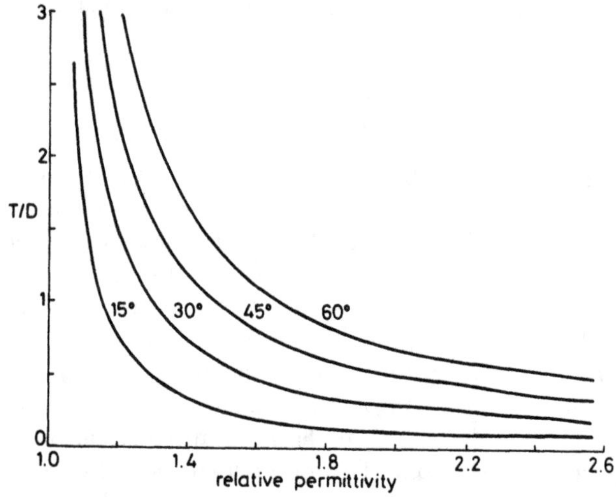

Figure 11.5 Normalised axial thickness of lens against relative permittivity
Parameter: horn semiflare angle

11.5 Zoned lenses

The thickness of a solid-dielectric lens can be reduced by zoning the lens to produce the profile shown in Figure 11.6 for a type 3 lens. The surface of the lens is divided into zones with the optical-path length differing by integral

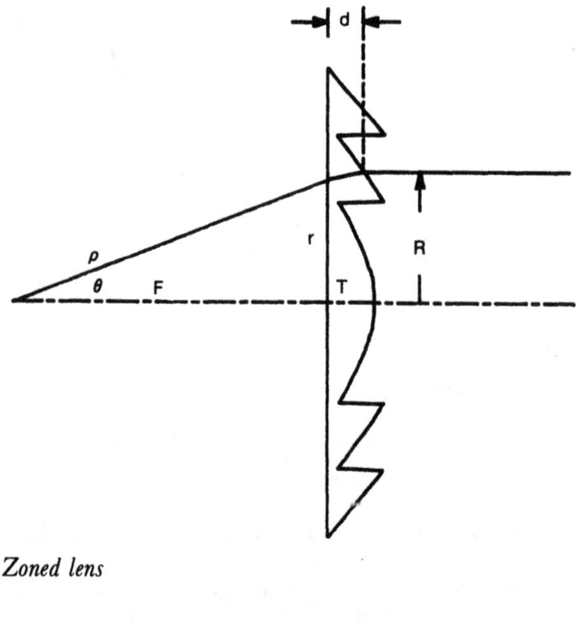

Figure 11.6 Zoned lens

multiples of a wavelength from one zone to the next zone. The design starts at the outer edge and moves towards the axis until the phase difference introduced by the lens is equal to just over a wavelength. At this point, the path length can be reduced by one wavelength and the output phase front will not be altered. This procedure then continues until the axis is reached. The maximum thickness of a zoned lens is approximately equal to $\lambda/(n-1)$ because of the phase delay introduced by the dielectric material. There will be a fixed thickness for mechanical stability, in addition to the zoned thickness. The design procedure is identical to that of the smooth-surface lens with the exception of the zoning. The equations of the shape of the surface are a modification of those for the smooth lens (eqns. 11.6 and 11.7) [6]. Using the notation of Figure 11.6,

$$z=\frac{(K-1)\lambda+(n-1)T+F-\rho}{\frac{1}{S}\left(\frac{n}{\sin\theta}\right)^{2}-1} \tag{11.17}$$

and

$$R=\rho\sin\theta+\frac{d}{S} \tag{11.18}$$

where $K=1, 2, 3\ldots$ from the axis of the lens and S is defined in eqn. 11.8.

A zoned lens is inherently a frequency-selective system because the design philosophy depends on wavelength. The bandwidth is inversely proportional to the number of zones across the lens surface. Zoned lenses have been quite widely used at microwave and lower-millimetre-wave frequencies where the mass of a smooth-profile solid lens would significantly increase the overall weight of the antenna.

11.6 Aperture amplitude distribution

The amplitude of the transverse field across the output aperture of the lens–horn combination is influenced by four factors:

(i) the horn aperture distribution without the lens
(ii) reflections at the surfaces of the lens
(iii) dielectric loss in the lens
(iv) the shapes of the lens surfaces.

The aperture distribution of the horn has a taper from the axis to the edge of the horn. The amount of amplitude taper depends on the horn dimensions and flare angle. In some cases it can be considerable, resulting in low aperture efficiency, particularly for wide-flare-angle conical corrugated horns.

The reflections from the surface of the lens are a function of the angle of incidence of the rays onto the lens surface (except for type 2 lenses). The amount of power reflected and transmitted can be computed using standard reflection formulae at a dielectric interface. The reflection of power at a surface for normal incidence is given by the well known formula $\rho=(n-1)/(n+1)$. For a typical solid dielectric with a relative permittivity of about 2.2, the maximum amount of power reflected is -14 dB, equivalent to a VSWR of 1.48. This is substantial, and the surface needs to be treated. For foam-dielectric lenses the reflections are much less and can generally be ignored in comparison with the other factors. For example if $\varepsilon_r=1.1$, the VSWR is 1.04 and, for $\varepsilon_r=1.4$ the VSWR is 1.18. A lens can be matched to free space by some form of quarter-wave transformer, such as adding corrugations to the lens surface. However, this makes the lens frequency selective.

The dielectric loss which takes place in the plastics is a function of three components: distance, the loss tangent of the dielectric and the refractive index of the dielectric. The distance travelled by power through the lens varies across the lens, being greatest at the centre and least at the edges. The maximum power loss in decibels in given by

$$\alpha=8.686\pi\frac{T}{\lambda}n\tan\delta \tag{11.19}$$

A typical attenuation for a solid-polystyrene lens of one-wavelength axial thickness is 0.004 dB, which is negligible. For a foam lens of length 5λ made from a polyurethane foam with $\varepsilon_r=1.4$ and $\tan\delta=0.003$, the attenuation is 0.48 dB. This is a worst-case value since foams can be obtained which have a tenth of this loss. However, the loss can be put to good use in some antennas, since the effect on the aperture amplitude distribution is to reduce the amplitude taper and hence improve the aperture efficiency.

The major factor influencing the amplitude distribution is the shape of the lens surface. This is because of the collimating influence of the lens profile. Energy diverging from the horn apex decreases in intensity as $1/\rho^2$. After refraction by the lens the rays become parallel, and no more spreading of the energy takes place. Referring to Figure 11.7, the power radiated by the feed through the solid angular ring formed by rotating $d\theta$ around the lens axis is given by

$$dP=2\pi\sin\theta\,P(\theta)\,d\theta \tag{11.20}$$

This power is equal to the power across the aperture ring

$$dP=2\pi rP(r)\,dr \tag{11.21}$$

Equating these powers gives

$$\frac{P(r)}{P(\theta)}=\frac{\sin\theta}{r}\frac{d\theta}{dr} \tag{11.22}$$

This equation gives the transverse amplitude distribution within the radius of the lens. It breaks down near the edge of the lens, where the geometric-optics assumption fails.

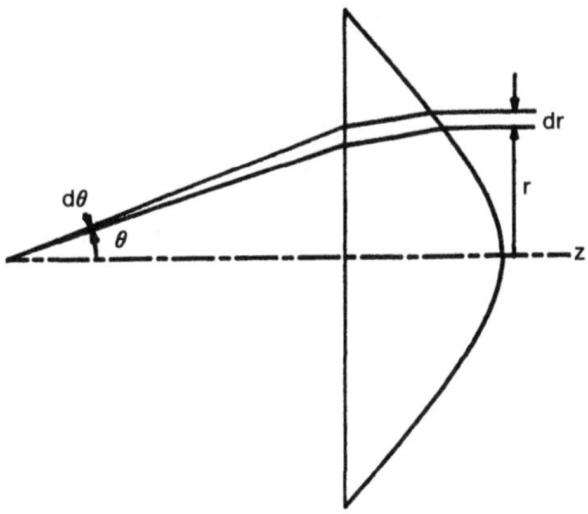

Figure 11.7 Influence of lens surface on aperture distribution

The transverse amplitude distribution for a type 1 lens can be found by inserting eqn. 11.2 into eqn. 11.22 and evaluating. This gives

$$\frac{P(r)}{P(\theta)} = \frac{(n \cos \theta - 1)^3}{F^2(n-1)^2(n - \cos \theta)} \tag{11.23}$$

The amplitude distribution for a type 2 lens is

$$\frac{P(r)}{P(\theta)} = \frac{(n \cos \theta - 1)^3}{F^2(n-1)^2(n \cos \theta - 1)} \tag{11.24}$$

For type 1 lenses, the refraction occurs at the first surface and the shape increases the amplitude taper. This is shown in Figure 11.8, and the effect of the lens is to concentrate the energy towards the axis at the aperture plane and hence reduce the efficiency of the antenna. For type 2 lenses, refraction occurs at the second surface and the amplitude taper is reduced, as shown in Figure 11.9 for a relative permittivity of 2.28. In this case the result of the horn plus lens is that the transverse amplitude distribution is more uniform after the lens.

For a type 3 lens, refraction occurs at both surfaces but the dominant surface is the second surface so that the amplitude distribution is similar to that of a type 2 lens. A closed-form expression cannot be obtained for the transverse distribution of a type 3 lens. The expressions in eqns. 11.6 and 11.7 must be used to compute eqn. 11.22 numerically. The factor $P(r)/P(\theta)$ increases as θ or r is increased; the amount depends on the relative permittivity of the lens. Figure 11.10 shows the combined result of this factor and the horn-amplitude distribution for a hybrid-mode horn with semiflare angle of 30°. As the relative permittivity is reduced from the value for a solid, the output-amplitude-aperture taper becomes less. At a particular value, it is nearly constant and as the relative permittivity further decreases the amplitude distribution has an inverse characteristic which peaks at some angle before decreasing to the horn edge. The value of relative permittivity which gives a nearly constant aperture

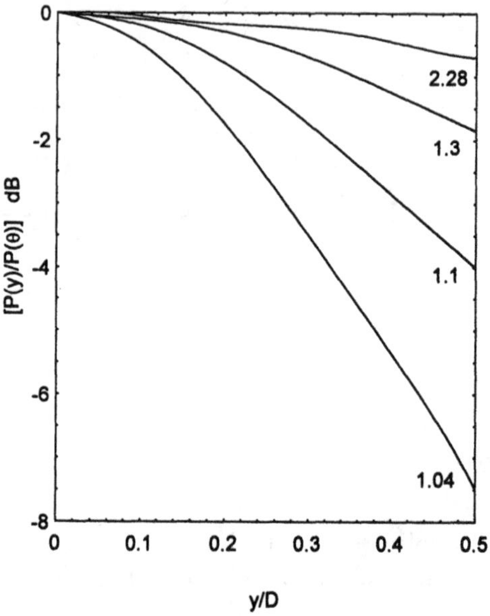

Figure 11.8 *Normalised transverse-power increase owing to type 1 lens*
$F/D = 2.69$
Parameter: relative permittivity

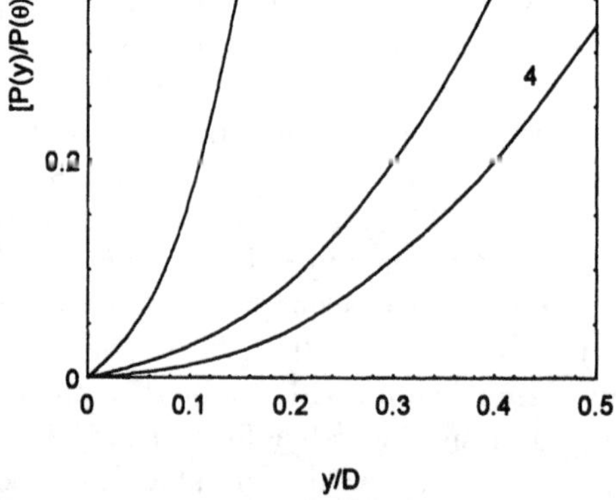

Figure 11.9 *Normalised transverse-power increase owing to type 2 lens*
$\varepsilon_r = 2.28$
Parameter: F/D

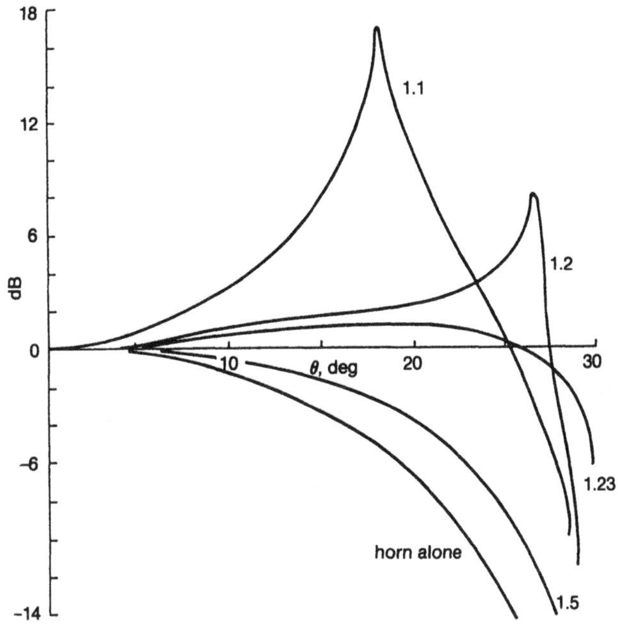

Figure 11.10 *Transverse aperture distribution for a type 3 lens with a 30°-semiflare-angle hybrid-mode horn*

Parameter: relative permittivity

distribution is of particular interest, since this value can be expected to give maximum aperture efficiency. It is plotted in Figure 11.11 as a function of horn semiflare angle. This situation is particularly notable, because there are very few horns, or indeed any other type of antenna, where it is possible to obtain a nearly constant field across the aperture.

11.7 Radiation characteristics

The far-field-radiation pattern can be calculated by evaluating the aperture distribution using the above information and then using the Kirchoff–Huygens formulae [6]. This method is appropriate because it assumes that the radiating aperture is in a surface of infinite extent. The fields over the surface are assumed to be zero everywhere except over the aperture area. While this is not completely true, since the field will exist outside the lens aperture, it is a good approximation. For a circular aperture with uniform phase and aperture power distribution $P(r)$, symmetric about the axis, the far field is given by

$$f(u) = \pi \frac{D^2}{2} \int_0^{D/2} \surd\{P(r)\} J_0(ur) r \, dr \tag{11.25}$$

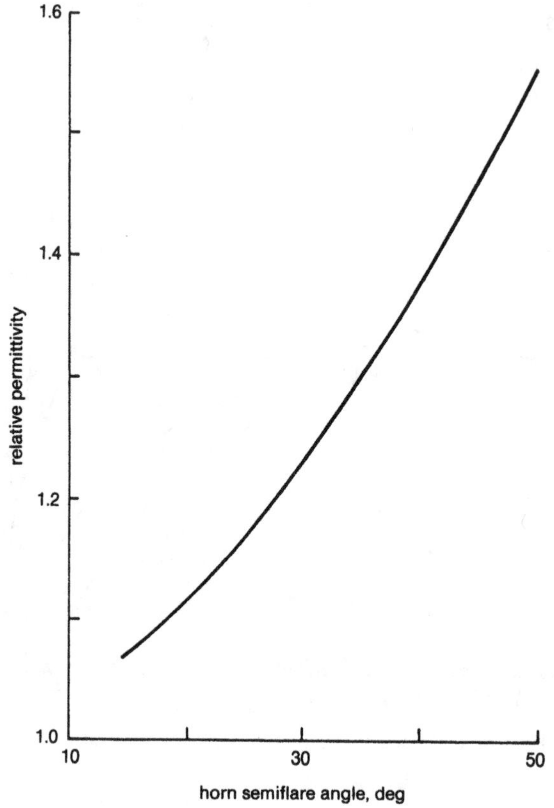

Figure 11.11 Value of relative permittivity which gives uniform transverse aperture distribution

where $u = (\pi D/\lambda) \sin \theta$.

The radiation patterns have been computed for a lens-corrected wide-angle corrugated horn using a type 2 solid lens and a type 3 foam lens. The aperture distribution of the horn alone was computed using a spherical-wave-expansion technique (References 7 and 8 and Chapter 4) with the assumption of zero crosspolarisation. Figure 11.12 shows typical patterns for a 30°-semiflare-angle horn of 5λ aperture diameter and a lens. The main beam is much narrower with the lens than without the lens, but sidelobes are present. The increase in gain is about 5 dB for both types of lens. The reason why both types give the same gain increase is that the high sidelobes in the type 3 lens detract from the narrower main beam.

A parametric study of the horn plus type 3 foam lens is shown in Figures 11.13–11.15. They show how the −3 dB halfbeamwidth, first-sidelobe level and directivity vary as the horn flare angle and relative permittivity of the lens are varied. The curves have been stopped at the minimum value of relative permittivity, where geometric optics break down. As the relative permittivity of the lens decreases, the beamwidth decreases but the sidelobe level increases.

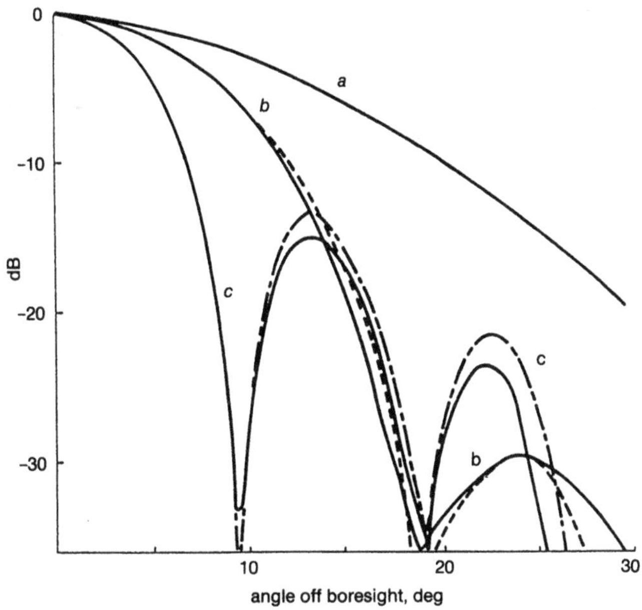

Figure 11.12 *Far-field-radiation pattern of 30°-semiflare-angle horn plus lens*

a Horn without lens
b Horn with type 2 lens
——— measured
- - - - theory
c Horn with type 3 lens
——— measured
- - - - theory

The minimum halfbeamwidth achievable is 3.8° when the sidelobe level is −15.5 dB. As expected, the uniform-aperture-distribution cases give a halfbeamwidth of 5.7° and sidelobe level of −17.6 dB. The directivity in all cases is substantially greater than the value appropriate to the horn alone, and peaks at the values of relative permittivity which give uniform aperture distribution.

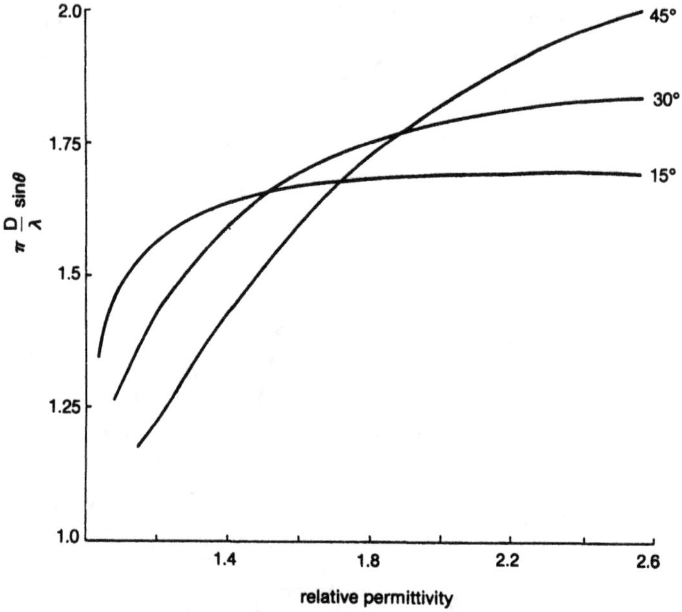

Figure 11.13 −3 dB halfbeamwidth against relative permittivity
Parameter: horn semiflare angle

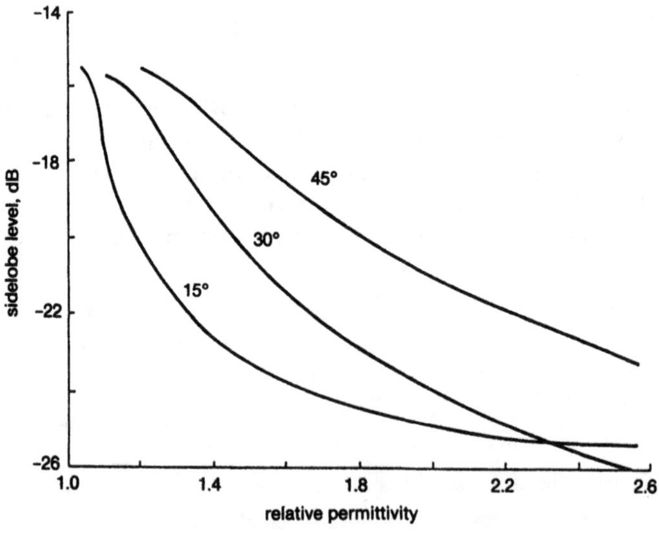

Figure 11.14 First-sidelobe level against relative permittivity
Parameter: horn semiflare angle

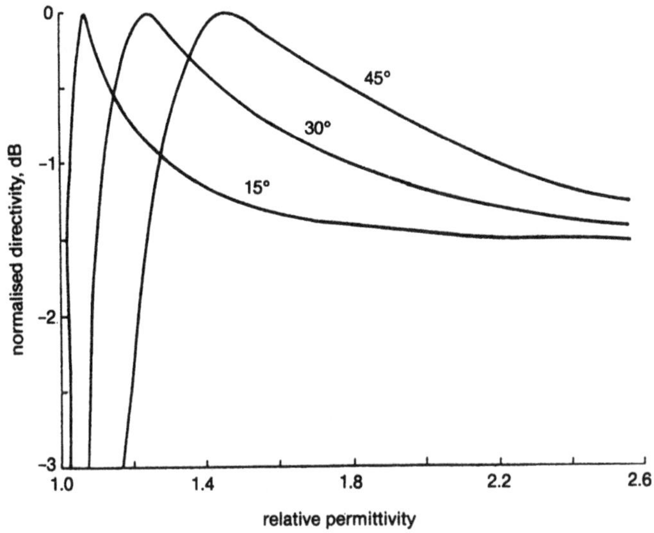

Figure 11.15 *Directivity relative to the directivity of a uniform illuminated aperture against relative permittivity*

Parameter: horn semiflare angle

11.8 Experimental lenses

The theory of lens-corrected horns has been verified by designing, manufacturing and testing two lenses. A type 3 lens was made from low-density expanded-polystyrene foam with an average relative permittivity of 1.13, and for comparison a type 2 lens was made from solid polyethylene with a relative permittivity of 2.26. Both lenses were designed at 10 GHz for a 30°-semiflare-angle conical corrugated horn with an aperture diameter of 15 cm, which has been shown by measurement to have a very low crosspolarisation. The type 3 lens had to be made from existing blocks of commercial foam. The available blocks had different densities, so a modification to the design procedure of Section 11.2 was used by calculating the refraction angle at each interface using geometric optics. The shape and composition of both lenses are sketched in Figure 11.16.

The measured radiation patterns for both experimental horns and lenses are shown in Figure 11.12. Agreement with theory is good, especially over the main-beam region. This seems to justify the use of geometric optics to predict the aperture-field distribution. The measured VSWRs of the horn alone and the horn plus lenses are shown in Figure 11.17. As expected, the solid lens increases the VSWR considerably, whereas the foam lens causes only a small increase in return loss. The crosspolarisation characteristics, in the 45° plane, of horn plus

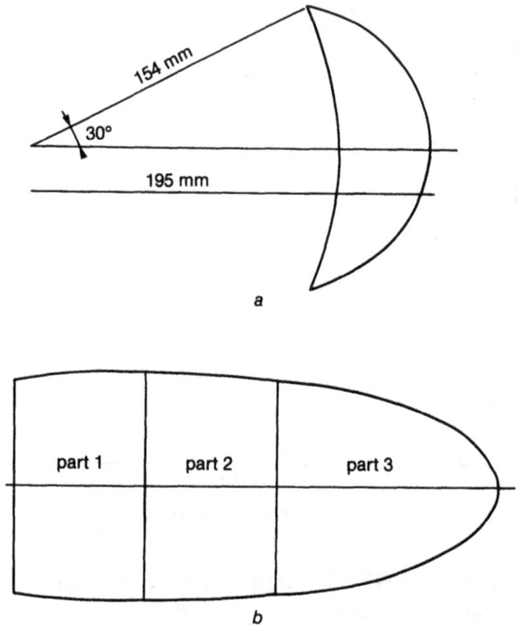

Figure 11.16 Type 2 and type 3 lenses

a Type 2
b Type 3

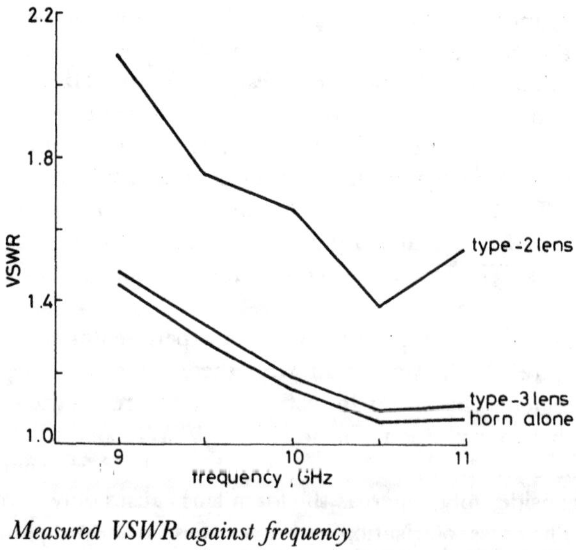

Figure 11.17 Measured VSWR against frequency

lenses were measured over a band of frequencies and compared with the values for the horn alone. The peak crosspolar level is plotted against frequency in Figure 11.18. The type 3 lens increases the peak crosspolar level by only a few decibels, thus maintaining the excellent performance of the corrugated horn without the lens. The type 3 lens has about a 22% bandwidth over which the peak crosspolar level is less than −40 dB. It is worth comparing the size of horn which would be required to obtain the same performance without the lens. A corrugated horn with a 5° semiflare angle, an aperture diameter of 10λ and a length of 57λ gives the same pattern performance as a 30° corrugated horn plus type 3 lens of aperture diameter 5λ with a length of 17λ. The lens generates a crosspolar component even if the horn is perfect. This is because the surface of the lens reflects and refracts orthogonal components of polarisations by differing amounts. The aperture efficiency of the 30° corrugated horn by itself is 52%, and this value does not allow for the area occupied by the flange. The addition of the type 3 lens raises the aperture efficiency to 93%, and with an optimally designed lens this could be raised to nearly 100%.

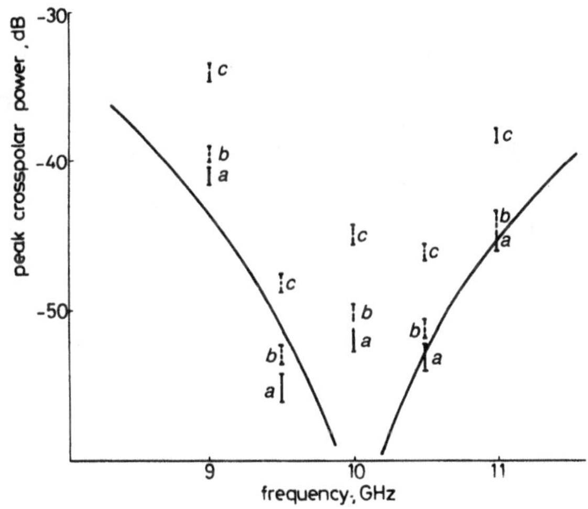

Figure 11.18 Measured and computed peak crosspolar radiated power

a horn alone
b horn plus type 3 lens
c horn plus type 2 lens
——— computed for horn alone

11.9 Uses of lens-corrected horns

There are many potential uses for a lens-corrected horn, particularly for the combination of a horn plus a foam lens. The lens enables the designer to control the copolar-radiation-pattern characteristics. A field with a uniform amplitude and phase can be generated just in front of the lens aperture. As a feed for a Cassegrain-reflector antenna, the horn plus lens gives a compact high-performance antenna. One of the most promising uses is at millimetre and submillimetre wavelengths. The construction of corrugated horns at wavelengths of a few millimetres is difficult and expensive. Short horns plus lenses are easier to manufacture. The surface tolerance required for a foam lens is much less stringent than for either a solid lens or the horn alone. At these wavelengths, the design principles discussed in this chapter should enable accurate designs to be produced.

11.10 References

1 RUST, N.M.: 'The phase correction of horn radiators', *J. IEE*, 1946, **93**(111A), pp. 50–51
2 OH, L.L., PENG, S.Y., and LUNDEN, D.C.: 'Effect of dielectrics on the radiation pattern of electromagnetic horn', *IEEE Trans.*, 1970, **AP–18**, pp. 553–556
3 COLLIN, R.E., and ZUCKER, F.J.: 'Antenna theory' (McGraw–Hill, 1969)
4 GUNDERSON, L.C., and HOLMES, G.T.: 'Lenses improve horn antennas' *Microwaves*, 1971, **10**, pp. 33–35
5 CLARRICOATS, P.J.B., and SAHA, P.K.: 'Radiation pattern of lens-corrected conical scalar horn', *Electron. Lett.*, 1969, **5**, pp. 592–593
6 SILVER, S. (Ed.): 'Microwave antenna theory and design' (McGraw-Hill, 1949; reprinted Peter Peregrinus, 1984), IEE Electromagnetic Waves series 19
7 CLARRICOATS, P.J.B., and SAHA, P.K.: 'Propagation and radiation behaviour of corrugated feeds', *Proc. IEE*, 1971, **118**, pp. 1167–1176
8 CLARRICOATS, P.J.B., and OLVER, A.D.: 'Corrugated horns for microwave antennas' (Peter Peregrinus, 1984), IEE Electromagnetic Waves series 18

Chapter 12
Dielectric feeds

L. Shafai and A.A. Kishk

12.1 Introduction

Introducing dielectric materials to the design of reflector feeds adds an additional parameter for optimisation. Several advantages can be gained, the most important of which is the fact that dielectric rods can support the HE_{11} mode and consequently improve the pattern symmetry. To benefit from this property of dielectrics, the feed horn must have a reasonable size, in aperture or length, to support the excitation of the HE_{11} mode efficiently. Since their properties are discussed in Chapter 10, this chapter will concentrate on small feeds which contain dielectrics. Two main types of feed will be considered here: the dielectric-rod feed, and waveguide and horn feeds which use dielectric for pattern shaping. In these feeds, the introduction of dielectric material helps with pattern symmetry and when used with waveguides reduces aperture size. The latter results in broadening of the radiation patterns and, consequently, low-cost small feeds can easily be designed that provide satisfactory performance.

Historically, the use of a dielectric rod as a reflector feed was first reported by Dombek [1]. Later the performance of a short dielectric rod, fed by a waveguide and enclosed by a small cavity, was experimentally studied by Kumar [2]. The use of short dielectric rods as reflector feeds was also studied by Ittipiboon *et al.* [3], who optimised the rod dimensions experimentally. Their result was later used by Shafai *et al.* in a coaxial waveguide to design a dualband feed for operation in the 4–6 GHz bands [4]. A general study of these antennas by a numerical method was carried out by Kishk and Shafai [5]. The moment method described in Chapter 5 was used for the study. Loading a waveguide with dielectric materials to improve the feed-radiation pattern was investigated by Raghavan *et al.* [6], who showed that satisfactory feed characteristics can be obtained using a variety of dielectric configurations.

In this chapter some of these feeds will be discussed and some representative data will be provided. The computed results are generated using the moment-method solution of Chapter 5. To indicate the range of solution accuracy, results using multiple formulations are also provided and discussed (see Chapter 5).

12.2 Dielectric-rod feed

The dielectric-rod antenna consists of a dielectric cylinder excited by a microwave guide such as a hollow waveguide, as shown in Figure 12.1. To solve the problem, a variety of approximate methods have been used to predict theoretically the experimentally observed radiation patterns. In 1967, James reviewed four different theoretical methods [7], namely Horton's analysis [8, 9], Bouix's analysis [10], Fradin's analysis [11] and Brown's analysis [12]. In Reference 8, Watson and Horton applied Schelkunoff's equivalence principle to the fictitious currents on the surface of a dielectric-rod antenna of rectangular cross-section, which gave an encouraging agreement with the experimental results. In Reference 9, Horton applied the theory to a uniform cylindrical antenna supporting a symmetrical transverse magnetic wave. In Reference 10, the radiation from the end of a uniform dielectric rod carrying HE_{11} mode was neglected, and only that from the radial surface was evaluated. In Reference 11, Fradin asserted that very little radiation occurs at the end of the rod, so it was neglected in his analysis, and he transformed the uniform dielectric rod into a fictitious volume of electric currents, using the second equivalence principle. In Reference 12, Brown and Spector asserted that the role of the uniform dielectric rod is to carry a surface wave to the end of the rod, where it radiates at wave discontinuities. Yaghjian and Kornhauser [13] used a hybrid-modal solution for a circular semi-infinite dielectric-rod antenna excited by the HE_{11} mode. Their results indicated an excellent pattern symmetry and increasing gain with increasing relative permittivity. However, the method is not useful for short dielectric rods. Using a twofold application of Kirchhoff integrals, Dombek [1] found an approximate solution. Initially, the waveguide aperture was assumed to be illuminated by the TE_{11} mode, which was excited by assuming that the dielectric is of infinite extent. This near-field distribution was subsequently used to compute the far field. For short rods, it is important to consider the effect of the waveguide surface which will help in predicting the sidelobes and the backlobes of the radiation patterns.

For a finite length of the dielectric rod, the pattern symmetry is not always possible, but can be controlled by its length or by tapering the dielectric rod at its ends. Experimental studies have shown that compact rod antennas can be designed to illuminate a paraboloid reflector efficiently [1,3,5]. For finite

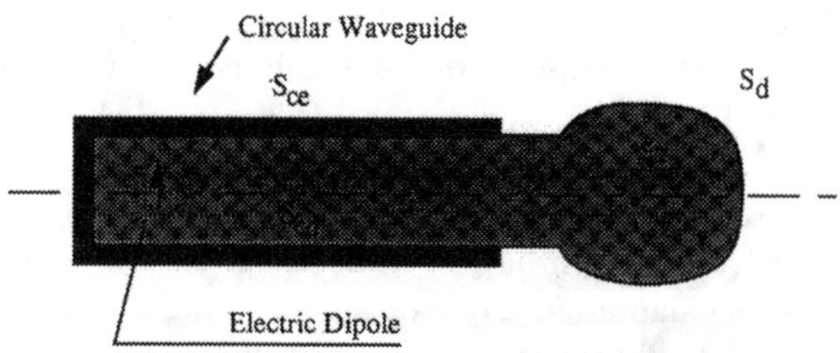

Figure 12.1 Dielectric-rod-antenna geometry

dielectric rods the problem is difficult to solve analytically, and a numerical method must be utilised.

There are many parameters to be studied in this problem and the existence of a numerical code allows their effect to be investigated. However, it is difficult to analyse or present the results for all possibilities. Here, only the radiation characteristics of a cylindrical dielectric rod with $\varepsilon_r = 2.5$, diameter $D = 0.5\lambda$ and extended a distance $L(\lambda)$ from the open end of the waveguide will be considered in detail. Other cases will be discussed briefly.

The example of a dielectric rod with $\varepsilon_r = 2.5$ and a diameter $D = 0.5\lambda$ is considered to compare its results with the experimental results of Dombek [1]. From the radiation patterns of the dielectric rod with different lengths L, the beamwidths are calculated at 20, 10, and 3 dB levels as shown in Figure 12.2, with L varying from 0 to 2λ. This Figure indicates excellent agreement with the experimental results. It is seen that the beamwidths are decreased by increasing

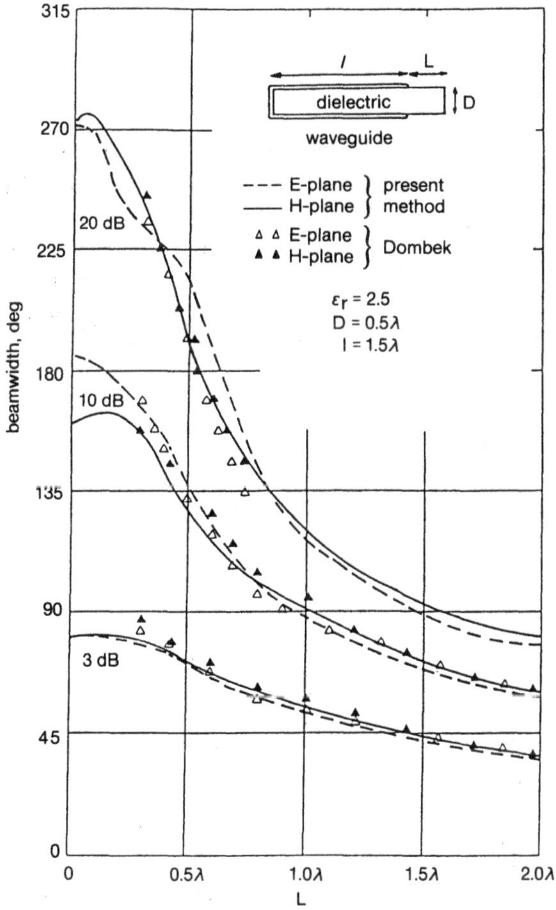

Figure 12.2 Beamwidths of dielectric-rod patterns at 3, 10 and 20 dB levels

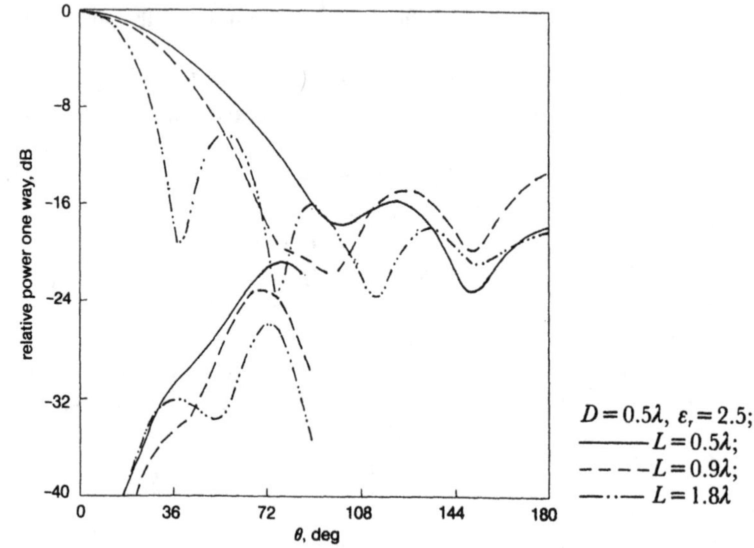

Figure 12.3 E-plane- and crosspolarisation-radiation patterns

L, owing to the excitation of the surface waves on the dielectric-rod surface. This means that the directive gain of this antenna increases as L increases. The disadvantage of this rod antenna is that its radiation pattern has a high backlobe level which may be reduced by a quarter-wavelength choke on the waveguide wall. Utilising this approach, it is found that the backlobe level reduces, on average, in excess of 10 dB. The radiation patterns are shown in Figures 12.3 and 12.4 in both E- and H-planes for different extensions L,

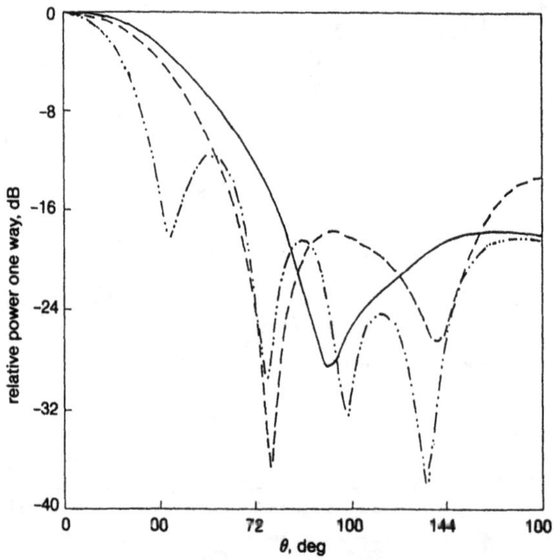

Figure 12.4 H-plane-radiation patterns for the examples in Figure 12.3

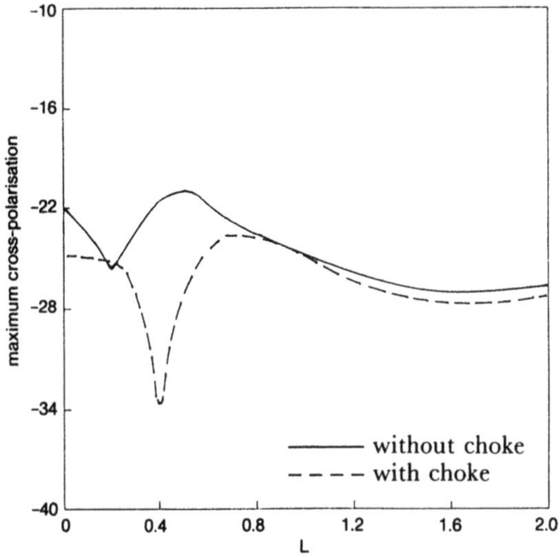

Figure 12.5 Peak crosspolarisation levels for the examples in Figure 12.2

respectively. One of the important results to consider is the peak crosspolarisation. This antenna type provides a symmetric radiation pattern, and the maximum crosspolarisation is expected to be low, below −22 dB, in Figure 12.5. However, despite the amplitude-pattern symmetry, the crosspolarisation level is high. This is due to the phase differences between the patterns of the principal planes, which are found to be large.

Considering this antenna as a feed, the gain factor and the spillover efficiency are calculated and shown in Figure 12.6. It is noticed that both the gain factor

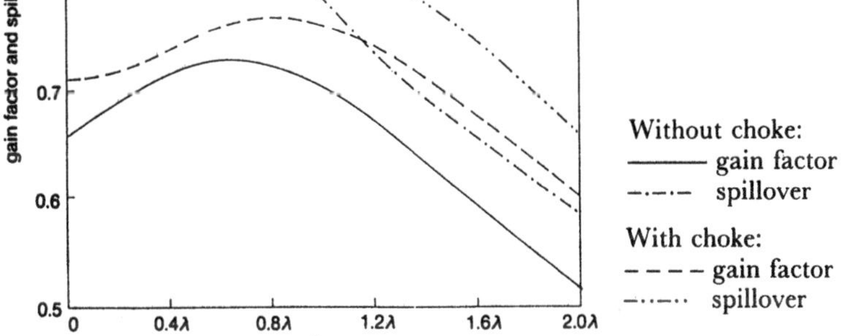

Figure 12.6 Gain factors and spillover efficiency for the examples in Figure 12.2

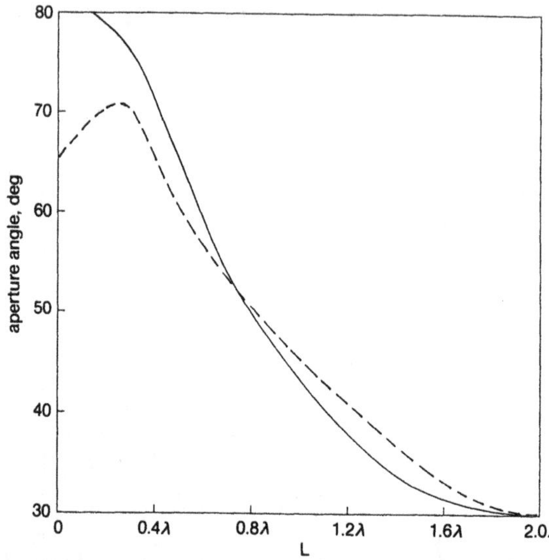

Figure 12.7 Aperture angles corresponding to the maximum gain factors in Figure 12.6
——— without choke; – – – – with choke

and spillover efficiency improve by the introduction of the choke on the waveguide wall. Both the dielectric extension *L* and the choke affect the required reflector-aperture angle for optimum gain or aperture efficiency. The computed results are shown in Figure 12.7.

Most popular feeds for paraboloid reflectors have a well defined phase-centre

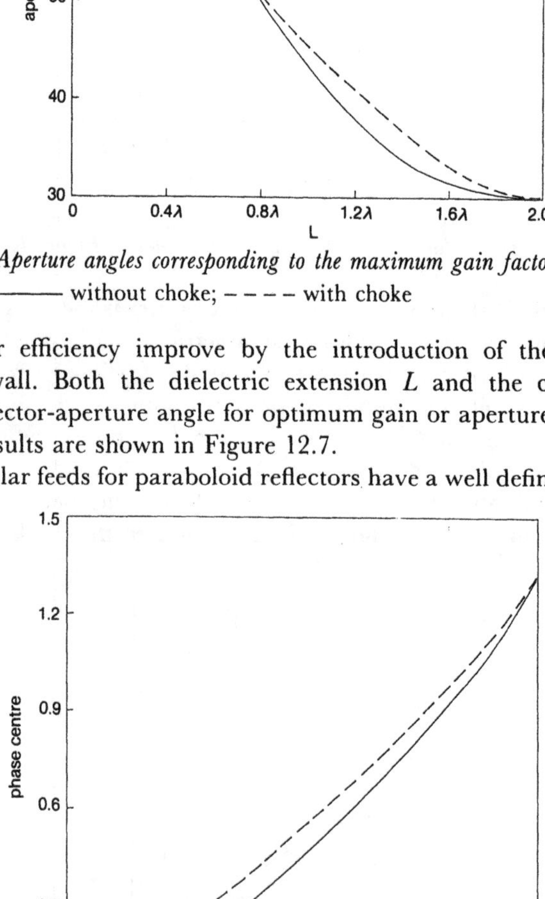

Figure 12.8 Phase centre locations for the examples in Figure 12.2 within the aperture of Figure 12.7
——— without choke; – – – – with choke

location. Knowledge of the phase-centre location of the feed is important since it must coincide with the reflector focus to achieve the optimum gain. The phase centre of the dielectric rod is calculated for $\theta \leqslant \psi_0$ as in Reference 15. Figure 12.8 shows the phase-centre separation distance from the waveguide aperture, calculated as mentioned above for $0 \leqslant \theta \leqslant \psi_0$, i.e. within the range of reflector-aperture angles shown in Figure 12.7. It can be seen that the phase centre is moved far from the waveguide open end by increasing L. Adding the choke also moves the phase centre away from the waveguide aperture. With increasing L, the influence of the choke becomes weaker and phase-centre locations with or without a choke approach each other.

12.2.1 Dielectric-diameter effect

We saw in Chapter 7 that, for hollow circular waveguides, the diameter has an important effect on the radiation patterns and the crosspolarisation level. For the dielectric rod, an example is selected where $L = 0.5\,\lambda$ and $\varepsilon_r = 2.5$. For different dielectric and waveguide parameters, the results are shown in Table 12.1. It can be seen that the peak crosspolarisation decreases as the diameter increases until $a = 0.35\,\lambda$, where it achieves the minimum peak crosspolarisation. Afterward, it increases with a. Also, as the diameter increases, the phase centre moves towards the dielectric-rod end and eventually outside. Of course, the performance may be different for other dielectric constants or extensions L.

Table 12.1 Effect of changing waveguide radius

a	ψ_0	Peak crosspolarisation $0 \leqslant \theta \leqslant \psi_0$	Peak crosspolarisation $0 \leqslant \theta \leqslant 90°$	Gain (dB)	η_p (%)	η_s (%)	$d_c(\lambda)$ $0 \leqslant \theta \leqslant \psi_0$
0.25	66	−21.98	−20.28	8.59	74.4	87.7	0.167
0.30	60	−27.98	−26.52	9.42	77.8	90.4	0.229
0.35	54	−41.50	−34.81	10.00	76.0	86.6	0.279
0.40	54	−31.00	−31.00	10.38	75.5	88.0	0.296
0.45	90	−13.10	−13.10	9.27	60.5	79.5	0.609

$\varepsilon_r = 2.5$, $L = 0.5\lambda$

12.2.2 Dielectric constant

For $L = 0.5\lambda$ and $a = 0.25\lambda$, the results were computed for ε_r changing from 2.5 to 5. Table 12.2 shows the results. This Table indicates that there is an optimum value of the dielectric constant to achieve an optimum crosspolarisation level. However, other parameters could not be improved. Also, the phase centre moves towards the dielectric end as ε_r increases.

Similar results are also obtained for a dielectric rod with a tapered end. Selecting a tapering angle of 90° and $L = 0.5\lambda$, the computed results are shown in Table 12.3. In general, the peak crosspolarisation deteriorates for $a \geqslant 0.35\lambda$ and $\varepsilon_r = 3.5$. Also, the phase centres move further towards the dielectric end as

Table 12.2 Effect of dielectric constant

ε_r	ψ_0	Peak crosspolarisation $0 \leq \theta \leq \psi_0$	$0 \leq \theta \leq 90°$	Gain (dB)	η_p (%)	η_s (%)	$d_c(\lambda)$ $0 \leq \theta \leq \psi_0$
2.5	66	−21.98	−20.28	8.59	74.4	87.7	0.167
3.0	60	−25.60	−23.56	9.00	77.8	88.5	0.232
3.5	60	−31.06	−28.90	9.10	75.0	86.6	0.299
4.0	54	−36.64	−29.49	8.96	67.5	74.7	0.356
5.0	60	−33.16	−28.05	8.40	55.8	65.5	0.480
$a=0.25\lambda$, $L=0.5\lambda$							

Table 12.3 Effect of tapering the dielectric end by 90°

a	ψ_0	Peak crosspolarisation $0 \leq \theta \leq \psi_0$	$0 \leq \theta \leq 90°$	Gain (dB)	η_p (%)	η_s (%)	$d_c(\lambda)$ $0 \leq \theta \leq \psi_0$
0.25	60	−23.80	−21.96	9.37	77.1	89.7	0.214
0.30	54	−29.58	−29.20	10.27	76.1	88.5	0.304
0.35	48	−32.87	−26.57	10.77	73.8	84.0	0.352
0.40	48	−26.29	−22.26	10.98	73.9	85.3	0.371
0.45	78	−09.93	−09.93	8.13	73.3	84.7	0.641
$\varepsilon_r=2.5$, $L=0.5\ \lambda$							
ε_r							
2.5	60	−23.80	−21.96	9.37	77.1	89.7	0.214
3.0	54	−27.97	−26.78	9.71	76.3	85.7	0.299
3.5	54	−31.29	−31.29	9.65	68.9	78.8	0.384
4.0	54	−34.00	−33.29	9.41	62.0	71.7	0.469
4.5	90	−23.80	−23.80	8.50	55.5	79.2	0.710
$a=0.25\lambda$, $L=0.5\lambda$							

compared with the untapered rod and the directive gain increases for all the examples. Again, these performance data can be changed by selecting different rod parameters. The best performance data as a feed were found for $a=0.35\lambda$ and $\varepsilon_r=2.5$ and when $\varepsilon_r=3$ and $a=0.25\lambda$. The radiation patterns for these two situations are shown in Figures 12.9 and 12.10, respectively.

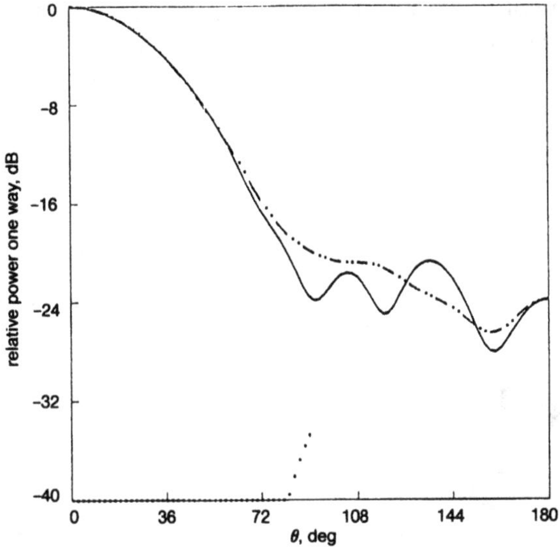

Figure 12.9 *E- and H-plane and crosspolarisation patterns for dielectric-rod antenna*
$a = 0.35\lambda$, $\varepsilon_r = 2.5$, $L = 0.5\lambda$
——— E-plane; —··— H-plane; ········ crosspolarisation

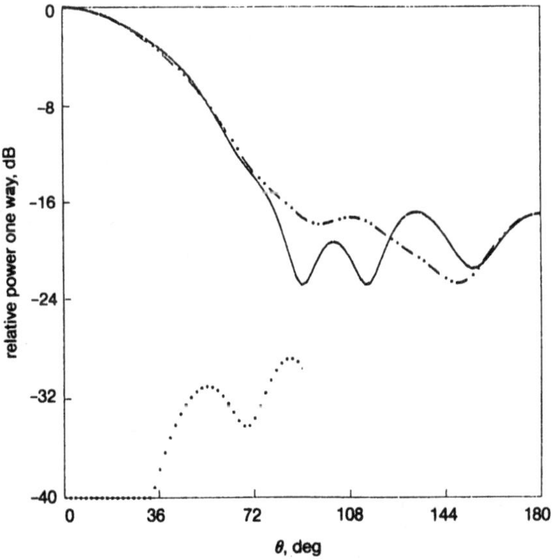

Figure 12.10 *E- and H-plane and crosspolarisation patterns for dielectric-rod antenna*
$a = 0.25\lambda$, $\varepsilon_r = 3.5$, $L = 0.5\lambda$; ——— E-plane; —··— H-plane; ········ crosspolarisation

Further examples were also studied where the dielectric shape was modified. They are shown in Table 12.4 and the corresponding radiation patterns are shown in Figures 12.11 and 12.12. The first two examples, give excellent pattern

Table 12.4 Performance of the shaped dielectric-rod antenna

Geometry		Peak crosspolarisation $0 \leqslant \theta \leqslant 90°$ (dB)	gain (dB)	Half beamwidths 3 dB E-plane	3 dB H-plane	10 dB E-plane	10 dB H-plane
a		−33.95	8.28	36.82	36.18	71.47	72.51
b	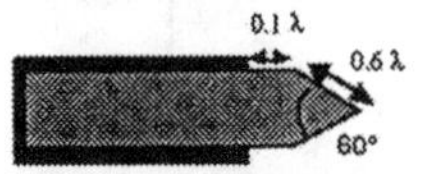	−24.74	8.11	37.21	38.32	73.42	71.35
c	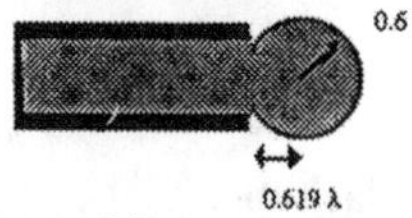	−24.43	13.47	19.43	20.25	33.13	35.17

$D = 0.6\lambda$, $\varepsilon_r = 2.5$

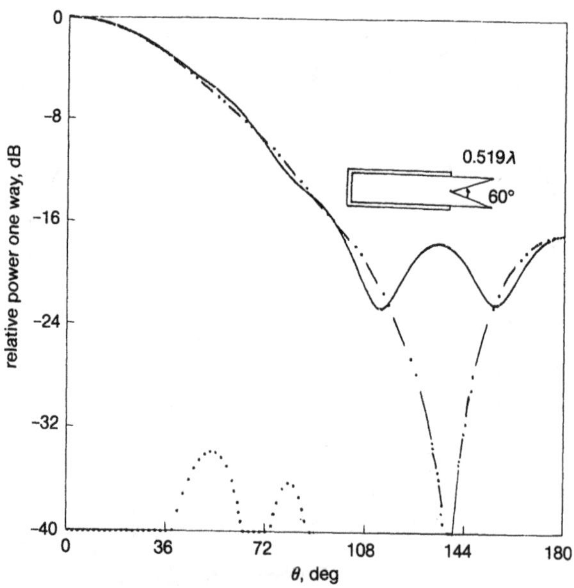

Figure 12.12 E- and H-plane and crosspolarisation patterns for dielectric-rod antenna for case b in Table 12.4

——— E-plane; —··— H-plane; ······· crosspolarisation

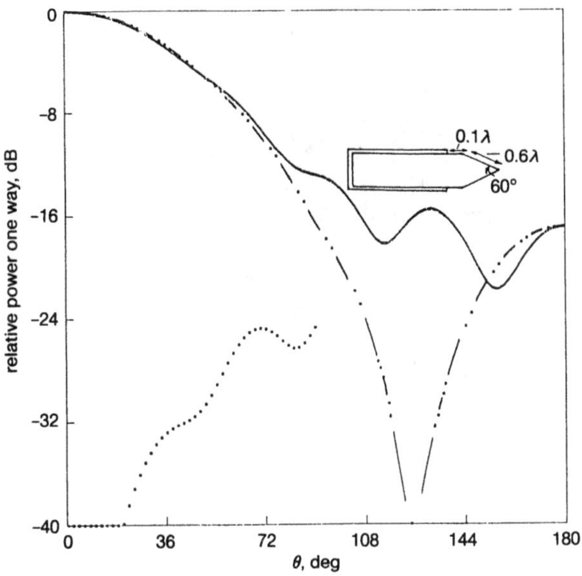

Figure 12.12 *E- and H-plane and crosspolarisation patterns for dielectric-rod antenna for case b in Table 12.4*

——— E-plane
—··— H-plane
········ crosspolarisation

symmetry, within the range of the reflector-aperture angle. The peak crosspolarisation is lower for the first example, since the phase-centre locations were nearly the same for both principal planes. Its principal-plane-pattern phase difference was less than 1°, while it was in excess of 4° for the second example. The shape of the radiation pattern for the third example was not suitable as a feed and its results are not shown. However, it gave the highest gain and may be used in such applications.

The radiation patterns shown in Figures 12.9–12.12 were all computed without a waveguide choke or flange. Their backradiation is therefore high. In the design of high-performance feeds one must therefore use additional measures to reduce the backradiation. Here we provide a few representative examples of possible feeds for shallow and deep reflectors.

12.3 Dielectric-rod feed with ring loading

A simple way of preventing backradiation of the dielectric rod is to place it within a cylindrical or conical flange. The example of the cylindrical flange, i.e. a ring loading of rod, is shown in Figure 12.13. The presence of the cylindrical cavity eliminates the backradiation and simultaneously improves further the

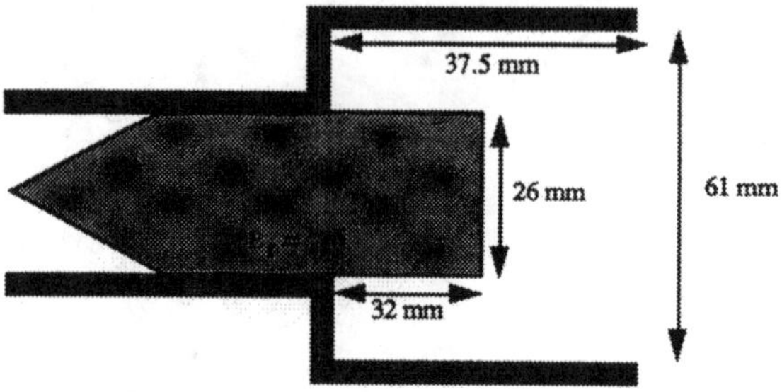

Figure 12.13 Dielectric-rod feed with ring loading

crosspolarisation. The computed pattern of this feed at the centre frequency is shown in Figure 12.14, which indicates excellent pattern symmetry and crosspolarisation below −40 dB. However, the ring loading is narrowband and the crosspolarisation quickly rises to that of the dielectric rod itself, i.e. around

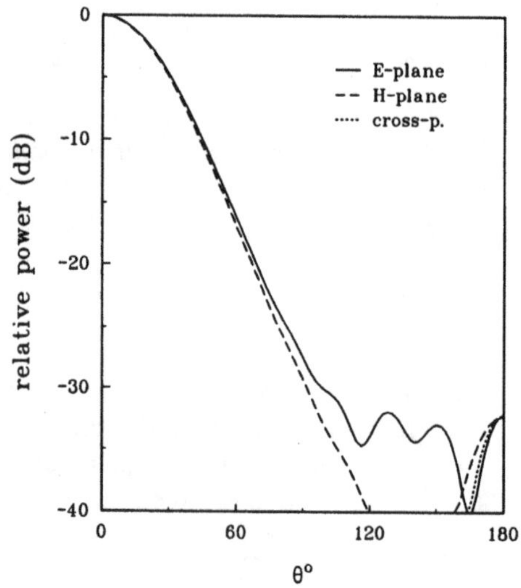

Figure 12.14 Radiation patterns of the ring-loaded feed illustrated in Figure 12.13 at $f=f_0$

Table 12.5 Performance of ring-loaded dielectric feed

	$0.95\,f_0$	f_0	$1.05\,f_0$
ψ_0	48	44	52
g_f	76.65	76.72	75.06
X_p	−25.41	−39.96	−23.3
$1/2\ \theta_E$(10 dB)	40.19	44.67	45.4
$1/2\ \theta_H$(10 dB)	45.67	43.53	53.53

−25 dB. The computed patterns within 10% band are shown in Figures 12.15 and 12.16. Also, the ring loading increases the effective aperture of the antenna and consequently its beamwidth is narrower than that of the dielectric rod itself. The resulting feed is therefore suitable for shallow reflectors with large f/D or offset-type reflectors. The performance data for this feed are shown in Table 12.5, which shows excellent results at the centre frequency, but deteriorate rapidly, within the 10% bandwidth, for crosspolarisation. However, its gain factor remains satisfactory.

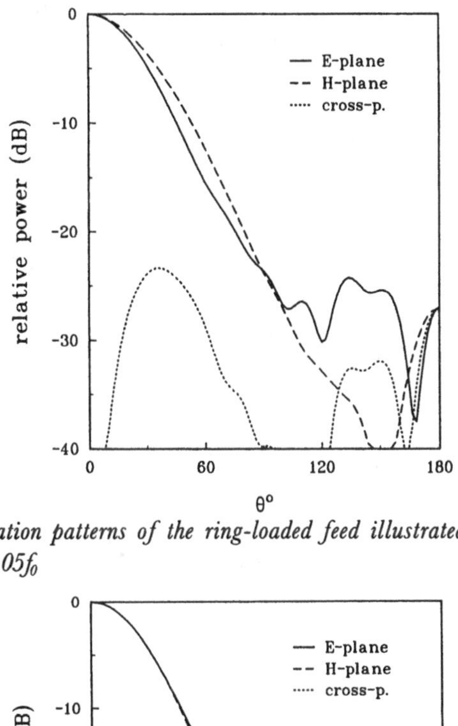

Figure 12.15 *Radiation patterns of the ring-loaded feed illustrated in Figure 12.13 at* $f = 1.05f_0$

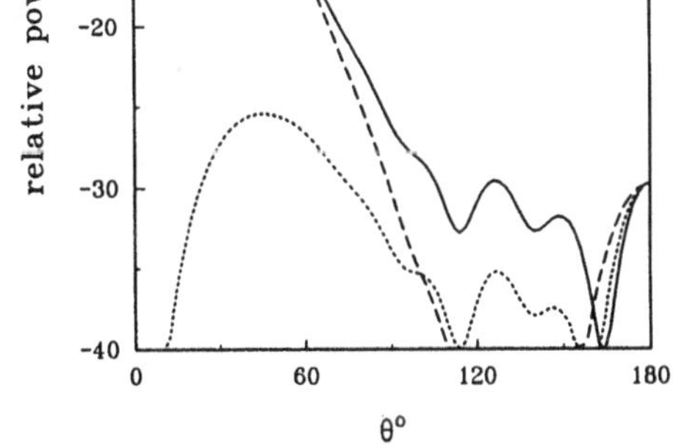

Figure 12.16 *Radiation patterns of the ring-loaded feed illustrated in Figure 12.13 at* $f = 0.95f_0$

12.4 Dielectric rod with cavity loading

Another method to reduce the backradiation of a dielectric rod is to use a cavity loading behind the waveguide, where an additional choke may be used. In this design the cavity dimensions can be optimised to place a dip on the pattern along the broadside to design the feed for deep reflectors. One such design is shown in Figure 12.17 [15]. Its performance within 10% bandwidth is shown in Table 12.6 and the computed patterns at the centre frequency are presented in Figure 12.18. As the results indicate, the introduction of the cavity has improved the feed performance and increased the required reflector aperture

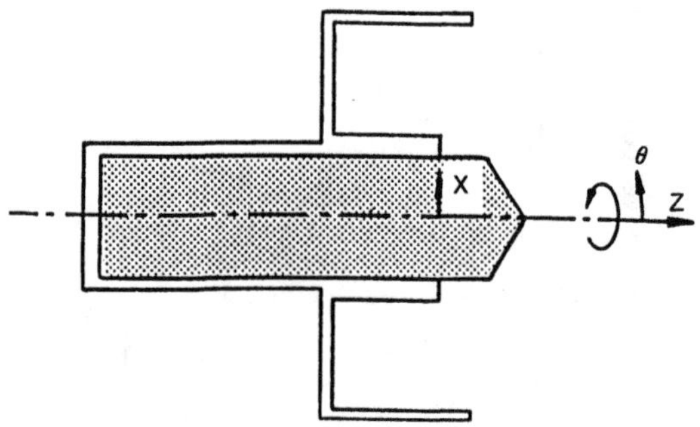

Figure 12.17 Geometry of a cavity-loaded dielectric feed

Waveguide diameter $=0.5\lambda$
Dielectric-rod diameter $=0.5\lambda$
Dielectric-rod extension $=0.15\lambda$
Rod and apex angle $=120°$
Cavity depth $=0.57\lambda$
Cavity radius $=0.88\lambda$

Table 12.6 Performance of feed in Figure 12.17

Frequency	$0.95\ f_0$	$0.975\ f_0$	f_0	$1.025\ f_0$	$1.05\ f_0$
ψ_0	46	76	74	74	74
η_g	51.50	81.56	80.57	80.36	80.37
η_s	64.10	89.70	88.20	88.76	89.39
η_i	76.19	90.90	91.35	90.55	89.39
X_p	−11.65	−25.51	−32.30	−25.97	−23.94
d_c			−0.80		

ψ_0 = aperture angle of reflector antenna at which gain factor is maximum

η_g = maximum gain factor at ψ_0

η_s = spillover efficiency at ψ_0

η_i = illumination efficiency at ψ_0

X_p – maximum crosspolarisation level ($0<\theta<90°$)

d_c = phase-centre location from the waveguide aperture

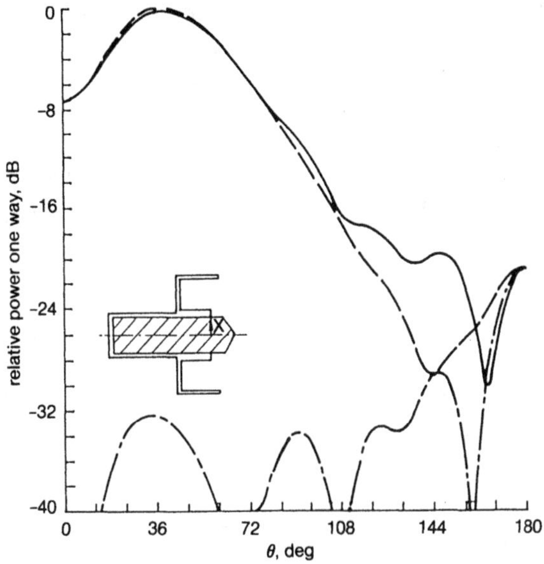

Figure 12.18 Radiation patterns of feed illustrated in Figure 12.17

——— E-plane
– – – – H-plane
—·— crosspolarisation

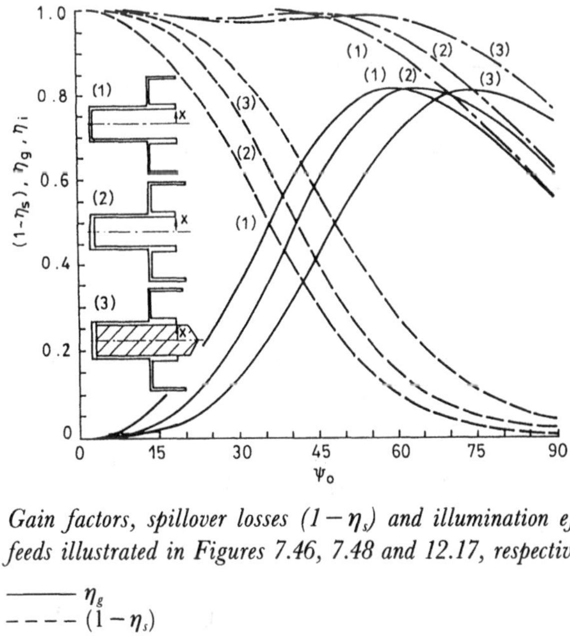

Figure 12.19 Gain factors, spillover losses $(1-\eta_s)$ and illumination efficiencies of the feeds illustrated in Figures 7.46, 7.48 and 12.17, respectively

——— η_g
– – – – $(1-\eta_s)$
—·— η_i

considerably. The performance of this feed is in Figure 12.19 compared with the corresponding feeds discussed in Chapter 7. It shows that the introduction of the dielectric-rod feed to the waveguide has further broadened its pattern and the resulting feed is ideal for very deep reflectors. Introducing a choke at the cavity periphery further improves its performance. Its geometry and radiation patterns are shown in Figure 12.20. Table 12.7 shows its performance data within 10% frequency band, which show further improvements over the results of Table 12.6.

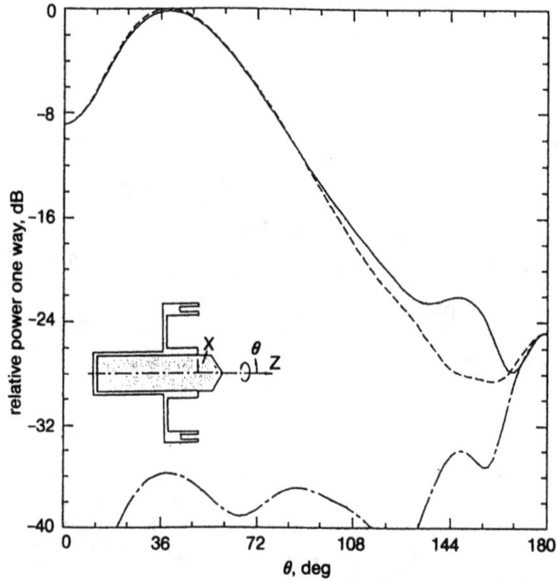

Figure 12.20 Radiation patterns of feed illustrated in Figure 12.17 with λ/4 choke

——— E-plane
- - - - H-plane
—·— crosspolarisation

Table 12.7 Performance of feed in Figure 12.17 with choke

Frequency	0.95 f_0	0.975 f_0	f_0	1.025 f_0	1.05 f_0
ψ_0	78	77	74	72	72
η_g	77.97	81.62	81.05	80.94	81.10
η_s	89.36	88.79	89.70	89.03	89.67
η_i	87.03	91.92	90.36	90.92	90.45
X_p	−17.44	−24.84	−35.77	−26.44	−23.91
d_c			−0.22		

12.5 Profiled dielectric rods

Shaping the dielectric rod adds an additional parameter for optimising the feed. For dielectric rods without a flange, the effect of shaping was discussed earlier. For the cavity loading discussed in Section 12.4, the end of the dielectric rod was made conical to improve crosspolarisation. Here, we consider an opposite case, where the dielectric-rod diameter outside the waveguide is gradually increased. A conical flange is used to reduce the backradiation, Figure 12.21. The radiation patterns of this feed at the centre frequency and within the 10% frequency band are shown in Figures 12.22–12.24, which show an excellent pattern shape. This feed is similar to the ring-loaded one, but has improved performance over a larger band. Since its beamwidth is quite narrow, it is an ideal feed for reflectors with large f/D or offset paraboloids. The performance of this feed is shown in Table 12.8.

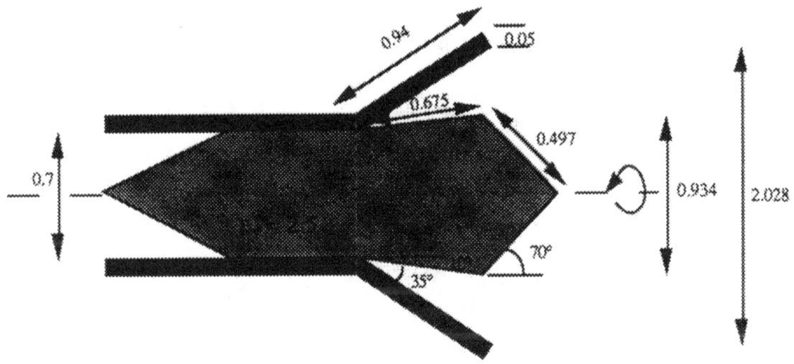

Figure 12.21 Geometry of a profiled dielectric feed (dimensions in wavelenghts)

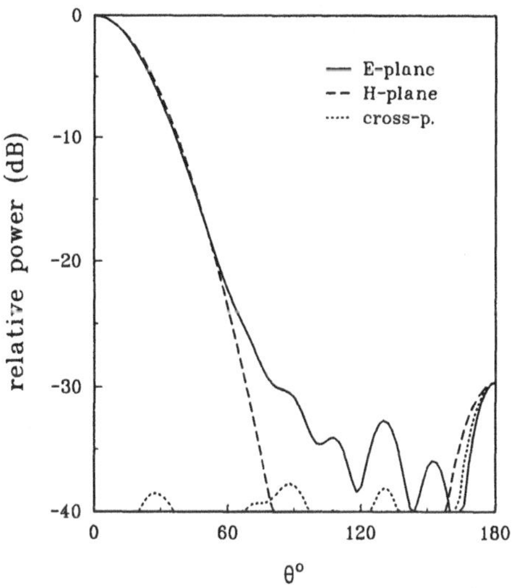

Figure 12.22 Radiation patterns of feed illustrated in Figure 12.21 at $f=f_0$

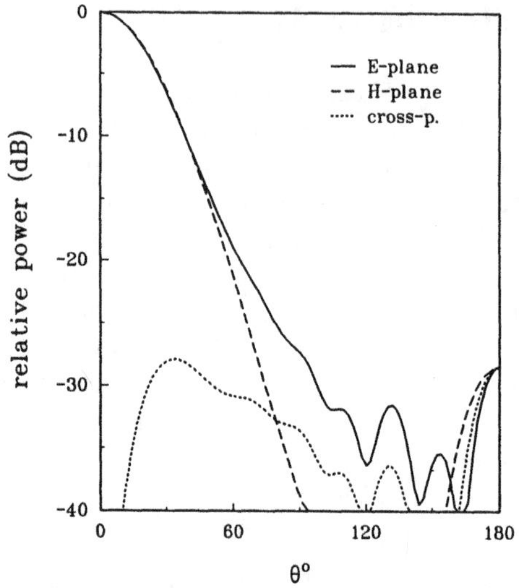

Figure 12.23 Radiation patterns of feed illustrated in Figure 12.21 at $f = 1.05f_0$

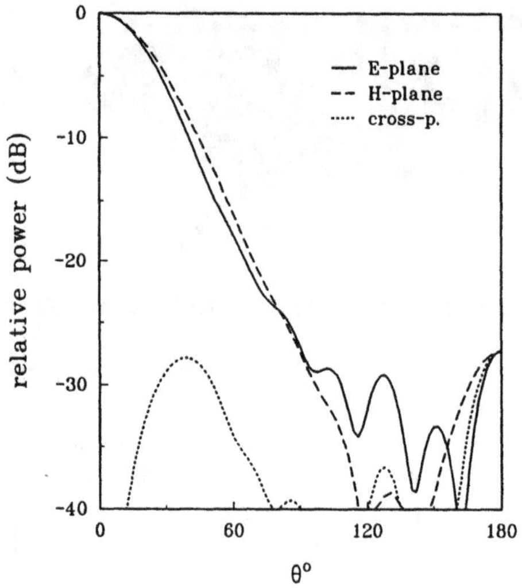

Figure 12.24 Radiation patterns of feed illustrated in Figure 12.21 at $f = 0.95f_0$

Table 12.8 Performance of profiled dielectric rod

	$0.95\ f_0$	$0.975\ f_0$	f_0	$1.075\ f_0$	$1.05\ f_0$
ψ_0	44	40	36	40	40
η_g	75.81	77.53	77.97	77.36	76.54
η_s	87.78	90.24	87.55	90.97	88.91
η_i	86.11	85.85	89.05	85.03	86.09
X_p	−23.7	−27.01	−36.88	−37.44	−29.16
d_c	2.52	2.42	2.32	2.26	2.24

12.6 Waveguides with dielectric rings

A simple way of improving the radiation pattern of a circular waveguide is to load its aperture with a dielectric ring (Figure 12.25). The ring section supports hybrid modes and, by optimising its dimensions, simple feeds with satisfactory performance may be obtained [6]. The performance of these feeds at different frequencies is shown in Table 12.9. Both gain factor and crosspolarisation increase with frequency. The corresponding radiation patterns at the high frequency end are shown in Figures 12.26–12.28. The results are very similar to those for waveguides, but improve faster with frequency for a ring-loaded waveguide. A simple way of explaining this result is to note that loading the waveguide with a dielectric increases its effective diameter. Consequently, the performance of a dielectric-ring-loaded waveguide is similar to that of a hollow waveguide, but at slightly higher frequency. However, the presence of a hybrid structure improves its crosspolarisation level over a wider bandwidth.

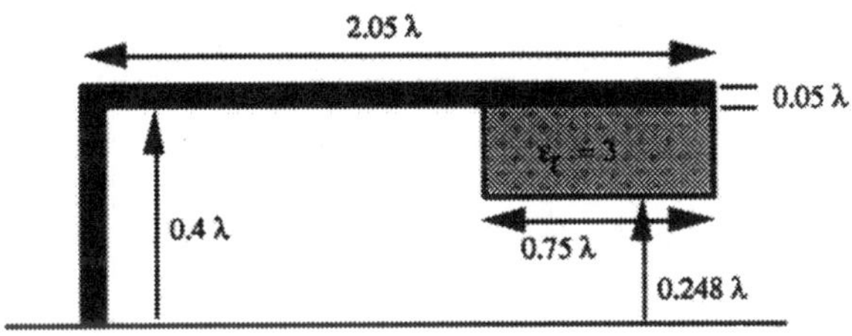

Figure 12.25 Geometry of a ring-loaded waveguide

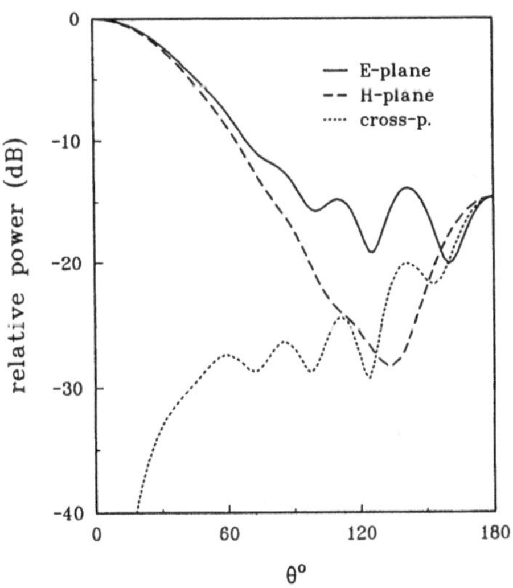

Figure 12.26 Radiation patterns of waveguide illustrated in Figure 12.25 at $f=f_0$

Table 12.9 Performance of dielectric-ring-loaded waveguide

	0.9 f_0	0.95 f_0	f_0	1.05 f_0	1.1 f_0	1.15f_0
ψ_0	68	64	64	60	60	56
g_f	69.3	70.58	71.6	72.55	73.7	74.7
X_p	−24.15	−25.23	−26.28	−27.39	−29.44	−34.48
1/2 θ_E(10 dB)	81.59	73.58	68.70	66.34	62.85	61.1
1/2 θ_H(10 dB)	67.65	65.46	64.0	62.16	60.15	58.86

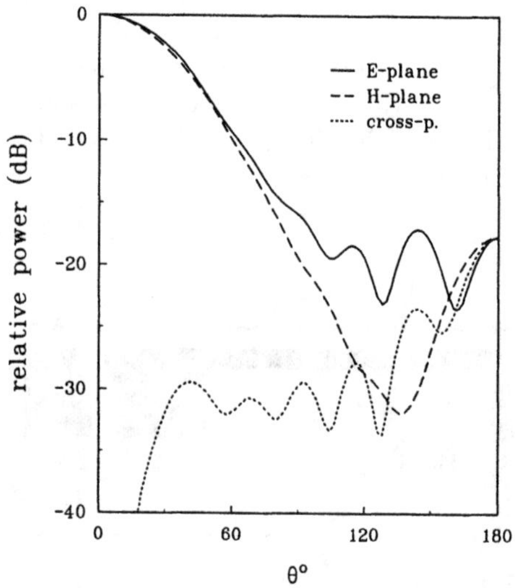

Figure 12.27 Radiation patterns of waveguide illustrated in Figure 12.25 at $f = 1.1f_0$

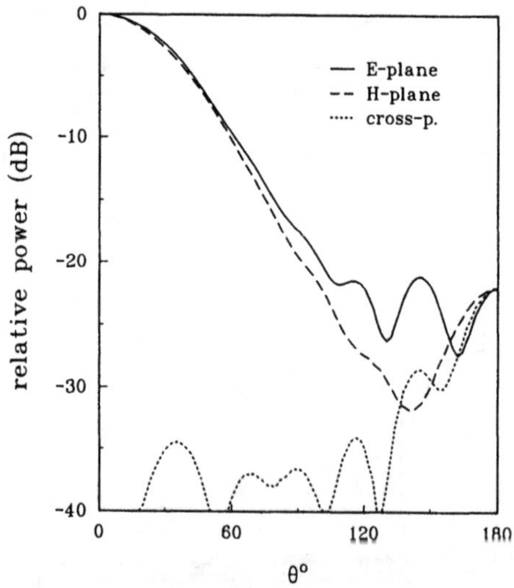

Figure 12.28 Radiation patterns of waveguide illustrated in Figure 12.25 at $f = 1.15f_0$

12.7 Dualband feed

In Chapter 7, it was shown that coaxial feeds can be designed for multiband operation, by exciting the TE_{11} mode separately in each coaxial waveguide. It was also shown there that, to equalise the principal pattern of the coaxial waveguide, the ratio of its radii must be optimised. In some applications, the operating-frequency bands are too close to each other and the optimum waveguide radii cannot be achieved. In such a case, one can load the central section of the guide with a dielectric to reduce its diameter to permit optimisation of the waveguide dimensions. An example of such a feed is shown in Figure 12.29 which was designed for operation in 4–6 GHz bands [4]. The 4 GHz portion of the antenna consists of a circular coaxial waveguide. The radii a and b of the conductors are chosen so that the coaxial waveguide can support the TE_{11} mode at 4 GHz, and optimised to give symmetric E- and H-plane patterns. The dimension b is also selected so that the coaxial aperture is impedance-matched to free space, but is constrained by the dimensional requirements of the 6 GHz antenna. The excitation of the TE_{11} mode in the coaxial waveguide is achieved by using an axial slot-type launcher, positioned on the outside wall of the guide. The circumferential location of this slot controls the polarisation.

The 6 GHz portion of the antenna consists of the circular waveguide formed by the thin-walled inside conductor of the coaxial structure and a dielectric rod which homogeneously fills the circular guide and extends beyond its end over an adjustable length. The radius c of the dielectric-filled circular waveguide is selected such that there is propagation of the TE_{11} mode at 6 GHz.

Using the numerical procedure, a coaxial feed with a quarter-wavelength choke was designed which had aperture dimensions of $a/\lambda = 0.4$ and $a/b = 2.5$. Its measured radiation patterns at 4 GHz are shown in Figure 12.30. The difference between the patterns in the two planes is always less than 1 dB, within the beam angle of 72°.

The 6 GHz antenna is essentially a dielectric-rod feed. For an isolated antenna, its performance can be optimised by selecting an appropriate extension of the dielectric rod beyond the waveguide open end. However, in the present coaxial geometry, it was found that the coupling between the antennas

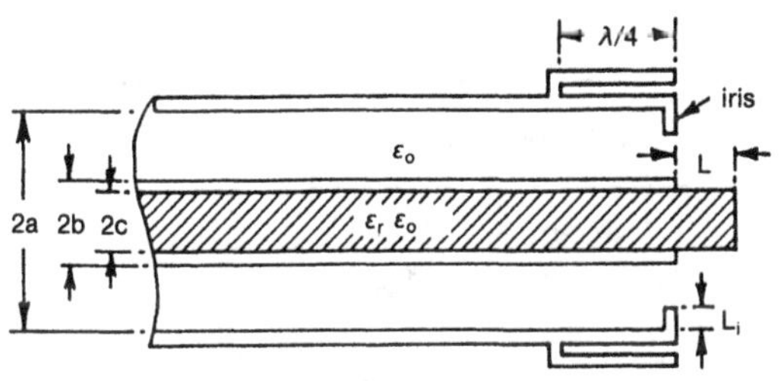

Figure 12.29 Coaxial dualband-feed geometry

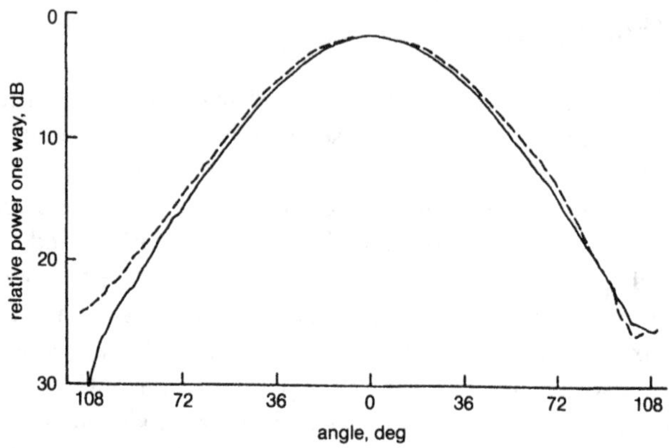

Figure 12.30 Measured radiation patterns of coaxial aperture with choke

——— E-plane
- - - - H-plane

adversely affects their radiation patterns. To reduce this coupling effect, an iris was placed on the outer conductor of the coaxial structure as shown in Figure 12.29. Subsequent investigations revealed that such an iris reduces both the coupling effects and the surface currents on the external surface of the antenna. However, the introduction of the iris may also affect the radiation patterns of the coaxial 4 GHz antenna. For different iris sizes, Tables 12.10 and 12.11 show the variation of the 10 dB beamwidth of the patterns in the principal planes of this antenna. It is evident that at 4 GHz the effect of the iris on the E-plane patterns is negligible, and for the H-plane patterns is small. This is because the field of the coaxial aperture is mostly concentrated around the inside conductor, and an iris on the external conductor does not affect the aperture-field distribution significantly. In addition, the quarter-wavelength choke suppresses the current flow on the iris, due to series infinite impedance of the choke. Figure

Table 12.10 Iris effect on 6 GHz antenna, 10 dB beamwidth

Iris width	Dielectric-rod length $L=0$		Dielectric-rod length $L=2.0$ cm	
L_i	E-plane	H-plane	E-plane	H-plane
(cm)	(deg)	(deg)	(deg)	(deg)
0	194	138	179	151
0.6	103	131	159	143
0.9	151	130	145	142

12.31 shows the measured patterns for the dielectric rod extension at $L=$ 2.0 cm, which indicates a pattern symmetry of better than 1 dB within the beam angle of 72°. Note that the extension of the dielectric rod beyond the flush position does not affect the 4 GHz-patterns significantly, which is evident from the results in Table 12.11 for the different iris widths. The maximum beamwidth variation is in the H-plane and for a 0.9 cm iris is about 3°. Since no corrugation on the inner conductor was used, the crosspolarisation of this feed was generally high and about −20 dB.

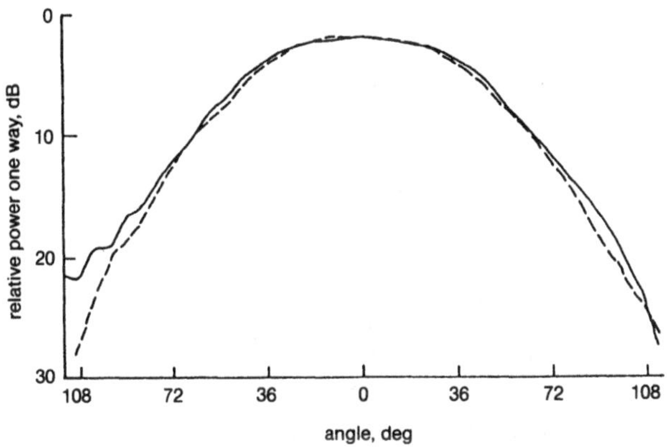

Figure 12.31 *Measured radiation patterns at 6.175 GHz*
$L=20$ mm, $L_i=9$ mm
——— E-plane
– – – – H-plane

Table 12.11 *Iris effect on 4 GHz antenna, 10 dB beamwidth*

Iris width L_i	Dielectric-rod length $L=0$		Dielectric-rod length $L=2.0$ cm	
	E-plane	H-plane	E-plane	H-plane
(cm)	(deg)	(deg)	(deg)	(deg)
0	118	122	119	124
0.6	118	124	120	127
0.9	119	126	120	129

Using the patterns of Figures 12.30 and 12.31, the gain factors of the antenna at 4 and 6 GHz were calculated and are shown in Figure 12.32. Its performance at 6 GHz is somewhat better than at 4 GHz. Thus, when this antenna is used as a feed in a prime focus reflector, the *f/D* of the reflector must be chosen to maximise the gain for the 4 GHz band.

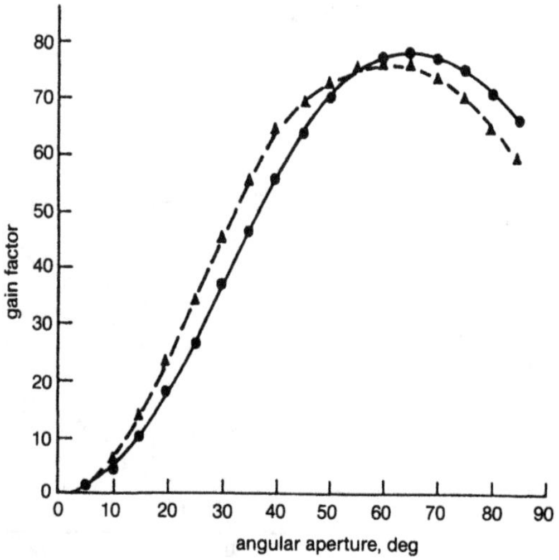

Figure 12.32 Gain factors of dualband antennas

—◆— 4 GHz
—•— 6 GHz

12.8 References

1 DOMBEK, P.K.: 'Dielektrische Antennen geringer querabmessungen als erreger für spiegelantennen', *Nach. tech. Z.*, 1975, **28**, pp. 311–315

2 KUMAR, A.: 'Experimental study of a dielectric rod enclosed by a waveguide for use as a feed', *Electron. Lett.*, 1976, **12**(25), pp. 666–668

3 ITTIPIBOON, A., SHAFAI, L., and BRIDGES, E.: 'A dielectric rod antenna as a high efficiency feed for reflector antennas', *Can. J. Elec. Eng.*, 1981, **6**(4), pp. 10–13

4 SHAFAI, L., ITTIPIBOON, A., BRIDGES, E., and HYJAZRI, F.: 'Dualband horn with inherent isolation between its transient and receive ports', *IEE Proc. H*, 1984, **131**(3), pp. 143–146

5 KISHK, A., and SHAFAI, L.: 'Radiation characteristics of the short dielectric rod antenna: A numerical method', *IEEE Trans.*, 1987, **AP–35**(2), pp. 139–146

6 RAGHAVAN, K., OLVER, A.D., and CLARRICOATS, P.J.B.: 'Compact dielectric feeds for prime focus reflector antennas'. Proceedings of IEEE symposium on Antennas and Propagation, Syracuse, 1988, vol. 1, pp. 350–353

7 JAMES, J.R.: 'Theoretical investigation of cylindrical dielectric-rod antenna', *Proc. IEE*, 1967, **114**, pp. 309–319

8 WATSON, R.B., and HORTON, C.W.: 'The radiation patterns of dielectric rods—experimental and theory', *J. Appl. Phys.*, 1948, **19**, pp. 661–670

9 HORTON, C.W., KARAL, F.C., and McKINNEY, C.: 'On the radiation patterns of dielectric rods of circular cross section-TM mode', *J. Appl. Phys.*, 1949, **21**, p. 1279
10 BOUIX, M.: 'Contribution a l'étude des antennas dielectrique'. Doctoral dissertation, University of Paris, 1952
11 FRADIN, A.Z.: 'Microwave antenna' (Pergamon, 1961)
12 BROWN, J., and SPECTOR, J.O.: 'The radiation properties of end-fire aerials', *Proc. IEE*, 1957, **104B**, pp. 27–34
13 YAGHJIAN, A.D., and KORNHAUSER, E.T.: 'A model analysis of the dielectric rod antenna excited by the HE_{11} mode', *IEEE Trans.*, 1972, **AP–20**, pp. 122–128
14 SHAFAI, L., and KISHK, A.A.: 'Phase centre of small primary feeds and its effects on the feed performance', *IEE Proc. H*, 1985, **132**(3), pp. 207–214
15 KISHK, A.A.: 'Simple primary feeds for deep reflectors', *IEE Proc. H*, 1989, **136**(2), pp. 169–171

Chapter 13
Dipole feeds

L. Shafai and A.A. Kishk

13.1 Introduction

The dipole antenna is the most fundamental and simplest antenna and was, historically, the first to be used as a reflector feed [1]. To illuminate the reflector efficiently, a small circular disc was used as a subreflector to direct the dipole radiation toward the reflector. An oversized dipole reflector has also been used in place of the disc reflector. While these designs are simple and are used extensively as low-cost or low-frequency feeds, the asymmetric dipole radiation in its E- and H-planes does not illuminate the reflector efficiently. In particular, the asymmetric radiation causes an excessive crosspolarisation. These are shown in Figure 13.1, where, a halfdipole is placed a quarter-wavelength in front of a small disc of 2λ diameter. Both backlobe and crosspolarisation are high, and the beamwidth ratio in two principal planes is excessive.

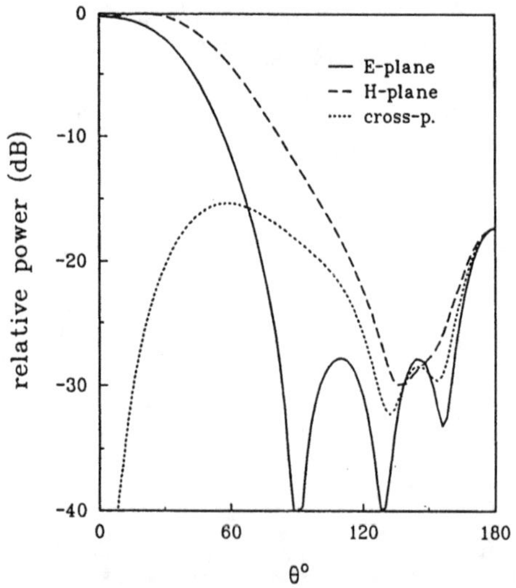

Figure 13.1 Far-field patterns of a halfwave dipole antenna placed a quarter-wavelength in front of a 2-wavelength-diameter disc

A numerical approach for optimising the dipole-disc antenna was used by Kildal and Skyttemyr [2]. They used a conducting ring in front of the dipole to shape the feed pattern. Here we shall consider other dipole-feed configurations and show that satisfactory feed patterns can usually be obtained by optimising the feed dimensions.

13.2 Ring-loaded dipole-disc feed

In this configuration the feed is essentially the original dipole-disc antenna, with a disc diameter of 2.2λ. However, a large ring, in the form of a short-circuited loop antenna, is placed about 0.2λ in front of the dipole. It couples strongly with the dipole field, and by optimising the loop diameter, symmetric radiation patterns can be obtained. The results reported in Reference 2 are shown in Figure 13.2. Comparing the copolar patterns with Figure 13.1, one observes that the backlobe level remains unchanged. Also, the beam-forming ring does not significantly affect the dipole-disc radiation pattern in the E-plane, but compresses its H-plane pattern to form a symmetric radiation pattern. The process reduces the beamwidth of the feed and, consequently, the reflector-aperture angle for peak efficiency, from 63° to 54°. The feed performance with a reflector is shown in Table 13.1 and Figure 13.3. This feed has a narrowband performance, since both the dipole and ring are resonant structures and their coupling that controls the radiation pattern varies with frequency.

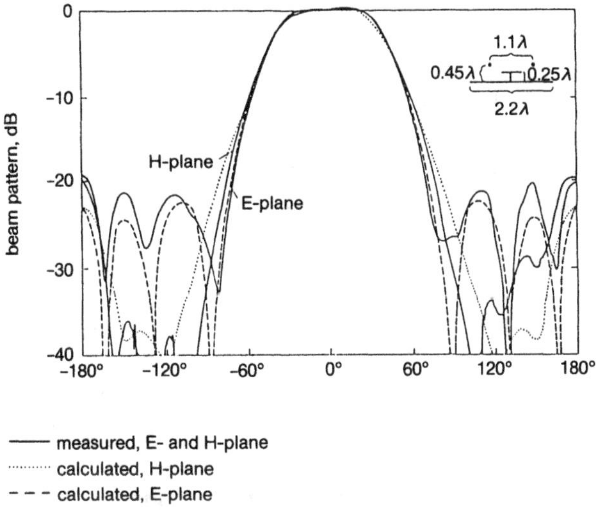

Figure 13.2 Computer and measured patterns of dipole-disc antenna with ring [2]

Table 13.1 Comparison of performance characteristics of directivity-optimised paraboloids and paraboloid with $\psi_0 = 72°$, which corresponds to 20 dB taper in E-plane

	Directivity-optimised paraboloids		Paraboloid with $\psi_0 = 72°$	
	Feed without ring	Feed with ring	Feed without ring	Feed with ring
Subtended halfangle ψ_0, deg	63	54	72	72
Gain of feed G_f, dBi	6.4	7.5	6.4	7.5
Aperture taper: E-plane, dB	13	6.6	20.5	20.7
H-plane dB	3.9	8.1	6.0	21.8
Spillover lobes: E-plane $A_1(\psi_0)$, dBi	−6.6	0.9	−14.1	−13.2
H-plane $C_1(\psi_0)$, dBi	2.5	−0.6	0.4	−14.3
Backlobe of paraboloid G^b, dBi	13.3	17.6	7.6	3.2
Subefficiencies:				
Spillover efficiencies, η^s	0.856	0.940	0.915	0.989
Polarisation efficiencies, η^p	0.983	1.000	0.975	1.000
Illumination efficiencies, η^i	0.868	0.898	0.775	0.626
Phase efficiencies, η^ϕ	0.999	0.997	0.996	0.089
Aperture efficiency, η^f	0.729	0.841	0.688	0.612

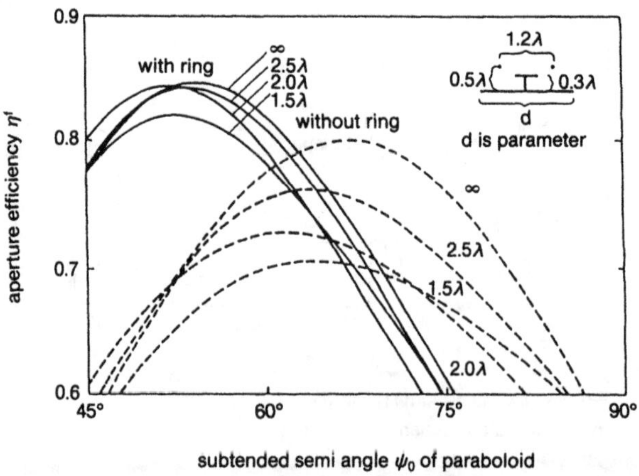

Figure 13.3 Aperture efficiency of a paraboloid reflector with a dipole-disc feed [2]

13.3 Cone-dipole feed

The asymmetry of the dipole-disc feed, as shown in Figure 13.1, is due to the flat subreflector which forms an approximate image of the dipole. However, since the dipole-radiation pattern is asymmetric, the resulting pattern also becomes asymmetric. With a flat-disc reflector, to generate a symmetric pattern one must use an antenna which also has a symmetric radiation pattern. This can be achieved with an electric dipole by bending its arms towards the disc and optimising the dimensional parameters. However, this solution is not attractive since it affects the antenna impedance and power-carrying capability. An alternative and more desirable solution is to keep the dipole in its standard form, but bend the disc. This modifies the disc to a cone reflector, as shown in Figure 13.4. A sample of the radiation patterns for this feed is shown in Figure 13.5. It shows good pattern symmetry and low backlobe level and crosspolarisation. These results are superior to those of many waveguide feeds. Also, since the pattern shaping is achieved by a geometrical correction, rather than mutual coupling, the feed performance is expected to be broadband. For this feed, key parameters are the cone angles, the separation of the dipole from the cone apex, and the cone size. To assist in design of such feeds, their performance with paraboloid reflectors is shown in Tables 13.2–13.4.

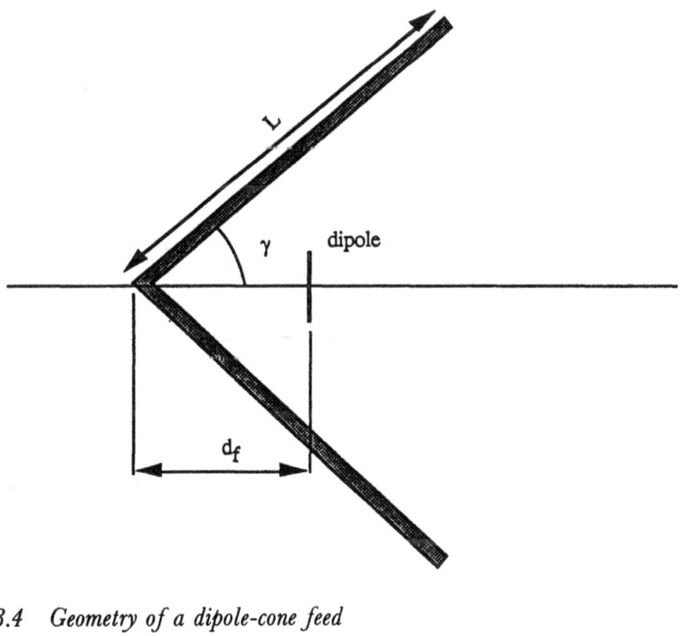

Figure 13.4 Geometry of a dipole-cone feed

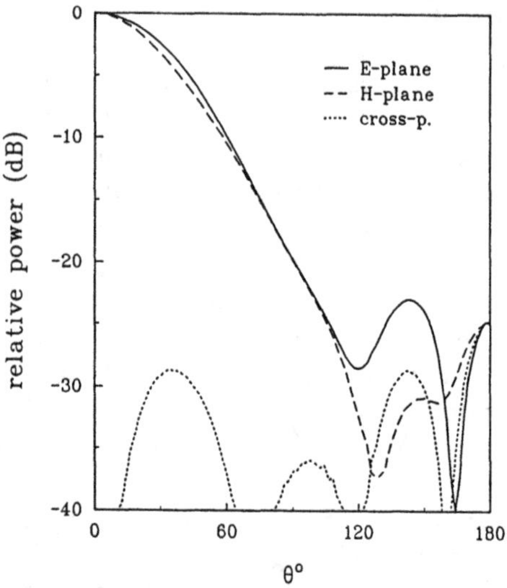

Figure 13.5 Far-field patterns of a cone-dipole feed
Cone size $L=1\lambda$, cone angle $\alpha=70°$, dipole distance $d_F=0.2$

Table 13.2 Effect of cone angle on cone-dipole-feed performance

α, deg	60	70	80	90
Crosspolarisation, $\alpha<90°$, dB	−20.45	−28.66	−21.95	−14.88
$1/2\theta_E(10\text{ dB})$	61.30	60.33	55.42	54.14
$1/2\theta_H(10\text{ dB})$	47.74	58.27	67.08	77.20
ψ_0	60	56	60	64
g_f	0.71	0.76	0.77	0.78
$D=1.0\,\lambda$, $d_f=0.2\lambda$				

Table 13.2 shows that, for the given parameters, the best cone angle for minimum crosspolarisation is 70°. At other angles the pattern symmetry is not optimum. Various computations have shown that, for other cone dimensions, the optimum cone angle remains around 70°, but may be changed slightly (by 2 or 3°) to minimise the crosspolarisation. The reflector-aperture angle for this example is small, similar to that of a ring-loaded dipole disc. The effect of cone size on the feed performance is shown in Table 13.3. For the best overall feed

Table 13.3 Effect of cone size on cone-dipole-feed performance

$D(\lambda)$	1	1.1	1.2	1.3	1.5
Crosspolarisation, $\theta<90°$, dB	−25.58	−23.42	−21.28	−18.66	−28.66
$1/2\theta_E(10\text{ dB})$	59.56	58.50	58.08	63.33	60.33
$1/2\theta_H(10\text{ dB})$	58.64	59.07	58.47	57.37	58.27
ψ_0	56	56	56	56	56
g_f	0.7740	0.7838	0.7858	0.7756	0.7605
$\alpha=70°$, $d_f=0.2\lambda$					

Table 13.4 Performance of cone-dipole-feed with frequency

f/f_0	0.95	1.0	1.05
Crosspolarisation, $\theta<90°$, dB	−19.18	−28.66	−18.58
$1/2\theta_E(10\text{ dB})$	61.69	60.33	62.91
$1/2\theta_H(10\text{ dB})$	57.20	58.27	57.81
ψ_0	56	56	56
g_f	0.7752	0.7605	0.7814
$D=1.0\lambda$, $\alpha=70°$, $d_f=0.2\lambda$			

performance, the cone size is between 1.0λ and 1.2λ. Finally, the performance of 1.0λ cone-dipole feed within a 10% frequency band is shown in Table 13.4. The most sensitive parameter is the crosspolarisation, which increases rapidly with frequency variation. An attempt to improve the performance of this feed with a choke was not successful, at least for the chosen dimensions. The computed patterns are shown in Figure 13.6, which shows a slight improvement in crosspolarisation, but the backlobe level has increased. A detailed optimisation was not carried out.

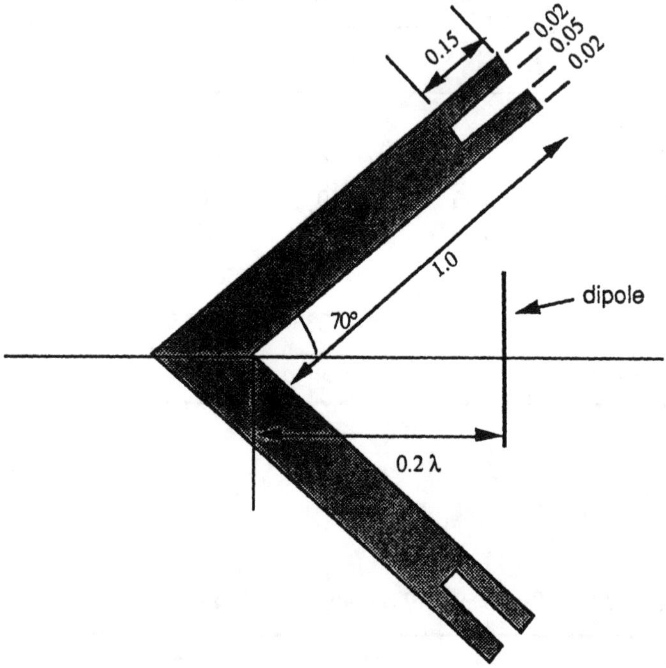

a

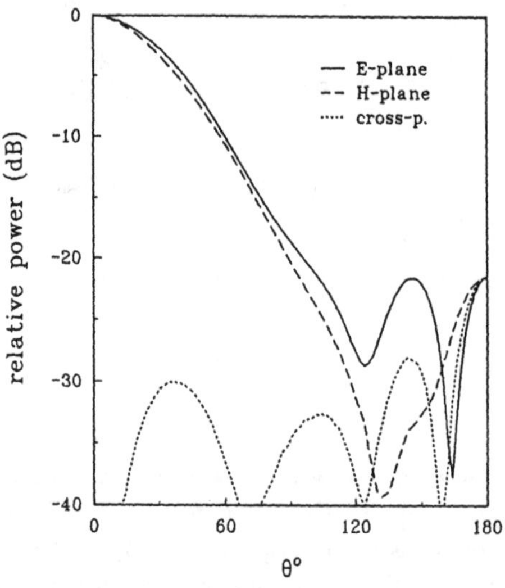

b

Figure 13.6 Far-field patterns of a cone-dipole feed with a choke at the cone periphery

a Geometry

b Field patterns

13.4 Cavity-dipole feed

As a second solution to dipole-pattern asymmetry, a short cylindrical section may be placed over the disc to form a cavity. Here, the cavity-dominant mode controls the pattern shape and, consequently, its behaviour is easier to understand. Figure 13.7 shows the geometry. Its performance for different cavity parameters is shown in Tables 13.5–13.7. In Table 13.5 the most sensitive parameter is the crosspolarisation, which reaches a minimum for a

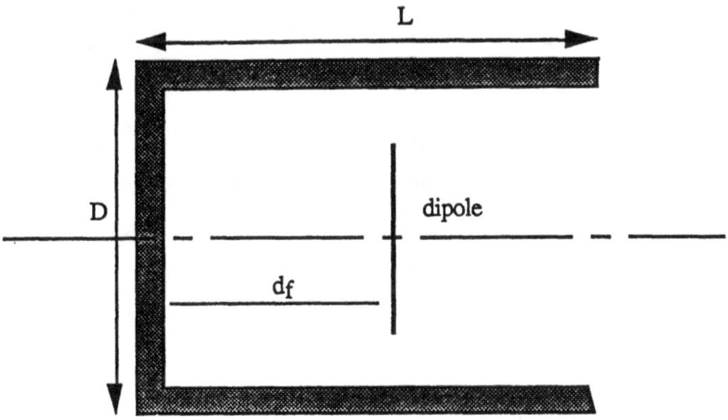

Figure 13.7 Cavity-dipole geometry

Table 13.5 Effect of cavity depth on feed performance

$L(\lambda)$	0.25	0.30	0.35	0.40	0.45	0.50
Cross-polarisation, dB						
$\theta<\psi_0$	−25.94	−29.18	−33.04	−38.83	<−40.00	−39.41
$\theta<90°$	−25.80	−28.94	−32.95	−38.77	<−40.00	−39.31
$1/2\theta_E$(10 dB)	53.41	53.71	53.76	53.57	53.19	52.69
$1/2\theta_H$(10 dB)	58.03	55.37	53.72	52.79	52.34	52.20
ψ_0	56	52	52	52	52	52
g_f	0.7498	0.7527	0.7554	0.7564	0.7565	0.7566

$D=1.2\lambda$, $d_f=0.25\lambda$, $t=0.02\lambda$

cavity depth of 0.45λ. For this depth, Table 13.6 shows that the separation of the dipole from the cavity floor is not a sensitive parameter. In this Table the crosspolarisation is below −40 dB, and is not shown. The performance of this feed at a range of frequencies is shown in Table 13.7. The crosspolarisation remains mostly below −40 dB, but the gain factor increases gradually with frequency. The computed radiation patterns at the centre frequency are shown in Figure 13.8. The crosspolarisation is negligible and the pattern symmetry is

Table 13.6 Effect of dipole distance d_f from cavity floor

df/λ	0.05	0.10	0.15	0.20	0.25
Cross-polarisation, dB					
$\theta < \psi_0$	⩽−40.00	⩽−40.00	⩽−40.00	⩽−40.00	⩽−40.00
$\theta < 90°$	⩽−40.00	⩽−40.00	⩽−40.00	⩽−40.00	⩽−40.00
$1/2\theta_E$(10 dB)	52.48	52.36	52.50	52.80	53.19
$1/2\theta_H$(10 dB)	52.69	52.63	52.57	52.51	52.34
ψ_0	52	52	52	52	52
g_f	0.7590	0.7592	0.7588	0.7580	0.7565

$L = 0.45\lambda$, $D = 1.2\lambda$, $t = 0.02\lambda$

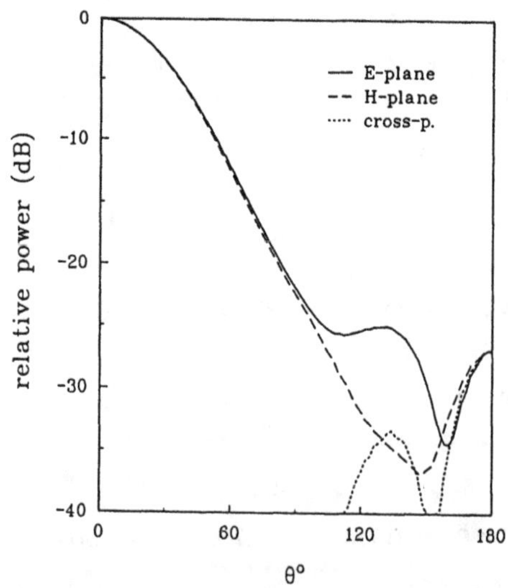

Figure 13.8 Far-field patterns of a cavity-dipole feed

Table 13.7 Performance of dipole-cavity feed with frequency

f/f_0	0.80	0.85	0.90	0.95	1.0	1.05	1.1
Crosspolarisation, dB							
$\theta<\psi_0$	−40.00	−40.00	−40.00	−40.00	−40.00	−40.00	−33.54
$\theta<90°$	−36.71	−39.76	−40.00	−40.00	−40.00	−40.00	
$1/2\theta_E$(10 dB)	63.20	60.65	58.1	52.80	53.19	50.176	46.12
$1/2\theta_H$(10 dB)	62.39	59.76	57.25	52.51	52.34	49.92	47.56
ψ_0	60	60	56	56	52	48	48
g_f	0.7336		0.7451	0.7495	0.7565	0.7639	0.7763

$L=0.45\lambda$, $d_f=0.25\lambda$, $t=0.02\lambda$, $D=1.2\lambda$
Crosspolarisation less than −40 dB is shown as −40 dB

excellent. Note that these results are obtained with a peripheral choke. The results of this Figure and Tables 13.5–13.7 indicate that this type of dipole feed has excellent pattern characteristics and that its performance is broadband.

An important piece of information from Table 13.6 is the invariance of the feed performance with d_f, the separation of the dipole from the cavity floor. This means one may replace the dipole with a microstrip antenna without reducing its performance, i.e. this dipole cavity feed is a flexible design for many applications. In Chapter 14 it will be shown that a similar cavity can also be excited with a dielectric resonator, giving excellent performance.

13.5 References

1 SILVER, S.: 'Microwave antenna theory and design', (McGraw–Hill, New York, 1949), p. 252

2 KILDAL, P.S., and SKYTTEMYR, S.A.: 'Dipole-disk antenna with beam-forming ring', *IEEE Trans.*, 1982, **AP–30**(4), pp. 529–535

Chapter 14

Microstrip and resonator type feeds

L. Shafai and A.A. Kishk

14.1 Introduction

The feed configurations discussed in Chapter 13 dealt with traditional designs which were based on using waveguide and horns and their variations. Their design and performance have been improved through the years and they currently form the most efficient and accepted feeds for reflector antennas. However, microstrip antennas are becoming increasingly popular because of their ease of fabrication and low cost. Progress in their design and analysis has also removed most of their earlier limitations in bandwidth and radiation characteristics. Consequently, they are also becoming acceptable candidates as reflector-antenna feeds. Early attempts used the microstrip-antenna advantages of low-profile geometry and low weight to design single or multiband operations in specific applications [1, 2]. However, recent progress in improving their performance extends their application to more general applications.

Two application areas are seeing microstrip antennas as reflector feeds: as a single unit for prime-focus reflectors; and in array feeds. In the latter they are used in multiple-beam applications, shaping the radiation patterns and compensating for the reflector-focusing imperfections. In recent years, in millimetre-wavelength applications, there have also been developments in their monolithic integration with electronics, which will further expand the microstrip applications through the possibility of signal processing in addition to passive array processing. These applications, however, are beyond the scope of this chapter and will not be addressed. Here, we shall limit our attention to the design of the microstrip-antenna element as a reflector-antenna feed for prime-focus reflectors.

In the past, microstrip antennas were not considered as feeds for reflector antennas, primarily because of the asymmetry of their radiation patterns and narrow impedance bandwidths. These problems have now, however, been overcome. The pattern symmetry can be controlled by the ground-plane size, and the bandwidth can be increased by parasitic coupling to adjacent patches. In this respect, microstrip patches are similar to tuned circuits. Since microstrip patches are resonant cavities, coupling two or more patches which have different but adjacent resonance frequencies provides an impedance behaviour that is sensitive to their coupling. Thus, by controlling their coupling coefficients, varieties of impedance forms from dual-frequency or multifrequency operation to a single wideband operation can be obtained. Properties of tuned circuits, can be used to assist in selection of appropriate patch dimensions and coupling coefficients to facilitate impedance matching. Thus, from the point of view of feed design it is more important to concentrate on the geometrical optimisation for radiation-pattern control.

The early use of microstrip antennas as reflector feeds was reported by Kerr [1]. The low-profile nature of microstrip was also used there to achieve a dualband operation. Feed optimisation and the achievable reflector characteristics in gain factor and crosspolarisation etc. were not, however, provided. Later in Reference 2, a microstrip feed was described which was supported centrally and resulted in a simple feed–support configuration. The use of single microstrip antennas and arrays of microstrip antennas as reflector feeds was investigated by Hall and Prior [3]. Good crosspolarisation performance was obtained with a patch–ring combination that had a broad radiation pattern. Array patterns were studied as a narrowbeam feed for application in offset reflectors or those with a large f/D. A numerically optimised feed was recently investigated by Kishk and Shafai [4], where a coupled-patch configuration was used to enlarge the feed impedance bandwidth above 10%.

One of the difficulties with microstrip antennas is the resistive losses due to the conductor and its substrate, which become significant at high frequencies. In an attempt to reduce these losses, Kishk *et al.* have studied the use of dielectric resonators as a candidate for reflector feeds and in lieu of microstrip patches [5, 6]. Feed performances similar to those of microstrip antennas are obtained.

In this chapter we provide computed and experimentally verified design results for both microstrip and resonator-type feeds. Their performance with reflector antennas is also discussed.

14.2 Microstrip-antenna feed

The configuration of Figure 14.1 was selected for analysis; it shows a lower microstrip patch excited by an electric dipole representing the feed probe, an upper patch which is parasitically coupled to the lower one to broaden the bandwidth and a ground plane with a peripheral conducting ring to support a quarter-wavelength choke. The antenna is analysed to optimise its various dimensions. In the following we provide radiation patterns for different variations of this geometry, starting with a standard circular patch [7].

Some computed results are presented here so as to examine the performance of such antennas. Initially, we investigated a microstrip antenna with a small ground plane of diameter $D \leqslant 0.8\lambda$ where λ is the wavelength in free space. With this diameter size it is possible to achieve radiation pattern symmetry [6]. For an ideal feed, the geometry must excite only the dominant TM_{11} mode in the cavity. However, with the finite structure all modes are excited and the patch

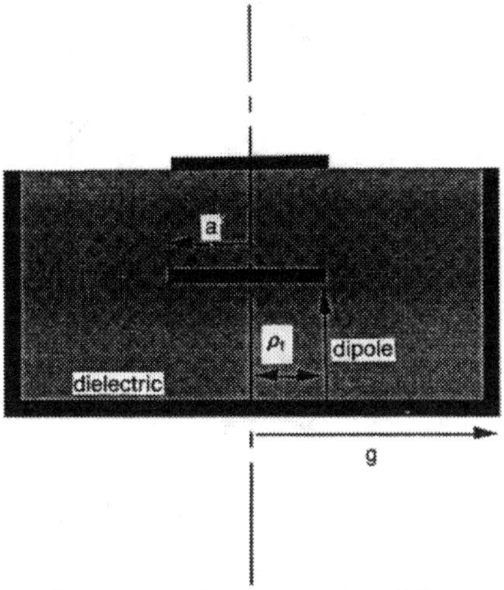

Figure 14.1 Geometry of microstrip for numerical modelling

and the ground-plane dimensions must be used to enhance the magnitude of the TM_{11} mode. Nevertheless, the contributions of the zero-order and the higher-order modes will exist, and these will cause the principal-plane patterns $E_{\theta|\phi=0}$ and $E_{\theta|\phi=180}$ to be different. Similarly, the crosspolarisation pattern in the $\phi=45°$ and $\phi=225°$ planes will not be the same. In the present calculations we have used the first seven modes (TM_{01}, $TM_{\pm 11}$, $TM_{\pm 21}$, and $TM_{\pm 31}$), which are sufficient for accurate representation of radiated fields [6]. Figure 14.2 shows

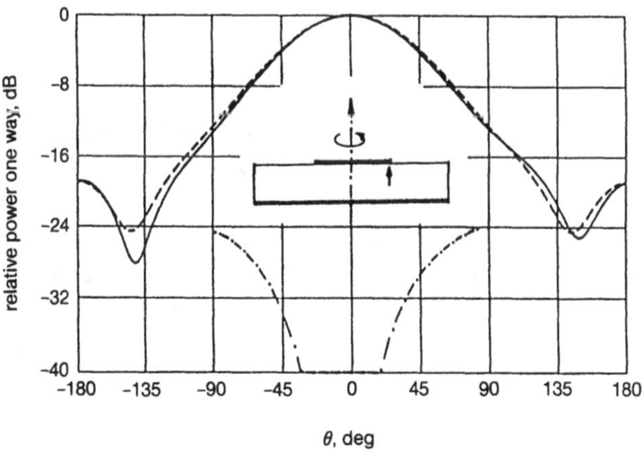

Figure 14.2 Radiation patterns of a circular microstrip patch

$a=0.71\lambda$, $t=0.05\lambda$, $g=0.4\lambda$, $p_f=0.17\lambda$, $\varepsilon_r=2.32$

——— E-plane

– – – – H-plane

–·–·– crosspolar

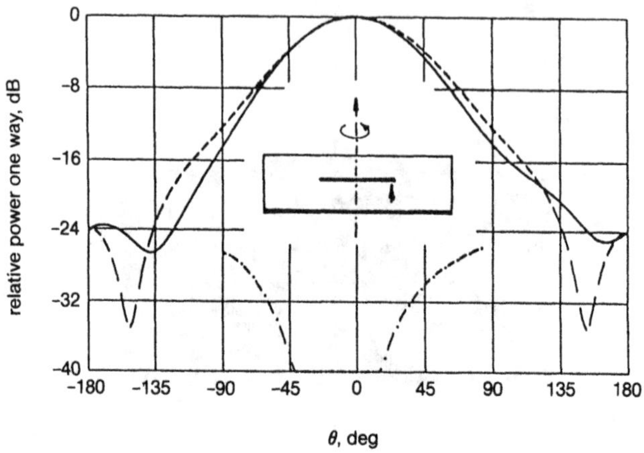

Figure 14.3 *As for Figure 14.2 for an embedded microstrip antenna of 0.1λ dielectric thickness and g = 0.4λ*

——— E-plane; - - - - H-plane; -·-·- crosspolar

the radiation patterns of a circular microstrip antenna with $a = 0.17\lambda$, $t = 0.05\lambda$, $g = 0.4\lambda$ and the feed at the patch edge. An excellent symmetry of the radiation is evident, and the peak crosspolarisation level is below -24 dB. However, the backlobe level is about -19 dB. The effect of covering the patch with a dielectric sheet of thickness 0.05λ was also studied, and the resultant patterns are shown in Figure 14.3. It is found that the dielectric cover reduces the radiation-pattern symmetry, but improves the backlobe level and the crosspolarisation. If the antenna is surrounded, from the sides, by a conducting cylindrical collar, the radiation patterns improve, as shown in Figure 14.4.

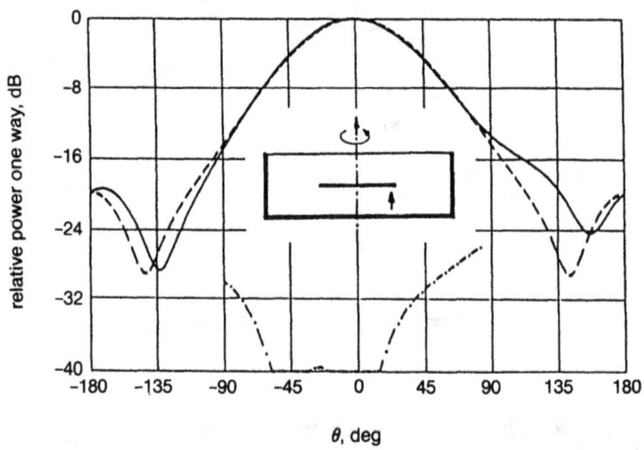

Figure 14.4 *As for Figure 14.2 but with a conducting collar surrounding the dielectric edges*

——— E-plane; - - - - H-plane; -·-·- crosspolar

The previous steps studied the possible improvement of the radiation-pattern symmetry. Now, to improve the bandwidth, we use a stacking technique. To examine the effect of stacking on the radiation pattern, the example of Figure 14.4 is selected and a second patch is added on top of the original patch. The radiation patterns of the new geometry are shown in Figure 14.5. Comparing Figure 14.5 with Figure 14.4, no significant change can be noticed. This illustrates that stacking the microstrip antenna with 0.05λ separation does not affect the radiation-pattern symmetry, although it improves the bandwidth. In Section 14.3, the effects of adjacent modes and quarter-wavelength chokes on the patterns are investigated.

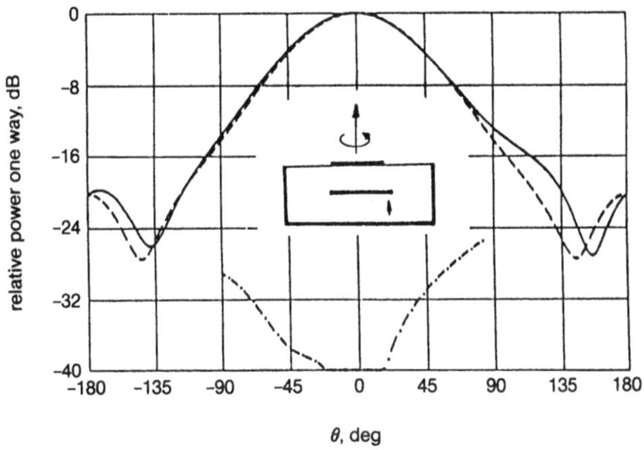

Figure 14.5 *Two-layer stacked microstrip*

Diameters: 0.34λ (bottom), 0.32λ (top)
$g = 0.4\lambda$
——— E-plane; – – – – H-plane; –·–·– crosspolar

14.3 Effect of mode excitation on the crosspolarisation

For reflector-feed applications, the dominant TM_{11}-mode excitation gives desirable pattern characteristics. However, as indicated previously, other modes are also present and contribute to the radiation patterns. Generally, with a TM_{11}-mode excitation the copolar patterns are adequately described by a consideration of the dominant TM_{11} mode. On the other hand, this mode is not adequate for accurate calculation of the crosspolarisation. To indicate the significance of the other modes to the crosspolarisation, the radiation pattern is calculated by inclusion of various modes, and the results are shown in Figure 14.6. It shows the crosspolarisation due to the first two modes and additional ones up to the TM_{31} mode. It is evident that at least four initial modes must be considered so that the crosspolarisation can be determined accurately.

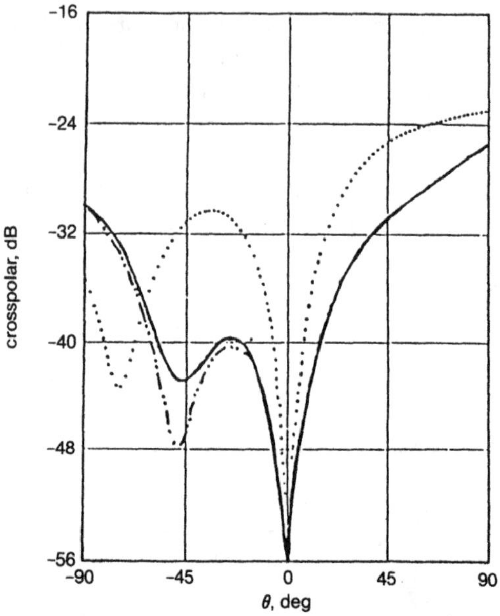

Figure 14.6 *Effect of excitation modes on the crosspolarisation level of the arrangement shown in Figure 14.4*

$\phi = 45°$
········ $TM_{01} + TM_{11}$
–··–··– $TM_{01} + TM_{11} + TM_{21}$
——— $TM_{01} + TM_{11} + TM_{21} + TM_{31}$

14.4 Effect of conducting chokes

Previous results indicated that, generally, the backradiation level of a microstrip antenna is high ($\geqslant -20$ dB). In waveguide feeds adding a quarter-wavelength choke reduces the backlobe level by about 10 dB or more (Chapter 7). This idea may be used with microstrip feeds to reduce the backlobe level. To investigate this possibility, we selected the geometry of Figure 14.5 and added a $\lambda/4$ choke (the distance between the central short-circuit and ground-plane edge is $\lambda/4$) behind the microstrip ground plane. The radiation patterns of the new geometry were recalculated and are shown in Figure 14.7. The result indicates an improvement in the backlobe level of about 4 dB, without disturbing the radiation-pattern symmetry or the bandwidths. Since this type of choke does not improve the radiation pattern significantly, the effect of a peripheral choke on the radiation pattern was studied. The computed patterns are shown in Figure 14.8. In this case the aperture diameter is increased by 0.2λ, and the backlobe level is improved by about 10 dB. The crosspolarisation pattern is also improved, in particular for the wide angular range. These improvements were noticed for antennas with ground-plane size $g = 0.4\lambda$. The peripheral choke was

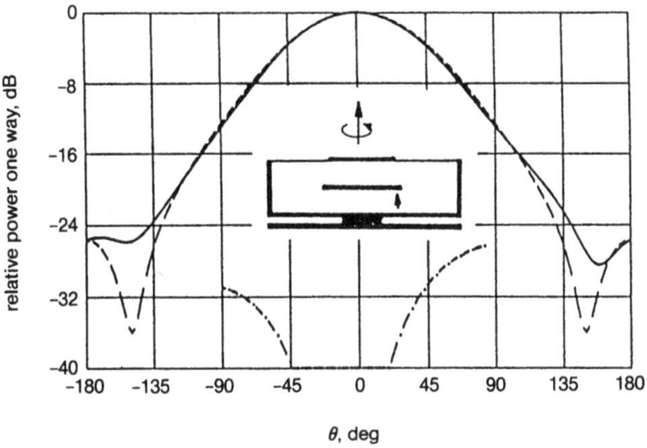

Figure 14.7 *Two-layer stacked microstrip as in Figure 14.5 with a λ/4 backchoke*

——— E-plane
– – – – H-plane
–·–·– crosspolar

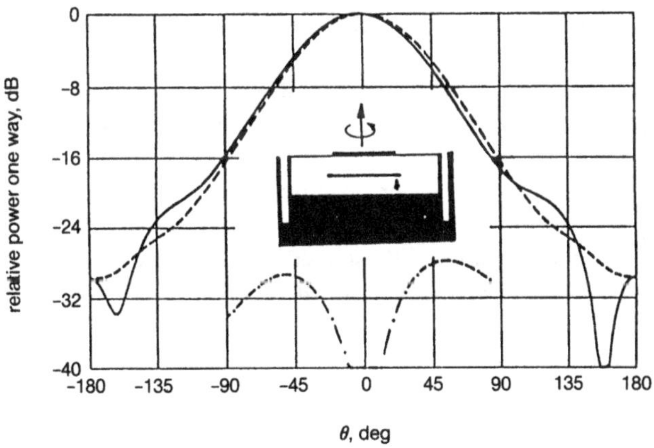

Figure 14.8 *Two-layer stacked microstrip as in Figure 14.5 with a λ/4 sidechoke*

——— E-plane
– – – H-plane
–·–·– crosspolar

found to be unsuitable for the smaller antennas. The measured radiation patterns of this antenna are shown in Figure 14.9. The crosspolarisation is higher than that computed, and shows a central peak owing to misalignment in the experimental setup.

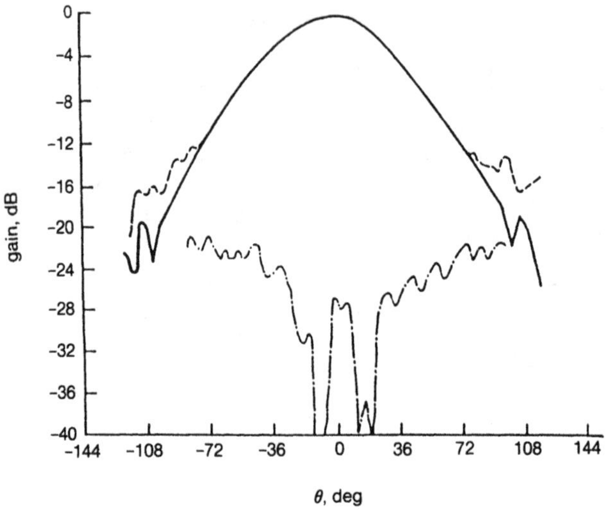

Figure 14.9 Measured patterns for the feed shown in Figure 14.8

——— E-plane
- - - - H-plane
—·—·— crosspolar at $\phi = 45°$

14.5 Feed performance with the reflector antenna

The return loss of the feed was also measured, and is shown in Table 14.1. Further improvement of these return losses can be achieved by modifying the feed-point location.

Table 14.2 summarises the performance of the above antennas, where the beamwidths at 3 dB and 10 dB levels and the peak crosspolarisation are given. To evaluate the performance of these feeds on a reflector antenna, the data on the gain factor, spillover efficiency and the corresponding aperture angles must be known. They are calculated and shown in Table 14.3. It shows that the aperture angle varies from 60° to 71° and the gain factor rises from 72.5% to 74.24%. The feed performance is therefore reasonable. The computed gain factors are somewhat smaller than those of waveguide feeds with corrugated flanges. However, the aperture blockage of microstrip feeds is small owing to their small size. Thus, when used on small reflectors, they should provide comparable performance. These microstrip feeds are, on the other hand, light weight and easy to fabricate and can readily be integrated with the receiving electronics.

Table 14.3 also shows the location of the phase centre of each antenna calculated over its aperture angle, given in column 2. The locations are measured from the ground plane, i.e. $z=0$, and are all positive, indicating that the phase centres are above the ground plane. However, it is interesting to compare the example of Figure 14.3 with the remaining examples, which have

Table 14.1 Measured return losses of feeds

Feed 1 (Figure 14.5)		Feed 2 (Figure 14.8)	
Frequency	Return loss	Frequency	Return loss
(GHz)	(dB)	(GHz)	(dB)
4.10	9.5	4.30	10.0
4.15	11.0	4.35	11.0
4.20	13.0	4.40	11.5
4.25	14.0	4.45	12.0
4.30	17.0	4.50	12.5
4.35	18.0	4.55	12.5
4.40	16.0	4.60	12.5
4.45	15.5	4.65	12.5
4.50	14.0	4.70	12.0
4.55	11.0	4.75	11.5
4.60	9.0	4.80	10.0

Table 14.2 Feed characteristics at various frequencies

Example of Figure	Peak cross-polarisation 0≤q≤90°	Gain	Beamwidths 3 dB E-plane	Beamwidths 3 dB H-plane	Beamwidths 10 dB E-plane	Beamwidths 10 dB H-plane
(*a*) Frequency f_0	(dB)	(dB)	(deg)	(deg)	(deg)	(deg)
14.3	−25.2	7.98	74.4	78.7	143.4	156.5
14.4	−25.2	8.37	72.3	72.5	141.8	141.9
14.5	−24.7	8.19	74.6	73.4	148.1	143.3
14.7	−26.1	8.36	74.2	73.4	142.9	142.2
14.8	−27.7	9.08	65.9	66.9	128.4	130.4
(*b*) Frequency $f=1.05\ f_0$						
14.3	−24.9	7.50	75.4	76.1	145.1	154.2
14.4	−28.7	7.13	80.3	85.9	170.1	171.1
14.7	−25.9	7.35	82.6	76.9	163.5	148.2
14.8	−25.3	9.05	66.1	67.2	129.5	131.1
(*c*) $f=0.95\ f_0$						
14.3	−26.1	7.68	73.1	78.1	144.3	158.1
14.4	−26.5	7.50	73.4	81.5	145.7	161.4
14.7	−25.1	8.24	72.7	74.4	141.1	143.8
14.8	−23.1	8.61	68.1	69.0	139.7	134.6

Table 14.3 Reflector-aperture angles, gain factors, spillover efficiencies and phase-centre locations above the ground plane for various feeds

Example of Figure	Aperture angle (deg)	Gain factor (%)	Spillover efficiency (%)	Phase centre (λ)
(a) $f=f_0$				
14.3	68	72.93	85.43	0.0075
14.4	66	73.14	85.23	0.1127
14.5	66	72.86	84.16	0.1202
14.7	66	74.29	86.24	0.1186
14.8	60	73.83	84.67	0.1827
(b) $f=1.05\,f_0$				
14.3	67	72.1	84.9	0.08
14.4	67	72.3	86.6	0.144
14.7	71	73.0	87.0	0.142
14.8	60	73.8	84.67	0.174
(c) $f=0.95\,f_0$				
14.3	69	73.2	85.8	0.065
14.4	68	73.3	87.1	0.085
14.7	66	74.3	86.8	0.114
14.8	66	71.2	83.9	0.416

the peripheral conducting collar. In the former cases, the phase-centre location is just above the ground plane. Since the total thickness of the dielectric is 0.1λ, the phase centres are inside the dielectric and under the lower patch. In the remaining cases, all phase centres are outside the dielectric. From these results the following important conclusion can be drawn: in a microstrip antenna as described by Figure 14.3, the radiation is mostly from the aperture between the patch and the ground plane. Incorporating the side collar raises the radiation zone to the periphery around the upper patch.

The performance of the above antennas listed in Tables 14.2*a* and 14.3*a* was also studied as a function of frequency. Within ±5% frequency variation, the computed results are shown in Tables 14.2*b*, 14.2*c*, 14.3*b* and 14.3*c*. An examination of these results reveals that the feed performance remains relatively constant within the band, in the magnitude of the peak crosspolarisation and the reflector-gain factor. The feed gain, however, decreases to some degree, regardless of whether the frequency is increased or decreased. This is primarily due to the increased excitation of the modes adjacent to the TM_{11} mode.

14.6 Patch-ring feed

The patch–ring configuration of Reference 3 (Figure 14.10) is a simple microstrip feed. Its numerically generated patterns are shown in Figure 14.11, where the crosspolarisation is computed in the 45° plane. These results agree

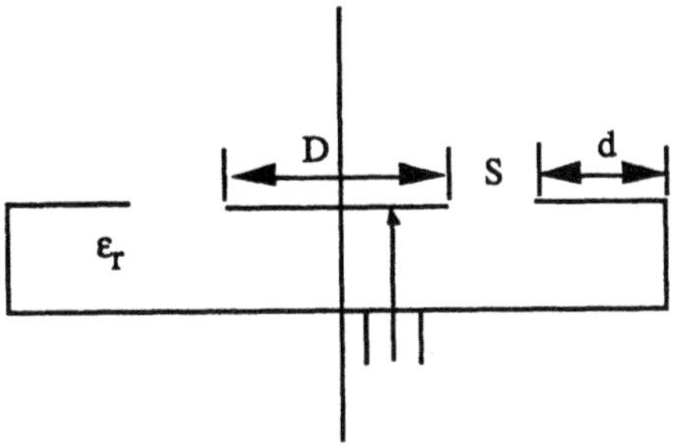

Figure 14.10 Patch-ring-feed geometry

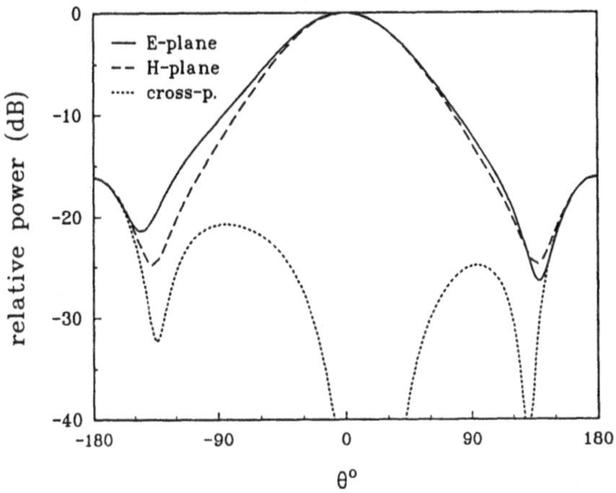

Figure 14.11 Computed far-field patterns of a patch-ring feed
$D=0.338\lambda$, $s=0.039\lambda$, $d=0.156\lambda$, $t=0.06\lambda$, $\varepsilon_r=2.5$

reasonably well with the measured data given in Reference 3. The backradiation is high, since the design does not incorporate a quarter-wavelength choke. A disadvantage of this feed is its closed-cavity configuration. With a cavity of this type the resistive losses are usually high and, even with an air-dielectric microstrip, cannot be eliminated. Consequently, the feed losses, and in turn its noise temperature, will be high. To reduce losses one may remove the end short-circuit to open the cavity (Figure 14.12). Since the radiation is mostly from the annular ring, opening the cavity does not affect its radiation patterns significantly. The new results are shown in Figure 14.13, which show some improvement over the patterns of the original cavity feed of Figure 14.10.

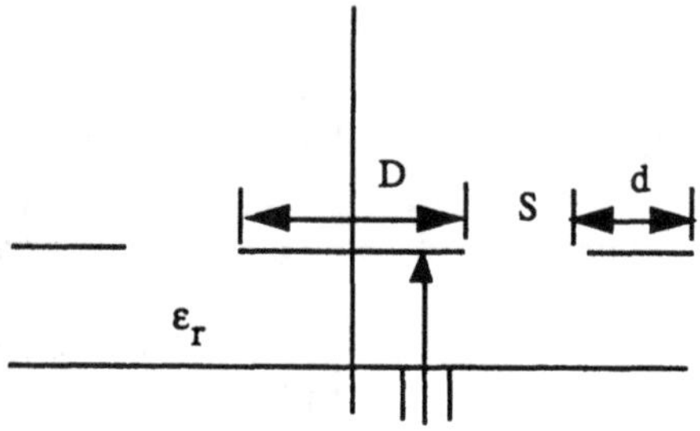

Figure 14.12 Modified patch-ring feed

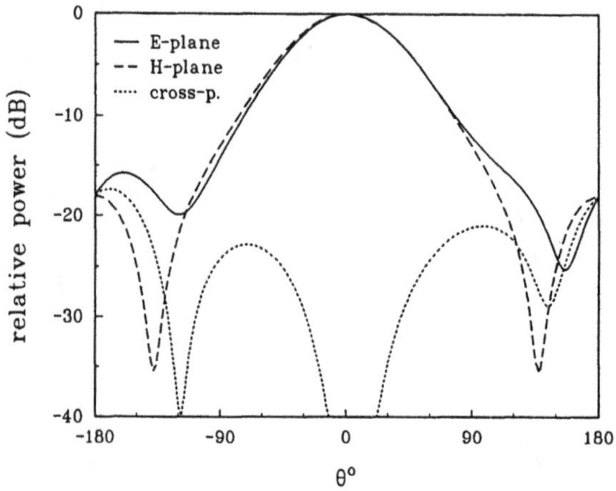

Figure 14.13 Computed far-field patterns of feed in Figure 14.12 with the data as in Figure 14.11

14.7 Dielectric-resonator feed

As indicated earlier, microstrip feeds tend to have low efficiencies at high frequencies owing to resistive losses of the conducting patch and substrate. Since microstrip antennas are resonant cavities, they can be replaced by a dielectric resonator. With a low-loss dielectric their efficiency could be better than microstrip patches at high frequencies. To show their potential as reflector feeds, the configurations of the microstrip feed from Section 14.5 is selected and is excited by a dielectric resonator. The dielectric relative permittivity is $\varepsilon_r = 8.9$. The antenna configuration without a peripheral choke is shown in Figure 14.14. The geometrical dimensions are: $H = 0.45\lambda$, $D = 1.0\lambda$ for the waveguide section and $d_c = 3.3$, $h = 0.3412\lambda$ for the resonator. The dimensions of the dielectric resonator are chosen to resonate at the TM_{110} mode. Assuming that the

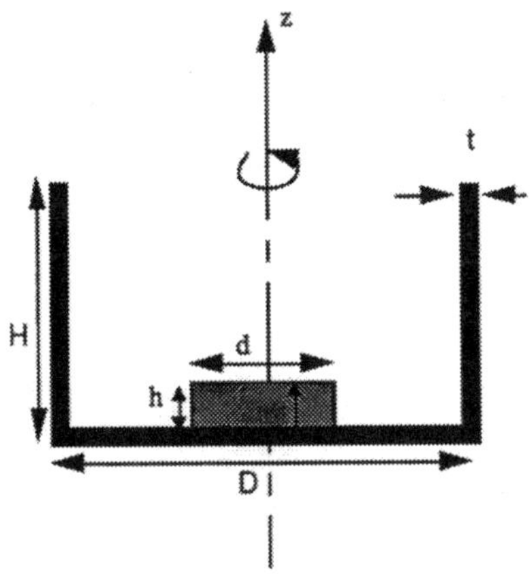

Figure 14.14 *Dielectric-resonator antenna housed inside a circular waveguide*

resonator is over a conductor and is covered with a perfect magnetic conductor, its diameter is calculated from

$$d = \frac{1.841}{2\sqrt{\varepsilon_r}} \sqrt{\left\{16 + \left(\frac{\pi}{1.841}\frac{d}{h}\right)^2\right\}}$$

As with microstrip antennas, since the dielectric resonator is an open cavity its excitation generates not only the dominant TM_{110} mode, but also the adjacent TM_{010}, TM_{210} and TM_{310} modes [6]. The far-field patterns of this feed are shown in Figure 14.15, which has high backlobes. With a larger waveguide section

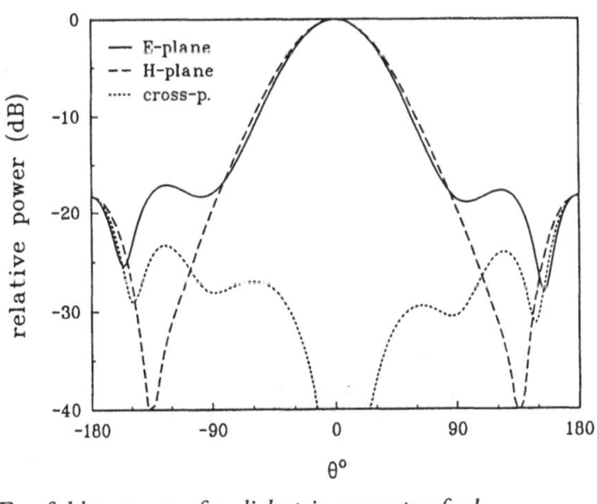

Figure 14.15 *Far-field patterns of a dielectric-resonator feed*
$D=\lambda$, $H=0.48\lambda$, $t=0.05\lambda$, $\varepsilon_r=8.9$

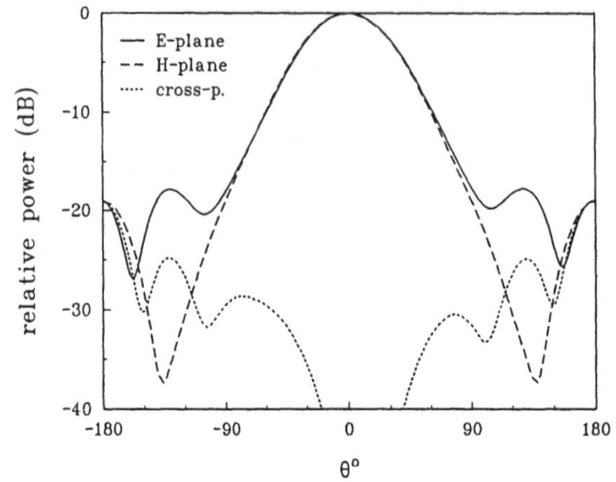

Figure 14.16 Far-field patterns of a dielectric-resonator feed with an enlarged cavity $D=\lambda$, $H=0.84\lambda$, $t=0.05\lambda$, $\varepsilon_r=8.9$

$H=0.84\lambda$, the patterns in Figure 14.16 show slight improvement in symmetry and reduced crosspolarisation. Since the waveguide section is cutoff for higher-order modes it further attenuates them and improves the aperture illumination by the dominant mode.

The above antenna, after incorporating a quarter-wavelength peripheral choke, is shown in Figure 14.17. Their far-field patterns are provided in Figures 14.18 and 14.19, which show improved pattern symmetry and reduced backlobe level. The peak crosspolarisation is very low at about −33 dB level, which is considerably better than that of the microstrip feed shown in Section 14.4. Here, the finite length of the waveguide attenuates the higher-order modes. A similar approach may also be used with microstrip feeds to improve their crosspolarisation.

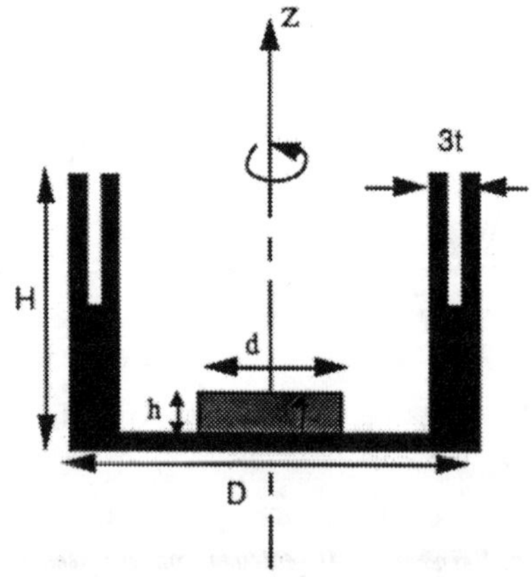

Figure 14.17 Dielectric-resonator feed with a choke

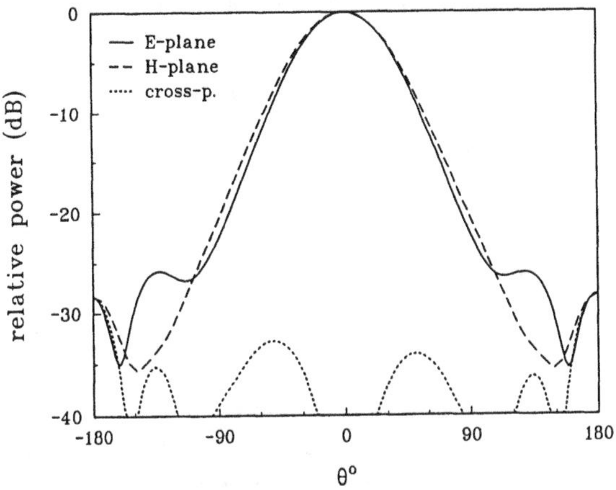

Figure 14.18 Far-field patterns of the dielectric-resonator feed of Figure 14.15 with a choke

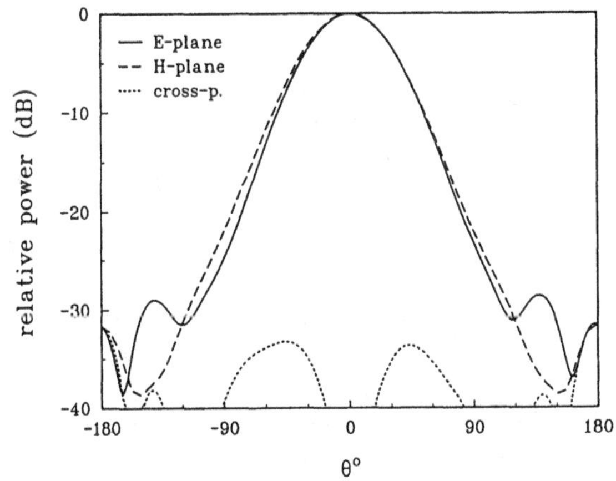

Figure 14.19 Far-field patterns of the dielectric-resonator feed of Figure 14.16 with a choke

The frequency characteristics of the above two antennas as a feed for a reflector are shown in Tables 14.4 and 14.5, which are similar to the performance characteristics of the microstrip feed shown in Tables 14.2 and 14.3.

Table 14.4 Frequency characteristics of antenna in Figure 14.15

f, GHz	9.50	9.75	10.00	10.25	10.50
10 dB beamwidth					
E-plane, deg	120.0	115.0	114.0	110.0	106.0
H-plane, deg	122.0	121.0	120.0	118.0	116.0
Crosspolarisation, dB	−25.20	−26.20	−27.00	−27.40	−27.60
Backlobe, dB	−17.30	−17.80	−18.30	−18.70	−19.20
d_c	−00.04	−00.04	−00.04	−00.04	−00.03
ψ_0^0	59.00	57.00	56.00	56.00	54.00
η_g	72.10	72.80	73.40	73.90	74.50
η_s	84.20	84.20	84.80	80.10	85.80

d_c = phase-centre location from waveguide aperture in free-space wavelengths (λ_0)
η_g = maximum gain factor at ψ_0
η_s = spillover efficiency at ψ_0

Table 14.5 Frequency characteristics of antenna in Figure 14.18

f, GHz	09.50	09.75	10.00	10.25	10.50
10 dB beamwidth, deg					
E-plane	116.0	112.0	108.0	105.0	102.0
H-plane	118.0	116.0	114.0	112.0	111.0
Crosspolarisation, dB	−27.80	−31.90	−32.80	−31.80	−31.30
Backlobe, dB	−24.20	−27.40	−28.30	−28.20	−27.60
d_c			00.06		
ψ_0	57.00	56.00	54.00	54.00	52.00
η_g	74.80	75.50	75.90	76.20	76.40
η_s	86.70	87.80	87.50	88.50	87.60

Definition of symbols are as in Table 14.4

An examination of the entries of these Tables reveals that the beamwidth, backlobe and crosspolarisation levels and the aperture angle all decrease with the wavelength. Further, the values of these parameters are less with the choke installed, which accounts for the improvement of the reflector-antenna efficiencies observed in each case. Finally, the phase centre is located inside the aperture with no choke, but moves outside when the choke is installed. In either case, the phase centre is stationary within the frequency range of operation. The excellent characteristics of this feed are basically a result of utilising the dielectric resonator as a probe. A resonator-excited feed can therefore be used in place of microstrip feeds with improved performance.

14.8 References

1 KERR, J.L.: 'Microstrip antenna developments'. Proceedings of workshop on Printed Circuit Antenna Technology, New Mexico State University, 1979, pp. 3.1–3.20

2 CONROY, P.J.: 'Disc antenna feed for parabolic reflector'. US patent 4 195 301, March 1980

3 HALL, P.S., and PRIOR, C.J.: 'Microstrip feeds for prime focus fed reflector antennas', *IEE Proc. H*, 1987, **134**(2), pp. 185–188

4 KISHK, A.A., and SHAFAI, L.: 'Optimization of microstrip feed geometry for prime focus reflector antennas', *IEEE Trans.*, 1989, **AP–37**(4), pp. 445–451

5 KISHK, A.A., AUDA, H.A., and AHN, B. C.: 'Accurate prediction of radiation patterns of dielectric resonator antennas', *Electron. Lett.*, 1987, **23**(25), pp. 1374–1375

6 KISHK, A.A., AUDA, H.A., and AHN, B.C.: Radiation characteristics of cylindrical dielectric resonator antennas with new applications', *IEEE Antennas Propag. Newsl.*, 1989, **31**(1), pp. 7–16

7 KISHK, A.A., and SHAFAI, L.: 'The effect of various parameters of circular microstrip antennas on their radiation efficiency and the mode excitation', *IEEE Trans.*, 1986, **AP–34**, pp. 959–977

8 KISHK, A.A., and SHAFAI, L.: 'Optimization of microstrip feed geometry for prime focus reflector antenna', *IEEE Trans.*, 1989, **AP–37**(4), pp. 445–451

Chapter 15

Array feeds

P.J.B. Clarricoats

15.1 Introduction

This chapter is concerned with the use of more than one element to form the feed for a reflector antenna. In Chapter 2, the design of shaped-coverage antennas using array feeds is described, but one of the fundamental design problems which arises when elements are combined, namely mutual coupling, is not addressed there.

Arrays, as antennas in their own right, have a long history extending back to the early years of radio broadcasting where they were used at HF and later at VHF. Their use at microwave frequencies emerged during the Second World War when they were deployed in radar systems, sometimes in conjunction with reflectors. Phased-array radars incorporating large numbers of elements were designed to provide inertialess beam scanning in the 1950s, and the need to consider mutual couping for general waveguide elements such as dipoles and slots led to very detailed studies being undertaken. Account of this work is to be found in many excellent texts [1, 2] and a very large number of papers (see bibliography in Reference 3). Our interest here is with the array as a feed for a reflector, so we shall not discuss some of the important complications, such as blind spots, which arise when an array is deployed as an individual scanning antenna and where the beam or beams are swung through large angles. These problems are discussed adequately elsewhere [1, 4, 5]. Brief mention is made here of scanning antennas where the array is fed so as to provide a limited beam scan which is magnified through reflector optics.

Array feeds for reflectors usually consist of waveguides operating in single- or dual-mode configurations, horns or microstrip radiators. In our discussion we concentrate on mutual coupling between waveguide and horn radiators, since these form a very important class and have been the subject of detailed analysis. The number of elements in an array feed can vary from a few, for example when the feed is to be used with a shaped reflector, to more than 100, as in the *Intelsat VI* C-band offset-paraboloidal antenna. This latter antenna was one of the first to incorporate a means to reconfigure the antenna pattern with the spacecraft in orbit, achieved through the use of a 3-way microwave switch in the beam-forming network (BFN). This permitted three different element excitations to be obtained, corresponding to the illumination, respectively, of the Atlantic,

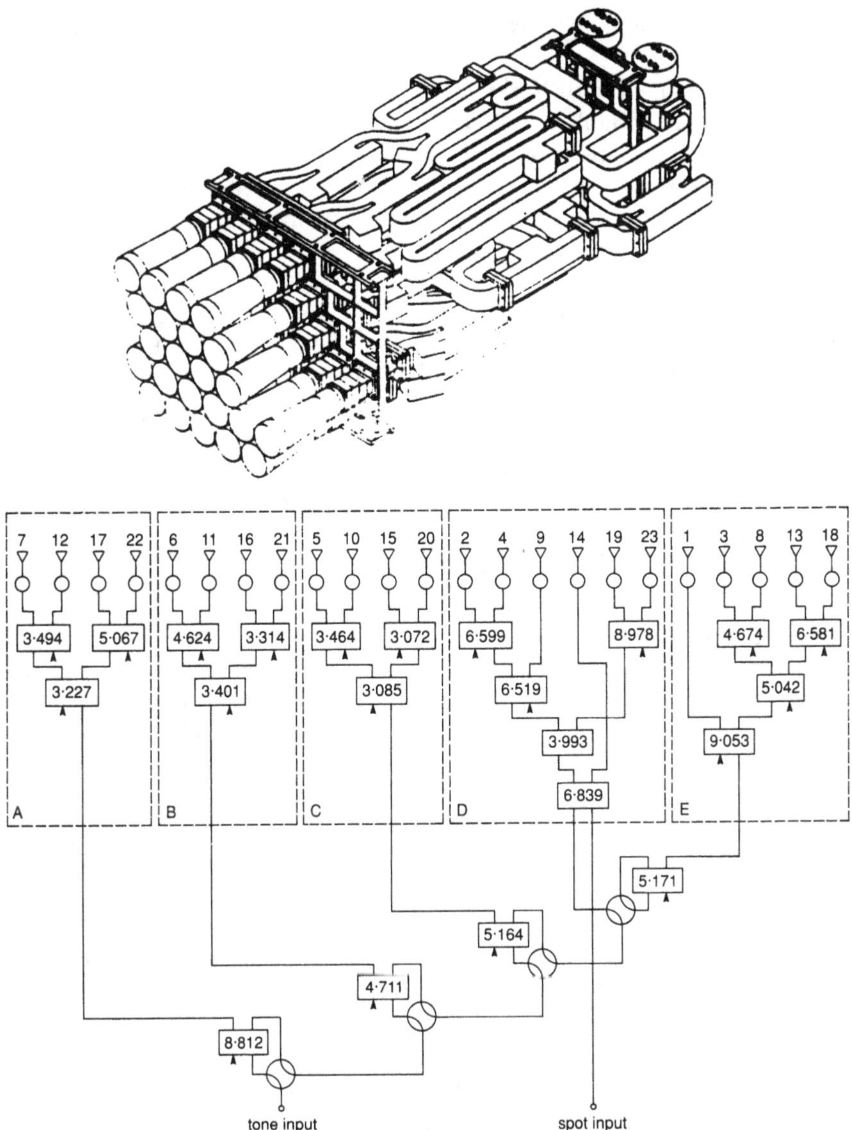

b

Figure 15.1 Switched reconfigurable Ku-band feed system and BFN topology [6]

a Ku-band feed system
b BFN topology, 307° coverage

Pacific and Indian Ocean regions. An example of a 23-element Ku-band horn subarray incorporating a switched BFN is shown in Figure 15.1 [6]. The detailed design of BFNs is a subject which lies beyond our scope, and only limited comments are presented here: suffice it to say that, for waveguide and horn arrays, both TEM and waveguides have been used to connect the elements and form couplers.

Reconfiguration of the BFN has been accomplished by the use of switches, as mentioned above, variable-power dividers (VPDs) and variable phaseshifters (VPSs) or by combinations of these. The example of Figure 15.1, which was designed to illuminate North America from four satellite locations, utilises limited reconfigurability, where a few VPDs are used on the input stages to connect groups of sub-BFNs.

The array excited by a classical BFN requires switching at high power. As an alternative, the use of the Butler-matrix coupler within a BFN has been explored extensively for satellite applications, and a survey has been published by Roederer [7]. There it is shown that reflectors fed from a beam-forming system with one feed and one power amplifier per beam do not provide power-to-beam allocation flexibility, unless complex high-power switching is employed, as above. One solution is to use overlapping feed clusters where each beam uses, typically, three, four, seven or nine feeds powered via a multiport amplifier. Identical amplifiers are inserted between two identical back-to-back multiport Butler-like couplers which perform two cascaded inverse transforms, so that each low-level input corresponds to a feed element and each beam is generated by proper excitation of the inputs. The disadvantage of the system is the need for two very large back-to-back matrices which, unless the coverage region contains a large number of scattered users, will have an adverse effect on amplifier efficiency. Spring and Moody have suggested [8] that the large multiport amplifiers be replaced by several smaller ones.

A similar multimatrix concept, independently proposed by Roederer [9], also implies overlapping feed clusters (Figure 15.2*a*). The feeding arrangement excludes half of the small matrices and includes fixed or variable phase shifters for each beam. Beams originate from three feeds A, and B and C powered via three 4 × 4 hybrid couplers from 12 equally excited amplifiers. Beam crossover is typically at − 3 dB below peak. Full flexibility in traffic-to-beam allocation is available by low-level switching of more or fewer channels into each beam. All amplifiers operate at maximum efficiency, and once the feed clusters create a − 10 dB taper at the reflector edge an aperture efficiency of 55% is achievable, with low sidelobes. Use of four, five, seven or nine feeds per beam provides better sidelobes and higher efficiency.

Multimatrix antennas can also provide fine beam hopping. Figure 15.2*b* shows the feed principle for scanning across Europe with a 30-wavelength reflector using 32 feeds. Depending on the user location, the phases are set to power four, six or nine feeds. Four 8 × 8 couplers and 32 2-bit phase shifters allow fine beam hopping with low sidelobes and give a minimum directivity corresponding to an aperture efficiency of 50%.

In another type of scanning antenna, the imaging array-fed reflector, a small array is magnified by optics. Single-reflector systems appear to have poor efficiency, but improved optics with shaped main reflectors and subreflectors, fed by a planar array of 32 elements, have recently been optimised.

Among the advantages which arrays offer over single elements is the opportunity to tailor the amplitude and phase of the field accurately over the reflector aperture. This is essential when a contoured coverage is required, or when beams are to be scanned. Limits are imposed by the element pattern, the number of elements and their spacing, the size and the total mass of the array and BFN, and ohmic losses, especially those arising in the BFN. The other complicating factor is mutual coupling, which will now be addressed.

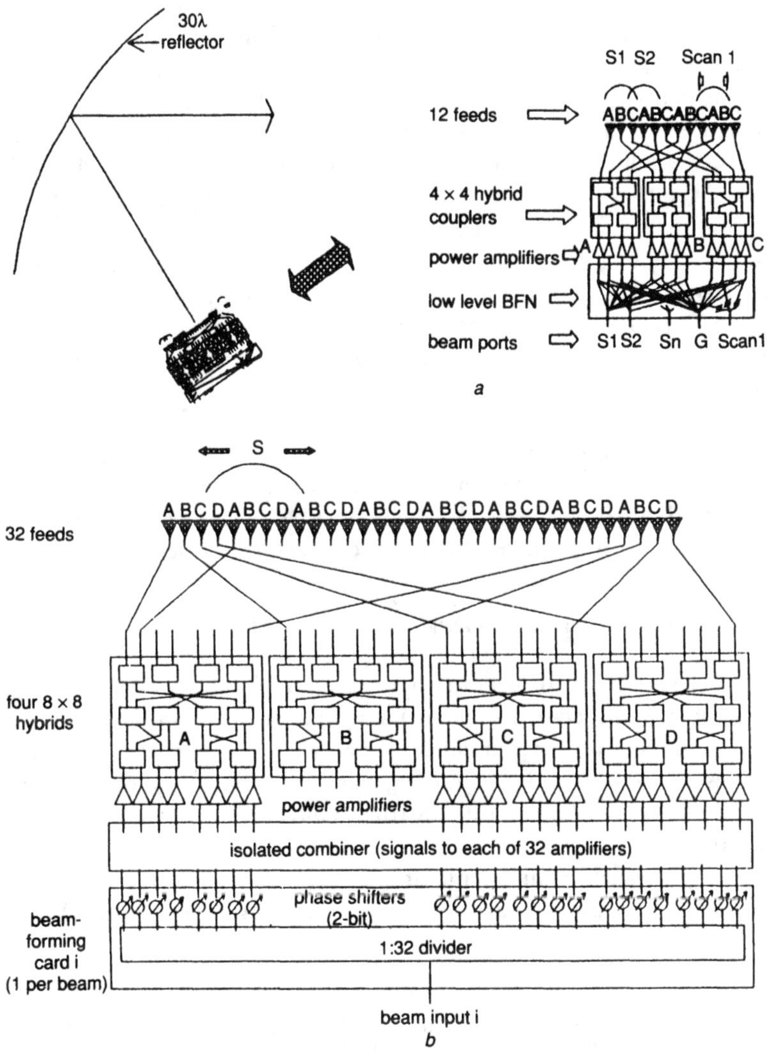

Figure 15.2 Multimatrix antennas [7]

a Multimatrix-antenna principle (ESA patents pending)
b Multimatrix scanning array

15.2 Mutual coupling in an array

15.2.1 Fundamentals

In an idealised array feeding a reflector, each element radiates as if it were isolated and the secondary radiation pattern of the reflector is a combination of the patterns arising from each element radiating alone. This is an idealisation; in practice radiation from each element excites secondary currents in the neighbouring elements, especially near neighbours, and these fields reradiate. There is a consequential influence on both the copolar and crosspolar secondary-radiation pattern of the antenna, its gain, beamwidth and the reflection of energy into the array BFN. Mutual coupling is a polarisation-dependent effect and is a decreasing function of the distance of separation of the elements, though with possible cyclic variations. It depends on the radiation pattern of the element, and is enhanced when, for example, two elements have their patterns and polarisations aligned.

Mutual coupling between circular apertures (waveguides and horns) has been analysed by, among others, Bird [10] and Clarricoats, Tun and Parini [11], and coupling between rectangular and square apertures by Bird [3] and Parini and Kitchener [12]. Elliptical apertures have also been studied by Bird [3].

An excellent summary of the techniques for analysing mutual coupling in arrays has been presented by Bird [3] and a summary is given in Figure 15.3. The integral-equation method he developed for coupled apertures, as used by Clarricoats, Tun and Parini [11], follows next. We chose coupled circular horns for our example, although results for other cases are presented later.

15.2.2 Mutual coupling in an array of conical horns

Figure 15.4 shows identical horn elements in an infinite perfectly conducting ground plane. In practice, the elements will be lodged in a finite ground plane, but the common use of parasitic elements at the periphery of the array minimises the effect of the infinite-ground-plane assumption. The present theory also gives useful insight into the behaviour of feed clusters where no ground plane is present.

The magnetic-field components at a distance R from the origin in a plane transverse to the z direction are related to the aperture electric field through the identity

$$\boldsymbol{H}_t(R)=\frac{j}{2\pi\omega\mu_0}\,(k_0^2+\nabla_t\nabla_t)\,.\,\hat{\boldsymbol{u}}_z\times\int\!\!\int_{-\infty}^{+\infty} \mathrm{d}S'\boldsymbol{E}_t(\boldsymbol{R}',0)G(\boldsymbol{R}-\boldsymbol{R}'\boldsymbol{I}) \quad (15.1)$$

where $|\boldsymbol{R}-\boldsymbol{R}'|=\sqrt{\{(x-x')^2+(y-y')^2+z^2\}}$, $G(R)=\exp(-jk_0R)/R$ is the free-space Green's function, and ∇_t is the transverse-gradient operator. $k_0=\omega/c$ is

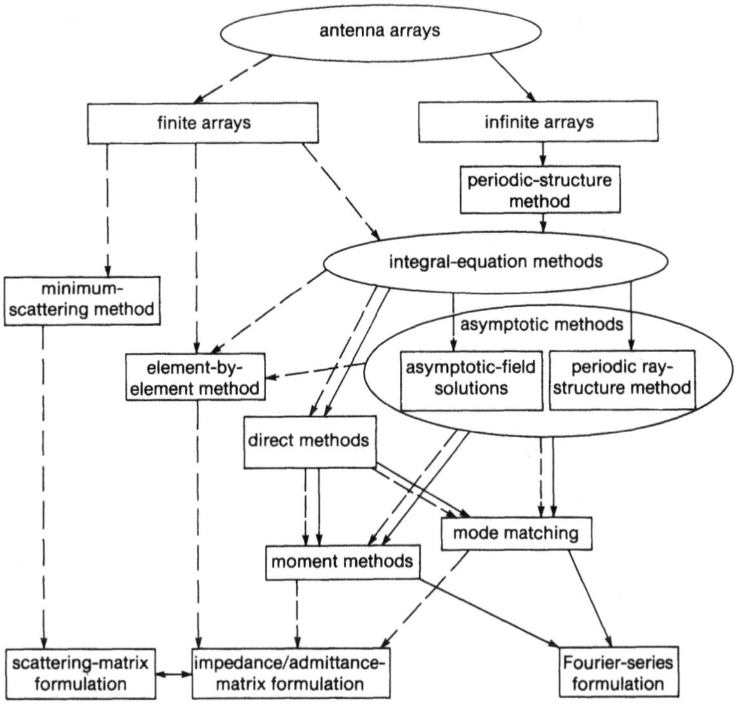

Figure 15.3 Relationship of methods for analysing mutal coupling [3]

the wave-propagation constant in free space. As $\boldsymbol{E}_t$ vanishes on the ground plane, the infinite integral reduces to a sum of integrals over the N horn apertures $D_i (i = 1, 2, \ldots, N)$.

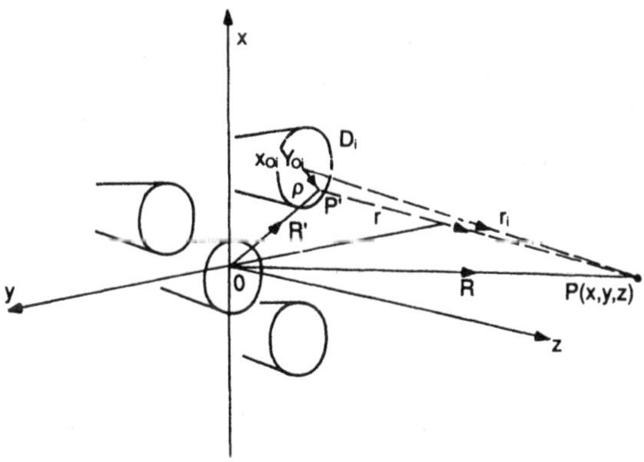

Figure 15.4 Planay array with the radiation field point P and the source point P′ in the horn aperture [11]

In horn i the tangential fields are given approximately by the M-mode expansions

$$\boldsymbol{H}_t^{(i)} = \sum_{m=1}^{M} \{a_m^{(i)} \exp(-j\gamma_m z) - b_m^{(i)} \exp(j\gamma_m z)\}\boldsymbol{h}_m \tag{15.2}$$

$$\boldsymbol{E}_t^{(i)} = \sum_{m=1}^{M} Z_m\{a_m^{(i)} \exp(-j\gamma_m z) - b_m^{(i)} \exp(j\gamma_m z)\}\boldsymbol{h}_m \times \hat{\boldsymbol{u}}_z \tag{15.3}$$

where $\boldsymbol{h}_m$ is the transverse magnetic field of mode m and Z_m is the wave impedance of mode m, which is given by

$$Z_m = \zeta_0 \begin{Bmatrix} k_0/\gamma_m \text{ TE modes} \\ \gamma_m/k_0\varepsilon_r \text{ TM modes} \end{Bmatrix} \tag{15.4}$$

ζ_0 is the intrinsic wave impedance of free space and ε_r is the relative dielectic constant of the material filling the horn, although in practice this value will normally be unity. $\gamma_m = \beta_m = j\alpha_m$ is the propagation coefficient of mode m. Mode m is associated with the mth-highest tranverse-wave number, or eigenvalue λ_m when the wave numbers of the TE and TM modes are arranged in order of ascending magnitude.

When the field point lies on the plane $z = 0$, eqn. 15.1 becomes an integral equation in the field components tangential to the array aperture. This equation is a constraint upon all possible trial solutions to the aperture fields. One possible set of trial solutions is the fields of the orthogonal modes inside the waveguides. We have found that the inclusion of 16 pairs of modes is sufficient to achieve very high accuracy at coupling levels as low as -40 dB.

By using the Galerkin procedure, $A_n^{(j)}$ and $B_m^{(i)}$ are obtained from the matrix equation

$$B_m^{(i)} = C_{mn}^{(ij)} A_n^{(j)} \tag{15.5}$$

where $A_n^{(j)} = a_n^{(j)} + b_n^{(j)}$ and $B_m^{(i)} = a_m^{(i)} - b_m^{(i)}$. Expressions for $C_{mn}^{(ij)}$, the diagonal elements of the matrix, are given in Appendix 15.7.

The scattering matrix is given by $\boldsymbol{S} = [\boldsymbol{I} + \boldsymbol{C}]^{-1}[\boldsymbol{I} - \boldsymbol{C}]$ where $\boldsymbol{C}$ is the admittance matrix. A general network representation of an array is in Figure 15.5(ii).

The crosscoupling integrals in Apendix 15.7 involve triple products of Bessel functions or their derivatives; however, they may be evaluated by contour integration provided that terms $J_{q+p}(k_0 R_{ij} \omega)$ are rephrased as the sum of Hankel functions of the first and second kind.

For conical horn elements, account must be taken of higher-order modes which are excited in the aperture but which are cutoff before reaching the the throat. These reflected modes not only alter the aperture field directly; they also cause further mutual coupling, which must be included in an accurate representation. Figure 15.5(i) shows, schematically, coupling between modes

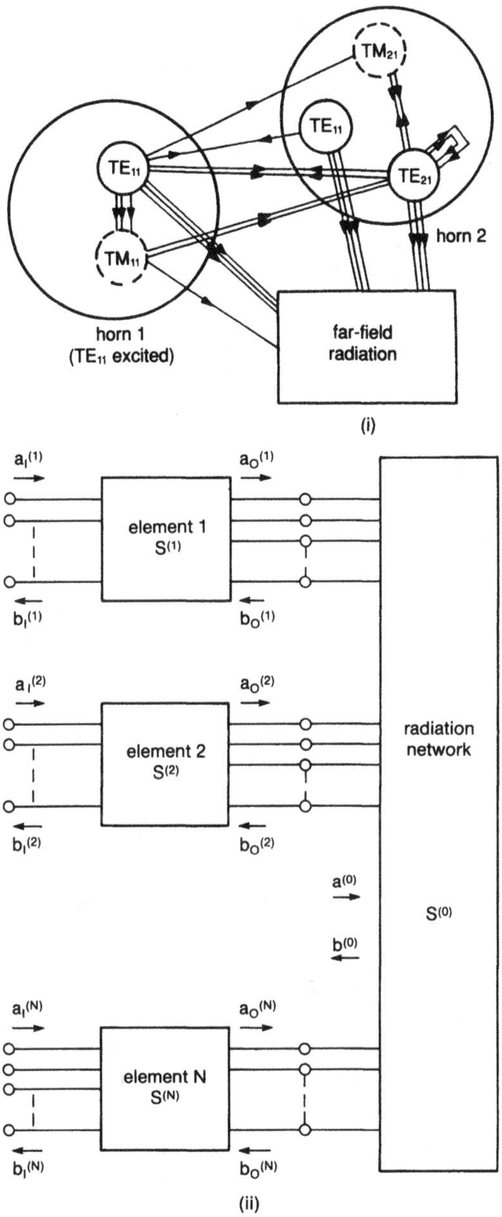

Figure 15.5 Conical-horn array

(i) Mode coupling between two adjacent elements in a conical-horn array [11]
→ weak
⇉ strong
⇶ very strong

(ii) Network representation of array [3]

in two horns of the array. Generally, nearest-neighbour coupling is strongest, but analysis and software allow for coupling extending beyond nearest neighbours. The strength of the coupling is qualitatively indicated by the number of lines forming the 'bond'; loops signify propagation part of the way into a horn of a mode which is cutoff in the horn throat. These modes have an associated phase factor which can be modelled for small flare angles by the addition of two terms. One corresponds to the phase change at the cutoff point, $+\pi/2$ for TE-modes and $-\pi/2$ for TM-modes; the other to the integrated phase change Φ along the horn, where

$$\Phi = \frac{2X_m(\beta - \tan\beta)}{\tan\alpha} \tag{15.6}$$

and

$$\beta = \cos^{-1}\frac{X_m}{ka}$$

X_m = characteristic root of mode, a = radius of horn aperture, α = horn flare angle and $k = 2\pi/\lambda$.

Once the aperture field is established, the radiation pattern is obtained using the Fourier-transform model, which is known to be accurate for an embedded array.

15.2.3 Mutual-coupling predictions for conical-horn arrays

15.2.3.1 Coupling between a pair of horns

Insight into the physics of mutual coupling is provided by considering first coupling between two horns as in Figure 15.6(i) and (ii). The choice of horn parameters corresponds to those used in an array for which experimental data are available. Figure 15.6(i) shows the scattering coefficient of coupling between TE_{11} modes in two horns as a function of horn radius with the apertures nearly touching. The results are obtained by including all significant propagating and evanescent modes in the computation. For comparison, results are also shown when higher modes are not included. For the most part, neglect of higher modes leads to an optimistic level for the mutual coupling. For the E-plane case, coupling falls monotonically until the horn supports the TM_{11}-mode, when a further increase in horn radius increases the coupling. In the H-plane case, the transition occurs when the TE_{12}-mode is supported. Figure 15.6(ii) enables comparison to be made between coupling in the E-plane between TE_{11}–TE_{21}, TE_{11}–TM_{01} and TE_{11}–TE_{11} combinations. Evidently, for horn sizes used in practice, i.e. with radius about 0.5λ, coupling to the TE_{21} mode is 20 dB stronger than coupling to the TE_{11} mode and 8 dB stronger relative to the TM_{01} mode. For the H-plane configuration, the level of coupling to the TE_{21} mode is similar to that for the E-plane, although the field orientation is rotated by 45°; for the TM_{01} mode the coupling is zero as a consequence of the relative field configurations.

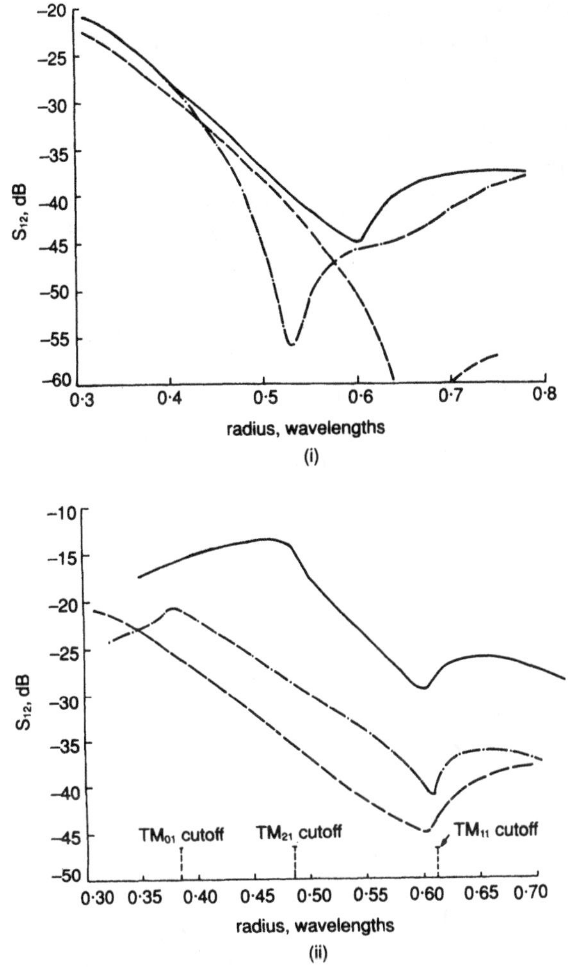

Figure 15.6 *E-plane-scattering coefficient for various mode couplings between two adjacent horns as functions of radius of aperture [11]*

Element spacing 2.1*a*

(i) TE_{11} mode

—— E-plane

–·–·– E-plane (single mode)

— — H-plane

Element spacing 2.1*a*

(ii) TE_{11} to TE_{11}, TM_{01} and TE_{21} modes

—— $TE_{11} - TE_{11}$

–·–·– TE_{11}–TM_{01}

— — TE_{11}–TE_{21}

15.2.3.2 Radiation patterns of array feeds

To establish a basis for comparison of the radiation patterns of various array feeds, we first show, in Figure 15.7, the copolar and crosspolar contours due to radiation from an isolated horn immersed in a ground plane and excited in the TE_{11} mode. The results are obtained by taking the Fourier transform of the electric field over the array aperture. Neglect of the phase curvature in the aperture of a 5° horn has been justified. As can be seen from Figure 15.7, the maximum level of crosspolarisation is very low, a consequence of a deliberate choice of horn radius. In Chapter 6 it is shown that the crosspolarisation of a circular horn depends quite critically on the horn radius in wavelengths.

Figure 15.8 shows the predicted copolar- and crosspolar-radiation contours for a 7-element array feed with the central element excited by the TE_{11}-mode, and with orthogonal orientations of the electric vector ***a*** and ***b***. The difference between the two polarisation directions stems from the rotational asymmetry of the array and the effect this has on mutual coupling.

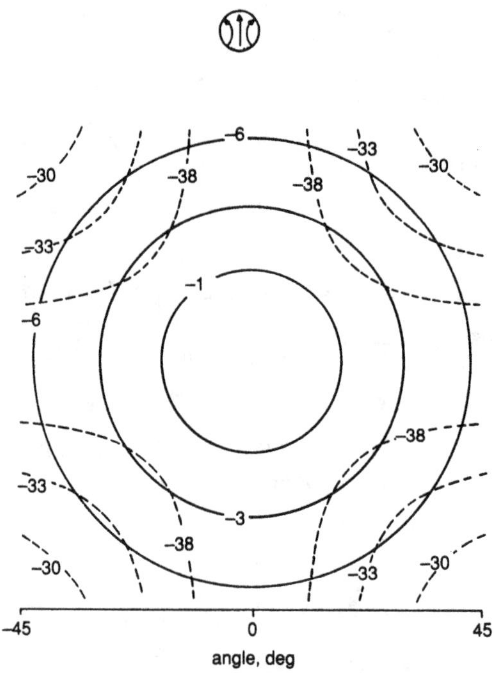

Figure 15.7 Predicted linearly polarised radiation fields from a single isolated conical horn [11]

Radius of aperture $a = 0.51\lambda$, *flare angle* = 5°

—— copolar

- - - crosspolar

Contours on θ and ϕ grid; scale depicts angular displacement in any ϕ direction from the boresight

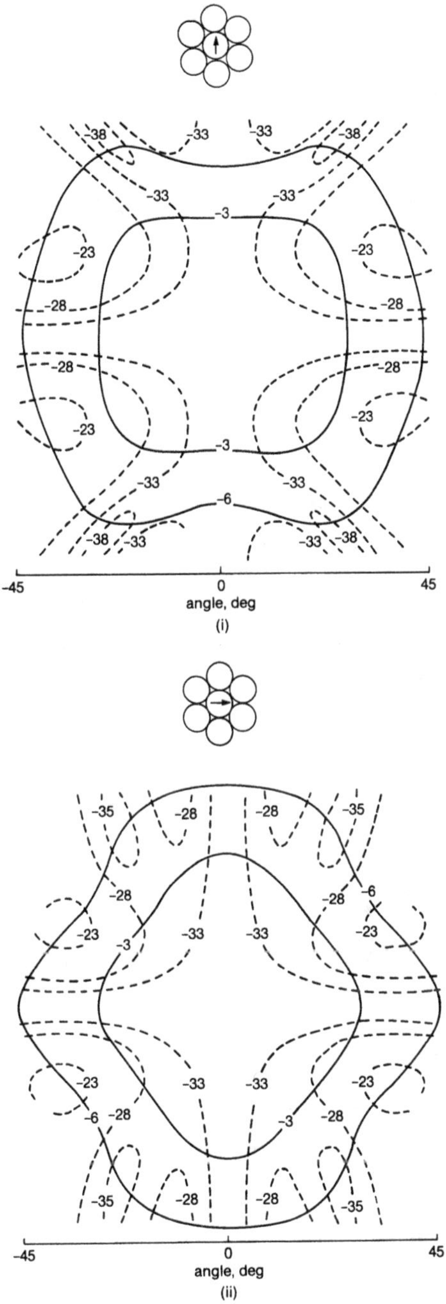

Figure 15.8 *Predicted linearly polarised radiation fields from an immersed conical element [11]. For details of element size and figures see Figure 15.4*

Element spacing = 2.01a; —— copolar; --- crosspolar
(i) Vertical polarisation; (ii) Horizontal polarisation

To interpret the results of Figure 15.8, we recall from Figure 15.7 that the crosspolarisation of the isolated element only rises above -40 dB in the 45° planes at observation angles well removed from boresight. By contrast, the crosspolarisation of the immersed element rises above -22 dB quite close to the boresight, which is due almost entirely to the effects of TE_{11}–TE_{21} and TE_{11}–TM_{01} mutual coupling. Both TE_{21} and TM_{01} modes are inherently rich in crosspolarisation; thus, their presence strongly disturbs the crosspolar pattern symmetry. Many designers have surrounded the 7-element cluster with rings of identical parasitic elements to reduce mutual coupling between subarrays and to reduce edge effects.

Figure 15.9 shows the influence on crosspolarisation of such elements, which are assumed to contain perfectly matched loads. On comparison with Figure

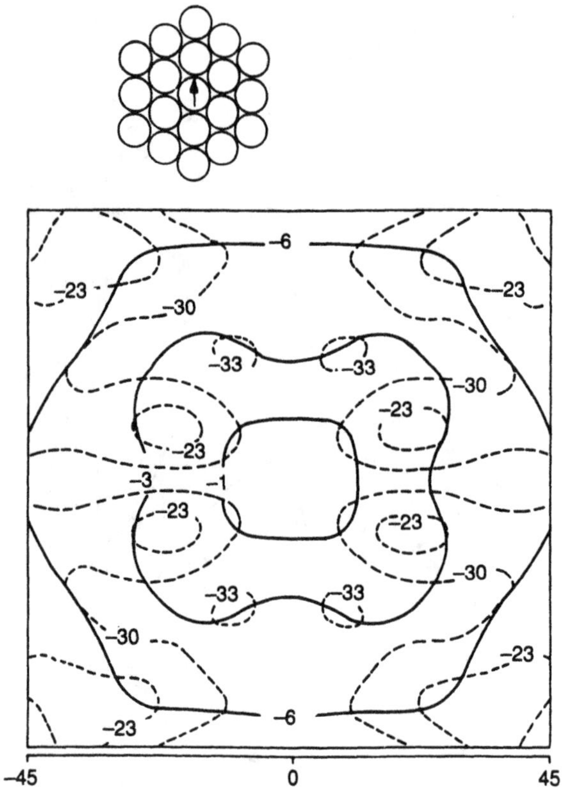

Figure 15.9 *Predicted linearly polarised radiation fields from an immersed conical element in a 19-element cluster [11]*

Element spacing = 2.01*a*
Vertical polarisation
—— copolar
--- crosspolar
For details of element size and figures see Figure 15.7

15.8 we see that the crosspolar lobes are slightly increased and their position has shifted. Thus, an assumption sometimes made that, only nearest neighbours contribute to the mutual coupling process, is clearly not entirely valid.

Next we turn our attention to so-called clean-beam excitation of arrays. Such arrays have been used in spacecraft antennas to reduce sidelobes and to improve crosspolarisation. Figure 15.10 shows a 7-element cluster with a 12-element passive guard ring. The central element is excited in phase and +10.5 dB above the six surrounding elements of the inner cluster. The array factor serves

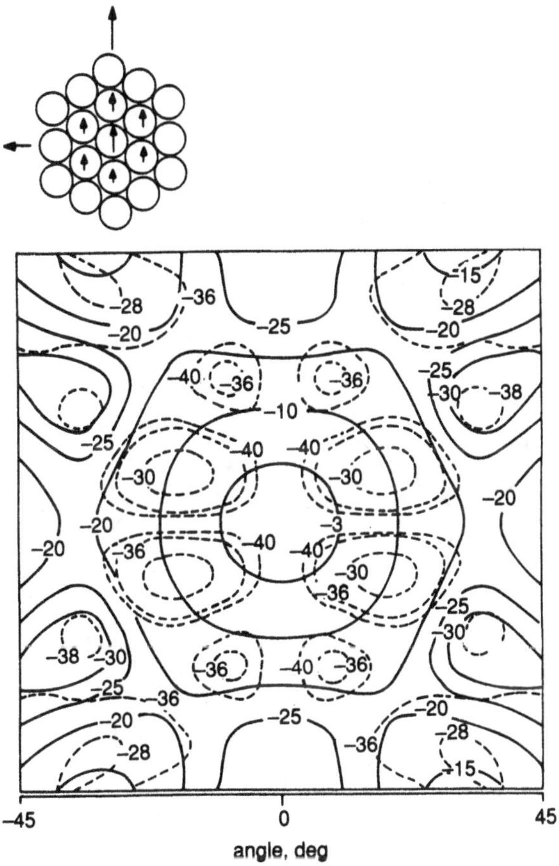

Figure 15.10 *Predicted linearly polarised radiation fields from a clean-beam cluster with mutual coupling [11]*

Element spacing = 2.01a
—— copolar
--- crosspolar
Inner ring of elements is excited in amplitude 10.5 dB below the central element; outer-ring elements are terminated in matched loads. For details of element size and figures see Figure 15.7

to suppress the crosspolarisation peak by about 5 dB, and there is a corresponding improvement in copolar pattern symmetry. As this array is one of the most important used in spacecraft-antenna design, it is worthwhile to compare the patterns of Figure 15.10 with those of Figure 15.11, where the effect of mutual coupling has been intentionally suppressed from the prediction process. Clearly, mutual coupling has a significant influence on the radiation-pattern contours, especially crosspolarisation, and it must be included in the design process.

From our earlier study in Figure 15.6(i) and (ii) of the dependence of crosspolarisation on horn-aperture size, it is to be expected that this parameter

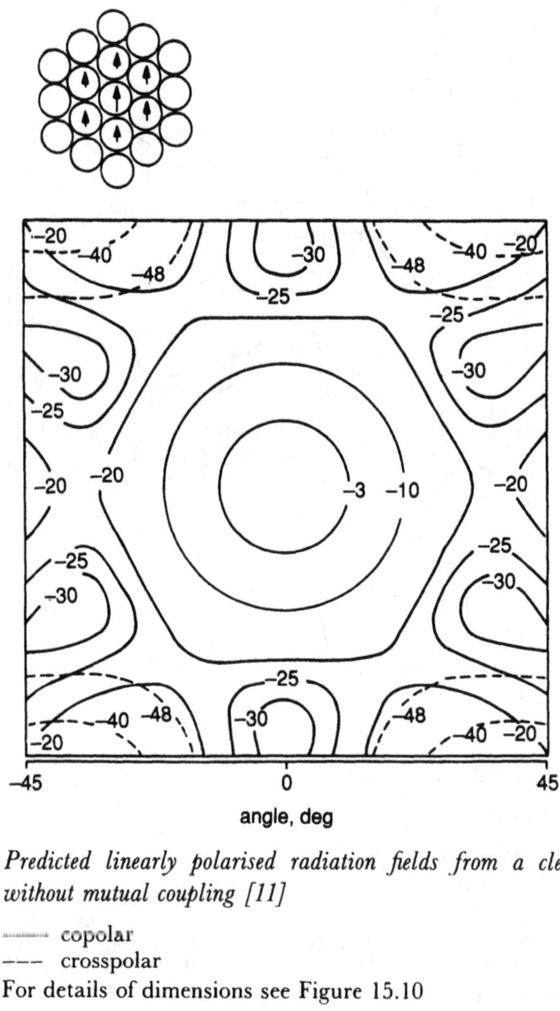

Figure 15.11 *Predicted linearly polarised radiation fields from a clean-beam cluster without mutual coupling [11]*

—— copolar
--- crosspolar
For details of dimensions see Figure 15.10

will also influence the crosspolarisation of a clean-beam array. Figure 15.12 shows that the peak crosspolar level falls below −32 dB when the radius exceeds about 0.5λ. Higher levels may be encountered in more complex arrays when element re-use is being deployed.

Finally, we turn to an example of a 29-element array containing three 7-element cells which, in conjunction with a reflector, could be used to generate three spot beams or, with suitable excitation, an elliptically shaped beam such as could be required for a European-coverage zone. The element dimensions and excitation levels are shown in the inset of Figure 15.13 for the elliptical-beam example.

This Figure shows the copolar and crosspolar contours, within the 10 dB copolar contour, the crosspolar level lies close to −40 dB. Furthermore, crosspolarisation is effectively suppressed by the array factor in the vertical plane. However, it should be noted that the above results are displayed at 11.2 GHz, which is in the middle of the 10.7–11.7 GHz range. Because of the rapid increase in crosspolarisation with decreasing horn-aperture size (in wavelengths), deterioration in performance by about 7.5 dB occurs at the lower band edge where the element radius is 0.52λ.

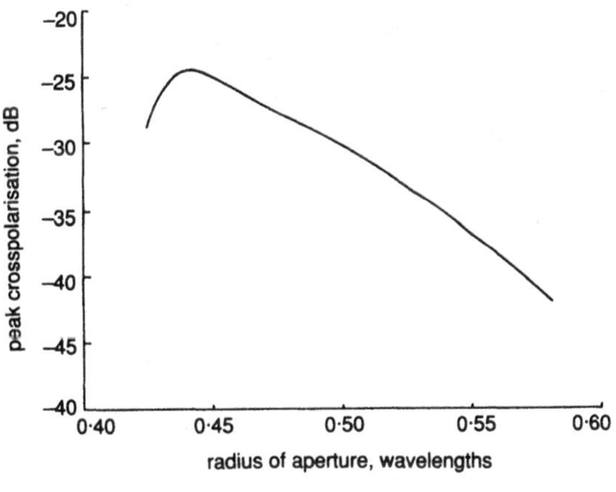

Figure 15.12 *Peak crosspolar variation as a function of aperture radius for a clean-beam cluster [11]*

For details of dimensions see Figure 15.10

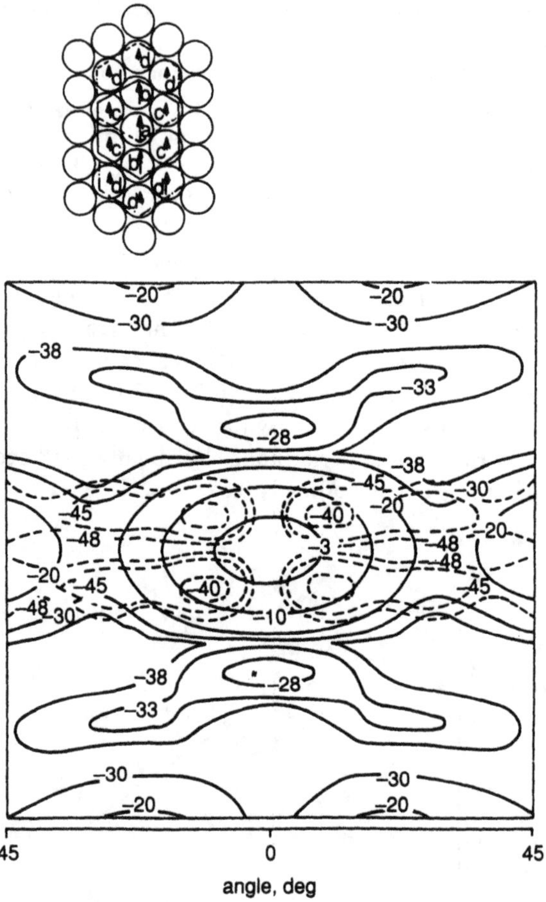

Figure 15.13 Predicted linearly polarised radiation fields from three overlapping clean-beam clusters [11]

$a = 0.55\lambda$, element spacing $= 1.2\lambda$ flare angle $= 5°$; for details of figure scales see Figure 15.7

—— copopolar

--- crosspolar

Overall excitation levels (vertical polarisation):

(i) 0

(ii) − 1.925 dB

(iii) − 8.01 dB

(iv) − 14.03 dB

All excitations are in phase, and the outer ring of dummy elements is terminated in matched loads

15.3 Effect of mutual coupling on secondary patterns

15.3.1 Prediction methods

We now consider, following the work of Clarricoats, Tun and Brown [13], the effect of mutual coupling in a conical-horn array on the secondary pattern of both single-offset and dual-offset-reflector antennas.

The array feed is shown in Figure 15.4. It comprises a finite number of conical-horn elements, which may either be excited from a matched input in the TE_{11} mode or passively terminated. The elements lie in a perfectly conducting ground plane which is assumed to have infinite extent. The assumption has only a slight effect on the array-feed pattern except at wide angles from boresight and, in general, has a negligible effect on the antenna secondary pattern. To determine the field due to the array at a point on the reflector surface, the contribution from every element of the array is summed after the effects of mutual coupling have been included.

Figure 15.14(i) shows an individual horn element. By using the E-field model, the radiation field at P for the mth mode with azimuthal dependence p in the ith aperture is given by

(a) TE modes:

$$(E_\theta)_{i,m}=jk_0^2 Z_0 a^3 A_m^{(i)} \frac{\exp(-jk_0 r_i)}{r_i} \times \left\{\frac{j^P Y_m P J_p(k_0 a \sin\theta_i)}{\chi_m'^2 k_0 a \sin\theta_i}\right\} \sin(p\phi_i-\psi_m) \qquad (15.7)$$

$$(E_\phi)_{i,m}= -jk_0^2 Z_0 a^3 A_m^{(i)} \frac{\exp(-jk_0 r_i)}{r_i} \cos\theta_i \times \left\{\frac{j^P Y_m J_p'(k_0 a \sin\theta_i)}{(k_0 a \sin\theta_i)^2-\chi_m'^2}\right\} \cos(p\phi_i-\psi_m) \qquad (15.8)$$

(b) TM modes:

$$(E_\theta)_{i,m}= -jk_0^2 Z_0 a^3 A_m^{(i)} \frac{\exp(-jk_0 r_i)}{r_i} \sin\theta_i \times \left\{j^P\left(\frac{\beta_m a}{\chi_m}\right) Y_m' \frac{J_p(k_0 a \sin\theta_i)}{(k_0 a \sin\theta_i)^2-\chi_m'^2}\right\} \cos(p\phi_i-\psi_m) \qquad (15.9)$$

$$(E_\theta)_{i,m}=0 \qquad (15.10)$$

where χ_m = root of the Bessel function $J_p(x)$, $Y_m'=J_p'(\chi_m)$, χ_m' = root of the Bessel function derivative $J_p'(x)$, $Y_m=J_p(\chi_m')$, ψ_m = polarisation angle, β_m = propagation coefficient, Z_0 = free-space wave impedance, and other symbols are defined in Figure 15.4.

Because the horn flare angle is small, it is unnecessary to allow for phase curvature over the horn aperture.

The spherical vector-field components $E_\theta E_\phi$ are converted to Cartesian components and the contributions from the individual horn apertures are

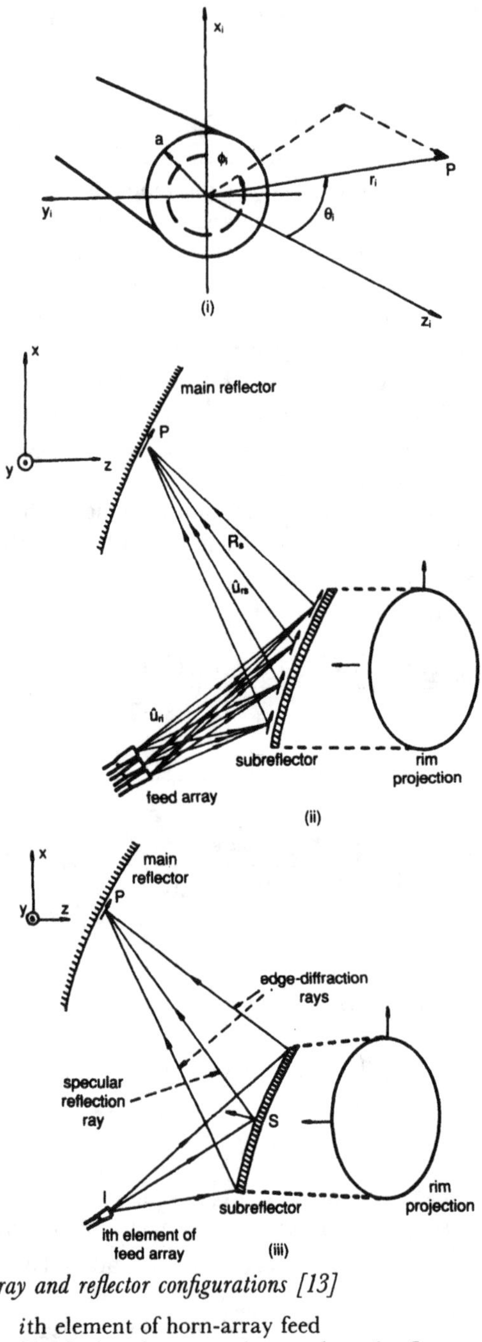

Figure 15.14 Array and reflector configurations [13]

(i) *i*th element of horn-array feed
(ii) Physical-optics scattering at the subreflector: configuration of the reflectors and array feed
(iii) Geometrical optics and diffraction at the subreflector: configuration of the reflectors and horn element

summed, taking account of the different amplitude and phase weightings and different space factors. We note that the point P is in the near field of the array since different values of r_i are used in the term $\exp(jk_0 r_i)/r_i$. The effect of mutual coupling is to modify the modal coefficients $A_m^{(i)}$.

For dual-reflector antennas, scattering of the feed radiation from the subreflector is evaluated either by application of physical optics or by a combination of geometric optics, GTD and slope diffraction.

In the physical-optics method, shown in Figure 15.14(ii), the PO surface currents are calculated over the subreflector surface using a local-plane-wave assumption. Thus, for a given incident ray $\hat{\boldsymbol{r}}_i$:

$$\boldsymbol{H}_{inc} = \frac{1}{Z_0} (\hat{\boldsymbol{r}}_i \times \boldsymbol{E}_{inc}) \tag{15.11}$$

and the surface current $\boldsymbol{J}_s$

$$\boldsymbol{J}_s = 2(\hat{\boldsymbol{n}}_s \times \boldsymbol{H}_{inc})$$

The contribution d$\boldsymbol{E}$ to the electric field at P from the surface element dx dy of the subreflector is given by

$$\mathrm{d}\boldsymbol{E} = C\{\boldsymbol{J}_s - \hat{\boldsymbol{r}}_s(\hat{\boldsymbol{r}}_s.\boldsymbol{J}_s)\} \frac{\exp(-jk_0 R_s)}{R_s}\, \mathrm{d}x\, \mathrm{d}y \tag{15.12}$$

There are three approximations in eqn. 15.1:

(*a*) no reactive fields are included
(*b*) $\boldsymbol{J}_s$ is calculated using a local plane wave
(*c*) fringe currents at the subreflector edge are neglected.

The method is well suited to arbitrarily shaped surfaces, but a fine *xy* grid (typically 5×10^3 points) is required since the phase is generally nonuniform and there is usually a stationary-phase point on the surface. However, note that $\boldsymbol{J}_s$ is computed only once for all the feed elements.

In the second method, GO, UGTD and slope diffraction are employed to calculate the field at P. Figure 15.14(iii) shows a sketch of the configuration. For a given point P on the main reflector and a given feed element *i*, a search is made to find the point of specular reflection on the subreflector surface. There, the distance IS + SP is stationary. The search is conducted using an *xy* grid, but for surfaces of revolution only a one-dimensional search is required. The GO procedure is supplemented by UGTD rather than GTD (which fails at shadow boundaries). Diffraction points are located by means of a stationary-phase search around the subreflector rim. Slope diffraction accounts for the field slope of the incident wave at the subreflector edge.

For each cell P of the main reflector, the process must be repeated for every feed element. For a generally shaped surface, 2-dimensional searching is time consuming but, for subreflectors which are formed from surfaces of revolution, GO and UGTD is faster than PO provided that the array contains only a small number of feed elements. For subreflectors with diameters larger than 20λ, we have found negligible differences in predicted patterns (including the sensitive crosspolar pattern) between the two methods.

The secondary-radiation pattern of the antenna is predicted by means of a PO integration over the main reflector using an *xy* grid. In our results, Ludwig's third definition is used to define the copolar and crosspolar components of the field. We have not included fringe currents at the main-reflector rim, so far-out sidelobes will lack accuracy. This is not a limitation in a spacecraft antenna, which is generally required to illuminate only a small region of the earth's surface. Our patterns are displayed in the form of contour plots with the convention that the vertical U axis represents $D/\lambda \sin\theta \cos\phi$ (where D is the main-reflector diameter and λ is the free-space wavelength), and the horizontal V axis represents $D/\lambda \sin\theta \sin\phi$.

15.3.2 Predicted primary and secondary patterns

15.3.2.1 Antenna configurations and feed patterns

Figure 15.15*a* and *b* shows the two reflector configurations investigated. The single-offset reflector of Figure 15.15*a* is representative of a spacecraft antenna with a 10 dB beamwidth of approximately 2°, while the dual-offset Gregorian antenna of Figure 15.15*b* has about half that beamwidth. The choice of reflector and subreflector sizes was made so that the same 19-element array feed could be used, thus enabling reflectors to be largely disassociated from feed effects.

The conical-horn array feed and its midband copolar and crosspolar patterns are shown in Figure 15.16. The central element is excited in phase with, and at an amplitude of 10.5 dB above, the surrounding cluster of six elements; these in turn are surrounded by a cluster of 12 elements terminated in matched loads. In the absence of mutual coupling, the feed exhibits a pattern with symmetry, very low sidelobes and a very low crosspolarisation (so-called clean beam). However, TE_{11}–TE_{21} and TE_{11}–TM_{01} mutual coupling slightly distorts the copolar pattern and raises the crosspolar level to -30 dB within the angle subtended by the reflector.

15.3.2.2 Single-offset reflector

Figure 15.17*a* shows the copolar and crosspolar patterns of the single-offset reflector of Figure 15.15*a*, fed from the array of Figure 15.16, but with neglect of mutual coupling, thereby virtually eliminating feed crosspolarisation because of the array factor. Figure 15.17*b* shows comparable results with mutual-coupling effects fully included. The characteristic high H-plane crosspolar lobes of Figure 15.17*a* are a well known feature of the offset-reflector antenna. Figure 15.17*b* shows that when mutual coupling in the feed is included the structure of the feed crosspolarisation is conveyed to the secondary pattern, but the four lobes are no longer symmetrical, as in Figure 15.16, because of interference. The copolar pattern is virutally unaffected down to -20 dB, but below this level there is marked asymmetry.

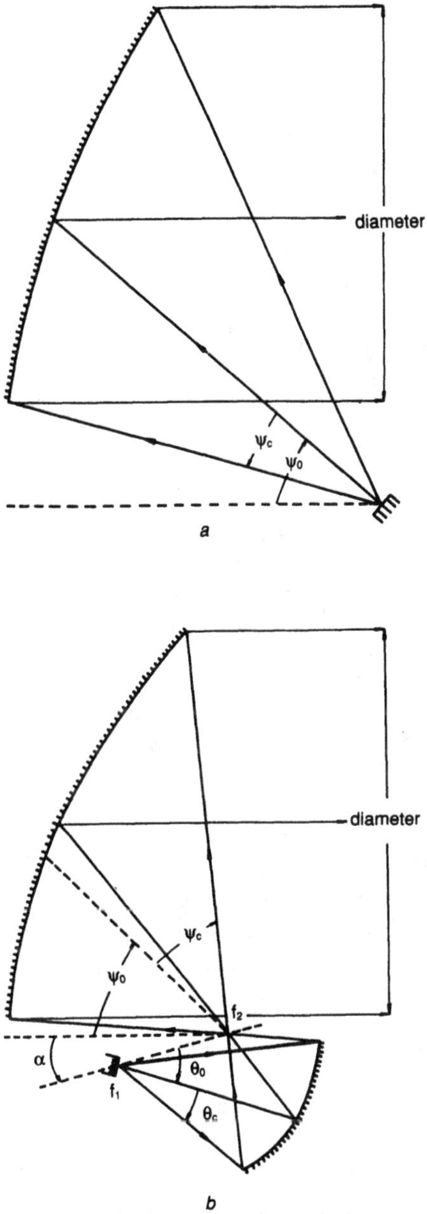

Figure 15.15 *Offset-reflector antennas with array feed [13]*

a Single-offset-reflector antenna with array feed: $D=50.5$, $\psi_0=40°$, $\psi_c=25°$

b Offset Cassegrain-reflector antenna with array feed: $D=100\lambda$, $\psi_0=45°$, $\psi_c=40°$, $x=13.74°$, $f_1 f_2=30\lambda$, $e=0.394$, $\theta_0=31°$, $\theta_c=22.8°$

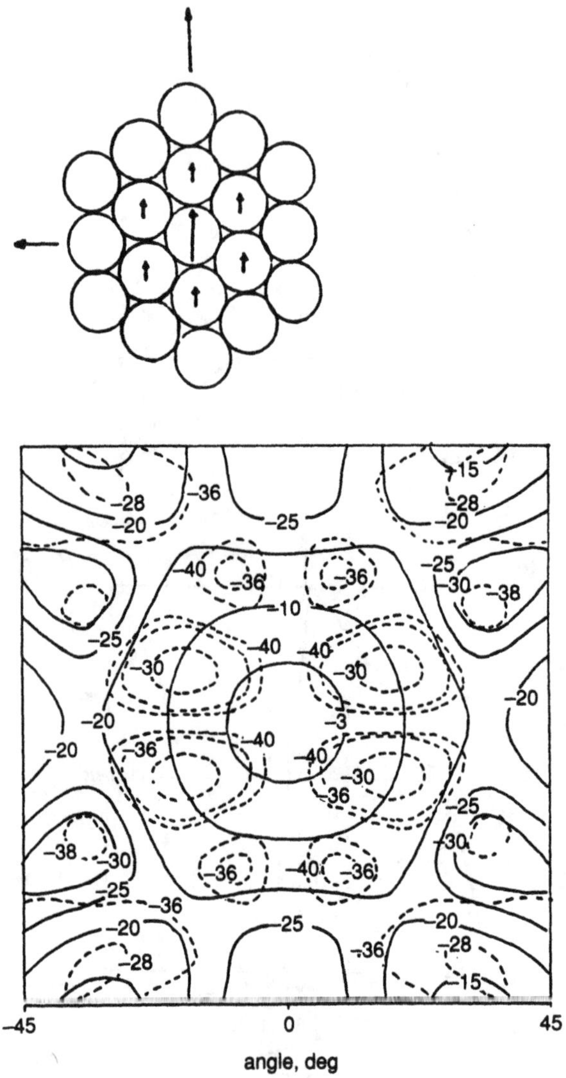

Figure 15.16 *Linearly polarised radiation fields from a 19-element clean-beam cluster [13]*

Contours on θ and ϕ grid; scale depicts angular displacement in any ϕ direction

—— copolar

--- crosspolar

Horn radius = 0.5λ element spacing = 1.005λ, flare angle = $5°$

Inner ring of elements is excited in amplitude 10.5 dB below the central element; outer ring of dummy elements is terminated in matched loads

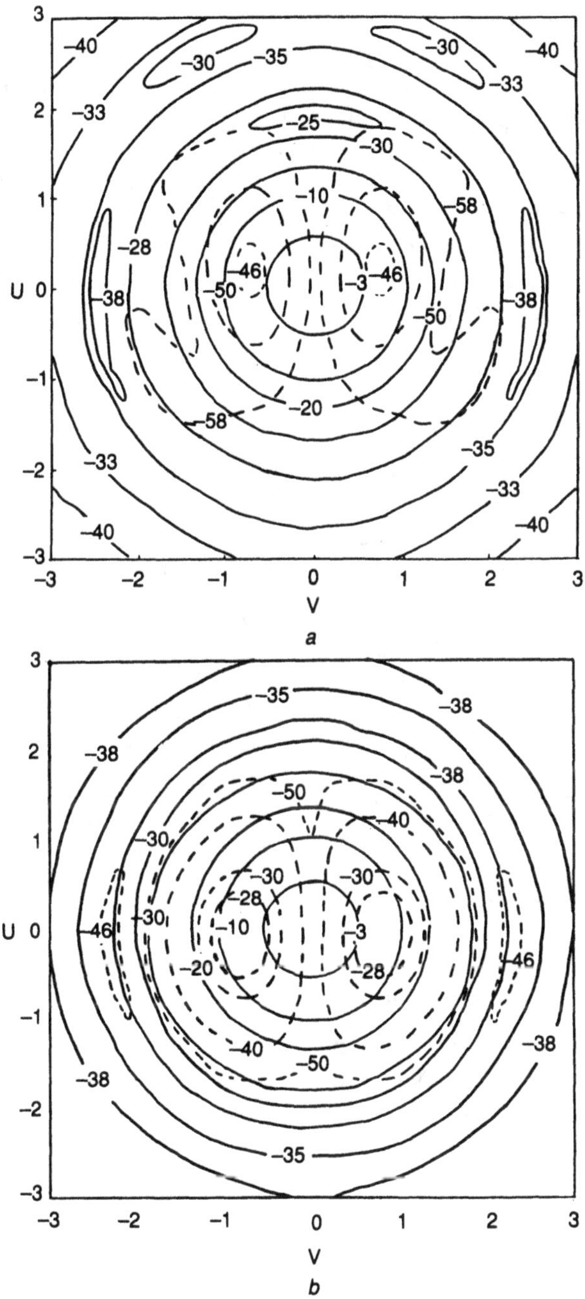

Figure 15.17 Far-field contours of single-offset-reflector antenna (Figure 15.15a) [13]

a Without mutual coupling
b With mutual coupling
—— copolar
– – – crosspolar

15.3.2.3 Dual-offset reflector

Figure 15.18*a* and *b* shows copolar and crosspolar patterns obtained using UGTD for the dual-offset reflector of Figure 15.15*b*, fed from the array feed of Figure 15.16 with, respectively, mutual-coupling effects excluded and included. As the parameters of the antenna were chosen to eliminate crosspolarisation within the limits of GO, Figure 15.18*a* reveals a very low level of crosspolarisation, even when diffraction is included. However, when mutual coupling is also included, as in Figure 15.18*b*, it becomes the dominant factor in determining the level of crosspolarisation. The secondary crosspolar lobes correspond closely to those of the feed, and they exhibit the same symmetry. They are noticeably lower than for the single-offset-reflector antenna, and the copolar pattern is symmetrical to -30 dB below the boresight value.

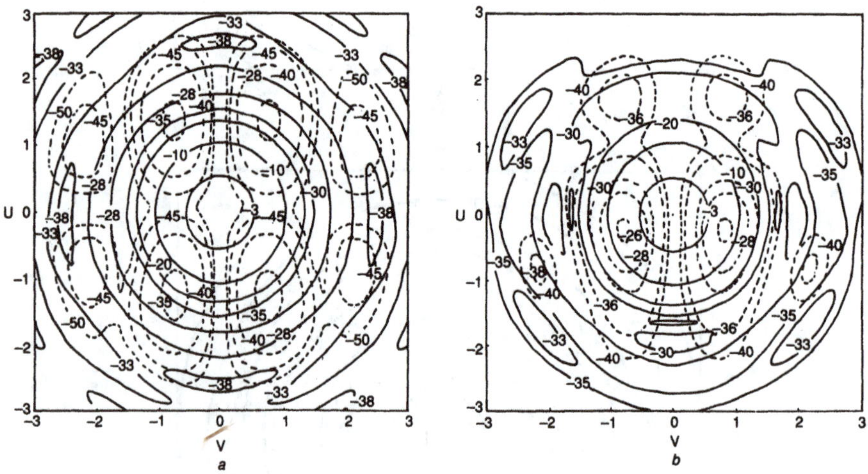

Figure 15.18 Far-field contours of offset Gregorian reflector antenna (Figure 15.15b) [13]

a Without mutual coupling
b With mutual coupling
—— copolar
--- crosspolar

15.3.3 Effect of mutual coupling on secondary patterns when the array feed is displaced

As discussed in Chapter 2, array-fed reflectors are used to generate contoured beams, and consequently elements are no longer at the reflector focus. Here we show the effect of mutual coupling on the secondary patterns for displaced array feeds. For the single-offset reflector, we have found little change in the level of crosspolarisation when the centre element of the cluster of Figure 15.16 is displaced by 4λ (causing approximately 3.5 beamwidths of beam displacement or 4° scan).

For a dual-offset antenna, which has been initially optimised for low crosspolarisation, the changes are more significant and mutual coupling produces a more marked influence on the predicted crosspolar pattern. In a practical array, the displaced feed elements could be repointed so as to improve copolar symmetry and reduce sidelobe levels but, for reasons of clarity and brevity, we display the effect of mutual coupling only for the crosspolar-contour patterns without feed repointing. The displacement of the feed in the E-plane is 4λ and an array, comprising subarrays of 19-element clusters, would contain 130 elements if this displacement were accorded in all directions about the reflector focus.

Figures 15.19*a* and *b* show the predicted crosspolar contours with, respectively, mutual-coupling effects excluded and included. On comparing Figure 15.19*a* with Figure 15.18*a*, we see that a prediction made excluding mutual coupling indicates an 11 dB increase in crosspolar level for this feed displacement. In fact, the peak crosspolar level rises monotonically with displacement and beam shift (see Figure 15.19*d*). Figure 15.19*b* shows that, with mutual coupling included, the crosspolar structure of the array feed (see Figure 15.16) is present but with symmetry disturbed. Figure 15.19*d* again shows the effect of beam shift.

When the feed is repointed the contours of Figure 15.19*c* show that symmetry is further disturbed, owing to interference effects, and the maximum level is now only slightly above -30 dB. However, the beneficial effects of repointing are manifest in Figure 15.20*a* and *b*, which shows, with mutual coupling included, the improvement in copolar symmetry and sidelobe levels.

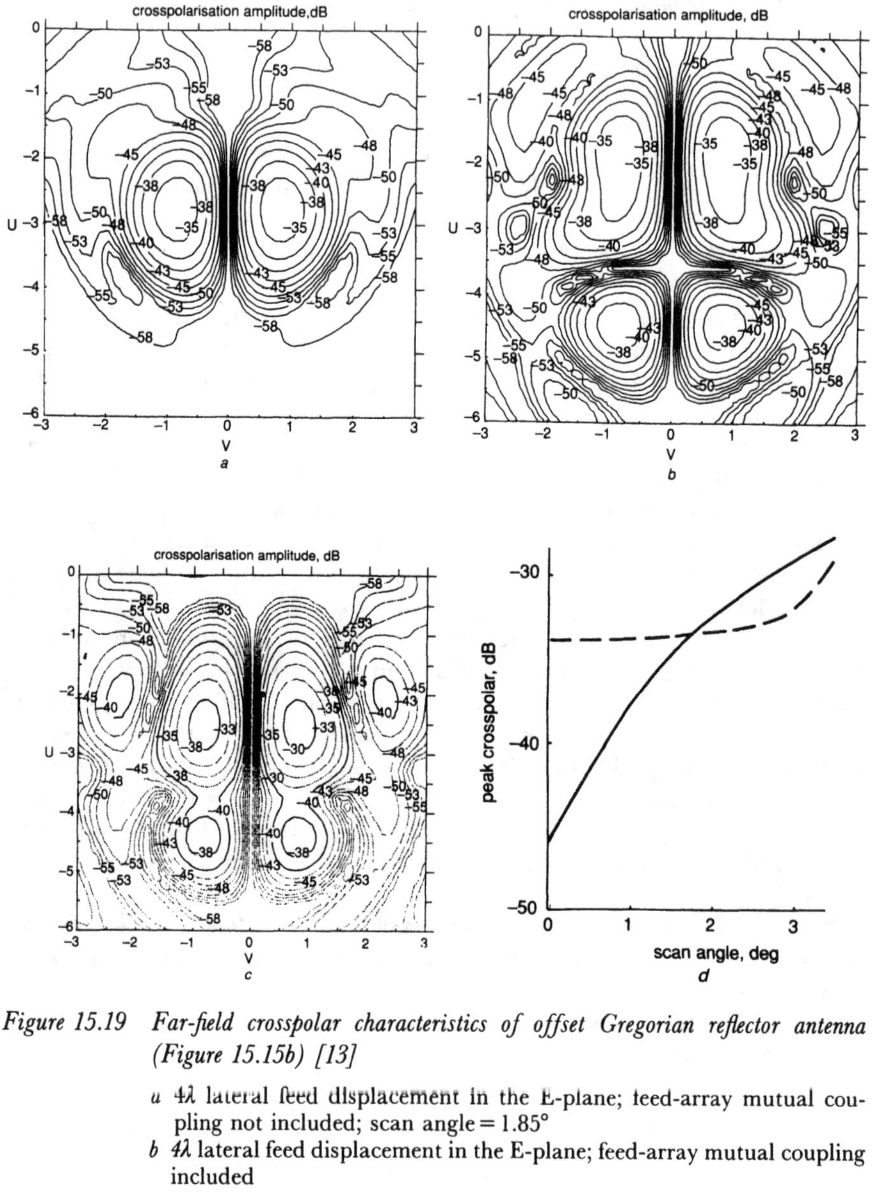

Figure 15.19 Far-field crosspolar characteristics of offset Gregorian reflector antenna (Figure 15.15b) [13]

a 4λ lateral feed displacement in the E-plane; feed-array mutual coupling not included; scan angle = 1.85°

b 4λ lateral feed displacement in the E-plane; feed-array mutual coupling included

c 4λ lateral feed displacement in the E-plane; feed-array mutual coupling included in the analysis and 19-element feed cluster repointed to the midpoint of the main reflector

d Far-field crosspolar level as a function of the main-beam scan angle

--- with mutual coupling

—— without mutual coupling

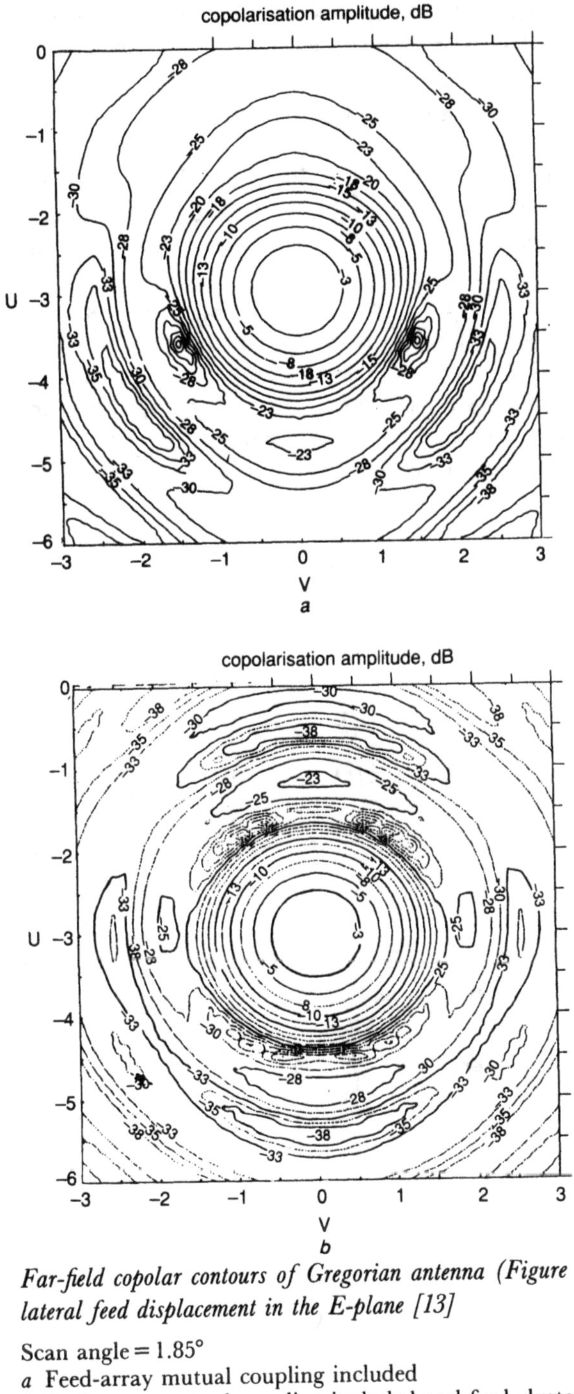

Figure 15.20 *Far-field copolar contours of Gregorian antenna (Figure 15.15b) with 4λ lateral feed displacement in the E-plane [13]*

Scan angle = 1.85°

a Feed-array mutual coupling included

b Feed-array mutual coupling included and feed cluster repointed as in Figure 15.19*c*

15.4 Mutual coupling for elements of rectangular and elliptical cross-sections

The method described in Section 15.2 has been extended by Bird [3] to apertures of rectangular and elliptical cross-section in an infinite conducting ground plane, and also by Parini, Kitchener and Law to square and rectangular apertures. In Figure 15.21, due to Bird, the radiation pattern is shown for an element at the centre of a 13 × 13 square array of square waveguides whose dimensions are given in the Figure. Five modes (TE_{10}, TE_{12}, TM_{12}, TE_{11} and TM_{11}) were specified in each aperture and the mutual coupling can be seen to alter the radiation pattern significantly when comparison is made with the pattern of the isolated element. The appearance of a grating lobe is noteworthy. To complement the results in Figure 15.6(i) and (ii), the coupling coefficients for different sizes of rectangular waveguide and for a pair of elliptical waveguides are shown in Figures 15.22 and 15.23.

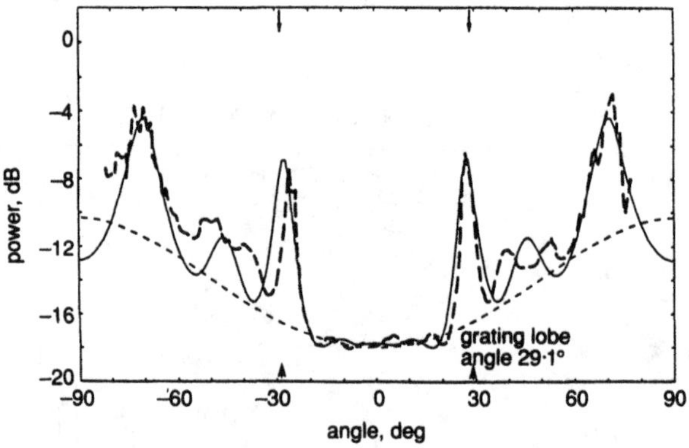

Figure 15.21 Radiation patterns of an element at the centre of a square grid array of square waveguides, E-plane [3]

—— 13 × 13 array (theory)
--- single-element (theory)
— — 19 × 19 array [1]
Square-grid spacing: 0.6729λ, square elements: $a = b = 0.6305\lambda$

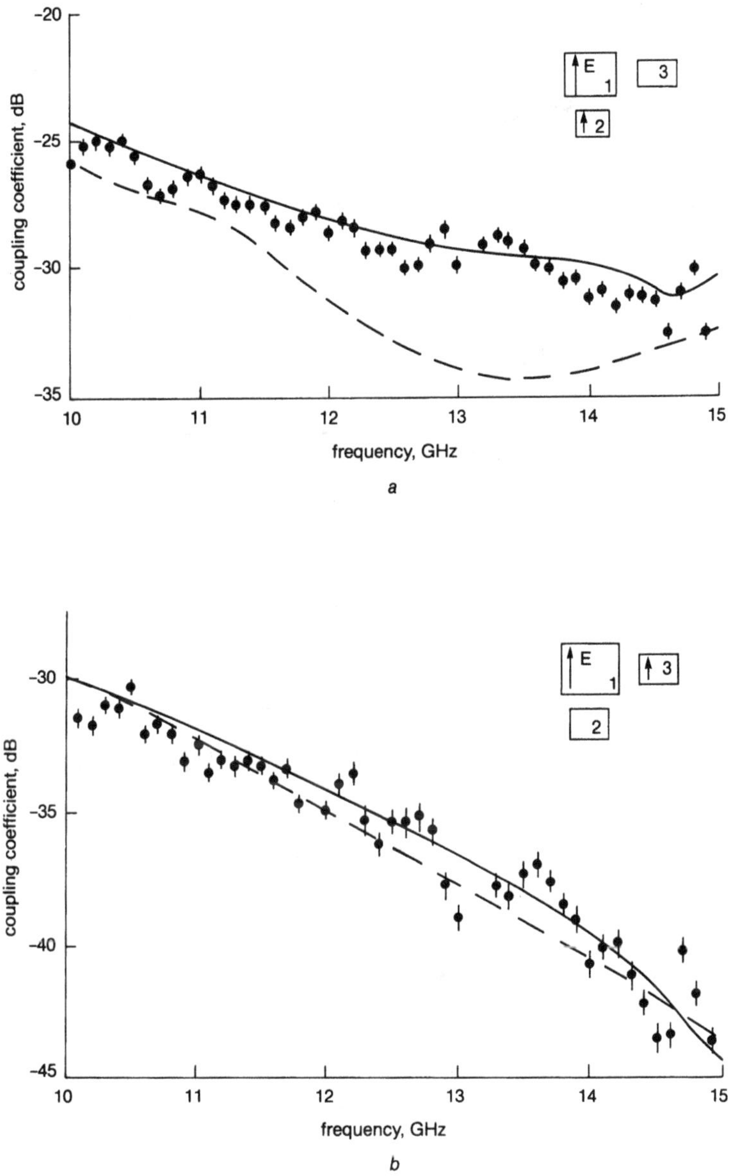

Figure 15.22 Coupling coefficient for different-sized rectangular waveguides [3]

Element 1 is square with side length 22.8 mm, elements 2 and 3 have width 15.7 mm and height 7.7 mm
a E-plane
b H-plane

●●● experiment
—— theory, seven modes
--- theory, single mode

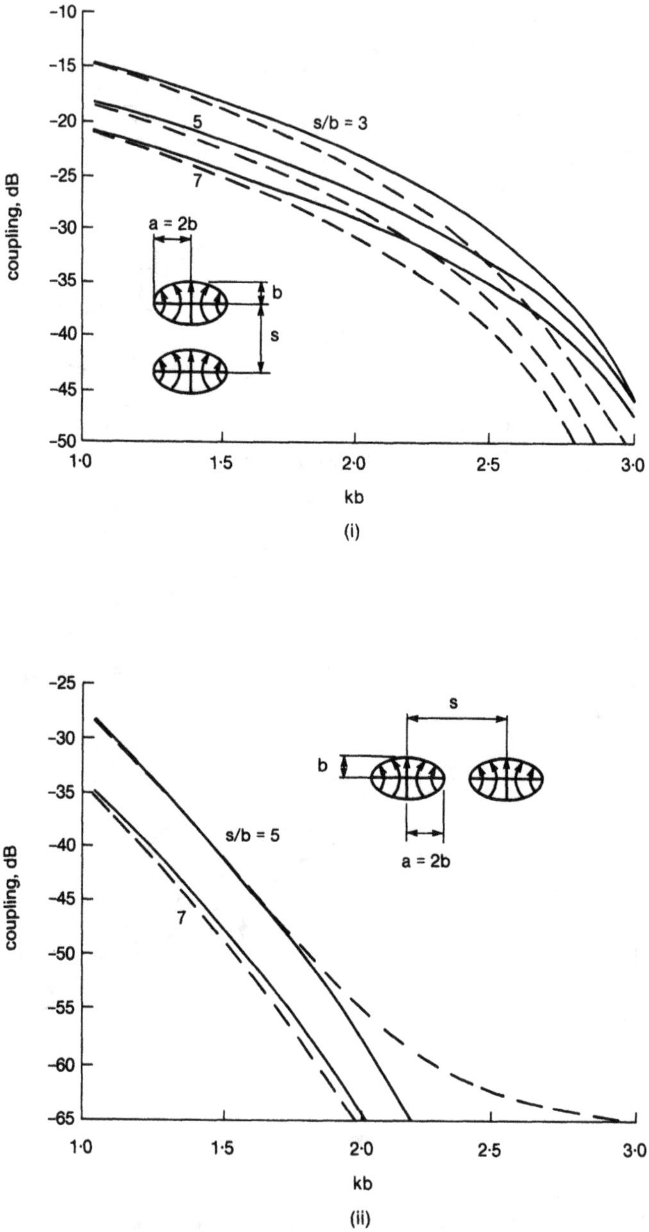

Figure 15.23 Coupling coefficient for identical rectangular and elliptical waveguides with $a/b = 2$ as a function of separation s [3]

(i) E-plane
(ii) H-plane
--- rectangular waveguide
—— elliptical waveguide

15.5 Mutual coupling for a microstrip patch array

Direct-radiating microstrip patch antennas are entering use in terrestrial, airborne and satellite applications, but they can also be used as a feed for a reflector. As with waveguide and horn apertures, when formed into an array, mutual coupling influences their performance and in this section a brief summary of a study by Shafai, Helaly and Bhattachatyya [14] for square patches is presented. Shafai and Bhattachatyya have also analysed an array of circular patches [15]. For either geometry, their basic model is shown in Figure 15.24, where each element is treated as a lumped load connected in series with the source internal resistance; the lumped loads represent the input impedances of the radiating elements. The curved arrows in Figure 15.24 denote the mutual coupling between the elements. Shafai treats individual patches by means of the cavity model of Figure 15.25, where two identical rectangular patches are shown. The input connections to each patch are indicated and, within the model, the patch is represented by a cavity with a magnetic wall (shown dotted) which is located near to the physical boundary of the patch [14]. The substrate thickness is *h*.

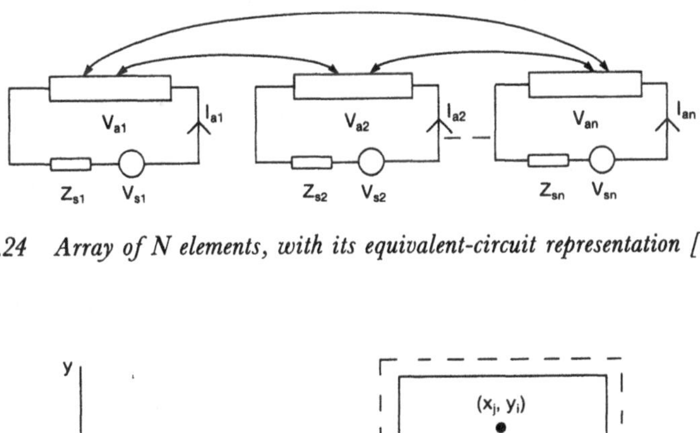

Figure 15.24 Array of N elements, with its equivalent-circuit representation [14]

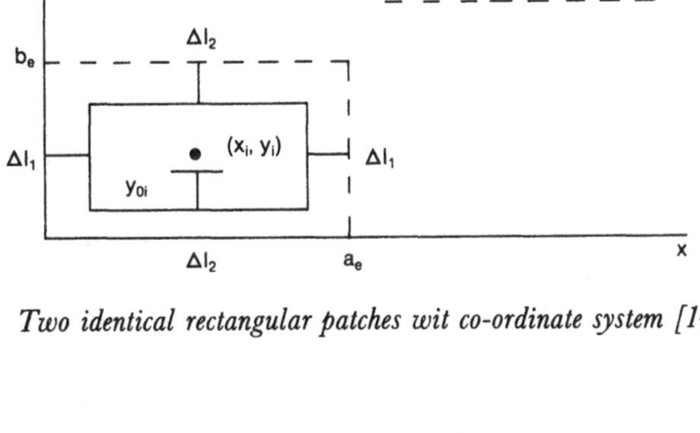

Figure 15.25 Two identical rectangular patches wit co-ordinate system [14]

In the model, the patch is represented as a grounded dielectic slab with magnetic-current distributions located at the open walls formed between the edge of the patch and the ground plane. Since the electric field is zero everywhere on the surface of the cavity except on the open walls, field equivalence may be used to replace the fields on the surface by equivalent magnetic currents in front of a perfect electric conductor. Mutual coupling occurs between these sources, and the magnetic field due to these equivalent sources is needed to calculate their effect. Central to the analysis is the determination of a dyadic Green's function of magnetic type for a grounded dielectric slab. From this, the magnetic field due to the magnetic currents on the open boundary may be calculated and the mutual admittance obtained. The details of this analysis are to be found in References 14 and 15. The principal equations for the array of square patches are given below.

Following Shafai [14], the mutual admittance Y_{ij} is given by

$$Y_{ij} = -\int_{cj} \boldsymbol{H}_{ij} . hEz\,(\hat{\boldsymbol{z}} \times \hat{\boldsymbol{n}})\ \mathrm{d}lj \tag{15.13}$$

where $\boldsymbol{H}_{ij}$ is the magnetic field of the ith antenna near the location of the jth antenna; and the second term in the integral of eqn. 15.13 is the equivalent magnetic current on the jth antenna. The mutual couping integral can be computed numerically [15].

The self admittance Y_{ii} can be derived in closed form and the susceptance jB is given by

$$jB = \frac{ja_e b_e}{2\omega\mu_0 h} \sec^2\left(\frac{\pi x_0}{a_e}\right)\left\{k_1^2 - \left(\frac{\pi}{a_e}\right)^2\right\} \tag{15.14}$$

and

$$k_1^2 = \omega^2 \epsilon \mu_0.$$

The patch radiates energy and the value of the conductance G can be obtained from the self reaction of the aperture through

$$G = \left\{R \int_{ci} \boldsymbol{H}_{ii} . hEz\,(\hat{\boldsymbol{z}} \times \hat{\boldsymbol{n}})\ \mathrm{d}l_i\right\} \sec^2\left(\frac{\pi x_0}{a_e}\right) \tag{15.15}$$

Table 15.1 shows the self impedance of a 2×2 subarray. The assumption is made of a single mode in the array shown in Figure 15.26, which has an element spacing of $2d$. The input impedance at the feed point is displayed for three values of separation on the assumption of coupling. As expected, the active impedance decreases with separation, and the corresponding result is given without coupling. In general, the effect of coupling on the radiation pattern of a microstrip array is no less important than for a waveguide or horn array.

Following Reference 15 and by way of illustration, we consider just the form of E_θ in the radiation field, and as a simplification consider a square patch of

Table 15.1 Self impedance of a 2×2 array of microstrip patches [15]

Patch	Z_{in} Mag.	<°	Patch	Z_{in} Mag.	>°
With coupling, d=0.4 cm			With coupling, d=10.0 cm		
	Ω	Ω	1	47.552	−5.93
1	35.908	−25.96	2	47.555	−5.93
2	36.147	−26.24	3	47.555	−5.93
3	36.266	−25.97	4	47.565	−5.92
4	36.643	−26.17			
With coupling, d=4.4 cm			No coupling		
1	55.184	−1.81	1	49.856	−4.25
2	55.171	−1.80	2	49.856	−4.25
3	55.173	−1.80	3	49.856	−4.25
4	55.160	−1.79	4	49.856	−4.25

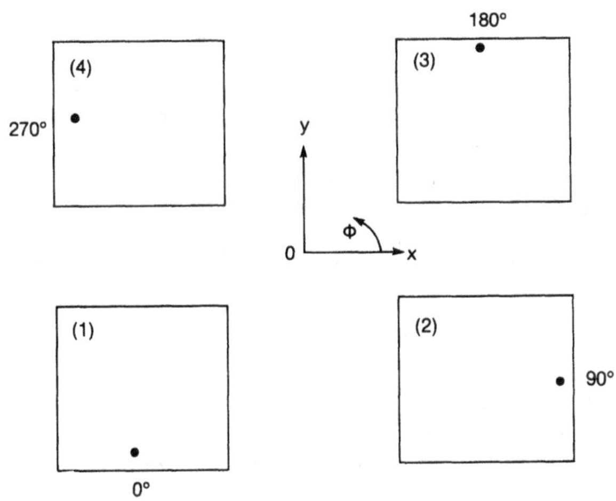

Figure 15.26 Geometrical layout of 2×2 subarrary [15]
Element spacing = 2d

side a. The radiation pattern of the ith element of the array is then given by

$$E_\theta = \frac{j4V_{ai}}{\lambda} \frac{\exp(-jkR)}{R} \exp(jka \cos\phi \sin\theta) \sin\left(\frac{ka}{2}\cos\phi \sin\theta\right)$$

$$\times \cos\left(\frac{ka}{2}\sin\phi \sin\theta\right) k \sin\phi \cos\phi \sin\theta$$

$$\times \left\{ \frac{1}{k^2 \cos^2\phi \sin^2\theta} + \frac{1}{k^2 \sin^2\phi \sin^2\theta - \left(\frac{\bar{u}}{a}\right)^2} \right\} \quad (15.16)$$

where $[V_a^i] = [B_{ij}]^{-1}[V_{sj}]$ and V_a and V_s are defined in Figure 15.24. E_ϕ is of similar form.

As with horn arrays, the influence of mutual coupling on the radiation pattern depends on the excitation of the array and the element separation. Figure 15.27 shows an array considered for a dual-polarised satellite-antenna application. The copolar- and crosspolar-radiation patterns are shown in Figure 15.28 (i) and (ii). For the example of Figure 15.28(i) where the elements are very closely spaced, there is a strong influence of mutual coupling, but this

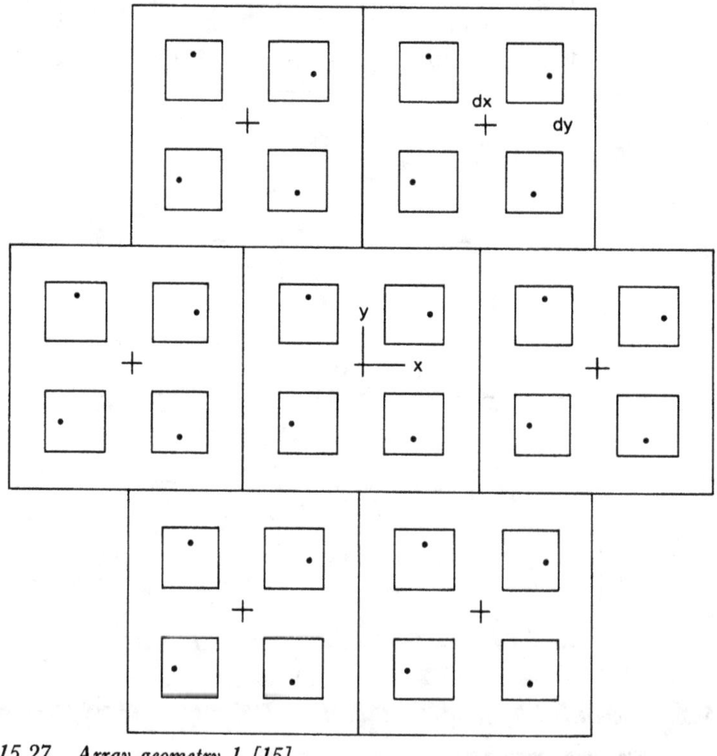

Figure 15.27 Array geometry 1 [15]

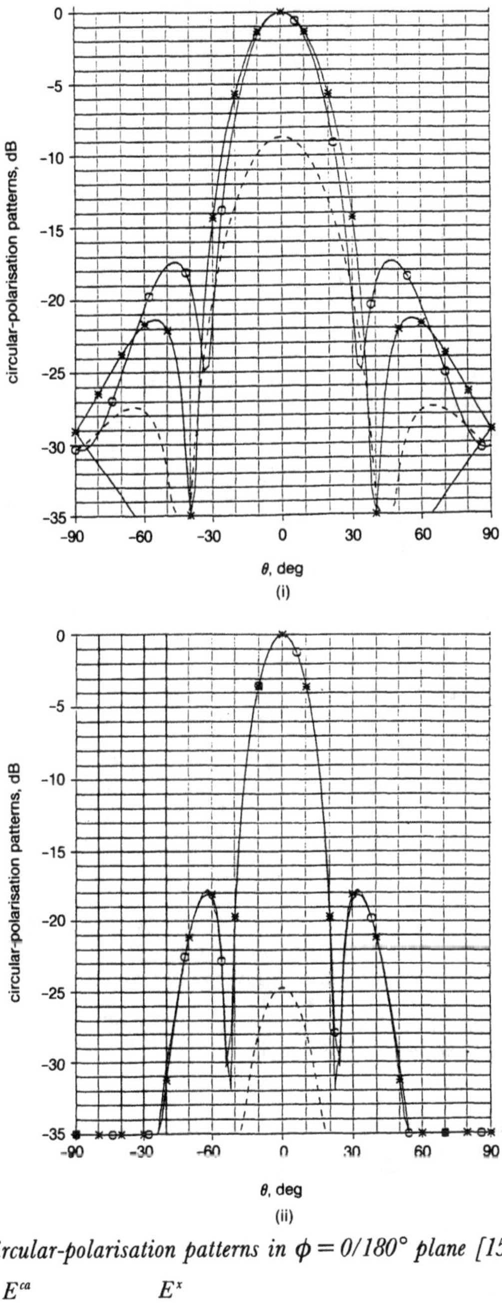

Figure 15.28 Circular-polarisation patterns in $\phi = 0/180°$ plane [15]

	E^{ca}	E^{x}
No coupling:	—*—	——
Coupling:	—○—	– – –

$f = 1.647$ GHz, $a = 58$ mm, $\varepsilon_r = 2.33$

(i) $dx = dy = 4$ mm

(ii) $dx = dy = 44$ mm

diminishes rapidly when the separation is increased to a representative value, as in Figure 15.28(ii). However, it is again important to recall that the effects of mutual coupling in an array are very sensitive to the manner of excitation of the array.

15.6 References

1 OLINER, A.A., and MALECH, R.G., *in* HANSEN, R.C. (Ed.): 'Microwave scanning antennas' (Peninsula Publishing 1985), Chapters 2–4
2 KING, R.W.P., *in* COLLIN, R.C., and ZUCKER, F.J. (Eds.): 'Antenna theory Part I' (McGraw-Hill Book Co., 1969), Chapter 9
3 BIRD, T.S.: 'Proceedings of the 1992 Queen Mary and Westfield College Antenna Symposium', lecture 5, pp. 1–22
4 LO, Y.T., and LEE, S.W.: 'Antenna handbook' (Van Nostrand Reinhold Publishing Co., 1988), Chapters 11–14
5 MAILLOUX, R.J., *in* RUDGE, A.W., MILNE, K., OLVER, A.D., and KNIGHT, P. (Eds.): 'The handbook of antenna design Vol. 1' (Peter Peregrinus, 1982), IEE Electromagnetic Waves Series 15, Chapter 5
6 CLARRICOATS, P.J.B., CRONE, G.E.A., JENSEN, N.E., and ROEDERER, A.G.: 'Antennas for satellite systems—a European view'. Proceedings of ISAP conference, 1992, pp. 965–980
7 ROEDERER, A.G.: 'Review of advanced multiport antennas'. ESA workshop on Advanced Beam Forming Networks for Space Applications, 1989, paper 2.2.1
8 SPRING, K., and MOODY, H.: 'Divided LLBFN/HMPA transmitted architectures'. US patent 4 901 085, February 1990
9 ROEDERER, A.G., and SABBADINI, M.: 'A novel semi-active antenna concept'. IEEE AP-S Symposium digest, 1990
10 BIRD, T.S.: 'Mode coupling in a planar circular waveguide array', *IEE J. Microw. Opt. Acoust.*, 1979, **3**, pp. 172–180
11 CLARRICOATS, P.J.B., TUN, S.M., and PARINI, C.G.: 'Effects of mutual coupling in conical horn arrays', *IEEE Proc. H*, 1984, **131**, pp. 165–171
12 PARINI, C.G., and KITCHENER, D.: 'The importance of mutual coupling in the analysis of finite arrays of rectangular apertures'. Proceedings of 19th European Microwave Conference, 1989, London (Microwave Exhibitions and Publishers Ltd., 1989)
13 CLARRICOATS, P.J.B., TUN, S.M., and BROWN, R. C.: 'Performance of offset reflector antennas with array feeds', *IEE Proc. H*, 1984, **131**, pp. 172–178
14 SHAFAI, L., HELALY, A., and BHATTACHATYYA, A.K.: 'Mutual coupling effects of microstrip array antennas'. University of Manitoba, report, June 1989
15 SHAFAI, L., and BHATTACHATYYA, A.K.: 'Input impedance and radiation characteristics of small microstrip phased arrays including mutual coupling, MSAT application', *Electromagnetics*, 1986, **6**(4), pp. 333–350

15.7 Appendix: Coefficients of the admittance matrix for an array of circular apertures [11]

TE^j–TE^i-mode coupling:

$$C_{mn}^{ij} = \frac{\pi a^3 \gamma_m y_m y_n}{N_m} \left\{ \frac{G_{mn}}{K} + \frac{pqKF_{mn}}{(x_m x_n)^2} \right\}$$

$$G_{mn} = \int_0^\infty \frac{\omega\sqrt{(1-\omega^2)}J_p'(K\omega)J_q'(K\omega)}{(\omega^2-u_m^2)(\omega^2-u_n^2)} \{S_{p,q}^{(1)} + D_{p,q}^{(1)}\}\, d\omega$$

$$F_{mn}=\int_0^\infty \frac{J_p(K\omega)J_q(K\omega)}{\omega\sqrt{(1-\omega^2)}}\{-S^{(1)}_{p,q}+D_{p,q}\}\,d\omega$$

TM^j–TM^i-mode coupling:

$$C^{ij}_{mn}=\frac{-\pi a^3 K\varepsilon_r\gamma_n y'_m y'_n}{x_m x_n N_m}L_{mn}$$

$$L_{mn}=\int_0^\infty \frac{\omega^3 J_p(K\omega)J_q(K\omega)}{\sqrt{(1-\omega^2)}\,(\omega^2-u_m^2)\,(\omega^2-u_n^2)}\{S^{(1)}_{p,q}+D^{(1)}_{p,q}\}\,d\omega$$

TE^j–TM^i-mode coupling:

$$C^{ij}_{mn}=\frac{-\pi a^4\gamma_m\gamma_n y'_m y_n Z_n}{x_m x_n^2 N_m Z_m}\,q\int_0^\infty \frac{\omega J_p(K\omega)J_q(K\omega)}{\sqrt{(1-\omega^2)(\omega^2-u_m^2)}}\{D^{(2)}_{p,q}-S^{(2)}_{p,q}\}\,d\omega$$

TM^j–TE^i-mode coupling:

$$C^{ij}_{mn}=\frac{+\pi a^4\gamma_m\gamma_n y_m y'_n}{x_n x_m^2 N_m}\,p\int_0^\infty \frac{\omega J_p(K\omega)J_q(K\omega)}{\sqrt{(1-\omega^2)\,(\omega^2-u_n^2)}}\{S^{(2)}_{p,q}+D^{(2)}_{p,q}\}\,d\omega$$

$$\mathrm{N_m}=\frac{-\pi a^4}{\varepsilon_{0p}}\begin{cases}\left(\dfrac{\gamma_m y_m}{x_m^2}\right)^2(x_m^2-p^2) & \text{TE modes}\\[2ex] \left(\dfrac{k_0\varepsilon_r y'_m}{x_m}\right)^2 & \text{TM modes}\end{cases}$$

$$\varepsilon_{0p}=\begin{cases}1 & p=0\\ 2 & p>0\end{cases}$$

ith aperture: 'm' mode with p azimuthal period
jth aperture: 'n' mode with q azimuthal period

$S^{(1)}_{p,q}=(-1)^p J_{q+p}(k_0\omega R_{ij})\cos\{(p+q)\phi_{ij}-\psi_n-\psi_m\}$
$D^{(1)}_{p,q}=J_{q-p}(k_0\omega R_{ij})\cos\{(p-q)\phi_{ij}+\psi_n-\psi_m\}$ $\quad\psi_m=0$ if $p=0$
$S^{(2)}_{p,q}=(-1)^p J_{q+p}(k_0\omega R_{ij})\sin\{(p+q)\phi_{ij}-\psi_n-\psi_m\}$ $\quad\psi_n=0$ if $q=0$
$D^{(2)}_{p,q}=J_{q-p}(k_0\omega R_{ij})\sin\{(p-q)\phi_{ij}+\psi_n-\psi_m\}$

ψ_m = polarisation angle defined relative to the y direction for TE modes and x direction for TM modes
ϕ_{ij} = angle between the line joining the centres of ith and jth apertures and the x direction from the jth aperture
R_{ij} = distance between the centres of ith and jth apertures
a = radius of aperture
$K=k_0 a$

$x_m = \lambda_m a$
$y_m = J_p(x_m)$
$y'_m = J'_p(x_m)$
$u_m = x_m / K$

The scattering matrix is given by $\boldsymbol{S} = [\boldsymbol{I} + \boldsymbol{C}]^{-1} [\boldsymbol{I} - \boldsymbol{C}]$ where $\boldsymbol{C}$ is the admittance matrix.

Index

Airy disc 47
Antenna efficiency 27
Antenna mass 18
Aperture antennas 25-28, 79-100, 448-452
  circular (see Circular apertures)
  compound 87-89
  cross polarisation 84
  directivity 96-97
  efficiency 27
  elliptical 448, 472
  gain 96
  Gaussian 95-96
  line source 85-87
  mutual coupling 448-452, 472-474
  near fields 97-98
  phase taper 90-91
  planar 24
  radiation equations 80-84
  radiation from 25-28, 79-99
  rectangular 85, 89, 241
  shaped 168-172
  sidelobes 86, 88, 94, 95
  taper on a pedestal 87
Aperture field method 22, 210
Arbitrary shaped horns 168-172
Array factor 457, 459, 464
Array feeds 18, 444-482
  conical horns 448-471
  crosspolarisation 448, 454-459
  elliptical 472
  microstrip (see Microstrip arrays)
  microstrip radiators 444, 475-480
  mutual coupling 448-482
  rectangular 472
Arrays
  direct-radiating 5
  feeds (see Array feeds)
  microstrip patch 475-480
  multimatrix 446
  multimode horn 251
  mutual coupling 448-482
  pure mode horns 150
  rectangular apertures 241
Astigmatism 38, 60
Astronomy 1, 3, 4, 24, 25, 61
Attenuation coefficient 102
Australian radiotelescope 47

Balanced hybrid condition 269, 272, 283-284, 298, 301, 325
Balanced hybrid mode 50, 260, 313, 324
Balanced hybrid model 266, 286
Beam deviation factor 38
Beam efficiency 9
Beam forming network 62, 70, 72, 73, 254, 444-448
Beam scanning 444
Beam squint 34
Beam waveguide feeds 43-44
Beamwaist 285-286
Bodies of revolution 129
Boundary conditions 104
Broadband horns
  corrugated horns 296-298
  definition 10
  ridge loaded 172-178
Broadcasting 61, 444
Butler matrix 446

Cassegrain antennas 2, 29, 31, 40, 42, 53, 56, 390
Cauchy's principal values 128
Cavity dipole feed 423-426
Choked waveguide feeds 187-189, 198-199, 212-218, 221-223
Circular apertures 25-28, 92-95
  admittance matrix of array 480-482
  beamwidth 27, 93
  blockage 27
  low sidelobe distributions 27
  mutual coupling between 448-452
  radiation equation 25, 92-93
  radiation pattern 25, 93
  sidelobe level 93
Circular polarisation 17, 18, 34, 66, 84, 178, 253
Circular waveguides
  contour patterns 232-240
  dielectric filled 321-327, 357-358
  fields 104, 154, 232
  multimode 232-240
  roots 115
  wavenumber 104
Circular waveguide feeds 61, 181-208
  beamwidths 191

choke with 187
crosspolarisation 193
dielectric filled 251, 321-327, 411
dielectric ring with 409
dielectric rod feed 392
efficiency 189, 201, 221, 225
gain 157, 189, 197-199, 201, 204, 206, 208
junctions 279
modes 153, 450
performance in reflector 195
phase centre 193-199, 201, 204-209
phase error effect 208
quarter wavelength choke 186
rectangular corrugated flange 200-202
rings in aperture 287
spillover efficiencies 189
trapezoidally corrugated flange 203-208
wall thickness 191
Clavin horn 251
Coaxial feeds 181, 209, 216, 225, 243, 251, 391
choke with 210
crosspolarisation 210
design data 213
dielectric 411
dual band dielectric 411
gain 209, 217
model 211
performance 215
phase centre 216-217
quater-wavelength choke 212
with choke 218
without choke 217
Coaxial probe 125
Coaxial radiation patterns 243
Coma 36, 38, 60
Communication systems 4, 18, 51, 53, 55, 62
Compact horns 159, 230, 251, 346
Cone dipole feed 419
Cone reflector 419
Conical horns 3, 92, 118, 149, 150, 152-160, 225, 230-232, 253
(see also Circular waveguide feeds)
bandwidth 159
crosspolarisation 152, 154, 156, 157, 159
efficiency 156, 157, 159
gain 156-159
large flare angles 159
lens corrected 370-390
mode conversion 159
mode expansions 450
mutual coupling 448-464
optimum designs 157
peak gain 157
phase factor 154
phase centre 159
radiation patterns 154
ridge loading 178
spherical phase cap 153
Copolarisation, definition 7, 151, 269
Corrugated flange 200, 203
Corrugated horns 258-317
admittance 260, 267, 272, 301
analytical techniques 265-286
aperture diameter 262, 263, 274, 276, 277, 286-288, 295, 302, 305
balanced hybrid condition 269
balanced hybrid model 260, 266
bandwidth 261, 263, 265, 276, 296, 298, 304
beamwaist 285
bibliography 258, 315-317
broadband 296-299
characteristic equation 268
characteristic impedance 277
comparison 13
copolar field 269
constant slot depth horns 289, 295
corrugation geometry 262
crosspolarisation 294, 295, 258-264, 269, 274, 277, 283, 286, 296-305, 309-315
design of the corrugations 286
design procedure 261-265
desirable properties 259
discrete section profile 310
dominant mode 260, 268, 272, 313
dual-depth 300-304
efficiency 305, 310, 312
electroforming 288
feed for offset 45
flare angle choice 262
Gaussian beam techniques 52, 282-285
geometry of corrugations 264
higher order modes 268, 279, 282, 289, 294, 298, 308, 312, 314
history 4, 258
impedance input 261-264, 276, 277, 296, 300, 305-307
impedance techniques (cylindrical) 266
impedance techniques (spherical) 271-273
junctions 275, 276, 278-281, 289, 293, 299, 307, 310-312, 314
lens corrected 384, 387-389
low mode conversion 310
maximum flare angle 306
millimetre-wave horns 288
modal matching technique 265, 275-282
multimode 250
narrow flare angle horns 269, 286-287
near-fields 273
numerical analysis 102-107, 113-121, 123
octave band horn 296
open ended waveguide 269
phase centre 305, 312
principles of operation 259-261
profile 305-313
rectangular horns 313 315
return loss 277, 279, 289, 292, 293, 308, 311, 315
ring-loaded slots 265, 276, 296, 298-300

sidelobes 259, 305, 308
sine-squared profile 309
slot depth 274, 277, 279, 282, 287, 289, 296, 300-302, 307, 308, 311
slot width 274, 286, 288, 296, 300, 302
small aperture horns 287
small horns 225-227
space harmonic techniques 264, 274
spherical impedance technique 271-273
spherical modal matching 278-282
spherical phase taper 261, 269
spherical wave expansions 271-273
surface admittance 260, 272
surface impedance 260
throat region 262, 264, 272, 276, 278, 280, 286, 289, 295, 296, 305-308, 313
tolerances 277
tracking horns 254
transmission line approximation 299
types 286-315
uses 258
waveguide feed 181-183
wide flare angle horns 60, 96, 261, 288-295
Corrugated surface 299
Cross guide 172
Crosspolarisation, definition 7, 151, 269
Cylindrical reflector 2, 160
Cylindrical waveguides 106, 113
fields 150-152
modal matching techniques 113-117
modes 150-152
multiple modes 230-243

DBS satellite receiving systems 4
Deep space antennas 43
Diagonal horns 168, 241
Diamond horn 172
Dichroic reflectors 74-77
Dielectric coated sphere scattering 145
Dielectric cone loaded horns (see Dielectric loaded horns)
Dielectric feeds 4, 391-415
dielectric rod (See Dielectric rod feed)
dual band feeds 411-414
history 391
pattern shaping 391
phase centre 396
waveguide with dielectric ring 409-410
Dielectric loaded horns 318-369
admittance 321
air gap 332
aperture diameter 331
attenuation 356-360
bandwidth 331
characteristic equation 324
circular waveguide 323-327
comparison 13
crosspolarisation 318-322, 325-340, 345, 347, 348, 350, 351-356, 361, 364-366
design considerations 328-330
design examples 336-346
dielectric foam cone in metal horn 337
dielectric losses 356-360
directivity 342, 344, 361, 363-365, 384, 385
efficiency 344, 345, 347, 350, 351, 354, 356, 359, 361, 364-366
exponential profile 354
flare angle 332
foam dielectric 330
frequency sensitivity 332
history 318
hybrid factor 325
hyperbolic profile 354
input guide radius 343
lens 364-367
lens to improve efficiency 365
lens to reduce crosspolarisation 366
losses 358-360
mass 329
metallised dielectric 344
metallised dual dielectric 338-341
millimetre-wave horns 342-346
mismatch 331
modal matching technique 322-328
parametric studies 330-336
permittivity of dielectric 330
phase centre 346
polynomial profile 354
power coupling integrals 328
principle of operation 321
profile horns 346-356
profile length 350
shape of profile 352
sine squared profile 353
throat region 332, 336
tolerance on dielectric 361-363
Dielectric resonator feeds 428, 438-442
crosspolarisation 440
phase centre 442
Dielectric ring feeds 248, 409
Dielectric rod feeds 225, 320, 391-414
cavity loading 404-406
choke 396
comparison 13
crosspolarisation 395
dielectric constant 397
dielectric diameter 397
efficiency 395, 396
gain 394-398, 401, 403, 409
phase centre 396
profiled 407
radiation characteristics 392
ring loading 401-403
shaped 400
spillover efficiency 396
tapered end 397
Dielectric rod loaded horn 320
Dielguide 4

Diffraction optimised reflector 31, 69
Diffraction synthesis 30, 68
Dipole disk feeds 416-418
  apertures 96
  conical horns 157
  definition 7-8
  lens corrected horns 384
Dipole feeds 416-426
  cavity dipole 423-426
  comparison 13
  cone dipole 419-422
  dipole disk 416-418
  efficiency 417
  ring loaded 417-418
Dipoles 1, 2, 12-14, 74-77, 125, 127, 142, 185, 210, 416-426, 428, 444
Dual band feeds 411-414
Dual mode horns 234-243
Dual offset reflector antennas 40-45, 66, 464, 468
Dual reflector antennas 4, 14, 29-32, 60
  advantages 30
  magnification 42
  offset 40-45, 66, 464, 468
Dual-depth corrugated horns 300-304
Dual-gridded reflector antennas 71

E-field model 81, 83, 84, 87, 461
Earth-station antennas 5, 17, 18, 24, 25, 30, 54, 230
Effective area 8
Efficiency, definition 8-9, 27
Eight-sided horn 172
Electroforming horns 288
Elliptical horns 168
Elliptical waveguide 472
Equivalence principle 80, 81, 127, 392
Eutelsat 2 satellite 69
Expansion function 131, 145

Far-field distance 24, 97, 98, 285
Feeds (See Array feeds, Circular waveguide feeds, Coaxial feeds, Dielectric feeds, Dielectric resonator feeds, Dielectric rod feeds, Dipole feeds, Dipole disk feeds, Microstrip antennas and feeds and Prime focus waveguide feeds
Feeds, choice of 14
Feeds, performance requirements 5
Feeds, quality assessment 13
Fermat's principle 372
Field equivalence principle 80, 81, 127
Finite difference method 133, 150, 169-172, 175, 178
Finite element method 150, 169
Floquet's theorem 274
Focal field analysis 46 51
Focal region field 181
Fourier mode 134
Fourier transforms 22, 80-95, 151, 168, 266

Fraunhofer region 97
Frequency selective surfaces 74-77
Fresnel integrals 91, 162, 165
Fresnel transform 98
Fresnel zone 56

G/T ratio 9
Gain definition 7-9
Gain factor 184
Galerkin procedure 450
Gaussian aperture distributions 95
Gaussian beam modes 284
Gaussian beam techniques 52, 282-285
Gaussian distribution 95
Gaussian pattern 52
Gaussian-Hermite modes 168
Gaussian-Laguerre modes 284
Geometric optics 4, 19, 22, 30, 40, 46, 56, 57, 62, 66, 166, 364-366, 371, 372, 380, 384, 387, 463
Geometric optics synthesis 68
Geostationary satellite 4
Grating lobe 472
Green's function 128, 133, 151, 448, 476
Gregorian antennas 2, 29, 40, 42, 56, 68, 464
GTD 19, 22, 31, 463

H-field model 83, 84
Helmholtz 101
Hertz cylindrical reflector 2
Hexagonal shaped horn 168
History
  array feeds 444
  corrugated horns 4, 258
  dielectric cone loaded horns 318
  dielectric feeds 391
  horns 3
  lens corrected horns 371
  microstrip feeds 427, 428
  modal matching 107
  prime focus feeds 181
  pure mode horns 149
  reflector and feeds 1
  ridge loaded horns 175
Horns (See Broadband horns, Compact horns, Conical horns, Corrugated horns, Dielectric loaded horns, Lens corrected horns, Multimode horns, Pure mode horns, Pyramidal horns, Ridge loaded horns and Sectoral horns)
Huygen's principle 80
Hybrid factor 268, 325
Hybrid feed 48
Hybrid mode horns 149, 373
Hybrid modes 12, 102, 149, 181, 242, 251, 318, 320, 321, 324, 357, 409
  spherical 121

Impedance characteristics 9-10
Intelsat 4, 5, 444

Jacobi-Bessel method 21
James Clerk Maxwell Telescope 5, 17, 55
Jansky, Karl 3
Jodrell Bank 4

Kirchhoff-Huygens principle 209, 272, 383

Laguerre polynomial 284
Laplacian operator 103
Laser 71
Lens antennas 79
Lens corrected horns 370-390
  aperture amplitude distribution 379-383
  aperture efficiency 389
  attenuation 380
  beamwidth 384
  crosspolarisation 373, 384, 387, 389
  design 371-372
  dielectric loaded horns 364-367
  directivity 384
  efficiency 376, 379-381, 383, 389
  experimental lenses 387
  history 371
  lenses (see Lenses)
  measured results 387
  metal plates 370
  phase centre 371, 377
  plastics foam lenses 373
  radiation characteristics 383-387
  shape of lens 373-378
  sidelobe level 384
  uses 389
  zoned lens 378
Lenses 12, 79, 283, 364-367, 370-390
  attenuation 380
  axial thickness 373
  design 371-372
  dielectric loss 379
  dual-surface 373
  equation 372
  foam plastic 371
  integrated with dielectric 364
  minimum relative permittivity 376
  power loss 380
  profile 364, 373
  reflections at the surfaces 379
  shape 371, 373, 379
  single-surface 373
  zoned 378
Line source apertures 85
  beamwidth 86
  radiation pattern 86
  sidelobe level 86
Loss tangent 342, 356-359, 373, 380
Ludwig definition of polarisation 7, 27, 84, 105, 151, 184, 218, 464

Marconi 2
Matched feed horns 35, 46, 230, 253
Metal plate lenses 370
Metallised dual dielectric cone loaded horn 338
Method of moments 11, 19, 74, 84, 123, 129, 131, 132, 145, 149, 277, 391
  application of 131
  bodies of revolution 129
  C-field formulation 144
  computation time 142
  conductor-dielectric combinations 147
  dielectric material 125, 128, 141-147
  E-field formulation 144
  excitation vectors 135
  expansion function 131
  generating curve 131
  H-field formulation 144
  integral equation formulations 143
  matrix symmetry 142
  mode symmetry 142
  Müller formulation 144
  pulse functions 132
  radiation patterns 139
  solution verification 145
  testing functions 131
  triangle function 133
Microstrip antennas and feeds 4, 5, 13, 14, 123, 147, 427-444, 475-480
  annular ring 437
  aperture blockage 434
  application 427
  arrays (see Microstrip arrays)
  back radiation 432
  bandwidth 431
  conducting chokes 432-434
  crosspolarisation 429-431, 434, 436
  dielectric sheet 430
  effect of mode excitation 431
  efficiency 434
  frequency performance 436
  gain 428, 434, 436
  history 427, 428
  impedance match 428
  monolithic integration 427
  patch 427, 475
  patch ring 428, 436
  performance 427, 434
  phase centre 434, 442
  prime focus reflectors 427
  quarter wavelength choke 428
  resistive losses 428
  self impedance 477
  stacked 147, 431
Microstrip arrays
  conductance 476
  mutual admittance 476
  mutual coupling 475
  radiation pattern 478
  self impedance 476
  susceptance 476
Microwave communication systems 53, 55
Microwave radio systems 24

Millimetre-wave horns 288, 342
Millimetre-waves 59, 168, 239, 282, 288, 319, 320, 342, 345, 379, 427
Modal matching technique 101, 105-122, 275-282, 306, 313, 322-328
 accuracy 115
 coaxial structures 107
 computational time 115
 conical waveguides 117
 corrugated horns 275-282
 cylindrical waveguides 113-117, 275-278
 dielectric loaded horns 107, 322-328
 history 107
 number of modes 115, 276
 numerical procedure 114
 power coupling integrals 112
 principles 106
 reflection coefficient 108
 scattering matrix 106, 109, 110
 spherical waves 117-122, 278-282
 tolerances 277
Monge-Ampere equation 67
Monopulse horns 239, 253, 254
Müller formulation 144
Multimatrix antennas 446
Multimode horns 229-257
 aperture efficiency increase 249
 bandwidth 229, 234, 245, 246, 248, 251
 circular waveguides 232
 coaxial waveguide 243, 251
 comparison 13
 conical horns 249
 contour patterns 232
 corrugated horn 242
 crosspolar cancellation 234
 crosspolarisation 229, 230, 234, 241, 242, 246-249, 253, 254
 diagonal horn 241
 dielectric ring 248
 efficiency 229, 234, 243, 249, 253
 excitation 229, 231, 234, 244, 246, 251
 junctions 244, 245, 247, 254
 matched feed horns 230
 multiple modes 230
 optimum modes 234
 pattern symmetry horns 246
 Potter horn 245
 pyramidal horns 248
 sidelobes 234
 rectangular waveguide 241
 tracking horns 230
Mutual coupling 63, 66, 448-482
 arrays 448-482
 conical horns 448-453
 crosscoupling integrals 450
 crosspolarisation in arrays 452
 elliptical apertures 472
 higher order modes 450, 452
 pair of horns 452
 rectangular apertures 472
 secondary patterns 461

Narrow flare angle horns 264, 286, 289
Near fields
 apertures 97
 corrugated horns 273
Newton, Isaac 1
Noise 4, 9, 24, 43, 218, 437
Noise temperature 9, 437
Numerical analysis 100-148
 axisymmetric feeds 123
 cylindrical wave analysis 102-104
 method of moments 123-148
 modal matching 106-121
 small feeds 123
 spherical waves analysis 105
Nyquist sample 66

Octave band horn 296
Offset reflector antennas 32-45, 253, 461-471
 array-fed 62
 beam-squint 34
 beam deviation factor 38
 blockage 32
 Cassegrainian 40
 crosspolarisation elimination 40
 dual 40-45, 468, 469
 focal field 50
 Gregorian 40, 464
 Intelsat VI 444
 matched feed horns 253
 single 32-40, 464, 468
 spherical 44
Open-ended waveguides 181
Optical path length 372
Optimisation 53, 64, 68, 70, 72, 185, 311, 332, 342, 350, 365, 366
Optimum gain horn 167
Orion satellite 71
Orthogonal mode sets 101
Orthogonality conditions 232
Overlapping feed clusters 446

Pencil beams 17
Phase centre 9, 60, 182-185
 circular waveguide feeds 193-199, 201, 204-209
 coaxial waveguide 216-217
 conical horns 159
 definition 183
 determination 182-185
 dielectric feeds 396
 dielectric loaded horns 346
 lens corrected horns 371, 377
 microstrip feeds 434, 442
 pyramidal horns 167
Phase efficiency 8, 9, 183, 184, 209, 312
Phase factor 154, 158, 163, 166
Phased-array radars 444

Physical optics 19, 24, 30, 51, 64, 67-70, 166, 463, 464
Physical optics synthesis 68
Plane waves 80, 91, 97
Plastics foam lenses 373
Polarisation definition 6-9, 151, 269
Potter horns 35, 117, 183, 229, 244, 246, 248
  design 245
  impedance 246
Power coupling integrals 112, 113, 121, 298, 314, 323, 326, 328
Prime focus waveguide feeds 181-228
  circular waveguide feeds 181-208
  coaxial waveguide feeds 209-217
  comparison 13
  dual band feeds 222-224
  efficiency 184
  gain 182-184
  high efficiency feeds 230
  history 181
  ring feeds 218
  small horn feeds 225
Profile horns
  corrugated 258, 259, 305-312
  dielectric loaded horns 346-355, 359, 370
Pure mode horns 13, 149-180
  conical horns (see Conical horns)
  history 149
  principles 149
  pyramidal horns (see Pyramidal horns)
  ridge loaded horns (see Ridge loaded horns)
  shaped aperture horns 168-171
Pyramidal horns 3, 11, 150, 152, 160-167, 174, 178, 248, 249
  crosspolarisation 165
  design 166
  efficiency 167
  gain 164-167
  impedance 167
  lengths 167
  optimum gain 167
  phase centre 167
  phase factor 163, 166
  quad-ridge 178
  radiation patterns 162
  rectangular 249, 253, 314, 319
  ridge loaded 178

Quasi-optics 283

Radar 1, 3, 4, 10, 14, 18, 19, 54, 61, 62, 172, 254, 258, 259, 296, 444
Radiation equation 84, 87, 90, 92, 93, 95, 162
Radiation from apertures 79-99
Radiation integral 80, 92, 151, 162, 272
Radiation patterns, definitions 5-6
Radio astronomy 1, 4, 24
Radio astronomy antennas 25
Radome covered reflectors 18, 53-57
  Cassegrain 56
  Gregorian 56
Rayleigh, Lord 2, 51
Rayleigh-Ritz procedure 169
Reciprocity 19, 48
Reconfigurable reflector antennas 10, 72-73
Reconfiguration 62, 446
Rectangular apertures 89
Rectangular corrugated horns 313-314
Rectangular waveguides 104, 152, 153, 160-162, 174-176, 241, 254, 293, 314, 472
  aperture fields 162
  double ridge 175
  fields 104, 160-162, 241, 314
  multimode 241
  mutual coupling 472
  ridge-loaded 174
  wavenumber 104
Rectangular waveguide feeds 90, 251, 253
Reflection coefficient 9, 18, 51-53, 56, 100, 108, 276, 280, 339
Reflector antennas 1-6, 9-12, 16-78, 181, 189, 195, 218-222, 225, 230, 243, 251, 253, 288, 396, 403-407, 427, 428, 434-438, 444, 446, 461-471
  aberrations 36
  analysis techniques 19
  aperture field method 22
  array fed 18, 62, 461-471
  array feeds 444, 446, 461-471
  axial defocusing 60
  beam deviation factor 38
  blockage 27
  Cassegrain 2, 29, 31, 40, 42, 53, 56, 390
  circular aperture 25
  classification 16
  contoured coverage 447
  crosspolarisation 18, 24, 33-38, 43, 46-50, 60, 61, 66, 68, 71, 77, 464, 468, 469, 478
  dichroic 74
  diffraction optimised 31
  directivity optimised 418
  dual (see Dual reflector antennas)
  dual gridded 71
  dual offset 40-45, 66, 464, 468
  efficiency 30-32, 51, 53, 85, 86
  feeds for deep reflectors 218
  focal field analysis 46-51
  focal region 34
  focal region field 181
  frequency selective surfaces 74
  gain 17, 18, 21, 22, 24, 25, 27, 30, 38, 46, 51, 53, 58-61, 64, 66, 69, 70, 77
  gain factor 184
  Gregorian 2, 29, 40, 42, 56, 68, 464
  imaging array-fed 446
  Jacobi-Bessel method 21
  matched feeds 30, 46, 230, 253
  offset (see Offset reflector antennas)

optimisation 53
performance requirements 5
phase centre 60
positional errors 60
radome covered 53
reconfigurable 72
reflection coefficient 18
reflection minimisation 53-57
return loss 18, 51-53, 56
scattered field 18, 20
secondary pattern 24-28
shaped beam antennas 61-70, 446
shaped reflector 18, 27, 62, 66, 70-73, 313, 320
sidelobes 27, 59-61, 64, 77, 446, 457, 464
spherical 17
subreflector 29-31, 40, 42-44, 53, 60, 68-70, 77, 416, 463, 464
surface errors 57-60
surface roughness 17
symmetrical paraboloidal 22-24
synthesis 61-70
tri-reflector 17
types 16
vertex matching 53
Refraction 370-373, 380, 381, 387
Refractive index 365, 370, 372, 380
Remote sensing 1, 9, 10, 14, 258, 259, 282
Return loss 9
Ridge loaded horns 172-178
bandwidth 173, 176-178
circular waveguide 174
crosspolarisation 172
impedance 176
input section 174
radiation patterns 174
rectangular waveguide 174
Ring feeds 218, 225, 401, 417
Ring-loaded slots 298

Satellite communications 4, 10, 14, 71
Satoh horn 246, 248
Scalar horn 261
Scanned feeds 18
Scanning antennas 444, 446
Scattering matrices 106, 108-113, 121, 450
Schelkunoff gain formula 166
Sectoral horns 160-163, 371
Shaped aperture horns 168-172
Shaped beams 4, 10, 14, 59, 61, 64, 66, 283, 313
elliptical 459
Shaped coverage antennas 444
Shaped pattern horns 229, 246-253
Slope diffraction 463
Small horn feeds 225-227
Snell's law 372
Space harmonics 274, 286
Spacecraft antennas 59, 61, 64, 74, 457, 458, 464
Spark erosion 288
Spherical Hankel function 127
Spherical modes 11, 106, 117, 271, 272, 278, 279, 289, 307
Spherical phase factor 90, 91, 118, 158, 168, 286
Spherical wave analysis 105
Spherical wave expansion 31, 273, 384
Spherical waves 5, 9, 11, 91, 97, 100, 102, 105, 118, 159, 266, 272
Spillover efficiency 8, 27, 189, 221, 395, 396
Spillover loss 25
Spot beams 459
Subarrays 456, 469
Syntactic foam 339, 362

TE modes 102, 103, 104, 114, 150, 151, 160, 171, 175, 232, 276, 314
Telecommunications 4, 24
Telescopes 1, 4, 5, 17, 18, 59
TELSTAR 4
TEM modes 175
Terrestrial communication antennas 17
Testing functions 136
TM modes 103, 114, 120, 151, 171, 232, 241, 242, 250, 267, 272, 276, 314, 366
Tracking feeds 4, 12, 87, 229, 230, 239, 241, 253, 254, 268, 325
Transmission lines 261, 299

Variable power divider 446
VSAT systems 5
VSWR 9, 167, 277, 297, 347, 350, 371, 380, 387

Wave equation 89, 101-103, 150, 151, 168-171, 272, 279, 284
Waveguide (see Circular waveguide, Rectangular waveguide)
Wavenumber 103-105, 115, 151, 154, 161, 171, 183, 244, 314, 321
Wide flare angle horns 261, 264, 278, 280, 288-295
Wideband horns
corrugated horns 296-298
definition 10
ridge loaded 172-178
Widebeam feeds 219-220

Zoned lenses 378

www.ingramcontent.com/pod-product-compliance
Lightning Source LLC
LaVergne TN
LVHW010630300926
849162LV00001B/115